Methods in Enzymology

Volume 389
REGULATORS OF G-PROTEIN SIGNALING
Part A

METHODS IN ENZYMOLOGY

EDITORS-IN-CHIEF

John N. Abelson Melvin I. Simon

DIVISION OF BIOLOGY
CALIFORNIA INSTITUTE OF TECHNOLOGY
PASADENA, CALIFORNIA

FOUNDING EDITORS

Sidney P. Colowick and Nathan O. Kaplan

Methods in Enzymology

Volume 389

Regulators of G-Protein Signaling

Part A

EDITED BY

David P. Siderovski

ELSEVIER
ACADEMIC
PRESS

AMSTERDAM • BOSTON • HEIDELBERG • LONDON
NEW YORK • OXFORD • PARIS • SAN DIEGO
SAN FRANCISCO • SINGAPORE • SYDNEY • TOKYO

Elsevier Academic Press
525 B Street, Suite 1900, San Diego, California 92101-4495, USA
84 Theobald's Road, London WC1X 8RR, UK

This book is printed on acid-free paper. ∞

For all information on all Academic Press publications
visit our Web site at www.academicpress.com

ISBN: 0-12-182794-1

PRINTED IN THE UNITED STATES OF AMERICA
04   05   06   07   08   9   8   7   6   5   4   3   2   1

Table of Contents

Section I. General Methods of RGS Protein Analysis
Subsection A. Expression and Post-translational Modification

Subsection B. Assays of RGS Box Activity and Allosteric Control

Subsection C. Electrophysiological Methods and RGS-Insensitive Gα Subunits

Subsection D. Mouse Models of RGS Protein Action

Subsection E. Methods of RGS Protein Inhibition

Section II. G-Protein Regulators of Model Organisms

Contributors to Volume 389

Article numbers are in parentheses and following the names of contributors.
Affiliations listed are current.

WOOIN AHN (8), *Department of Physiology, University of Texas Southwestern Medical Center, Dallas, Texas 75390-9040*

SEENA AJIT (17), *Neuroscience Discovery Research, Wyeth Research, Princeton, New Jersey 08543*

COREY BENDER (17), *Neuroscience Discovery Research, Wyeth Research, Princeton, New Jersey 08543*

DANIEL L. CHASE (18), *Department of Molecular Biophysics and Biochemistry, Yale University School of Medicine, New Haven, Connecticut 06520*

SCOTT A. CHASSE (24), *Department of Biochemistry and Biophysics, The University of North Carolina at Chapel Hill, Chapel Hill, North Carolina 27599-7260*

HUANMIAN CHEN (12), *Department of Pharmacology and Toxicology, Medical College of Georgia, Augusta, Georgia 30912-2300**

JIN-GUI CHEN (20), *Department of Biology, The University of North Carolina at Chapel Hill, Chapel Hill, North Carolina 27599-3280*

MARY J. CLARK (10), *Department of Pharmacology, University of Michigan Medical School, Ann Arbor, Michigan 48109-0632*

MICHAEL A. CLARK (12), *Department of Pharmacology and Toxicology, Medical College of Georgia, Augusta, Georgia 30912-2300*

HENRIK G. DOHLMAN (23, 24), *Department of Biochemistry and Biophysics, The University of North Carolina at Chapel Hill, Chapel Hill, North Carolina 27599-7260*

CRAIG A. DOUPNIK (9), *Department of Physiology and Biophysics, University of South Florida College of Medicine, Tampa, Florida 33612*

TIMOTHY C. ELSTON (23), *Department of Mathematics, University of North Carolina at Chapel Hill, Chapel Hill, North Carolina 27599-3250*

YING FU (14), *Department of Pharmacology, University of Michigan, Ann Arbor, Michigan 48105*

ADAM GILBERT (17), *Chemical Screening Sciences, Wyeth Research, Pearl River, New York 10965*

STEPHEN J. GOLD (13), *Deptartment of Psychiatry, University of Texas Southwestern Medical Center, Dallas, Texas 75390*

SEBASTIAN GRANDERATH (21), *International Graduate School in Genetics and Functional Genomics, University of Cologne, Köln D-50931, Germany*

MELINDA D. HAINS (5), *Department of Pharmacology, The University of North Carolina School of Medicine, Chapel Hill, North Carolina 27599*

NAN HAO (23), *Department of Biochemistry and Biophysics, The University of North Carolina at Chapel Hill, Chapel Hill, North Carolina 27599-0812*

Current Affiliation: National Institute on Alcohol Abuse and Alcoholism, National Institutes of Health, Bethesda, Maryland 20892-8115

T. KENDALL HARDEN (5), *Department of Pharmacology, The University of North Carolina School of Medicine, Chapel Hill, North Carolina 27599*

KATHLEEN HARRISON (2), *B Cell Molecular Immunology Section, Laboratory of Immunoregulation, National Institute of Allergy and Infectious Diseases, National Institutes of Health, Bethesda, Maryland 20892*

JIE HUANG (1), *Department of Pharmacology, University of Texas Southwestern Medical Center, Dallas, Texas 75390*

XINYAN HUANG (14), *Department of Pharmacology, University of Michigan, Ann Arbor, Michigan 48105*

STEPHEN R. IKEDA (11), *Laboratory of Molecular Physiology, National Institute on Alcohol Abuse and Alcoholism, National Institutes of Health, Bethesda, Maryland 20892 USA*

MASARU ISHII (7), *Department of Pharmacology, Osaka University Graduate School of Medicine, Osaka 565-0871, Japan*

CRISTINA JAÉN (9), *Department of Physiology and Biophysics, University of South Florida College of Medicine, Tampa, Florida 33612*

SEONG-WOO JEONG (11), *Department of Physiology, Yonsei University Wonju College of Medicine, Kangwon-Do 220-701, Republic of Korea*

YAFEI JIN (16), *Department of Medicinal Chemistry, College of Pharmacy, University of Michigan, Ann Arbor, Michigan 48109*

CHRISTOPHER A. JOHNSTON (4), *Department of Pharmacology, The University of North Carolina at Chapel Hill, Chapel Hill, North Carolina 27599-7365*

ALAN M. JONES (20), *Department of Biology, The University of North Carolina at Chapel Hill, Chapel Hill, North Carolina 27599-3280*

TERESA L. Z. JONES (3), *Division of Diabetes, Endocrinology and Metabolism, National Institute of Diabetes and Digestive and Kidney Diseases, National Institutes of Health, Bethesda, Bethesda, Maryland 20892*

JOHN H. KEHRL (2), *B Cell Molecular Immunology Section, Laboratory of Immunoregulation, National Institute of Allergy and Infectious Diseases, National Institutes of Health, Bethesda, Maryland 20892*

ADAM J. KIMPLE (4), *Department of Pharmacology, The University of North Carolina at Chapel Hill, Chapel Hill, North Carolina 27599-7365*

RANDALL J. KIMPLE (4), *Department of Pharmacology, The University of North Carolina at Chapel Hill, Chapel Hill, North Carolina 27599-7365*

CHRISTIAN KLÄMBT (21), *Institut für Neurobiologie, Universität Münster, Münster, 48149 Germany*

MICHAEL R. KOELLE (18), *Department of Molecular Biophysics and Biochemistry, Yale University School of Medicine, New Haven, Connecticut 06520*

TOHRU KOZASA (15), *Department of Pharmacology, University of Illinois at Chicago, Chicago, Illinois 60612*

YOSHIHISA KURACHI (7), *Department of Pharmacology, Osaka University Graduate School of Medicine, Osaka 565-0871, Japan*

DEBORAH M. KURRASCH (1), *Department of Pharmacology, University of Texas Southwestern Medical Center, Dallas, Texas 75390-9040[†]*

[†]*Current Affiliation: Department of Physiology, University of California San Francisco, San Francisco, California 94143-2611*

NEVIN A. LAMBERT (12), *Department of Pharmacology and Toxicology, Medical College of Georgia, Augusta, Georgia 30912-2300*

KENGLI LAN (14), *Department of Pharmacology, University of Michigan, Ann Arbor, Michigan 48105*

MIN LIU (15), *Department of Pharmacology, University of Michigan, Ann Arbor, Michigan 48109*

XIANG LUO (8), *Department of Physiology, University of Texas Southwestern Medical Center, Dallas, Texas 75390-9040*

CHANTAL MORATZ (2), *B Cell Molecular Immunology Section, Laboratory of Immunoregulation, National Institute of Allergy and Infectious Diseases, National Institutes of Health, Bethesda, Maryland 20892*

RICHARD M. MORTENSEN (14), *Department of Physiology, University of Michigan, Ann Arbor, Michigan 48105*

HENRY I. MOSBERG (16), *Department of Medicinal Chemistry, College of Pharmacy, University of Michigan, Ann Arbor, Michigan 48109*

SHMUEL MUALLEM (8), *Department of Physiology, University of Texas Southwestern Medical Center, Dallas, Texas 75390-9040*

MASAKATSU NANAMORI (14), *Department of Pharmacology, University of Michigan, Ann Arbor, Michigan 48105*

RICHARD R. NEUBIG (14, 15, 16), *Department of Pharmacolgy, University of Michigan, Ann Arbor, Michigan 48109*

BART W. NIEUWENHUIJSEN (17), *Neuroscience Discovery Research, Wyeth Research, Princeton, New Jersey 08543*

JOHN R. OMNAAS (16), *Department of Medicinal Chemistry, University of Michigan, Ann Arbor, Michigan 48109*

FERNANDO RAMIREZ (17), *Chemical Screening Sciences, Wyeth Research, Princeton, New Jersey 08543*

JOYCE J. REPA (1), *Departments of Physiology and Internal Medicine, Touchstone Center for Diabetes Research-8.212B, University of Texas Southwestern Medical Center, Texas 75390-8854*

JEFFREY D. ROTHESTEIN (15), *Department of Neurobiology, Johns Hopkins University, Baltimore, Maryland 21287*

DAVID P. SIDEROVSKI (4, 5, 19), *Department of Pharmacology, Lineberger Comprehensive Cancer Center, UNC Neuroscience Center, The University of North Carolina at Chapel Hill, Chapel Hill, North Carolina 27599-7365*

PAUL C. STERNWEIS (15), *Department of Biochemistry, University of Texas Southwestern Medical Center, Dallas, Texas 75207*

JOHN R. TRAYNOR (10), *Department of Pharmacology, University of Michigan Medical School, Ann Arbor, Michigan 48109-0632*

YAPING TU (6), *Department of Pharmacology, University of Texas Southwestern Medical Center, Dallas, Texas 75390-9041‡*

QIN WANG (15), *Department of Pharmacology, University of Michigan, Ann Arbor, Michigan 48109*

YUREN WANG (17), *Neuroscience Discovery Research, Wyeth Research, Princeton, New Jersey 08543*

THOMAS M. WILKIE (1, 6), *Department of Pharmacology, University of Texas Southwestern Medical Center, Dallas, Texas 75390*

‡*Current Affiliation: Department of Pharmacology, Creighton University School of Medicine, Omaha, NE 68178*

FRANCIS S. WILLARD (4, 19), *Department of Pharmacology, The University of North Carolina at Chapel Hill, Chapel Hill, North Carolina 27599-7365*

NECMETTIN YILDIRIM (23), *Department of Mathematics, University of North Carolina at Chapel Hill, Chapel Hill, North Carolina 27599-3250*

KATHLEEN H. YOUNG (17), *Neuroscience Discovery Research, Wyeth Research, Princeton, New Jersey 08543-8000*

FENGWEI YU (22), *Temasek Life Sciences Laboratory, National University of Singapore, Singapore 117604, Singapore*

VENETIA ZACHARIOU (13), *Department of Pharmacology, University of Crete, Heraklion 71003, Crete*

WEIZHONG ZENG (8), *Department of Physiology, University of Texas Southwestern Medical Center, Dallas, Texas 75390-9040*

QINGLI ZHANG (9), *Department of Physiology and Biophysics, University of South Florida College of Medicine, Tampa, Florida 33612*

HUAILING ZHONG (14, 16), *Department of Pharmacology, University of Michigan, Ann Arbor, Michigan 48105*

Preface

Heterotrimeric G protein signaling has been the subject of several previous volumes of this august series, most recently with the "G Protein Pathways" volumes 343 to 345. Volume 344, in particular, contains several notable chapters devoted to emergent methods of analysis of the "regulator of G-protein signaling" (RGS) proteins. This superfamily of G-protein modulators first came to prominence in late 1995/early 1996 with a flurry of papers, spanning from yeast and worm genetics to T- and B-lymphocyte signaling, that heralded the discovery of a missing fourth component to the established receptor > G-protein > effector signaling axis. The ability of RGS proteins, via their hallmark RGS domain or "RGS-box", to accelerate the intrinsic guanosine triphosphatase (GTPase) activity of G-protein alpha subunits helped explain the timing paradox between GTPase activity seen with isolated G-alpha subunits *in vitro* and the rapid physiological timing of G protein-coupled receptor (GPCR) signaling observed *in vivo*.

The nomenclature used in the field since the discovery of the RGS protein superfamily has certainly drifted from the first efforts in 1996 of Druey, Horvitz, Kehrl, and Koelle to establish the name "regulators of G-protein signaling" for all proteins containing the RGS-box. Thus, before considering the breadth of content within volumes 389 (Part A) and 390 (Part B), I will try to clarify some of my own editorial decisions on issues of nomenclature. I greatly prefer "GTPase-*accelerating* protein" rather than "GTPase-*activating* protein" as the definition underlying the acronym "GAP" that describes RGS-box action on G-alpha subunits. This distinction helps highlight that the intrinsic GTPase activity of G-alpha subunits is much greater than that of Ras-superfamily monomeric GTPases – distant cousins of G-alpha whose minimal GTPase activity is truly activated by GAPs (distinct from the RGS proteins) that contribute a key catalytic residue. I also greatly prefer the use of "RGS-box" or "RGS domain" to describe the conserved alpha-helical bundle that comprises the G-alpha binding site and catalytic GAP activity common to the superfamily. Undoubtedly, with such a large superfamily, there will be structural variations on the central theme of the RGS-box nine-helical bundle; however, the proliferation of alternate names for subgroups of RGS domains (such as "rgRGS" for the RhoA guanine nucleotide exchange factors [Rho-GEFs], or "RH" for the N termini of the GPCR kinases [GRKs]) does a disservice to those of you looking in from outside the field.

Clearly, there is now no recourse to the fact that the numbering system for RGS proteins, first implemented in 1996 by Koelle and Horvitz, has gone awry and no longer allows facile cross-species comparisons. For example, three "RGS1" proteins are discussed in volume 389: mouse/human RGS1 (chapters 2, 6, 8), *Caenorhadbitis elegans* RGS-1 (chapter 18), and Arabidopsis AtRGS1 (chapters 19 & 20), yet the mammalian, nematode, and plant "RGS1" proteins share no structural (nor functional) relatedness. Also, whereas *Drosophila* RGS7 (vol. 389, ch. 21) and mammalian RGS7 proteins (vol. 390, ch. 10–14) share the same multi-domain architecture of DEP, GGL, and RGS domains, the *C. elegans* RGS-7 does not (yet the *C. elegans* EAT-16 and EGL-10 proteins do; vol. 389, ch. 18). Therefore, reader beware.

In my choices regarding content for these volumes, I have read the mandate "regulators of G-protein signaling" broadly. Thus, not only does Part B (vol. 390) contain 22 consecutive chapters spanning seven subfamilies of RGS-box proteins, but both Parts A and B also contain chapters devoted to G-protein signaling regulators outside the RGS-box superfamily. For example, four chapters deal with individual components of an emerging, non-GPCR-related nucleotide cycle for G-alpha subunits underlying microtubule dynamics, mitotic spindle organization, and chromosomal segregation: namely, the GoLoco motif-containing proteins Pins and RGS14 (vol. 389, ch. 22; vol. 390, ch. 16), the G-alpha exchange factor Ric-8A (vol. 390, ch. 23), and the microtubule constituent tubulin (vol. 390, ch. 24). Chasse & Dohlman (vol. 389, ch. 24) describe their adroit exploitation of the "awesome power of yeast genetics" to identify a multitude of novel modulators of the GPCR signaling archetype: the *Saccharomyces cerevisiae* pheromone response pathway. In the preceding chapter (vol. 389, ch. 23), Dohlman, Elston, and co-workers describe mathematical modeling of the actions of pheromone signaling components and modulators, like the RGS protein prototype SST2, that are known in great detail. It is my hope, and surely of the authors as well, that the modeling techniques detailed in this chapter will be employed by others to elicit understanding of the interplay between regulators of G-protein signaling and their targets in other systems and organisms.

Indeed, the overall intent of all of the chapters in Parts A and B is to facilitate further investigations of these newly-appreciated G-protein signaling modulators in their physiological contexts. My personal belief is that, over time, RGS proteins will prove valuable drug discovery targets. RGS-insensitive ("RGS-i") G-alpha subunits, described most thoroughly here by Neubig and colleagues (vol. 389, ch. 14), have facilitated several unequivocal demonstrations that endogenous RGS proteins control the dynamics of GPCR signal transduction *in vivo* (e.g., vol. 389, chapters 10–12, 14). Chapters 16 and 17 of volume 389 represent initial forays into developing peptidomimetic and small molecule inhibitors of RGS protein action that should provide "proofs-of-principle" required to energize the actions of "Big Pharma" in considering RGS proteins as legitimate drug targets. Real-time fluorescence and surface plasmon resonance assays of the RGS protein/G-alpha interaction, as described in chapters 4 and 19 of volume 389, should help accelerate this pursuit.

I remain ever grateful to my many colleagues in the G-protein signaling community, and members of my own laboratory, who readily contributed chapters for these two volumes. I also thank Ms. Cindy Minor for her organizational assistance, and my family for being patient with me during the completion of this enormous task. Indeed, it must be formally stated to my children that, although I wrote and edited chapter contributions during your ballet, baseball, and basketball practices, I *never* did during your games or performances!

DAVID P. SIDEROVSKI

METHODS IN ENZYMOLOGY

Section I

General Methods of RGS Protein Analysis

Subsection A

Expression and Post-translational Modification

[1]   Quantitative Real-Time Polymerase Chain Reaction Measurement of Regulators of G-Protein Signaling mRNA Levels in Mouse Tissues

By DEBORAH M. KURRASCH, JIE HUANG,
THOMAS M. WILKIE, and JOYCE J. REPA

Abstract

Regulators of G-protein signaling (RGS) play a critical role in G-protein-coupled receptor signaling in mammalian cells. RGS proteins are GTPase-accelerating proteins (GAPs) for α subunits of heterotrimeric G proteins of the G_i and G_q class. RGS GAPs can modulate the frequency and duration of G-protein signaling and may constitute a new family of therapeutic targets. Identifying the tissue distribution and cellular localization of RGS proteins has been hindered by the lack of effective antibodies for immunodetection. The measurement of mRNA levels for RGS proteins, however, can provide insight into their tissue specificity and regulation. This article describes the use of a highly sensitive and rapid method for measuring RGS mRNA in mouse tissues. This quantitative real-time polymerase chain reaction method is established for the 19 reported mouse RGS genes and is used to study the tissue distribution of the R4 family of RGS genes and the diurnal regulation of RGS16 in mouse liver.

Introduction

RGS Regulation of G-Protein Signaling

Regulators of G-protein signaling (RGS) are critical components of G-protein-coupled receptor (GPCR) signal transduction in eukaryotic cells (Chen *et al.*, 2003; reviewed in Hollinger *et al.*, 2002; Neubig *et al.*, 2002). The activity of heterotrimeric G proteins is regulated by a cycle of GTP binding and hydrolysis on the Gα subunit. In its inactive state, Gα binds GDP in a complex with the G$\beta\gamma$ subunits. When the associated GPCR binds agonist, the Gα subunit releases GDP, binds GTP, and dissociates from G$\beta\gamma$, rendering both Gα and the G$\beta\gamma$ heterodimer available to modulate downstream effector proteins. Intrinsic GTPase activity of the Gα subunit hydrolyzes Gα_{GTP} to Gα_{GDP}, terminates signaling, and allows the quiescent heterotrimeric G protein (G$\alpha_{GDP}\beta\gamma$) to reassemble.

Metazoan organisms express four classes of Gα subunits: G$_i$, G$_q$, G$_{12}$, and G$_s$. RGS proteins regulate G$_i$ and G$_q$ class proteins (Dohlman *et al.*, 1997; Ross *et al.*, 2000). A distantly related family of proteins, called RhoGEF (rg)RGS, regulate G$_{12}$ α subunits (Kozasa *et al.*, 1998). RGS regulation of the G$_s$ class is uncertain. RGS proteins are GTPase-accelerating proteins (GAPs) for G$_i$ and/or G$_q$ class α subunits. RGS proteins stabilize the transition state during Gα_{GTP} hydrolysis, thereby accelerating the rate of inactivation (Tesmer *et al.*, 1997). Because RGS proteins can control the frequency and duration of G-protein signaling (Chatterjee *et al.*, 2003; Doupnik *et al.*, 1997; Luo *et al.*, 2001; Popov *et al.*, 2000; Zeng *et al.*, 1998; Zhong *et al.*, 2003), the distribution (Gold *et al.*, 1997; Hollinger *et al.*, 2002) and regulation (Garnier *et al.*, 2003; Krumins *et al.*, 2004) of RGS proteins likely contribute to the specificity of G-protein-signaling pathways.

Surprisingly little is known about the regulation of RGS levels *in vivo*. The paucity of effective antibodies has hindered the measurement of endogenous RGS proteins. A further complication is that alternatively spliced mRNAs of several RGS genes encode protein isoforms that may be regulated differentially by posttranslational modifications and/or subcellular localization. However, the detection of RGS mRNA is possible by *in situ* hybridization, northern analysis, or RNase protection assay. Interestingly, many studies indicate that the regulation of RGS gene expression responds to physiological stimuli transduced by G proteins (Dohlman *et al.*, 1996; Garnier *et al.*, 2003; Zahariou *et al.*, 2003). Because RGS mRNA expression can be induced by G-protein-coupled agonists (Dohlman *et al.*, 1996), it may be possible to monitor known or unknown G-protein-signaling cascades by measuring RGS gene expression in cells and tissues. To that end, this article focuses on the ease and accuracy with which quantitative real-time polymerase chain reaction (qRT-PCR) can be employed to study RGS gene regulation and G-protein signaling *in vitro* and *in vivo*.

Background, Quantitative Real-Time PCR

Although many methods exist to detect mRNA, including northern blot analysis, RNase protection assay, *in situ* hybridization, and traditional reverse transcription PCR (RT-PCR), qRT-PCR is becoming the method of choice (Bustin, 2000; Heid *et al.*, 1996). The advantage of qRT-PCR over other conventional assays is that qRT-PCR is an extremely robust method, allowing detection of a dynamic range of up to 10^7-fold as opposed to agarose gel-based methods that enable approximately a 10-fold resolution. Instead of detecting PCR amplification at the end point of the PCR reaction, as is the case with traditional RT-PCR, qRT-PCR measures the kinetics of the reaction in the early geometric phase of amplification, thereby providing a more accurate quantitation of initial levels of cDNA. Very minute amounts of RNA (typically a microgram of total RNA) are required, thereby allowing one to measure transcript levels in limiting tissue sources (e.g., retina or pancreatic islet). Finally, qRT-PCR is a rapid, highly reproducible technique.

It is important to point out the limitations of qRT-PCR. First, because visualizing RNA integrity is not practical, great care must be taken in isolating clean, intact RNA. Close observation of the amplification characteristics of a normalizing gene (also referred to as a housekeeping gene) for each sample is the only way to judge RNA quality. Second, proper primer design is essential to measure RNA accurately by this method (described in detail later). In addition, if multiple transcripts exist for a given gene, a particular set of primers may not recognize all forms. Finally, as with many RNA detection methods, the selection of an appropriate normalizing gene is important. Housekeeping genes are commonly used for this purpose and include cyclophilin, β-actin, GAPDH, 36B4, RPL19, HPRT, and 18S ribosomal RNA (Thellin *et al.*, 1999).

Principle of Method for qRT-PCR

The principle of qRT-PCR is the measurement of an amplified product by capturing emission signals (for the protocol described in this article, an intercalating dye is the detection chemistry employed). Thus, the PCR reaction can be followed in "real time" by detecting the fluorescence intensity of the PCR product after every cycle instead of at the end point, as is the case with traditional RT-PCR. The reliability of qRT-PCR lies with the proper design of primers for the gene of interest; they must be highly specific and exhibit amplification of the same efficiency as the primers for the housekeeping gene used for normalization. With each qRT-PCR cycle, the primers anneal to the cDNA, the product is replicated, and the SYBR green dye intercalates with the double-stranded DNA product to produce a fluorescent signal. Thus, within each cycle, during the exponential phase, the increasing fluorescence signal is proportional to the exponentially increasing number of copies of the amplified gene product. Further, when the fluorescent signal increases above background, the PCR cycle time (C_T) is proportional to the initial amount of starting material in the sample well; the lower the C_T, the higher the concentration of initial starting material. Data for each sample are analyzed by utilizing an arithmetic formula (explained in detail later) to establish the amount of target gene.

Methods

Materials

DNase I (RNase free), random hexamers (pdN$_6$), and MgCl$_2$ are from Roche Molecular Biochemicals (Indianapolis, IN). Superscript II reverse transcriptase and dNTPs are from Invitrogen (Carlsbad, CA). Oligonucleotide primers are from GIBCO-Invitrogen and Integrated DNA Technologies (Coralville, IA). RNA is prepared using RNA-STAT60 (Tel-Test, Inc., Friendswood, TX). 2X SYBR green PCR master mix, 96-well optical reaction plates, and optical adhesion covers are from Applied BioSystems (Foster City, CA). Chloroform and isopropanol are HPLC grade, and nuclease-free water is generated by pretreatment with diethyl pyrocarbonate (DEPC). Liquid handling supplies are dedicated for RNA use only (polypropylene centrifuge tubes, pipettes, and aerosol-barrier pipette tips). All qRT-PCRs are performed in triplicate on an Applied Biosystems Prism 7000 sequence detection system.

RNA Purification

Tissues are dissected rapidly from mice, frozen immediately in liquid nitrogen, and then stored at $-85°$. The frozen sample is dropped into a polypropylene centrifuge tube containing the appropriate volume of RNA-STAT60. For most organs, 1 ml of RNA-STAT60 per 100 mg of tissue is optimal. However, due to the extreme lipid content of adipose tissue and the high RNase activity inherent in pancreas, a higher volume of RNA-STAT60 (10 ml/100 mg) is recommended to improve extraction efficiency and RNA stability, respectively. The tissue is then homogenized using a polytron (e.g., PowerGen 125, Fisher Scientific) for 20 s. The homogenized sample is incubated at room temperature for 5 min. To initiate phase separation, chloroform is added (200 μl/100 mg tissue), and the sample is shaken vigorously by hand for 15 s. The sample is incubated for 2 min at room temperature and is then placed on ice until all samples are processed. Samples are centrifuged at $12,000g$ for 15 min at $4°$. The mixture will separate into three phases: (lower) phenol–chloroform layer that contains DNA and proteins, (middle) a white interphase, and (upper) the clear aqueous phase containing the RNA. The aqueous phase is transferred to a fresh tube and the RNA is precipitated by the addition of isopropanol (500 μl per 100 mg tissue). After incubating the samples for 10 min at

TABLE I
TYPICAL YIELDS OF TOTAL RNA ISOLATED FROM VARIOUS ADULT MOUSE[a] TISSUES

Tissue	Approximate wet weight (mg)	Quantity of RNA (μg)
Adipose, white (epididymal)	280	200
Adipose, brown	70	175
Adrenal gland (pair, male)	4	25
Cerebrum	215	325
Cerebellum	60	100
Colon	370	800
Heart	125	200
Intestine (scraped mucosa)		
Duodenum	300	1,300
Jejunum	300	1,000
Ileum	175	700
Kidney (pair)	350	1,600
Liver	1000	10,000
Lung (pair)	145	360
Muscle	250	130
Ovaries (pair)	10	75
Pancreas	150	2,500
Skin	250	400
Spleen	60	1,000
Testis (pair)	200	1,500
Uterus	80	600

[a] Three-month-old Agouti 129/SvEv, mice, all tissues from males except uterus and ovaries. A pair of adrenal glands from the female mouse will weigh about 8–10 mg.

room temperature, they are centrifuged at 12,000g for 15 min at 4°. The supernatant is discarded and the RNA pellet is washed once with 75% ethanol (1 ml/100 mg tissue). Samples are centrifuged at 7500g for 5 min at 4°. The ethanol is aspirated, the pellet is dried briefly, and the RNA is resuspended in nuclease-free H_2O. The RNA concentration is determined spectrophotometrically at 260 nm, and the purity of RNA is assessed by its 260/280-nm ratio (typically >1.7). Samples should be diluted in ddH_2O for quantitation, as DEPC-treated H_2O is sufficiently acidic to give an inaccurate 260/280 ratio. RNA samples should be stored at −85° and, when handled properly, will retain their integrity for greater than 3 years. Typical RNA yields for a variety of mouse tissues are given in Table I.

Conversion of RNA to cDNA

RNA samples are first treated with DNase to eliminate any contaminating genomic DNA. Two micrograms of RNA is placed into a 0.2-ml PCR tube to which the following is added: 3.36 μl of 25 mM MgCl$_2$

(final concentration of 4 mM), 0.32 μl of DNase I previously diluted 1:5 (0.64 units), and nuclease-free H_2O to a final volume of 20 μl. Also prepare at least one "no-template control" tube that contains H_2O in place of RNA. Tubes are placed into a traditional thermocycler and are processed at 37° for 30 min and at 75° for 10 min.

Immediately following DNase treatment, the RNA is reverse transcribed to cDNA. To the 20-μl DNase reaction the following is added: 20 μl 5 × first strand buffer (final concentration of 1×), 10 μl 0.1 M dithiothreitol (DTT; 10 μM), 20 μl 10 mM dNTPs (2 mM), 10 μl 0.8 mg/ml pdN$_6$ (0.08 μg/μl), 1 μl Superscript II reverse transcriptase (2 units), and nuclease-free H_2O to a total volume of 100 μl per tube. The samples are then returned to the thermocycler and exposed to the following conditions: 25° for 10 min, 42° for 50 min, and 72° for 10 min. The samples are then stored at −20° until use. The final concentration of cDNA in these samples is assumed to be 50 ng per 2.5 μl (standard volume used in the qRT-PCR reaction).

Using master mixes whenever possible is advisable in order to reduce pipetting errors. Thus, a master mix is prepared for the DNase treatment (containing $MgCl_2$ and DNase I) and another for the RT reaction [containing buffer, DTT, dNTPs, p(dN)$_6$, reverse transcriptase, and DEPC-treated H_2O]. These master mixes allow all of these reagents to be introduced to each RNA sample with one pipetting stroke, thus reducing intersample variation.

Primer Design

mRNA sequences are obtained from the National Center for Biotechnology Information database (http://www.ncbi.nlm.nih.gov). Each mRNA sequence is aligned to its corresponding genomic sequence using the BLAST application (NCBI) to determine exon/intron boundaries. Every effort is made to identify a prime pair that spans an intron. This will minimize the chance of amplifying a product from genomic DNA contamination of the RNA sample because the extension conditions favor short products. Primer Express software, version 2.0 (Applied Biosystems) is used to determine optimal primer sequences for use with the ABI Prism 7000 sequence detection system. Each primer should be compared to the entire nonredundant GenBank nucleotide database using the BLAST application (NCBI) to ensure that it recognizes only the gene of interest. Primers designed by this method are optimal for this system as they amplify short products of 60–80 bases in length under the standard two-step PCR cycling conditions employed by the ABI Prism system (1 cycle at 50°, 2 min; 1 cycle at 95°, 10 min; 40 cycles of 95° – 0.15 min, 60° – 1 min, followed by a temperature ramp to display product dissociation characteristics).

Primers are resuspended in PCR-grade H_2O to 100 μM and stored at $-20°$. From these stocks, the primers are diluted to 2.5 μM and each pair (forward/reverse) is mixed at a 1:1 ratio to make a working concentration of 1.25 μM. The final concentration of primers in each qRT-PCR reaction is 150 nM.

All primer sets must be validated before routine use (described later).

qRT-PCR Reaction and Primer Validation

A 96-well optical reaction plate is prepared with each well containing a 20-μl qRT-PCR reaction. The 20 μl consists of 2.5 μl cDNA (assumed to be 50 ng), 10 μl 2 $\times$ SYBR green PCR master mix, 2.4 μl primer mix, and 5.1 μl H_2O. Samples are run in triplicate for each gene (primer mix). Because the SYBR green dye will intercalate with all double-stranded PCR products, multiple genes cannot be distinguished within a single reaction well and thus must be measured independently. Therefore each cDNA sample is evaluated for target genes of interest and a housekeeping gene that serves as a normalizing control in independent wells. In addition, a no-template control sample is included for each primer set to verify that none of the reagents are contaminated by DNA or that the primers do not produce a signal by dimer formation.

Before routine use, all primer sets must be validated to ensure amplification of a single product with appropriate efficiency. To do this one performs a template dilution assay using a cDNA sample from a tissue source expected to contain abundant RNA of interest. The identification of an appropriate tissue source usually can be found in the literature, often in the first report of cloning the gene. Alternatively, the tissue expression profile for a gene may be found at http://expression.gnf.org/cgi-bin/index.cgi (Gene Expression Atlas website of the Genomic Institute of the Novartis Research Foundation). The cDNA sample is serially diluted 1:5 so that analyses are performed for 50 (full-strength), 10, 2, 0.4, 0.08, 0.016, 0.0032 ng, and the no-template control sample (0 ng). qRT-PCR is run on these samples for the new primer set as well as a primer set for a housekeeping gene. Upon completion of the run, the amplification plot for each gene is viewed and a horizontal threshold line is set through the geometric progression phase of amplification. The cycle number determined at threshold (C_T) is plotted against the cDNA concentration (log ng) and should give a linear plot with a slope of -3.3 ± 0.1. The postrun dissociation curve should also be evaluated to ensure a single peak (product) and no amplification by the no-template control. If these conditions are not met, additional primer sets must be designed and tested until one is found to meet these criteria. In summary, a "failed" primer set exhibits any one of the following characteristics: does not amplify a product; does not achieve a C_T vs log[cDNA, ng] slope of -3.3 ± 0.1; produces amplification of the no-template control (C_T of NTC $= 7 + C_T$ of sample); or does not exhibit a single, sharp peak in the dissociation curve analysis that follows each run. Based on the hundreds of primer sets validated by this method, we have an approximately 60% success rate with the first set tested for a given gene.

Once primers are validated they can be used to measure the RNA content of experimental samples. Fifty nanograms (2.5 μl) cDNA of each sample is tested in triplicate, and with accurate pipetting the coefficient of variance for the triplicates will be less than 4%. Fifteen samples can be assessed (a target gene and normalizer gene) per 96-well plate; PCR amplification is complete in just over 2 h.

Quantitative Analysis

Data obtained from the PCR reaction are analyzed using the comparative C_T method (User Bulletin No. 2, Perkin Elmer Life Sciences). The C_T for each sample is manipulated first to determine the ΔC_T [(average C_T of sample triplicates for the gene of interest) – (average C_T of sample triplicates for the normalizing gene)] and second to determine the $\Delta\Delta C_T$ [(ΔC_T sample)-(ΔC_T for the calibrator sample)]. The calibrator is a sample run concurrently on the same plate and designated as an external control, usually the sample showing the lowest expression level (highest ΔC_T) or a sample from the basal treatment group. As all of these values are expressed in log scale, the relative mRNA levels are then established by conversion to a linear value using $2^{-\Delta\Delta CT}$. Statistical evaluation is generally based on biological variation reflecting the mean and standard deviation of multiple samples per treatment group.

Application of qRT-PCR to the Measurement of RGS mRNA Levels

Prior to assaying the tissue distribution and regulation of RGS expression *in vivo*, primer sets were validated for R4, R7, R12, and RZ subfamily members (Table II). Each primer set was validated using a tissue with high expression of the designated RGS gene product and cyclophilin was used as the normalizing gene.

To demonstrate that these primer sets were selective for a particular RGS mRNA, the tissue distribution for the R4 subfamily members was explored using a variety of organs obtained from C57BL/6 mice maintained on a chow diet and sacrificed at the third hour of the light cycle (Fig. 1). Data represent the relative mRNA levels for each RGS in these tissues. In support of reports in the literature, the expression profiles of these RGS

TABLE II

QUANTITATIVE REAL-TIME PCR PRIMERS FOR MEASURING RGS mRNA LEVELS IN THE MOUSE.

RGS family	Gene	Primer sequence (5' → 3') Forward	Primer sequence (5' → 3') Reverse	Position Forward	Position Reverse	Amplicon length (bp)
R4	RGS1 (AF215667)	CGGCCAAGTCCAAAGACATAC	TGTCTGGTTGGCAAGGAGTTT	198–218	280–260	83
	RGS2 (NM_009061)	ATCAAGCCTTCTCCTGAGGAA	GCCAGCAGTTCATCAAATGC	241–261	299–280	58
	RGS3 (AF215669)	TCACACGCAAATGGGAACCT	GCCAGCTTATTCTTCATGTCCTT	109–127	178–156	70
	RGS4 (AB004315)	GGGCTGAATCGTTGGAAAAC	ATTCCGACTTCAGGAAAGCTTT	176–195	250–229	75
	RGS5 (NM_009063)	GCGGAGAAGGCAAAGCAA	GTGGTCAATGTTCACCTCTTTAGG	406–423	477–454	72
	RGS8 (AK044337)	GAGCAAGCATCTCTGACCATCT	ACTGTCGTTGACTGGTCCTCAT	211–232	276–255	66
	RGS13 (AF498319)	ATTATACAGACACAGTGGCGTGTTTT	AAATTGCACGAGACACCAAGTAC	20–44	84–62	65
	RGS16 (NM_011267)	CCTGGTACTTGCTACTCGCTTTT	AGCACGTCGTGGAGAGGAT	36–58	103–85	68
	RGS18 (NM_022881)	CTTCTCCTACAGAGGCCTGACTT	GGCCAAGAGGGCAGATCTAC	329–351	391–372	63
R7	RGS6 (NM_015812)	CGGAGCGAGTCGAGACAAT	TGAGCCATCCTGAGTGTCTTC	314–332	375–355	62
	RGS7 (NM_011880)	GGGAGCCAGAAAACACAGACT	TTGCCTTGTTTTGCATTGTTC	538–558	600–580	63
	RGS9 (NM_011268)	CACCCAGCGCCCTTTGG	GACAATAAGCCCTTGTTTTTGGA	320–335	377–357	58
	RGS11 (XM_128488)	TGCTGAGCTAGACTATGCCATCT	CTAAAGCCCCTTGTTTTTGGA	382–404	443–423	62
R12	RGS10 (NM_026418)	TTGGCTAGCGTGTGAAGATTTC	TGGCCTTTTCTCTGCATCTG	269–290	330–312	62
	RGS12 (NM_173402)	GCTGGATCTCGTTCCGATTAA	CCGCAACACTTCCGTGACT	3284–3304	3359–3341	76
	RGS14 (NM_016758)	GCCCCGGCCAGATATGT	GCGCATAGCTGTCGAACTTC	579–595	649–630	71
RZ	RGS17 (NM_019958)	GGAGAGCATCCAGGTCCTAGA	TGAGACCAGGACAAGACTTCATC	323–343	388–366	66
	RGS19 (NM_026446)	GCAACCCCTCCACTCA	TGTGCTGTTTCTCAGCCTCAT	104–120	157–137	54
	RGS20 (NM_021374)	GCCTGCTGCTTCTGTTGGT	GGTCTTCTTGGTTTCTGACAGTGA	222–240	288–265	67

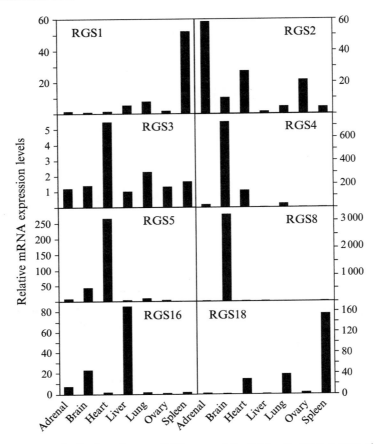

FIG. 1. Tissue distribution of mouse RGS genes of the R4 family. Tissues were obtained from C57BL/6 male mice maintained on a basal diet (Teklad 7001) at the third hour of the light cycle. RNA was prepared and analyzed by qRT-PCR; cyclophilin was used as the normalizer, and the lowest expressing tissue for each RGS was used as the calibrator.

mRNAs are highly region specific, with very little tissue overlap of high mRNA expression between R4 subfamily members (Gold *et al.*, 1997).

Finally, to demonstrate that the regulation of RGS mRNA expression could be monitored *in vivo* using qRT-PCR, the circadian regulation of RGS16 was explored. Northern analysis of pooled samples ($n = 3$) had demonstrated previously that RGS16 mRNA was regulated diurnally in liver (Fig. 2, top). These same samples were utilized to explore the relative expression of RGS16 mRNA using qRT-PCR (Fig. 2, bottom). Although the general trend is remarkably similar between the northern analysis and

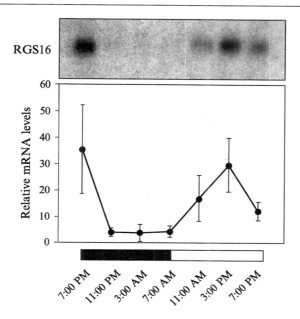

FIG. 2. Comparison of Northern analysis and qRT-PCR. Liver samples were obtained from Agouti 129/SvEv female mice ($n = 3$) at 4-h intervals and total RNA was prepared. (*Top*) Four hundred twenty micrograms of total RNA from each mouse of a given time point was pooled for the preparation of poly(A)$^+$ RNA. Five micrograms of poly(A)$^+$ RNA was analyzed by Northern analysis using a ^{32}P-labeled cDNA probe for RGS16. The completion of this analysis required 1 week. (*Bottom*) Two micrograms of total RNA from each mouse was analyzed by qRT-PCR for the expression of RGS16 and β-actin. This procedure did not involve the use of radioisotopes and was completed in an afternoon. Values depict mean $\pm$ SEM for $n = 3$.

the qRT-PCR result, qRT-PCR data demonstrate a limitation of the prior technique and an advantage of the latter technique, namely quantitation of relative RGS16 mRNA expression by qRT-PCR is more accurate, as demonstrated by the wide variation observed between samples—a detail that was not apparent in the pooled Northern blot. Further, qRT-PCR was faster, more reliable, and able to generate circadian curves for multiple genes of interest, in addition to RGS16, on the same plate (data not shown).

Future Applications

The development of this method to determine the expression of RGS genes will now allow the following questions to be addressed: Can RGS levels be regulated by metabolic state, diet, drug, or GPCR agonist treatment? If so, how does RGS regulation affect G-protein-coupled signaling pathways? Does the circadian rhythm exhibited by RGS16 in liver result from an internal clock mechanism or by a feeding state? Can the measurement of RGS levels serve as a marker for G-protein-coupled receptor action in a given tissue? qRT-PCR should prove to be a useful tool to answer these and other questions concerning the complex biology of G-protein signaling.

Acknowledgments

DKO was supported by NIH HL072551, and TMW receives support from NIH GM61395, PAR-98-057, and the WELCH Foundation (I 1382). We thank Jeff Cormier and Angie Bookout for technical assistance.

References

Bustin, S. A. (2000). Absolute quantification of mRNA using real-time reverse transcription polymerase chain reaction assays. *J. Mol. Endocrinol.* **25**, 169–193.

Chatterjee, T. K., Liu, Z., and Fisher, R. A. (2003). Human RGS6 gene structure, complex alternative splicing, and role of N terminus and G protein γ subunit like (GGL) domain in subcellular localization of RGS6 splice variants. *J. Biol. Chem.* **278**, 30261–30271.

Chen, J.-G., Willard, F. S., Huang, J., Liang, J., Chasse, S. A., Jones, A. M., and Siderovski, D. P. (2003). A seven transmembrane RGS protein that modulates plant cell proliferation. *Science* **301**, 1728–1731.

Dohlman, H. G., and Thorner, J. (1997). RGS proteins and signaling by heterotrimeric G-proteins. *J. Biol. Chem.* **272**, 3871–3874.

Dohlman, H. G., Song, J., Ma, D., Courchesne, W. E., and Thorner, J. (1996). Sst2, a negative regulator of pheromone signaling in the yeast *Saccharomyces cerevisiae*: Expression, localization, and genetic interaction and physical association with Gpa1 (the G-protein α subunit). *Mol. Cell. Biol.* **16**, 5194–5209.

Doupnik, C. A., Davidson, N., Lester, H. A., and Kofuji, P. (1997). RGS proteins reconstitute the rapid gating kinetics of $G\beta\gamma$ activated inwardly rectifying K^+ channels. *Proc. Natl. Acad. Sci. USA* **94**, 10461–10466.

Garnier, M., Zaratin, P. F., Ficalora, G., Valente, M., Fontanella, L., Rhee, M., Blumer, K. J., and Scheideler, M. A. (2003). Up-regulation of regulator of G-protein signaling 4 expression in a model of neuropathic pain and insensitivity to morphine. *J. Pharmacol. Exp. Ther.* **304**, 1299–1306.

Gold, S. J., Ni, Y. G., Dohlman, H. G., and Nestler, E. J. (1997). Regulators of G-protein signaling (RGS) proteins: Region specific expression of nine subtypes in rat brain. *J. Neurosci.* **17**, 8024–8037.

Heid, C. A., Stevens, J., Livak, K. J., and Williams, P. M. (1996). Real time quantitative PCR. *Genome Res.* **6**, 986–994.

Hollinger, S., and Hepler, J. R. (2002). Cellular regulation of RGS proteins: Modulators and integrators of G-protein signaling. *Pharmacol. Rev.* **54**, 527–559.

Kozasa, T., Jiang, X., Hart, M. J., Sternweis, P. M., Singer, W. D., Gilman, A. G., Bollag, G., and Sternweis, P. C. (1998). p115 RhoGEF, a GTPase activating protein for $G\alpha_{12}$ and $G\alpha_{13}$. *Science* **280**, 2109–2111.

Krumins, A. M., Barker, S. A., Huang, C., Sunahara, R. K., Yu, K., Wilkie, T. M., Gold, S. J., and Mumby, S. M. (2004). Differentially regulated expression of endogenous RGS4 and RGS7. *J. Biol. Chem.* **279**, 2593–2599.

Luo, X., Popov, S. G., Bera, A. K., Wilkie, T. M., and Muallem, S. (2001). RGS proteins provide biochemical control of agonist-evoked [Ca^{2+}]$_i$ oscillations. *Mol. Cell* **7**, 651–660.

Neubig, R. R., and Siderovski, D. P. (2002). Regulators of G-protein signalling as new central nervous system drug targets. *Nature Rev. Drug Disc.* **1**, 187–197.

Popov, S. G., Krishna, U. M., Falck, J. R., and Wilkie, T. M. (2000). Ca^{2+}/calmodulin reverses phophatidylinositol 3,4,5-triphosphate-dependent inhibition of regulators of G-protein signaling GTPase activating protein activity. *J. Biol. Chem.* **275**, 18962–18968.

Ross, E. M., and Wilkie, T. M. (2000). GTPase-activating proteins for hetrodimeric G proteins: Regulators of G protein signaling (RGS) and RGS-like proteins. *Annu. Rev. Biochem.* **69**, 795–827.

Tesmer, J. J. G., Berman, D. M., Gilman, A. G., and Sprang, S. R. (1997). Structure of RGS4 bound to AlF$_4^-$-activated G$_{i\alpha1}$: Stabilization of the transition state for GTP hydrolysis. *Cell* **89**, 251–261.

Thellin, O., Zorzi, W., Lakaye, B., De Borman, B., Coumans, B., Hennen, G., Grisar, T., Igout, A., and Heinen, E. (1999). Housekeeping genes as internal standards: Use and limits. *J. Biotechnol.* **75**, 291–295.

Zachariou, V., Georgescu, D., Sanchez, N., Rahman, Z., DiLeone, R., Berton, O., Neve, R. L., Sim-Selley, L. J., Selley, D. E., Gold, S. J., and Nestler, E. J. (2003). Essential role for RGS9 in opiate action. *Proc. Natl. Acad. Sci. USA* **100**, 13656–13661.

Zeng, W., Xu, X., Popov, S., Mukhopadhyay, S., Chidiac, P., Swistok, J., Danho, W., Yagaloff, K. A., Fisher, S. L., Ross, E. M., Muallem, S., and Wilkie, T. M. (1998). The N-terminal domain of RGS4 confers receptor selective inhibition of G-protein signaling. *J. Biol. Chem.* **273**, 34687–34690.

Zhong, H., Wade, S. M., Woolf, P. J., Linderman, J. J., Traynor, J. R., and Neubig, R. R. (2003). A spatial focusing model for G-protein signals. *J. Biol. Chem.* **278**, 7278–7284.

[2] Regulation of Chemokine-Induced Lymphocyte Migration by RGS Proteins

By Chantal Moratz, Kathleen Harrison, and John H. Kehrl

Abstract

G-protein-coupled receptors (GPCRs) activate heterotrimeric G proteins by inducing the G-protein α (Gα) subunit to exchange guanosine diphosphate for guanosine triphosphate. Regulators of G-protein signaling (RGS) proteins enhance the deactivation of Gα subunits, thereby reducing the activation of downstream effectors. Several members of the RGS family are expressed in lymphocytes. Among RGS proteins with the highest levels of expression are RGS1, RGS2, RGS10, RGS13, RGS14, RGS16, and RGS19. Perhaps the most important G-protein-coupled receptors in lymphocytes potentially subject to regulation by RGS proteins are the chemokine receptors. By signaling through these receptors, chemokines help orchestrate immune cell trafficking both during the development of the immune system and during responses to exogenous or infectious agents. Thus, the level and regulation of RGS proteins in lymphocytes likely significantly impact lymphocyte migration and function. This article provides some tools for the analysis of RGS protein expression in lymphocytes and outlines a number of methods for the analysis of the effects of RGS proteins on lymphocyte migration and chemokine receptor signaling.

Introduction

The migration of developing lymphocytes into lymphoid tissues, lymphocyte recirculation, and the organization of secondary immune tissues depend on a regulated configuration of cell surface adhesion molecules, chemoattractant receptors, and a spatial overlay of chemoattractant gradients (Ansel and Cyster, 2001). Most chemoattractant receptors and all chemokine receptors signal through heterotrimeric G proteins. Activated chemokine receptors function as an exchange factor for $G\alpha_i$ subunits, thereby causing the substitution of GTP for GDP, which leads to the dissociation of $G\alpha$ subunits from $\beta\gamma$ heterodimers (Neer, 1995). This causes the activation of downstream effectors and eventually leads to directed cell migration. G_i activation is necessary, but is probably not sufficient for chemotaxis, whereas the release of $\beta\gamma$ heterodimers is required (several reviews discuss the downstream signaling pathways involved in cell migration) (Devreotes and Janetopoulos, 2003; Fukata et al., 2003).

Two major mechanisms limit the duration that the $G\alpha$ subunits remain GTP bound, thereby limiting the duration that $G\alpha$-GTP can activate downstream effectors. First, $G\alpha$ subunits possess an intrinsic GTPase activity, which hydrolyzes the bound GTP, allowing the heterotrimer to reform and signaling to cease (Neer, 1995). Second, members of the regulator of G-protein signaling (RGS) family increase $G\alpha$ GTPase activity dramatically, a property that defines them as GTPase-activating proteins (GAPs) (Kehrl, 1998). Approximately 25 human RGS proteins have now been identified. When tested in standard in vitro GAP assays, most RGS proteins possess GAP activity for α subunits of the G_i and G_q subfamilies. Thus the amount of RGS proteins available to interact with GTP-bound $Gi\alpha$ present at the plasma membrane of a lymphoid cell may significantly alter its response to chemokine stimulation. Previous studies have shown that the expression of RGS1, RGS3, RGS4, and RGS13 in B lymphocytes and RGS16 in T cells impairs chemokine-induced signaling (Bowman et al., 1998; Lippert et al., 2003; Moratz et al., 2000; Reif and Cyster, 2000; Shi et al., 2002). Conversely, B cells from mice that lack Rgs1 have a heightened response to the chemokines CXCL12 and CXCL13 (Moratz and

Kehrl, 2004). This article discusses the RGS proteins expressed in lymphocytes, some of the current methodologies for analyzing lymphocyte migration, and some strategies for exploring the role of RGS proteins in lymphocyte migration.

RGS Proteins in Lymphocytes

Analysis of RGS Protein Expression in Lymphocytes

Because of the limited availability of antibodies with sufficient sensitivity to recognize endogenous RGS proteins, compounded by the difficulties in obtaining a sufficient number of highly purified lymphocyte cell populations, reverse transcription polymerase chain reaction (PCR) has been used predominantly to examine RGS protein mRNA expression. While some preliminary information on the expression of RGS proteins in lymphocytes is available (Table I), the expression of RGS proteins during lymphocyte development; in specific B-cell subsets such as marginal zone B cells, germinal center B cells, B1 B cells, memory B cells, and plasma cells; in T-cell subsets such as CD4 Th1 cells, CD4 Th2 cells, and CD8 cytotoxic T cells; and in natural killer cells requires further study. A survey of RGS mRNA expression in various mouse lymphoid tissues, bone marrow cells, fetal liver hematopoietic cells, and splenocytes is shown in Table I along with the expression pattern in two human Burkitt's lymphoma cell lines, which are likely derived from germinal center B cells. Based on this limited survey, known expressed sequence tags isolated from lymphocytes, plus additional unpublished data, the major RGS proteins expressed in lymphocytes are from the R4 and R12 subfamilies and include RGS1, RGS2, RGS10, RGS13, RGS14, and RGS16. Two other RGS proteins from the R4 class are found at lower levels, RGS3 and RGS18. Marginal zone B cells express significant levels of RGS3, and expressed sequence tags for RGS18 have been isolated from germinal center B cells. One R4 member, RGS8, and one R12 subfamily member, RGS12, have not been reported in lymphocytes with the exception of a single EST for RGS12 from a B-cell line. R7 subfamily members are largely absent from lymphocytes, as are RZ family members, with the exception of RGS19, GAIP. Based on initial analyses, RGS proteins expressed in human and mouse lymphocyte are remarkably similar. Two sets of PCR primers capable of amplifying mRNAs for those RGS proteins found in lymphocytes (mouse and human) are listed in Table II. Also listed are web sites for some of the commercial sources of antibodies purported to react with different RGS proteins found in lymphocytes. The authors cannot vouch for the reliability of the commercial antibodies.

TABLE I
RGS mRNA EXPRESSION IN MOUSE IMMUNE TISSUES AND TWO BURKITT'S
LYMPHOMA CELL LINES[a]

RGS	Peyer's patches	Lymph node	Thymus tissue	Spleen cells	Spleen cells[b]	Bone marrow	Fetal liver	Ramos cells	GA10 cells
RGS1	+2	+1	ND	+2	+3	+2	ND	+1	+2
RGS2	+3	+3	+3	+3	+3	+1	+1	+1	+2
RGS3S	+1	+1	ND	ND	ND	ND	ND	+1	ND
RGS3L	ND	ND	+1	+1	+1	ND	ND	ND	ND
RGS4	ND	+1	ND	ND	ND	ND	ND	ND	ND
RGS5	+3	+3	+2	ND	ND	ND	ND	ND	ND
RGS6	ND	ND	ND	ND	ND	ND	ND	ND	ND
RGS7	ND	ND	ND	ND	ND	ND	ND	ND	ND
RGS8	ND	ND	ND	ND	ND	ND	ND	ND	ND
RGS9	+1	+1	ND	+1	+2	ND	ND	ND	ND
RGS10	+3	+3	+3	+3	+3	+1	+1	+2	+2
RGS11	+1	+1	ND	+1	+1	ND	ND	ND	ND
RGS12	ND	+1	ND	+1	+1	ND	ND	ND	ND
RGS13	+2	+3	+1	+1	+2	ND	ND	+3	+1
RGS14	ND	+1	+1	+2	+2	ND	ND	+1	+1
RGS16	+1	+1	+1	+2	+2	ND	ND	+2	+2
RGS17	+1	+1	+2	+1	ND	ND	ND	ND	ND
RGS18	ND	+1	ND	+1	+1	+2	+2	ND	ND
RGS19	+3	+3	ND	+3	+3	ND	+1	+2	+2
RGS20	ND	ND	ND	ND	ND	ND	ND	ND	ND

[a] RT-PCR analysis of RNA extracted from mouse Peyer's patches, peripheral lymph node, thymus, purified spleen cells, spleen cells (purified spleen cells 1 week postimmunization), bone marrow cells, or fetal liver cells. Ramos and GA10 are human Burkitt's lymphoma cell lines. Each reverse-transcribed RNA was subjected to 33 cycles of PCR with appropriate PCR primers. Results scored as not detected (ND), minimal (+1), moderate (+2), or high (+3) expression.

[b] Purified spleen cells 1 week postimmunization.

Preparation of Cells for the Analysis of RGS Proteins

Standard protocols for reverse transcription PCR or western blotting can be used with the aforementioned PCR primers or RGS antibodies, respectively. Specific lymphocyte subsets from mice (or humans) can be isolated as followed for RNA preparation, for western blotting, and for studies of chemokine signaling. The focus is on the preparation of lymphocytes from mice. Primary and secondary lymphoid tissues in mice can be identified with the help of a basic immunology textbook such as "Fundamental Immunology," "Roitt's Essential Immunology," or "Current Protocols in Immunology."

TABLE II
HUMAN AND MOUSE PRIMER SETS FOR RGS PROTEINS IN LYMPHOCYTES
AND AVAILABLE ANTIBODIES[a]

RGS1 ACCTGAGATCTATGATCCCACATCTGG and GGCTATTAGCCTGCAGGTCAT	460 bp
Rgs1 GATCCCACATCTGGAATCTGG and GCTGTCGATTCTCGAGTATGG	310 bp
http://www.abcam.com/index.html?datasheet=2715	
http://www.novus-biologicals.com/data_sheet.php/4862/RA/33	
http://cbi.swmed.edu/computation/pga/west/sp10_Homo_sapiens/RGS1.htm	
RGS2: CAGACCCATGGACAAGAGCGC and TAGCATGAGGCTCTGTGGTGA	590 bp
Rgs2: AGTGCAGGCAACGGCCCCAAG and TGGGGCTCCGTGGTGATCTGT	570 bp
http://www.abcam.com/index.html?datasheet=9963	
http://www.novus-biologicals.com/data_sheet.php/3543	
RGS3L: TTGGCTGTCAGGGCAGCTGTACAATAGTGG and CACTGAACTCAGTGCGAAGGAAGGCTTGGA	1300 bp
Rgs3L: GTGCTTATTCACTTTGGAGGCACA and TGGGTGGGAGGTCTTGTCCTACAG	410 bp
http://www.abcam.com/index.html?datasheet=2564	
http://www.aviva-antibody.com/catalog.php?action=124&item_id=1978	
RGS3S: CTCCGAGGCATGTACCTCACTCGCAACGGG and CACTGAACTCAGTGCGAAGGAAGGCTTGGA	200 bp
Rgs3S: TCCTGAGTCTCAAGGTGGGGGAC and CAGAGCGGAGGAAGCGAGGGTAAGAGT	562 bp
Same as RGS3L	
RGS10: AGCCTCAAGAGCACAGCCAAATGG and TGCTTCTTAAAGCTGCCAGTCC	494 bp
Rgs10: AGATGGGAGCTCAAGCAGCGG and GGAGGCTCGCTTAGCTGCGGT	470 bp
http://www1.scbt.com/catalog/detail.lasso?-token.order_id= &-database=catalog2003&-layout=web_detail&-Op= eq&catalog_number=sc-6206&-search	
RGS13: ATGAGCAGGCGGAATTGTTGGA and GAAACTGTTGTTGGACTGCATA	480 bp
Rgs13: TCTACACCATTGTACCAGCATGAG and CCTTCTAAGGTCAATCATGGACATGCTGCT	1080 bp
None available, possible noncommercial sources Dr. John Kehrl (jkehrl@niaid.nih.gov); Dr. Kirk Druey (kdruey@niaid.nih.gov)	
RGS14: GGGCACAGCAGCTTCAGATCTTCA and GCCCTGAGACTCTCGGCGCAAGGC	320 bp
Rgs14: AACGGGCGCATGGTTCTGGCTGTCTCAGATGG and CCGACTCTCGCTCTCACTCTC	870 bp

(continued)

TABLE II *(continued)*

None available, possible noncommercial sources Dr. John Kehrl	
(jkehr1@niaid.nih.gov); Dr. David Siderovski (dsiderov@med.unc.edu)	
RGS16: CACCTGCCTGGAGAGAGCCAA and	540 bp
TGGCAGAGGCGGCTGAGGCTT	
Rgs16: GTTCAAGACGCGGCTGGGAAT and	540 bp
AAGGCTCAGCTGGGCTGCCGC	
None available, possible noncommercial sources	
Dr. Kirk Druey (kdruey@niaid.nih.gov)	
RGS18: ATGGAAACAACATTGCTTTTCT and	708 bp
TTATAACCAAATGGCAACATCTGA	
Rgs18: CAGAATATGGATATGTCACTGGTTTTCTTCTCTCA and	720 bp
TACTCATAACCAAATGGCAACA	
TCTGACTTTACATC	
None available, possible noncommercial sources Dr. Yuka Nagata	
(nagata@rtc.riken.go.jp); Dr. Michael Clarke (mclarke@umich.edu)	
RGS19: AGCCGCAACCCCTGCTGCCTG and	540 bp
AGGAGGACTGTGATGGCCCCT	
Rgs19: ACCTCCCAGTCGCAATCCCTG and	550 bp
GACTGTGGGGCCCCCTGGAGT	
http://www1.scbt.com/catalog/detail.lasso?-token.order_id=	
&-database=catalog2003&-layout=web_detail&-Op=	
eq&catalog_number=sc-6207&-search	

[a] The 5' primer is listed first followed by the 3' primer. The size of the PCR product generated is indicated after each primer set. For each RGS protein the human PCR primer set is listed first followed by the murine primers. Internet web sites for RGS specific antibodies are listed below the PCR primers.

1. Sacrifice mice, and lymphoid tissues such as thymus, peripheral lymph nodes, mesenteric lymph nodes, Peyer's patches, or spleen can be harvested.

2. Dissociate the tissues by gentle teasing and disruption with forceps, needles, or scissors in phosphate-buffered saline (PBS)/1% bovine serum albumin (BSA) and then filter through 70-mm Nylon mesh (Falcon) to remove the connective tissue.

3. Peritoneal cavity cells can also be harvested by injecting 7–10 ml of cold PBS into the peritoneal cavity using a 27-gauge needle and removing the cell wash with a 19-gauge needle.

4. Bone marrow cells can be harvested by dissecting the femur and tibia, removing excess tissue, cutting the ends off the bones, and then flushing the marrow out with PBS containing 1% BSA through both ends of the bones.

5. Bone marrow and spleen should be depleted of red blood cells by treatment with ACK buffer (0.15 M NH_4CL, 1 M $KHCO_3$, 0.1 M Na_2EDTA, pH to 7.3) and subsequent washing in 1× PBS.

6. The cell suspensions should be centrifuged at 1000 rpm for 5 min at 4°, the supernatants removed, and the pellet suspended in PBS/1% BSA. A count of the suspensions should be made with a hemocytometer to determine cell number per milliliter.

7. Specific lymphocytes can be isolated using a fluorescent-activated cell sorter (FACS). Pellet cells isolated from tissues and suspend in FACS buffer (PBS pH 7.4/1% BSA fraction V). Stain cells for specific cell markers that identify a subset of cells, such as follicular (CD21$^+$/CD23$^+$) and marginal zone B cells (CD21$^+$/CD23$^-$). After a brief incubation at 4°, wash cells three times with FACS buffer and suspend at 1×10^7/ml in FACS buffer. The cells can then be sorted by flow cytometry. Obtaining significant numbers of cells with this method can be problematic. However, if you are interested in a population that represents a small percentage of cells, it may be the best way to obtain significant numbers of the population to use. After sorting, the cells should be resuspended in RPMI 1640/10% fetal calf serum (FCS) and incubated at 37° for 1 h prior to any signaling or migration assay.

8. Alternatively, cell populations can be obtained by negative selection by magnetic separation. Purify cells from bulk populations by staining cells with specific monoclonal antibodies and negatively selecting those cells, leaving an enriched cell population. For example, to obtain B cells, suspend the cell population in FACS buffer (PBS pH 7.4/1% BSA fraction V) and stain with biotinylated Mab CD4, CD8, GR-1, Mac-1, Ter119, CD11c, and DX5 to remove T cells, macrophages, erythrocytes, natural killer cells, granulocytes, and dendritic cells. Incubate cells at 4° for 15 min, wash with PBS, and suspend in FACS buffer. Add Dynabeads M-280 streptavidin beads (Dynal) washed twice in PBS and twice with FACS buffer to the cell suspensions and incubate at 4°, rotating slowly for 15 min. The ratio of beads to cells is determined according to the manufacturer's protocol. Attach the suspensions to the magnet (Dynal) and allow to separate. Collect the nonadherent suspension, reapply to the magnetic source, and collect the nonadherent cells in suspension again. Wash, count, suspend in RPMI 1640 containing 10% FCS, and incubate this cell population at 37° for 1 h prior to any assay.

Assessment of Chemokine Signaling in Lymphocytes and Alterations by RGS Proteins

Available Assays: Known and Expected Effects of RGS Proteins

Lymphocytes express a broad array of chemokine receptors, whose expression levels determine whether a specific lymphocyte can respond to a specific chemokine. Exposure of lymphocytes to the appropriate chemokine elicits a number of measurable responses, including a transient increase in intracellular calcium levels; the activation of small GTPases such as Rac; transient increases in the phosphorylation of numerous proteins, including extracellular receptor kinases (ERKs), AKT, Src family kinases, and Pyk2; transient dephosphorylation of ERM proteins, which facilitates loss of microvilli and polarization; increased adhesiveness; actin remodeling and cell polarization; and finally directed cell migration. Assays for some of these responses are detailed later. The ability to measure the consequences of RGS protein expression in lymphocytes in these assays depends in large part on a means of manipulating RGS protein expression. Overexpression studies have shown that RGS1, RGS3, RGS13, RGS16, and RGS18 can limit chemokine receptor signaling in lymphocytes (predominantly ERK activation, calcium flux, and chemotaxis have been measured) (Bowman et al., 1998; Lippert et al., 2003; Moratz, et al., 2000; Reif and Cyster, 2000; Shi et al., 2002). These studies have relied on the use of transfected lymphocyte cell lines or transgenic expression in mouse lymphocytes. No published information is available on whether RGS10 or RGS19 affects chemokine receptor signaling. However, the low specificity of RGS19 for Giα2, likely the major mediator of chemotaxis, suggests that it may not have a major role in regulating chemokine signaling (Woulfe and Stadel, 1999). RGS2 minimally affects chemokine receptor signaling presumably because of its poor activity as a Giα GAP. Despite these studies, major questions remain about the roles of RGS proteins in normal lymphocyte function and in the regulation of chemotaxis. Additional studies using RGS-specific siRNAs in lymphocyte cell lines may provide insights into whether RGS proteins possess any chemokine receptor selectivity. The analysis of lymphocytes from mice that have a single disrupted RGS gene will provide some assistance, but such an analysis will likely be complicated by relatively subtle phenotypes and problems of compensation. Because the genes for *Rgs1, Rgs2, Rgs13,* and *Rgs18* are clustered on chromosome 1, interbreeding individual knockout animals will not be possible (Sierra et al., 2002). Several assays used to assess the responses of lymphocytes to chemokines are listed. These assays can be adapted to study the consequences of either over- or underexpression of RGS proteins.

Measurement of Ca^{2+} Influx

Several methods are used to examine changes in intracellular Ca^{2+} in lymphocytes. The major advantage of the following method is that it allows examination of the responses of specific subtypes of lymphocytes within a cell population (Griffioen et al., 1989). Expression of RGS proteins in lymphocyte cell lines reduces the magnitude of the response to chemokines and increases the rate at which Ca^{2+} levels return to baseline. Increased concentrations of chemokine have not reversed the effect of RGS protein overexpression (Moratz, unpublished data). B cells from *Rgs1* -/- mice demonstrate a modest increase in the magnitude of the Ca^{2+} response to chemokine stimulation and a prolongation of the time required to return to a baseline level (Moratz and Kehrl, 2004).

1. Wash cells in Hank's buffered saline solution (HBSS) with calcium (1.26 mM) and magnesium (1 mM) (Biosource International), 10 mM HEPES, and 1% FCS and then suspend at 1×10^7 cells/ml.

2. Add the fluorescent calcium probe Indo-1 (indo-1/acetoxymethyl ester) (Sigma or Molecular probes) at a final concentration of 2 μg/ml plus pluronic detergent (Molecular probes) at a final concentration of 300 μg/ml.

3. Incubate the cells 30 min at 30° while protected from light.

4. Wash cells with HBSS buffer and suspend at 1×10^7/ml in HBSS buffer. Cells can be stained with mAbs to cell surface markers for 10 min and washed three times in HBSS buffer.

5. Warm the cells at 37° for 3 min prior to stimulation.

6. To stimulate, load cells into the Time Zero module (Cyteck, Fremont, CA) and run at 1000 cells/s. Collect baseline for 30 s, inject a sham of 50 μl of HBSS, and at 60 s, inject the stimulant, such as a chemokine.

7. Perform the measurements on a FACSVantage flow cytometer (Becton Dickinson) equipped with an argon laser tuned to 488 nm and a krypton laser tuned to 360 nm. Analyze Indo-1 fluorescence at emission settings 390 and 530, with a slit width of 20, for bound and free probes, respectively. Data can be analyzed using FlowJo software (Tree Star Inc.). Results are best shown as ratio fluorescence (violet/blue).

Induction of ERK and Akt Phosphorylation

The phosphorylation of ERK and Akt occurs within minutes of exposure of lymphocytes to chemokines (Tilton *et al.*, 2000). Overexpression of RGS proteins in lymphocyte cell lines reduces the magnitude of the response substantially, which again cannot be rescued by increasing amounts of chemokine (Moratz, unpublished observation). Because of the difficulties in transiently transfecting lymphocytes, other more transfectable cell lines are often used for this assay, including COS-7, HEK 293, CHO, and HeLa cells. In some instances, the chemokine receptor of interest needs to be cotransfected. Pertussis toxin treatment for 6 h prior to chemokine stimulation at a concentration of 100 ng/ml can serve as a control. Pertussis toxin triggers the ADP-ribosylation of Giα, thereby preventing GDP/GTP

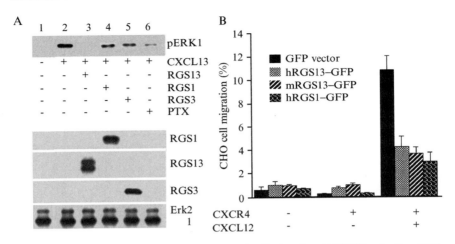

FIG. 1. Chemokine-induced ERK activation and cell migration. (A) ERK activation. COS-7 cells were transfected with expression vectors for CXCR5 and RGS1, RGS3, or RGS13. A day after the transfection, cells were treated with 1 μg/ml of CXCL13 for 15 min prior to harvest and preparation of cell lysates. An immunoblot of cell lysates with a pERK-specific antibody reveals a reduction in pERK levels in RGS1-, RGS3-, and RGS13-transfected cells as well as in cells treated with pertussis toxin. Levels of expression of RGS1, RGS2, RGS13, and Erk1/2 are shown. (B) CXCL12-induced cell migration. CHO cells were transfected with CXCR4 in the presence of expression constructs for GFP, RGS1–GFP, RGS3–GFP, or RGS13–GFP. The following day the cell cultures were harvested and loaded into the insert (5-μm pore size) of a Transwell chamber. CXCL12 was added into a fibronectin-coated bottom chamber (100 ng/ml). After 6 h incubation at 37° in a 5% CO_2 incubator, the migrated cells were harvested and resuspended in 300 μl FACS buffer. GFP-positive cells (included both weak and strongly positive cells) were counted, and the distributions of GFP expression in loaded cells, migrated cells, and nonmigrated cells were determined. The percentage of migrating cells was calculated by dividing the number of GFP-positive cells in the migrated sample by the total number of GFP-positive cells loaded in the upper chamber. Data modified from Shi *et al.*, 2003.

exchange and the dissociation of the heterotrimer and markedly reduces chemokine-induced ERK or Akt phosphorylation. An example of the effects of RGS proteins on chemokine-induced ERK phosphorylation is shown in Fig. 1A.

1. Isolate the lymphocytes as described previously and make a single cell suspension of 1×10^7 cells/ml in PBS/1% BSA media. Cells should be incubated for 60 min at 37° prior to stimulation. Genetically or pharmacologically manipulated lymphocytes can be compared to wild-type cells.

2. Alternatively, transfect COS (or other easily transfectable cell type) cells with the appropriate chemokine receptor expression construct

(0.5–1 μg) in the presence or absence of a tagged RGS protein expression vector (0.5–3 μg) using Superfect (Qiagen) following the manufacturer's protocol. Twenty-four hours after transfection, starve the cells with fresh media without FCS for 6 h prior to the assay.

3. Add the chemokine, i.e., CXCL12 (100 ng/ml), CXCL13 (1000 ng/ml), or CCL19 (250 ng/ml) (R&D Research, Inc.), to the culture at a specific time point. Harvest, centrifuge, and use the cells for making lysates by adding RIPA buffer (150 mM NaCl, 50 mM Tris, pH 7.5, 5 mM EDTA, and 1% Triton X-100, plus protease inhibitors and 2 mM sodium pervanadate) on ice for 20 min.

4. Centrifuge lysates for 20 min at 14,000 rpm, collect the lysate supernatant, and determine the protein concentration (Bio-rad protein assay). Lysates can be used directly or kept at $-80°$ for later use.

5. Fractionate the samples (25–100 μg protein) by SDS–PAGE and immunoblot. We typically transfer to pure nitrocellulose membranes, block the membranes with 10% milk in Tris-buffered saline (TBS), and incubate with the appropriate dilution of antiphosphospecific antibody (Cell Signaling Technology) in 5% nonfat dry milk in TBS with 0.1% Tween 20 at 4° overnight with gentle shaking. By carefully choosing antibodies, both pAKT and pERK can be examined simultaneously (protocols from Alliance for Cellular Signaling: http://www.cellularsignaling. org).

6. Wash the blot three times with TBS/0.1% Tween 20 for 20 min. Use either a directly HRP-conjugated secondary antibody for 1 h at room temperature or a biotinylated secondary with a third strepavidin–HRP-conjugated antibody to detect the primary antibody. Between steps wash as described earlier.

7. Detect proteins by a standard chemoluminescence assay.

8. Subsequently strip the blot and reprobe with ERK and Akt-specific antibodies to ensure equal loading of the lanes.

9. To verify the expression levels of transfected RGS proteins, antibodies directed against the Tag epitope (i.e., HA, myc, or FLAG) can be used for Western blotting the cell lysates.

Measurement of F-Actin Polymerization

The exposure of lymphocytes to chemokines induces a rapid formation of F-actin (Liao *et al.*, 2002). The effects of RGS proteins on chemokine-induced F-actin polymerization have not been reported, although they would be predicted to interfere. The following method provides a simple assay, which utilizes a flow cytometer, thereby allowing the examination of individual cells.

1. Stain the cells for phenotypic markers with either R-PE (phycoery-thrin) or APC (allophycocyanin)-labeled monoclonal antibodies at room temperature, wash, and warm cells to $37°$ for 10 min at 1.25×10^6 cells/ml in HBSS/1% FCS with Ca^{2+} (1.26 mM) and Mg^{2+} (1 mM).

2. Add the chemokine of interest at appropriate concentrations for signaling and vortex the cells.

3. At various time points, dilute a solution of FITC-phalloidin (0.8 U/ml), unconjugated phalloidin (5 μg/ml, Molecular Probes), saponin 0.1%, and 23% paraformaldehyde in PBS 1:5 into the sample. Some protocols do not recommend using unconjugated phalloidin and it is not necessary.

4. Incubate the fixed and permeabilized cells on ice for 10 min, wash twice with PBS containing 1% BSA and 0.1% saponin, and resuspend in PBS.

5. Collect data on a FACSCalibur (BD Bioscences); FITC-phalloidin staining should be quantified on a linear scale.

Measurement of Chemotaxis using a Transwell Chamber

The most common method for measuring lymphocyte chemotaxis uses a chamber based on Boyden's original design (Boyden, 1962), but which has been adapted so that multiple samples can be assayed at one time. This method provides a rapid and easily quantifiable means for assessing the response of lymphocytes to a chemokine. By employing flow cytometry, the chemotatic responses of subpopulations of cells within an input population can be measured. In addition, if an assay for an individual cell characteristic exists, this method can be adapted to examine the migratory ability of functional subsets of cells. For example, ELISPOT assays for immunoglobulin secretion or flow cytometric assays for intracellular cytokine production can be used in conjunction with the modified Boyden chamber to assess these functional response in chemokine-responsive or nonresponsive cells. Despite these advantages, there are several disadvantages of using a Transwell chamber to assess lymphocyte migration. Foremost, the cells cannot be observed during their migration. Second, and perhaps equally important, the assay does not distinguish between chemotaxis, the movement toward or away from a soluble factor, and chemokinesis, enhanced locomotion coincident with exposure to a soluble factor. The expression of RGS1, RGS3, RGS4, RGS13, or RGS16 markedly interferes with the chemokine-induced cell migration of lymphocytes in this assay. A comparison of expressing RGS1, RGS3, or RGS13 on the chemotaxic response to CXCL12 is shown in Fig. 1B. We have also used lymphocyte cell lines that constitutively overexpress an RGS protein using retroviral or lentiviral expression vectors (Moratz, 2000) and, more recently, we have begun to analyze lymphocyte cell lines that have a reduced level of an endogenous RGS protein using RNA knockdown strategies. In addition, transiently transfected lymphocyte cell lines can also be used to evaluate the effects of RGS proteins on lymphocyte chemotaxis. However, for these transient assays it is necessary to use constructs that express an RGS–GFP fusion protein. This allows gating only on those cells that express the RGS protein. Jurkat cells, an immature T-cell line, are readily transfectable by electroporation and migrate well to CXCL12 (40–60% of input cells). Human B-cell lines can also be transfected, although transfection efficiency is usually only on the order of 1–10%. Again, electroporation has proven to be the best transfection method. The cell lines with which we have had the best experience are HS-Sultan and B-JAB, both of which migrate in response to CXCL12 (10–20% of input cells). Some care needs to be taken in choosing B lymphocyte cell lines, as many of the commonly used human lymphoma cell lines express high levels of endogenous RGS proteins and migrate very poorly to chemokines, despite adequate receptor expression. The following method is used for analyzing the effect of an RGS protein on Jurkat cell migration.

1. Transfect Jurkat cells with the RGS–GFP fusion protein construct or a GFP expression vector. We generally use the Amaxa Systems nucleofector protocol for Jurkat cells (http://www.amaxa.com/products/immunology_hematology/cell_lines/vca1003/), although other electroporators can also be used successfully. Culture the transfected Jurkat cells overnight in RPMI 1640 plus 10% FCS. Harvest the transfected cells and suspend at 1×10^7/ml.

2. For the assay we use a 5-μm pore size chamber for both primary lymphocytes and lymphocyte cell lines (Costar transwell chambers #3421, Corning Inc.). In the bottom of the chamber add 600 μl of media either with or without chemoattractant. For Jurkat cells, 100 ng/ml of CXCL12 is sufficient for a maximal response.

3. Place the insert into the chamber and add 100 μl of cells prewarmed to 37° to the inserts and incubate cultures for 2–3 h at 37°.

4. After the incubation, remove the insert, collect cells from the lower chamber, spin them down, and suspend them in 300 μl of FACS buffer.

5. On a flow cytometer, collect data for 1 min for each sample at the high flow rate (important for the calculations). Set the gate on the flow cytometer so that the number of GFP-positive cells can be determined. By analyzing the migratory response of the Jurkat cells on the basis of the RGS–GFP expression level, the effect of expressing a low, intermediate, or a high level of the RGS protein can be assessed.

6. In order to calculate the percentage migration, a 100-μl aliquot from the input cell sample should be run on the flow cytometer like the other samples.

7. The percentage specific migration can be calculated by the following formula: GFP-positive cells in the bottom of the chemoattractant–containing well minus GFP-positive cells in the bottom without chemoattractant divided by GFP-positive cells in the input sample. Compare the GFP control to the RGS–GFP protein.

Use of the Zigmond Chamber to Examine Cell Morphology Changes in Response to a Chemokine Gradient

A Zigmond chamber allows the establishment of a precise chemokine gradient and is useful for observing the orientation of lymphocytes in such a gradient (Zigmond, 1989). In contrast to neutrophils, where such a chamber can be used to measure their directed movement, lymphocytes move poorly, although dramatic membrane changes can be observed. The use of RGS proteins fused to green fluorescent protein allows the detection of any alteration in the intracellular localization of the RGS–GFP fusion protein that occurs during the lymphocyte polarization. To date we have not detected any major alteration in the localization of RGS1–GFP or RGS18–GFP following chemokine exposure; however, other RGS proteins may be more responsive to chemokine signaling. The consequence of exposing a B lymphocyte cell line to CXCL12 is shown in Fig. 2.

1. Coat the coverslip of a Neuro Probe Z02 Zigmond chamber with fibronectin (Sigma F0895, 1:200 dilution of 1% fibronectin in PBS) for 1 h at room temperature

2. Rinse the coverslip with PBS and add cells to the coverslip in a humidified chamber at $37°$ 5% CO_2 for 30 min (cells in RPMI 1640 with 2% FCS) at a concentration of 1×10^6 cells/ml prewarmed to $37°$ for 1 h.

3. Rinse the cell layer on the coverslip with RPMI 1640 plus 2% FCS and drain the excess liquid off. Invert the coverslip onto the chamber so that the cells line over the bridge and then tighten clamps. Add media to open ends of the grooves.

4. Observe cells on the bridge and evaluate cell morphology and movement in the absence of a chemokine gradient for a short duration.

5. Wick away liquid in one groove of the chamber while simultaneously adding media containing chemokine to the other end of the groove to replace media.

6. Observe cells on the bridge and evaluate cell movement as the chemokine gradient forms. The chamber must remain at $37°$ during observation. Cell movement and cell polarization should be observed via an inverted microscope (Zeiss Axiovert 200 or equivalent) with an attached video camera or a charge-coupled device (CCD) digital camera (Hamamatsu Orca-ER or equivalent).

No chemokine   2 min

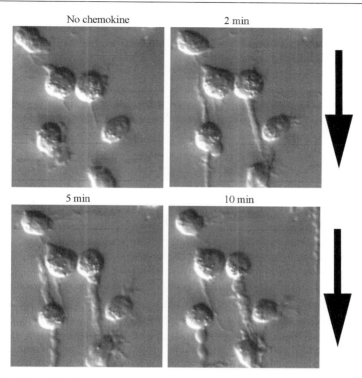

5 min   10 min

FIG. 2. Chemokine-induced morphology change. Still-frame images from a time-lapse experiment with the Zigmond chamber examining the response of a highly migratory subclone variant line of the HS-Sultan human B-cell line. The still frame on the upper left is of cells on the bridge of the chamber when both grooves of the chamber contain media alone. The time after addition of the chemokine CXCL12 is indicated. The four images are of the same group of cells. The arrow indicates the direction of the chemokine gradient.

7. Do not let the coverslip or cells dry, make sure chamber is clean and dry before use, and avoid air bubbles when inverting coverslip onto the chamber and when replacing media with media plus chemokine. Care needs to be taken not to damage the cells by inadvertently sliding the coverslip while placing it on the chamber.

In Vivo *Cell Trafficking*

A number of approaches have been used to study lymphocyte trafficking *in vivo*, including imaging of explanted lymph nodes or intravital imaging using two-photon or confocal microscopes (Cahalan *et al.*, 2003;

Delon *et al.*, 2002) One approach, which does not require expensive imaging equipment, but is limited to animal studies, has been to *ex vivo* label mouse lymphocytes and then reintroduce them into a donor animal for tracking using either the flow cytometric analysis of cells derived from lymphoid tissues or immunohistochemistry. This approach is most useful for examining the *in vivo* trafficking of genetically modified lymphocytes or lymphocytes treated with pharmaceutical agents, whose effects are not readily reversible. The consequences of manipulating RGS protein levels of *in vivo* lymphocyte trafficking should provide some interesting insights into the functional importance of these proteins. Two types of dyes can be used to trace cells *in vivo*. They are intracellular dyes such as CFSE [5-(and-6)-carboxyfluorescein diacetate succinimidyl ester] or lipophilic fluorescent tracers such as the dialkylcarbocyanines DiI, DiO, DiD, and DiR. CFSE is stably incorporated into the cell and can be used to monitor short- and long-term (up to several months) cell migration and cell division. In contrast, lipophilic dyes are incorporated into the cell membrane and are not useful for long-term experiments. The trafficking of CFSE-labeled B lymphocytes 1–7 days after transfer is shown in Fig. 3.

1. To label high numbers of cells with CFSE (Sigma or Molecular Probes), resuspend the cells at a concentration of 50×10^6/ml in HBSS

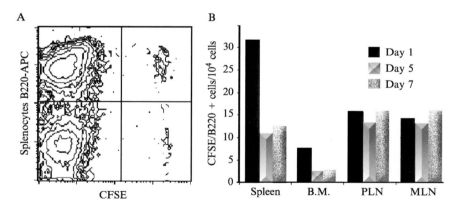

FIG. 3. Localization of CFSE-labeled B lymphocytes following transfer to recipient mice. (A) Flow cytometry analysis of cells isolated from spleens 5 days posttransfer of CFSE-labeled cells. Spleen cells were isolated from recipient mice and stained with B220 conjugated directly to APC. Analysis shows B220+/CFSE+ cells in contour plots of lymphocyte-gated cells, located in the upper right quadrant. (B) Quantifying CFSE cells after transfer in recipient lymphoid tissues. Data from flow cytometry analysis were used to calculate the number of CFSE B cells present in spleen, bone marrow (B.M.), peripheral lymph node (PLN), and mesenteric lymph node (MLN) at 1, 5, and 7 days posttransfer.

(without serum). For lower cell numbers, resuspend the lymphocytes in PBS containing 5% FCS at 0.5 to 10 $\times$ 10^6 cells/ml. FCS is needed at the low cell concentration to buffer the toxic effect of the concentrated dye.

2. Dilute the stock CFSE dye to 50 μM in PBS and add 110 μl of dye solution (final concentration is 1–5 μM) per milliliter of cells and mix rapidly by pipetting.

3. Incubate for 5 min at room temperature and wash with 10 volumes of PBS 5% FCS; centrifuge cells for 5 min at 300g at 20°. Remove supernatant and wash three additional times or until wash buffer is clear.

4. To label lymphocytes with lipophilic dyes, resuspend the cells at 5–10 $\times$ 10^6 cells/ml in buffered salt solution (HBSS).

5. Add the dye to the cell solution so that its final concentration is 0.5–50 μM. Incubate the cells at room temperature for 5 min.

6. Remove excess dye by centrifuging the cells for 5 min at 300g at 20°. Remove the supernatant and add 10 volumes of HBSS with 5% FBS. Repeat the wash three times.

7. To transfer the labeled lymphocytes, suspend them in PBS and inject 10–40 $\times$ 10^6 cells in 0.2 ml intravenously into the lateral tail vein of a mouse.

8. Lymphoid organ cells of recipient mice are typically harvested 30–90 min after transfer, counted, and subjected to flow cytometric analysis. Because CFSE acquires the identical spectral characteristics as fluorescein when carboxyl groups are cleaved by cytoplasmic esterases, it is excited efficiently by the 488-nm argon laser standard in most flow cytometers. This allows the concurrent use of other fluorochromes such as phycoerythrin (PE) or allophycocyanin (APC). Block single cell suspensions of 1 $\times$ 10^6 cells per sample with the appropriate blocking reagents (such as 5% BSA/PBS or CD16/CD32 Fcγ block; if a secondary monoclonal is needed to detect the primary monoclonal antibody, the cells should be blocked with 5% serum of the source of the secondary antibody in a protein-blocking solution) on ice for 15 min, wash, and then incubate with PE- or APC-labeled mAb in FACS buffer. After a 15-min incubation on ice, wash the samples three times with FACS buffer and perform analysis on a flow cytometer. If the analysis is to occur more than 24 h after transfer, a higher concentration of CFSE should be used (5 μM) in the initial labeling. Despite their loss of fluorescence with time, CFSE-labeled cells can be detected for several months following transfer. The accumulation of wild-type cells in various lymphoid organs can be compared to those of genetically modified cells.

9. The positioning of the transferred cells in lymphoid tissues can be determined by fluorescence microscopy of tissue sections. Use harvested lymphoid tissues to make frozen tissue blocks in Tissue Tek OCT compound (Fisher, Inc.). Fix tissue sections in 4% paraformaldehyde in

0.1 M phosphate buffer, pH 7.4, wash twice in PBS, and then incubate in 20% sucrose PBS at 4° for 1 h. After rinsing in cold PBS, snap freeze the tissues in Tissue Tek OCT compound (Fisher). Tissue sections (8 μm) can be cut from the snap-frozen block using a cryostat. Rehydrate thawed sections in PBS, fix again in 4% paraformaldehyde as described earlier, wash in PBS, and then quench with 50 mM NH$_4$Cl in PBS (quenching of debatable value). The tissue sections can then be stained with the APC or PE-labeled antibodies of interest. Note that no organic fixatives can be used (alcohol, acetone, xylene), as they will remove the dye from the tissue. Only water-based fixatives can be used such as paraformaldehyde. Additionally, paraffin embedding is not acceptable.

References

Ansel, K. M., and Cyster, J. G. (2001). *Curr. Opin. Immunol.* **13,** 172–179.

Bowman, E. P., Campbell, J. J., Druey, K. M., Scheschonka, A., Kehrl, J. H., and Butcher, E. C. (1998). *J. Biol. Chem.* **273,** 28040–28048.

Boyden, S. (1962). *J. Exp. Med.* **115,** 453–466.

Cahalan, M. D., Parker, I., Wei, S. H., and Miller, M. J. (2003). *Curr. Opin. Immunol.* **15,** 372–377.

Delon, J., Stoll, S., and Germain, R. N. (2002). *Immunol. Rev.* **189,** 51–63.

Devreotes, P., and Janetopoulos, C. (2003). *J. Biol. Chem.* **278,** 20445–20448.

Fukata, M., Nakagawa, M., and Kaibuchi, K. (2003). *Curr. Opin. Cell. Biol.* **15,** 590–597.

Griffioen, A. W., Rijkers, G. T., Keij, J., and Zegers, B. J. (1989). *J. Immunol. Methods* **120,** 23–27.

Kehrl, J. H. (1998). *Immunity* **8,** 1–10.

Liao, F., Shirakawa, A. K., Foley, J. F., Rabin, R. L., and Farber, J. M. (2002). *J. Immunol.* **168,** 4871–4880.

Lippert, E., Yowe, D. L., Gonzalo, J. A., Justice, J. P., Webster, J. M., Fedyk, E. R., Hodge, M., Miller, C., Gutierrez-Ramos, J. C., Borrego, F., Keane-Myers, A., and Druey, K. M. (2003). *J. Immunol.* **171,** 1542–1555.

Moratz, C., Hayman, R., Hua, G., and Kehrl, J. (2004). *Mol. Cell. Biol.* **24**(13), 5767–5775.

Moratz, C., Kang, V. H., Druey, K. M., Shi, C. S., Scheschonka, A., Murphy, P. M., Kozasa, T., and Kehrl, J. H. (2000). *J. Immunol.* **164,** 1829–1838.

Moratz, C. unpublished data.

Neer, E. J. (1995). *Cell* **80,** 249–257.

Reif, K., and Cyster, J. G. (2000). *J. Immunol.* **164,** 4720–4729.

Shi, G. X., Harrison, K., Wilson, G. L., Moratz, C., and Kehrl, J. H. (2002). *J. Immunol.* **169,** 2507–2515.

Sierra, D. A., Gilbert, D. J., Householder, D., Grishin, N. V., Yu, K., Ukidwe, P., Barker, S. A., He, W., Wensel, T. G., Otero, G., Brown, G., Copeland, N. G., Jenkins, N. A., and Wilkie, T. M. (2002). *Genomics* **79,** 177–185.

Tilton, B., Ho, L., Oberlin, E., Loetscher, P., Baleux, F., Clark-Lewis, I., and Thelen, M. (2000). *J. Exp. Med.* **192,** 313–324.

Woulfe, D. S., and Stadel, J. M. (1999). *J. Biol. Chem.* **274,** 17718–17724.

Zigmond, S. H. (1989). *Curr. Opin. Cell. Biol.* **1,** 80–86.

[3] Role of Palmitoylation in RGS Protein Function

By Teresa L. Z. Jones

Abstract

Palmitoylation, the reversible, post-translational addition of palmitate to cysteine residues, occurs on several regulators of G-protein signaling (RGS) proteins. Palmitoylation can occur near the amino terminus, as for RGS4 and RGS16, but can also occur on a cysteine residue in the $\alpha 4$ helix of the RGS box, which is conserved in most RGS proteins. For some of the RGS proteins, palmitoylation is required to turn off G-protein signaling by accelerating GTP hydrolysis on the $G\alpha$ subunit. This article discusses the role of palmitoylation in RGS function and protocols are given for metabolic and *in vitro* labeling of RGS proteins with [^{3}H]palmitate and measurement of GTP hydrolysis in membranes.

Introduction

Regulators of G-protein signaling (RGS) proteins are potent regulators of cell signaling. Their activity can be harnessed by regulating their protein concentration through changes in synthesis and degradation, but also more rapidly and efficiently through post-translational modifications. Many RGS proteins undergo palmitoylation, the post-translational addition of palmitate to cysteine residues through a thioester bond. This reversible modification can change the conformation, membrane localization, and orientation of proteins. Phosphorylation, another post-translational modification, also occurs on a subset of RGS proteins and has been discussed elsewhere (Hollinger and Hepler, 2004).

Protein Lipidation

Protein lipidation occurs on many extracellular and intracellular proteins, including G-protein signaling proteins (Casey, 1995). The best characterized modifications are myristoylation, isoprenylation, and palmitoylation (Table I). Myristoylation is the addition of a 14 carbon myristate group to an amino-terminal glycine through an amide bond. The $G\alpha_i$ subclass of $G\alpha$ subunits undergoes this modification. Isoprenylation is a multistep process on proteins ending with a CAAX motif (C-cys, A-aliphatic amino acid, X-any amino acid) that culminates in the addition

TABLE I
PROTEIN LIPIDATION

Modification	Myristoylation	Isoprenylation	Palmitoylation
Lipid	14C myristic acid	15C or 20C isoprenoid group	16C palmitic acid
Consensus sequence	Amino terminus MGXXXSX[a]	Carboxy-terminus CAAX[b]	No identified consensus sequence
Linkage	Stable amide bond to glycine	Stable thioether bond to cysteine	Reversible thioester bond to cysteine
Cellular site of modification	Cytosol	Cytosol	Membranes
Cellular sites of lipidated proteins	Cytosol and membranes	Cytosol and membranes	Membranes
Examples of lipidated proteins	$G\alpha_i$ subunits, src family kinases	$G\gamma$ subunits, *ras* G proteins	G-protein-coupled receptors, $G\alpha$ subunits, RGS proteins

[a] For the myristoylation consensus sequence in position 3, charged residues, proline, and large hydrophobic residues are not allowed; in position 6, small uncharged residues are allowed (Ser, Thr, Ala, Cys, Asn, and Gly with Ser favored); and in position 7, proline is not allowed.

[b] A is an aliphatic residue and X is any residue.

of an isoprenoid group to the cysteine residue, cleavage of the −AAX peptide, and methylation of the cysteine residue. $G\gamma$ subunits undergo isoprenylation with either a 15 carbon farnesyl group on the $G\gamma$ of transducin or a 20 carbon geranylgeranyl group on the other $G\gamma$ subunits.

Palmitoylation occurs on most G-protein-coupled receptors, $G\alpha$ subunits, and many RGS proteins and has conspicuous differences with myristoylation and isoprenylation (Linder and Deschenes, 2003; Qanbar and Bouvier, 2003). Palmitoylation occurs at the membrane, and palmitoylated proteins generally have other lipid modifications or membrane-targeting signals to get them to the membrane. In addition to the absolute requirement of a cysteine residue near the membrane, no consensus amino acid sequence has been identified. Compared to myristoylation, the two extra carbon groups on palmitate increase the membrane affinity to a point that strongly inhibits translocation off the membrane. Studies with lipidated peptides indicate that the 14 carbon myristoyl group leads to marginal membrane affinity, but a 16 carbon palmitoyl group tightly attaches peptides to membranes and essentially stops random dissociation (Peitzsch and McLaughlin, 1993). From a functional standpoint, the most significant difference between palmitoylation and other modifications is the reversibility of palmitoylation compared to the stability of myristoylation and isoprenylation, which remain for the life of the protein. Palmitoylated proteins undergo a thioacylation cycle between their palmitoylated and depalmitoylated states. For example, the β-adrenergic receptor (Mouillac *et al.*, 1992) and $G\alpha_s$ (Degtyarev *et al.*, 1993) undergo a thioacylation cycle after agonist binding to the receptor. The palmitoylated state favors some interactions such as for $G\alpha_s$, the binding to $G\beta\gamma$ (Iiri *et al.*, 1996), and the coupling to the receptor and effector (Ugur *et al.*, 2002). However, the depalmitoylated state is also important for the β-adrenergic receptor to bind to its kinase (Moffett *et al.*, 1993) and for $G\alpha$ subunits to bind to RGS proteins (Tu *et al.*, 1997).

Enzymology of Palmitoylation

Our knowledge of the enzymes involved in adding and removing palmitate from proteins is at an early stage compared to other lipid modifications. Two enzymes have been identified in yeast that can transfer palmitate to proteins—the Erf2p/Erf4p complex is a palmitoyl transferase for the yeast Ras2 protein (Lobo *et al.*, 2002) and Akr1p is a palmitoyl transferase for Yck2p (Roth *et al.*, 2002). Both of these enzymes share a DHHC cysteine-rich domain, and residues in this domain are critical for transferring palmitate. The DHHC cysteine-rich domain is found in about 20 mammalian proteins, but the function of these proteins has not been reported (Putilina *et al.*, 1999). AcyI protein thioesterase 1 (APT1) is a strong candidate for the enzyme that removes palmitate (Duncan and Gilman, 1998, 2002). Although this cytosolic enzyme was first isolated as a lysophospholipase, deletion of the gene in yeast leads to a dramatic reduction in palmitate turnover on the yeast $G\alpha$ subunit, Gpa1 (Duncan and Gilman, 2002). APT1 is a member of the α/β hydrolase family of acylhydrolases based on its crystal structure (Devedjiev *et al.*, 2000). At least one other isoform of this enzyme exists, although the specificity of the enzymes for different palmitoylated proteins has not been determined (Wang and Dennis, 1999).

Palmitoylation of RGS Proteins

[^{3}H]Palmitate incorporation into RGS proteins was first shown for RGS–GAIP (De Vries *et al.*, 1996), with several others subsequently shown to undergo palmitoylation (Fig. 1 and Table II). Based on sequence homology, other RGS proteins are likely to undergo palmitoylation (Tables II and III). Palmitoylation occurs in different patterns for RGS proteins— multiple sites at the amino terminus, cysteine string motifs, and internal palmitoylation. At a basic level, the regulation of palmitoylation is a function of the proximity of the side chain of the cysteine residue to the

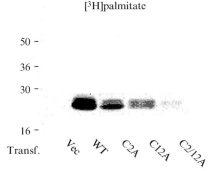

[³H]palmitate

FIG. 1. Palmitoylation of RGS16 after metabolic radiolabeling. COS cells transfected with vector alone or with cDNAs for the wild-type RGS16 or the indicated cysteine mutants were labeled metabolically with [³H]palmitate and were then homogenized and separated into particulate and soluble fractions by centrifugation. Particulate fractions were immunoprecipitated with a polyclonal antibody to RGS16 followed by separation on SDS–PAGE on 12% acrylamide gels and fluorography. Gels were exposed to Kodak MS film for 13 days at -80°. Molecular mass markers in kilodaltons are shown to the left. Reprinted with permission from Druey *et al.* (1999).

protein acyl transferase and palmitoyl CoA in the membrane or the exposure of the thioester bond between cysteine and palmitate to a cytosolic acyl protein thioesterase. For RGS3 and RGS10, receptor activation leads to an increased incorporation of [³H]palmitate (Castro-Fernandez *et al.*, 2002). For RGS4 and RGS16, palmitoylation at the internal site is dependent on amino-terminal palmitoylation, possibly by stabilizing membrane binding or targeting the RGS protein to a protein acyl transferase in lipid rafts to allow palmitoylation of the poorly accessible cysteine residue (Hiol *et al.*, 2003; Tu *et al.*, 1999). The function of multiple acylation sites may be to ensure the persistence of membrane localization, even during turnover of the palmitate at individual sites, although the stoichiometry of palmitoylation for any RGS protein is not known.

Function of RGS Protein Palmitoylation

Prevention of palmitoylation by mutation of cysteine residues leads to an inability of RGS16 to turn off G-protein signaling in cells (Druey *et al.*, 1999). RGS16, like other RGS proteins, does not need palmitoylation for GTPase-accelerating protein (GAP) activity in assays with soluble $G\alpha$ subunits and RGS proteins (Druey *et al.*, 1999). However, palmitoylation increases the GAP activity of RGS4 and RGS16 greatly when measured in a membrane-based assay, after overexpression of the RGS protein or after addition of the palmitoylated RGS protein to the membranes (Fig. 2).

TABLE II
PUTATIVE ACYLATION SITES OUTSIDE OF THE RGS BOX[a]

RGS protein	Type of acceptor site	Cysteine residues crucial for acylation or possible sites of acylation[b]	Reference
RGS–GAIP	Cys string motif	8 Cys near the aminoterminus	De Vries *et al.*, 1996
RET–RGS1	Cys string motif	12 Cys between the TM domain and the RGS box	
RGSZ1	Cys string motif	9 Cys near the aminoterminus	
RGS17	Cys string motif	9 Cys near the aminoterminus	
RGS2	Amino terminus	Cys-13	
RGS3	[c]	[c]	Castro-Fernandez *et al.*, 2002
RGS4	Amino terminus	**Cys-2, Cys-12**	Srinivasa *et al.*, 1998
RGS5	Amino terminus	Cys-2, Cys-12	
RGS8	Amino terminus	Cys-19, Cys-27	
RGS13	Amino terminus	Cys-6, Cys-9, Cys-12	
RGS16	Amino terminus	**Cys-2, Cys-12**	Druey *et al.*, 1999
RGS6	Internal sites	4 Cys between DEP and GGL domains	
RGS7	Internal sites	3 Cys between DEP and GGL domains	Rose *et al.*, 2000
RGS11	Internal sites	3 Cys between DEP and GGL domains	

[a] RGS proteins chosen for this table have either been shown to undergo palmitoylation (in boldface) or share homology with palmitoylated proteins. They are arranged in subclasses based on sequence homology (De Vries *et al.*, 2000). RGS10 undergoes palmitoylation (Castro-Fernandez *et al.*, 2002; Rose *et al.*, 2000), but like RGS1, has no cysteine residue outside of the RGS box.

[b] Residues shown in boldface are critical for [³H]palmitate incorporation based on mutagenesis studies. The other sites are speculative based on sequence homology.

[c] No obvious site of palmitoylation with three cysteine residues in a proline-rich region from residues 88 to 135 and then three other cysteines before the RGS box.

For RGS10, palmitoylation also increases agonist-stimulated GAP activity greatly in receptor–G_i protein proteoliposomes (Tu *et al.*, 1999).

An understanding of the function of palmitoylation for RGS proteins can be based on the premise that palmitoylation acts by changing the relationship of a protein with the membrane. The function of palmitoylation for interactions between a palmitoylated protein and the membrane can be divided into three interrelated categories: membrane attachment and targeting, membrane microdomain localization, and protein orientation at the membrane.

TABLE III
MAMMALIAN RGS PROTEINS WITH A CYSTEINE RESIDUE IN THE $\alpha 4$ HELIX OF THE RGS BOX[a]

RGS proteins with a cysteine aligned with C98 in RGS16		Cysteine residues in other positions in the $\alpha 4$ helix[b]	No cysteine residues in the $\alpha 4$ helix[c]
RGS1	RGS12	PDZRhoGEF (−1)	RGS6 (Val)
RGS2	RGS13	p115RhoGEF (−3, −4)	RGS7 (Val)
RGS3	RGS14		RGS-PX1 (Val)
RGS4	RGS16		LARG (Ser)
RGS5	RGS–GAIP		D-AKAP2 (Ala)
RGS8	RGSZ1		GRK2 (Ile)
RGS9	RET-RGS1		GRK3 (Ile)
RGS10	Conductin (Axil)		
RGS11	Axin		

[a] The sequence of this helix for RGS16 is **EEN**LE**FW**LA**C**EE**FK**. The underlined residues are in contact with the switch region of $G\alpha_{i1}$ in the crystal structure of the RGS4–$G\alpha_{i1}$ complex. Boldface residues are highly conserved among RGS proteins.

[b] Position of cysteine relative to Cys-98 in RGS16 is shown in parentheses.

[c] The residue in the Cys-98 position is shown in parentheses. Reprinted with permission from Hiol *et al.* (2003b).

Membrane Attachment and Targeting. Proteins that undergo palmitoylation have some affinity for membranes independent of palmitoylation because the modification occurs at the membrane. Palmitoylation often occurs within or adjacent to membrane-targeting signals that direct proteins to different intracellular membranes (Ugur and Jones, 2000). Palmitoylation is not a targeting signal itself, but acts in conjunction with a protein sequence to stabilize the targeting. The affinity of the protein sequence for the acceptor at the membrane will determine the relative importance of palmitoylation for maintaining the membrane localization. For RGS4 and RGS16, a mutation of cysteine residues that eliminate [^{3}H]palmitate incorporation did not change the membrane attachment or the plasma membrane targeting of these proteins, suggesting that the targeting sequence is adequate for membrane targeting (Chen *et al.*, 1999; Hiol *et al.*, 2003b; Srinivasa *et al.*, 1998). RGS4 and RGS16, as well as RGS5, have an amphipathic amino-terminal sequence with structural homology to the CTP:phosphocholine cytidylyltransferase that may lead to its intercalation into the membrane and stabilization of membrane attachment (Bernstein *et al.*, 2000; Chen *et al.*, 1999).

Lipid Raft Localization. A growing body of evidence indicates that biological membranes are not homogeneous, but contain microdomains enriched in cholesterol and sphingolipids (Brown and London, 2000; Simons and Toomre, 2000). These microdomains, called lipid rafts, are

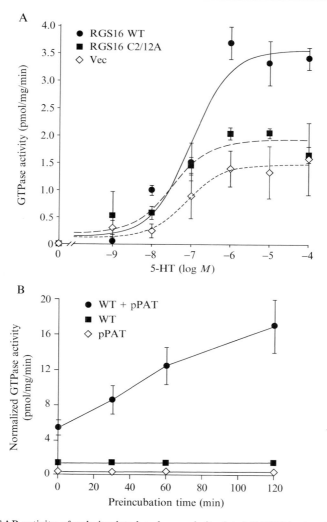

Fig. 2. GAP activity of palmitoylated and nonpalmitoylated RGS16 in transfected cell membranes or after addition to membranes. (A) HEK293 cells stably expressing a fusion protein between the 5-HT$_{1A}$ receptor and Gα_{o1}(C351G) were transfected transiently with plasmids containing *LacZ* as a control (Vec), RGS16 (WT), or RGS16 with Cys-2 and Cys-12 mutated to alanine (C2/12A). Cells were pretreated with pertussis toxin to block the activity of the endogenous Gα_i and Gα_o proteins. Forty-eight hours after transfection, agonist-stimulated steady-state GTPase activity was determined after addition of the indicated concentrations of 5-HT and incubation for 20 min at 37°. (B) Purified, recombinant RGS16 WT was treated with and without partially purified PAT (pPAT) and palmitate. These preparations and partially purified PAT were preincubated for various time periods with

the home to a number of acylated proteins because saturated fatty acids have a high affinity for the liquid-ordered nature of these lipid domains created by cholesterol and the saturated fatty acids on the sphingolipids. However, not all palmitoylated proteins are found in lipid rafts because, for example, RGS7 and Gα13 undergo palmitoylation, but do not reside in lipid rafts (Rose et al., 2000; Waheed and Jones, 2002). Lipid rafts are also the home to signaling proteins where they can segregate proteins to enhance or inhibit signaling based on the function of proteins in or out of the lipid rafts (Miura et al., 2001; Simons and Toomre, 2000). The RGS9-1–Gβ5L complex translocates to lipid rafts along with the Gα subunit, transducin, and an isoform of arrestin upon activation of the receptor (Nair et al., 2002). (The palmitoylation of RGS9 has not been reported.)

RGS16 is found in lipid rafts and palmitoylation is critical for this targeting (Hiol et al., 2003b). The lipid raft localization of RGS16 may have several different effects. RGS16 inhibits G13 signaling, independent of effects on GTP hydrolysis (Johnson et al., 2003). Overexpression of RGS16 translocates Gα13 into lipid rafts and this change in membrane localization may at least partially explain its inhibitory function on G13 signaling. Most Gα subunits reside in lipid rafts, but the function of the lipid raft targeting for RGS proteins may not be a simple matter of colocalization with the Gα subunit. Disruption of lipid rafts by cholesterol depletion does not diminish the ability of RGS16 to accelerate the GTP hydrolysis of a Gα subunit and even accentuated the agonist-stimulated GAP activity (Hiol et al., 2003b). Instead, lipid raft localization may act to bring RGS16 to a microdomain enriched in protein acyltransferase activity in order to promote palmitoylation of a conserved cysteine buried in the RGS box and facing away from the membrane (Dunphy et al., 2001; Osterhout et al., 2003) (Fig. 3).

Orientation of Proteins at the Membrane. The membrane attachment of proteins increases their effective concentration, but the price is a decrease in their degree of freedom. Palmitoylation can act to orient proteins at the membrane to facilitate interactions with other membrane-bound proteins. RGS4 and RGS16 do not require palmitoylation for their GAP activity in solution, but do need this modification for GAP activity in cell membranes (Fig. 2). RGS4, RGS10, and RGS16 undergo palmitoylation on a conserved cysteine residue in the RGS box (Hiol et al., 2003b; Tu et al.,

---

HEK293 membranes expressing the 5-HT$_{1A}$/Gα$_{o1}$ fusion protein. GTPase activity of the membranes was determined after a 20-min incubation with 5-HT (10^{-6} M). Values shown are the mean ± SEM of four experiments. Reprinted with permission from Hiol et al. (2003b) and Osterhout et al. (2003).

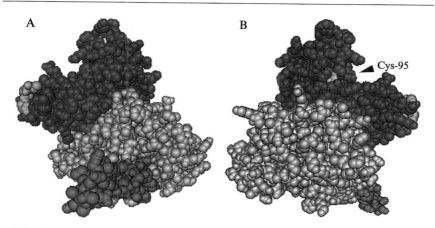

FIG. 3. Structure of the RGS box binding $G\alpha_i$. A space-filling model of the RGS box of RGS4 (dark green) binding $G\alpha_{i1}$ (gray) based on the crystal structure of this complex (Tesmer *et al.*, 1997). (A) The presumed membrane-facing surface of the complex. (B) Rotation by about 120° through the vertical axis of the structure shown in A. Cys-95 (yellow), which corresponds to Cys-98 in RGS16, is only visible on this surface. The amino-terminal end of the RGS box, which starts at residue 51, is shown in aqua, and the amino- and carboxyl-terminal ends of $G\alpha_{i1}$ are shown in pink. Reprinted with permission from Hiol *et al.* (2003b). (See color insert.)

1999) (Table III). This residue is buried within the structure of the RGS box with the only surface accessibility on a face of the structure a 120° rotation from the likely membrane-facing surface of the RGS4/$G\alpha_i$ complex (Fig. 3) (Tesmer *et al.*, 1997). Whereas palmitoylation in general enhances the GAP activity of RGS4 and RGS16, palmitoylation of the internal cysteine residue is particularly important (Fig. 4). In our model, the $\alpha 4$ helix of the RGS box must rotate about 90° in order for the side chain of the cysteine residue to undergo palmitoylation and the palmitate to insert in the membrane (Fig. 5). The proximal end of the $\alpha 4$ helix is in direct contact with the switch region of the $G\alpha$ subunit. Therefore, a change in the orientation of the $\alpha 4$ helix is likely to affect the GAP activity of the protein. The crystal structure of the RGS4/$G\alpha_i$ complex provides a foundation for understanding the interactions between these proteins, but the structure is based on their interactions in solution where these contacts are not limited. However, these proteins act at the membrane and their membrane orientation undoubtedly affects their ability to form a tight fit for the RGS protein to stabilize the transition state of the $G\alpha$ subunit in GTP hydrolysis. Palmitoylation of the conserved cysteine residue in the $\alpha 4$ helix may be necessary to optimally orient the RGS protein for the membrane-bound $G\alpha$ subunit.

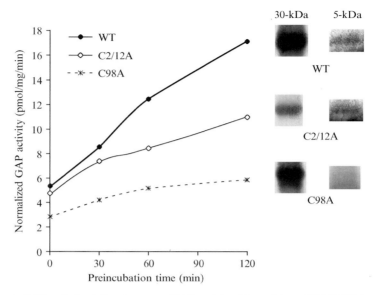

FIG. 4. RGS palmitoylation increases GAP activity. Recombinant WT RGS16 and the palmitoylation-defective mutants (C2/12A and C98A) were treated with partially purified PAT and palmitate and then preincubated for various time periods with HEK293 membranes expressing the 5-HT$_{1A}$/Gα_{o1} fusion protein. GTPase activity of the membranes was determined after a 20-min incubation with 5-HT (10^{-6} M). Normalized GTPase activity is calculated as raw activity divided by the activity of 5-HT-treated membranes not preincubated with RGS proteins. Values represent the mean ± SEM of four independent experiments. Activity correlates with palmitoylation of the Cys-98 residue in RGS16 as revealed by [^{3}H]palmitate incorporation into a 5-kDa band generated by clostripain cleavage of full-length (30-kDa band) RGS16 WT or C2/12A, but not C98A. Reprinted with permission from Osterhout *et al.* (2003).

Metabolic Incorporation of Palmitate into RGS Proteins

Metabolic Labeling of Cells

Source of Radiolabel. For metabolic labeling, [^{3}H]palmitate is used frequently because it enters the cell readily and is converted to [^{3}H]palmitoyl-CoA. [^{3}H]Palmitate is available through a number of commercial sources, such American Radiolabeled Chemicals, Amersham Biosciences, Moravek Biochemicals, Perkin-Elmer, and Sigma-Aldrich. The obvious problem with the use of tritium is its low energy, which requires special conditions for fluorography and long exposures to film. [^{14}C]Palmitate is also available commercially, but its significantly higher cost makes it prohibitive for metabolic labeling of cultured cells. An alternative is the

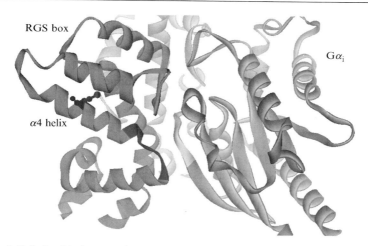

FIG. 5. Relationship between internal palmitoylation and the RGS–G protein interface. A ribbon representation of the crystal structure of RGS4 (green) bound to $G\alpha_i$ (gray) demonstrating the close proximity of Cys-95 in RGS4 (brown) to the residues between the $\alpha3$ and the $\alpha4$ helices of RGS4 (purple) that interact with switch region I (yellow) of $G\alpha_i$. The membrane-facing side of the complex is predicted to be in the foreground. Reprinted with permission from Osterhout *et al.* (2003). (See color insert.)

use of iodinated fatty acid analogs that behave like their respective fatty acids (Berthiaume *et al.*, 1995). The fatty acids are synthesized with ^{125}I in the omega position of the linear saturated fatty acid chain, resulting in shorter exposure times for fluorography.

Intracellular Conversion of Palmitate. A problem with metabolic labeling is the conversion of [^{3}H]palmitate into other fatty acids and amino acids. The former is not necessarily a significant problem because palmitoylation is known to include the incorporation of other long chain fatty acids (Qanbar and Bouvier, 2003). (The proper term for palmitoylation should be *S*-thioacylation.) To differentiate palmitoylation from myristoylation, radiolabeling can be done in the presence of a protein synthesis inhibitor that will block myristoylation that occurs cotranslationally. To avoid the problem of conversion of the tritium signal into amino acids, cell labeling can be done with short labeling times or in the presence of cycloheximide to stop new protein synthesis. In our experience with a number of cell lines, incubations longer than 1 h did not increase the level of [^{3}H]palmitate incorporation into $G\alpha$ or RGS proteins because of the degradation of the radiolabel and palmitate turnover on the protein.

Preparation of [^{3}H]Palmitate. [^{3}H]Palmitate is stored in ethanol for maintaining the stability of the fatty acid, but ethanol is toxic to cells.

Before metabolic labeling, the concentration of ethanol should be <0.3% of the total volume of the labeling medium. Therefore, the ethanol must be removed to achieve a high concentration of [³H]palmitate in the labeling medium. The volume of labeling medium is the minimum amount needed to cover the cells during a short 1-h incubation period. For 75-cm² flasks, we use 5 ml of medium containing 500 μCi/ml of [³H]palmitate. Shortly before metabolic labeling, 500-μl aliquots of [³H]palmitate (American Radiolabeled Chemicals, 5 mCi/ml, specific activity 30–60 Ci/mmol) of the calculated total amount of radioactivity needed for the experiment are placed into 1.5-ml microcentrifuge tubes. The volume is reduced in a Speed-Vac concentrator (Thermo Savant) to about 50–100 μl. The radioactive liquid is combined into one tube and reduction is continued until a volume <0.3% of the volume of the labeling medium is achieved. The aliquots should not be reduced to dryness because of a significant loss of activity. Dimethyl sulfoxide (DMSO) in a volume equal to 1% of the volume of the labeling medium is added to the [³H]palmitate to increase its solubility and entry in the cell. The [³H]palmitate/DMSO mixture is added to serum-free cell culture medium within an hour of use and is kept at 37°.

Incubation with the Radiolabel. The cells should be fairly confluent to maximize the yield of radiolabeled proteins. The cell culture medium is removed and the cells are washed one time in serum-free medium. The cells are then incubated for 1 h in serum-free medium for depletion of intracellular fatty acids. This medium is replaced with the labeling medium and the cells are incubated for 1 h. During both incubation periods, the cells are returned to the incubator. Care should be taken that the cells are covered with the small volume of labeling medium. After the incubation period, the labeling medium is removed and the cells are harvested quickly by scraping into cold phosphate-buffered saline. A cell pellet is obtained by centrifugation at 2000g for 10 min at 4° and stored at −80° until use.

Isolation and Detection of the Radiolabeled RGS Protein

Cell Fractionation. To evaluate their incorporation of [³H]palmitate, RGS proteins need to be separated from other cellular proteins, as many cellular proteins undergo palmitoylation. Because palmitoylated RGS proteins, like other intracellular palmitoylated proteins, are only found in membranes, the membrane fraction should be used for immunoprecipitation or other isolation techniques. In addition, acyl protein thioesterase 1 is a cytosolic protein, so separation of the membranes from the cytosol may prevent some removal of the palmitate. All steps are performed on ice. The indicated volumes for this protocol are for a cell pellet from a 75-cm² flask that generally yields 0.8–1.2 mg of membrane protein.

i. The cell pellets are removed from the $-80°$ freezer and thawed slowly in an ice bucket.

ii. Four hundred microliters of cold homogenization buffer (50 mM mannitol, 5 mM HEPES, pH 7.4, 1 μM leupeptin, 0.3 μM aprotinin, 100 μU/ml of α_2-macroglobulin, 1 μM pepstatin, and 0.1 mg/ml of soybean trypsin inhibitor) is added. The cells are homogenized by trituration with 12 up/down strokes through a 23-gauge needle and 1-ml syringe and are transferred to a 1.5-ml microcentrifuge tube.

iii. Step ii is repeated with an additional 400 μl of cold homogenization buffer to obtain any remaining cells from the cell pellet.

iv. Trituration is repeated with four up/down strokes on the combined sample.

v. The homogenized cells are centrifuged at 700g for 3 min at $4°$ to remove nuclei and any unbroken cells.

vi. The supernatant is transferred to a Beckman 1.5-ml polyallomer ultracentrifuge tube. Four hundred microliters of cold homogenization buffer can be added to the pellet and step v repeated after gentle mixing to increase the yield of membranes.

vii. Fractions are separated by centrifugation in a Beckman TL-100 centrifuge in a TLA-45 rotor at 45,000 rpm (125,000g_{max}) for 1 h at $4°$.

viii. The supernatant is removed and the pellet is homogenized in 200 μl of cold homogenization buffer using a Kontes disposable pestle.

ix. The protein concentration is determined using the Bio-Rad protein assay kit and immunoglobulin G as the standard.

Immunoprecipitation of RGS Proteins. Procedures and reagents that increase the yield of radiolabeled proteins are necessary after metabolic labeling with [^{3}H]palmitate because of the low specific activity of the radiolabel in the cell and the weak tritium signal. An antibody that works well for immunoblotting may not work for immunoprecipitation because of the high sensitivity of immunoblotting and immobilization of the protein on a support. Figure 1 shows an example of immunoprecipitation of RGS16 after metabolic labeling with [^{3}H]palmitate. Antibodies to epitope tags, which are required because of the limited number of quality antibodies to native RGS proteins, are highly variable in their ability to immunoprecipitate proteins. Our laboratory has had better results with rabbit polyclonal antibodies compared to monoclonal antibodies.

i. One milligram of radiolabeled membrane protein is solubilized in 900 μl of IP buffer (50 mM Tris-HCl, pH 7.5, 150 mM NaCl, and 1 mM EDTA) with 0.2% (w/v) sodium dodecyl sulfate (SDS) for 1 h on a rotator in the cold room. The microcentrifuge tubes are taped or placed in a covered box to prevent problems with radioactive contamination if a tube opens.

ii. One hundred microliters of 10% (w/v) Triton X-100 is added for solubilization and prevention of denaturation of the antibody by the SDS. The samples are vortexed gently. Then, 5 μl of the antiserum or 10 μg of the affinity-purified antibodies is added. The samples are incubated overnight on the rotator at 4°.

iii. A 50% slurry of protein A–Sepharose CL-4B (Amersham Biosciences) is added through a cut-off pipette tip. Due to rapid settling of the Sepharose, the volume of protein A–Sepharose needed for an experiment is vortexed before transferring 35 μl to each tube. Protein A–Sepharose is prepared by adding 1.5 g of the powder and about 10 ml of IP buffer without detergents to a 15-ml conical tube and allowing hydration to occur over several hours with occasional gentle vortexing. The buffer is poured off and replaced with IP buffer without detergents to give 50% (v/v) of protein A–Sepharose. This preparation of protein A–Sepharose can be stored at 4° for at least a month.

iv. The samples are returned to the rotator for 2 h and are then centrifuged at 4° for 10 min at 8000g in a tabletop centrifuge.

v. The supernatant is removed carefully by vacuum aspiration through a 23-gauge needle and a 1-ml syringe.

vi. The samples are washed two times in 1 ml of cold IP buffer containing 0.2% SDS and 1% (w/v) Triton X-100 and one time with 1 ml of cold IP buffer without detergents. The samples are centrifuged at 8000g for 3 min at 4° between washes except for the last centrifuge, which is for 10 min.

vii. The supernatant is removed and 43 μl of 1× Laemmli SDS–PAGE loading buffer is added. The samples are heated to 90° for 5 min, and the Sepharose is packed tightly by centrifugation at 8000g for 5 min. Forty microliters of the supernatant is added to a 10-well, 1.5-mm-thick gel for SDS–PAGE.

Detection of Radiolabeled RGS Proteins. Visualization of a signal from gel bands containing tritium-labeled proteins requires some form of enhancement. For several reasons we have changed from using a liquid fluor, such as En³Hance (DuPont), to a wax method (EA wax, EA Biotech Ltd). The liquid fluor required separate disposal of a radioactive chemical waste, which often led to cracking of the gel during drying and needed exposure times of up to 1 month. The EA wax is easier and simpler and decreases exposure times by one-third.

i. The gel containing the tritium-labeled proteins is transferred to a nitrocellulose membrane. The adequacy of transfer can be checked by staining the proteins with ponceau S. The membranes are incubated in 0.2% ponceau S in 3% acetic acid for about 15 min and then in water until the bands are seen. The membrane should be dried overnight because any moisture leads to a blotchy appearance on the film.

ii. For application of the EA wax, glass plates are heated to 80° for 30 min. The nitrocellulose membrane is placed on top of wax paper supplied by the manufacturer and the wax paper is placed on one glass plate. A generous amount of wax is applied to the nitrocellulose membrane and then the second wax paper and second glass plate are placed on top. (This creates a sandwich of a glass plate, wax paper, nitrocellulose membrane, wax, wax paper, and glass plate). The hot plates are compressed to spread the wax over the surface of the membrane. The wax-coated membrane is removed from the sandwich of the wax paper and allowed to cool at room temperature. The membrane is taped to aluminium foil that lines a film cassette. Kodak MS film is exposed to the membrane at −80° for exposure times between 3 days and up to 2 weeks. Exposure times longer than 2 weeks do not lead to darker bands.

In Vitro Incorporation of [³H]Palmitate into RGS Proteins

General Considerations

For experiments using purified RGS proteins, palmitate can be incorporated in vitro in the absence or presence of an enzyme, protein acyltransferase (Fig. 6). The advantage of nonenzymatic palmitoylation is the simplicity of the conditions and the retention of RGS protein purity. The problem is the low stoichiometry of palmitate incorporation because of the slow kinetics of the reaction. Conditions that increase the level of palmitoylation create their own problems, such as a long incubation time, leading to a loss of activity of the RGS protein, and pH values >8.0, leading to palmitoylation in unnatural sites (Duncan and Gilman, 1996). The advantages of enzymatic palmitoylation are the greater speed, completeness, and specificity of the reaction. However, the disadvantages are (1) the complexity of partially purifying protein acyltransferase (PAT) activity from membranes and (2) the loss of purity of the RGS protein by the addition of a membrane preparation containing detergents.

Enzymatic Palmitoylation of RGS Proteins

We may soon have a recombinant PAT, but at this time PAT activity must be purified partially from cell or tissue membranes. Our laboratory has published a purification of PAT activity from rat liver plasma

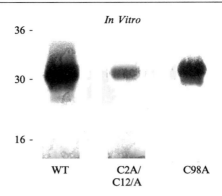

FIG. 6. Palmitoylation of RGS16 after *in vitro* enzymatic radiolabeling. Recombinant wild-type and mutant RGS16 were purified from *E. coli* and treated with partially purified PAT in the presence of [³H]palmitate. Proteins were then separated by SDS–PAGE, and gels were transferred to nitrocellulose membranes and prepared for fluorography. Gels were exposed to Kodak MS film for 3 days at −80°. Molecular mass markers in kilodaltons are shown to the left. Reprinted with permission from Osterhout *et al.* (2003).

membranes (Hiol *et al.*, 2003a). This preparation leads to a high level of [³H]palmitate incorporation into RGS proteins and has no GAP activity when added to cell membranes (Fig. 2B) (Osterhout *et al.*, 2003). Depending on the application, incubation of the RGS protein in a reaction buffer with a crude membrane fraction or an intermediate purification from this protocol may be adequate. The membrane preparations contain acyl-CoA synthetase activity that can produce palmitoyl-CoA from palmitate, coenzyme A, and ATP.

i. Partially purified PAT in buffer [20 mM Tris–HCl, pH 7.4, 1 mM EGTA, 10 μg/ml leupeptin, and 0.15% (v/v) Triton X-100] can be stored as aliquots at −80° and retain activity for 1–2 months.

ii. A glutathione S-transferase–RGS fusion protein is produced in *Escherichia coli* and is purified on glutathione–Sepharose (Amersham Biosciences) in a similar manner as described elsewhere (Garrison *et al.*, 2002; Ishii and Kurachi, 2004; Willard and Siderovski, 2004). The glutathione S-transferase is cleaved from the RGS protein with biotinylated thrombin that is cleared using streptavidin-agarose (Novagen). The purified proteins are dialyzed against buffer [50 mM Tris–HCl, pH 8, 100 mM NaCl, 1 mM EDTA, and 5% (v/v) glycerol] and are stored as aliquots at −80°.

iii. In a microcentrifuge tube, 10 μg of the partially purified PAT is added to 40–100 μg of purified RGS protein in IB buffer (20 mM Tris–HCl, pH 7.4, 150 mM KCl, and 1 mM EDTA). Coenzyme A and ATP are added to give final concentrations of 200 μM and 2 mM. The reaction is started with the addition of 1 μl (5 μCi) of [^{3}H]palmitate (American Radiolabeled Chemicals, 30–60 Ci/mmol). The final volume of the reaction is made up to 100 μl with IB buffer. One hundred micromolar of cold palmitate or 100 μM [^{3}H]palmitoyl-CoA can be substituted for the [^{3}H]palmitate.

iv. The samples are incubated for 45 min at 30°, and the reaction is stopped by adding 25 μl of 1% (w/v) of SDS or Laemmli SDS–PAGE loading buffer for analysis by SDS–PAGE.

Nonenzymatic Palmitoylation of RGS Proteins

The assay is performed under the same conditions as described earlier except that the membrane preparation is omitted and 200 μM [^{3}H]palmitoyl-CoA is added. [^{3}H]Palmitoyl-CoA can be synthesized with a high efficiency from [^{3}H]palmitate, coenzyme A, and ATP in the presence of acyl-CoA synthetase (Hiol *et al.*, 2003a). A longer incubation time is needed to show a good level of [^{3}H]palmitate incorporation, e.g., 2 to 6 h at 30°.

Functional Assays with Palmitoylated RGS Proteins

General Considerations

Most studies addressing the function of protein palmitoylation rely on comparisons between the wild-type protein and mutants in which cysteine residues critical for palmitoylation have been changed to alanine residues. These studies have the advantage of assessing the role of palmitoylation for protein localization and function within cells. In addition, mutagenesis is relatively easy to perform. However, mutagenesis brings up the problem that amino acid substitutions themselves may alter protein function, as has been reported for cysteine mutations in Gαq (Hepler *et al.*, 1996). Also, the function of individual palmitoylation sites cannot be tested if they are dependent on palmitoylation on other sites. For RGS4 and RGS16, palmitoylation on cysteine-95 and cysteine-98, respectively, is dependent on amino-terminal palmitoylation (Hiol *et al.*, 2003b; Tu *et al.*, 1999). *In vitro* studies comparing palmitoylated versus nonpalmitoylated forms of a protein offer the advantage of specifically testing the effects of the modification. The best-characterized function of RGS proteins is the acceleration of GTP hydrolysis. The GAP activity of RGS proteins has been investigated in solution, after reconstitution in artificial and cell membranes, and after overexpression in cell membranes. Palmitoylated RGS proteins can be used in these assays with a few caveats. The stoichiometry of palmitoylation must be high enough to see differences between palmitoylated and nonpalmitoylated forms (see following protocol). For reconstitution studies, extended preincubation times should be included because the membrane attachment of RGS proteins is a slow, multistep process (Tu *et al.*, 2001). The GAP activity of palmitoylated RGS proteins increased with preincubation times up to 120 min (Figs. 2B and 4) (Hiol *et al.*, 2003b). Protocols for determining the GAP activity in solution and with reconstituted artificial membranes can be found elsewhere (Hooks and Harden, 2004; Krumins and Gilman, 2002; Martemyanov and Arshavsky, 2004; Ross, 2002). The following protocol is used for GAP activity in cell membranes used with the reconstitution of palmitoylated proteins or after overexpression of palmitoylation-defective mutants (Fig. 2).

Stoichiometry of Palmitoylation

This assay is performed in triplicate. The filtration and washing steps are used to remove any unbound [^{3}H]palmitate.

i. RGS proteins undergo [^{3}H]palmitoylation as described earlier. Aliquots of 260 pmol of the RGS protein, calculated from the amount of pure RGS protein in the palmitoylation assay, are transferred to microcentrifuge tubes. The samples are diluted with 400 μl of F buffer [20 mM Tris–HCl, pH 7.4, 150 mM KCl, 1 mM EDTA, 0.1% (w/v) bovine serum albumin, and 1% (w/v) sucrose].

ii. Ultrafree filtration units (Amicon) are preequilibrated with F buffer by adding 400 μl of buffer and then centrifuging at 2000g in a tabletop centrifuge until the buffer has passed through the membrane.

iii. The samples are added to the filtration units and are centrifuged at 2000g for 20 min.

iv. The filters are washed once with 400 μl of 70% ethanol and then twice with 400 μl of F buffer.

v. The insert cups are transferred to vials for liquid scintillation counting and 10 ml of liquid scintillation fluid is added. The radioactivity is determined by liquid scintillation spectrometry.

vi. The specific activity (cpm/pmol) of the [^{3}H]palmitate was determined by spotting known quantities of [^{3}H]palmitate onto the filter membrane of the insert cup and then directly determining the radioactivity as in step v. Stoichiometry is calculated as the ratio of the picomoles of palmitate to the picomoles of protein.

Preparation of Membranes for the High-Affinity GTPase Assay

This assay uses HEK293 cells expressing a fusion protein of the serotonin (5-HT$_{1A}$) receptor and a mutant Gα_{o1} that is resistant to pertussis toxin because cysteine-351 is mutated to glycine (Kellett *et al.*, 1999). Endogenous

$G\alpha_i$ and $G\alpha_o$ proteins that could couple to this receptor are inactivated by treatment with pertussis toxin. The receptor–$G\alpha$ fusion protein ensures a 1:1 stoichiometry of activation between receptor and G protein following agonist stimulation and allows measurement of steady-state GTPase activity.

i. HEK293 cells stably expressing the 5-HT$_{1A}$/$G\alpha_{o1}$ (C351G) fusion protein are grown and maintained in Dulbecco's modified Eagle's medium containing 10% fetal bovine serum, 2 mM L-glutamine, 400 μg/ml G418 sulfate (Calbiochem), and antibiotics in a humidified 5% CO_2 incubator.

ii. For transfection studies, HEK293 cells expressing the 5-HT$_{1A}$/$G\alpha_{o1}$ (C351G) fusion protein are grown to near confluence in 75-cm^2 flasks. Ten micrograms of plasmid DNA containing the cDNA for the wild-type or mutated RGS protein are transfected into the cells using Superfect (Qiagen) per the manufacturer's instructions. Thirty hours after transfection, the cells are treated overnight with pertussis toxin (List Biological Laboratories) that is added to the medium for a final concentration of 50 ng/ml. Forty-eight hours after transfection the cell are harvested (see step iii).

iii. For membrane reconstitution studies, the cells are treated overnight with pertussis toxin, as described earlier. The next day the cells are harvested by scraping into cold phosphate-buffered saline and then centrifuging at 2000g for 10 min at 4°. Cell pellets can be stored at $-80°$ for 1–2 months.

iv. The cell pellet is resuspended in ice-cold TE buffer (10 mM Tris–HCl, pH 7.4, 1 mM EDTA) by gentle vortexing and is then homogenized with 30–50 passages of a Dounce homogenizer followed by 15 passages through a 25-gauge needle and 1-ml syringe. The sample is centrifuged at 4000g for 5 min at 4° to remove unbroken cells and nuclei. The supernatant is centrifuged at 45,000 rpm (125,000g_{max}) for 30 min at 4° in a TLA-45 rotor in a Beckman TL100 centrifuge to pellet the membrane fraction.

v. The pellet is resuspended in TE buffer to $\sim$2 mg/ml and can be stored as aliquots at $-80°$ for 1–2 months.

High-Affinity GTPase Activity of Cell Membranes

This assay can be performed on membranes from HEK293 cells stably expressing the 5-HT$_{1A}$/$G\alpha_{o1}$ (C351G) fusion protein that include RGS proteins from either transient transfection or reconstitution into the membranes. Cells are treated overnight with pertussis toxin (50 ng/ml) prior to membrane harvest to ablate the GTPase activity of endogenous $G_{i/o}$ proteins.

i. For reconstitution of RGS proteins into the membranes, palmitoylated RGS proteins (0–1 μM) are preincubated with the membrane preparations (10 μg total protein) for 0 to 120 min at 37° in a final volume of 50 μl.

ii. 5-HT (1 μM gives a near maximal response) in an ATP-regenerating buffer system (20 mM creatine phosphate, 0.1 U/μl creatine kinase, 200 μM AMP-PNP, 500 μM ATP, 2 mM ouabain, 200 mM NaCl, 10 mM MgCl$_2$, 4 mM dithiothreitol, 200 μM EDTA, pH 7.5, 80 mM Tris–HCl, pH 7.5, 1 μM GTP) spiked with 50,000 cpm [γ-^{32}P]GTP (Amersham Biosciences, 3000 Ci/mmol) in a final volume of 50 μl is added to the reconstituted membranes for a total reaction volume of 100 μl. The samples are incubated for 20 min at 30°. The assay is performed in triplicate. Nonspecific GTPase activity is determined in simultaneous reactions containing 100 mM nonradioactive GTP.

iii. Reactions are stopped by adding 750 μl of ice-cold 10% (w/v) Norit A-activated charcoal (Sigma-Aldrich) in 50 mM phosphoric acid followed by centrifugation at 10,000g for 20 min at 4°. The charcoal binds the free or protein-bound [γ-^{32}P]GTP. Radioactivity in the supernatants containing the liberated γ-^{32}P phosphate is determined by liquid scintillation spectrometry.

iv. Specific GTPase activity is calculated as pmol/mg/min followed by subtraction of the nonspecific activity in the presence of 100 mM GTP. Picomoles are calculated as the ratio of cpm from the free γ-^{32}P phosphate in the sample and the specific activity in the assay (cpm/pmol of [γ-^{32}P] GTP determined by counting 1 μl of the reaction buffer and dividing by the pmol of GTP based on 1 μM GTP in 50 μl of buffer). Milligrams are the amount of membrane protein used in the GTPase assay (10 μg).

Acknowledgments

I thank Drs. Kirk M. Druey and Abel Hiol for help with the manuscript. I thank Drs. Penelope C. Davey, Michael Y. Degtyarev, Abel Hiol, Sergey R. Kusnetsov, Yukiko Miura, James C. Osterhout, Ozlem Ugur, and Abdul A. Waheed who contributed in the development of these protocols and our understanding of protein palmitoylation.

References

Bernstein, L. S., Grillo, A. A., Loranger, S. S., and Linder, M. E. (2000). RGS4 binds to membranes through an amphipathic α helix. *J. Biol. Chem.* **275,** 18520–18526.
Berthiaume, L., Peseckis, S. M., and Resh, M. D. (1995). Synthesis and use of iodo-fatty acid analogs. *Methods Enzymol.* **250,** 454–466.

Brown, D. A., and London, E. (2000). Structure and function of sphingolipid and cholesterol rich membrane rafts. *J. Biol. Chem.* **275**, 17221–17224.

Casey, P. J. (1995). Protein lipidation in cell signaling. *Science* **268**, 221–225.

Castro-Fernandez, C., Janovick, J. A., Brothers, S. P., Fisher, R. A., Ji, T. H., and Conn, P. M. (2002). Regulation of RGS3 and RGS10 palmitoylation by GnRH. *Endocrinology* **143**, 1310–1317.

Chen, C., Seow, K. T., Guo, K., Yaw, L. P., and Lin, S. C. (1999). The membrane association domain of RGS16 contains unique amphipathic features that are conserved in RGS4 and RGS5. *J. Biol. Chem.* **274**, 19799–19806.

Degtyarev, M. Y., Spiegel, A. M., and Jones, T. L. Z. (1993). Increased palmitoylation of the Gs protein α subunit after activation by the β-adrenergic receptor or cholera toxin. *J. Biol. Chem.* **268**, 23769–23772.

Devedjiev, Y., Dauter, Z., Kuznetsov, S. R., Jones, T. L. Z., and Derewenda, Z. S. (2000). Crystal structure of the human acyl protein thioesterase I from a single X-ray data set to 1.5.A. *Struct. Fold Des.* **8**, 1137–1146.

De Vries, L., Elenko, E., Hubler, L., Jones, T. L. Z., and Farquhar, M. G. (1996). GAIP is membrane-anchored by palmitoylation and interacts with the activated (GTP bound) form of G αi subunits. *Proc. Natl. Acad. Sci. USA* **93**, 15203–15208.

De Vries, L., Zheng, B., Fischer, T., Elenko, E., and Farquhar, M. G. (2000). The regulator of G-protein signaling family. *Annu. Rev. Pharmacol. Toxicol.* **40**, 235–271.

Druey, K. M., Ugur, O., Caron, J. M., Chen, C.-K., Backlund, P. S., and Jones, T. L. Z. (1999). Amino terminal cysteine residues of RGS16 are required for palmitoylation and modulation of Gi and Gq-mediated signaling. *J. Biol. Chem.* **274**, 18836–18842.

Duncan, J. A., and Gilman, A. G. (1996). Autoacylation of G-protein α subunits. *J. Biol. Chem.* **271**, 23594–23600.

Duncan, J. A., and Gilman, A. G. (1998). A cytoplasmic acyl protein thioesterase that removes palmitate from G-protein α subunits and p21(RAS). *J. Biol. Chem.* **273**, 15830–15837.

Duncan, J. A., and Gilman, A. G. (2002). Characterization of Saccharomyces cerevisiae acyl protein thioesterase 1, the enzyme responsible for G-protein α subunit deacylation *in vivo*. *J. Biol. Chem.* **277**, 31740–31752.

Dunphy, J. T., Greentree, W. K., and Linder, M. E. (2001). Enrichment of G-protein palmitoyltransferase activity in low density membranes: *In vitro* reconstitution of Gαi to these domains requires palmitoyltransferase activity. *J. Biol. Chem.* **276**, 43300–43304.

Garrison, T. R., Apanovitch, D. M., and Dohlman, H. G. (2002). Purification of RGS protein, Sst2, from *Saccharomyces cerevisiae* and *Escherichia coli*. *Methods Enzymol.* **344**, 632–647.

Hepler, J. R., Biddlecome, G. H., Kleuss, C., Camp, L. A., Hofmann, S. L., Ross, E. M., and Gilman, A. G. (1996). Functional importance of the amino terminus of Gqα. *J. Biol. Chem.* **271**, 496–504.

Hiol, A., Caron, J. M., Smith, C. D., and Jones, T. L. Z. (2003a). Characterization and partial purification of protein fatty acyltransferase activity from rat liver. *Biochim. Biophys. Acta* **1635**, 10–19.

Hiol, A., Davey, P. C., Waheed, A. A., Osterhout, J. C., Fischer, E., Chen, C.-K., Milligan, G., Druey, K. M., and Jones, T. L. Z. (2003b). Palmitoylation and RGS16 signaling. I. Mutation of amino-terminal cysteine residues on RGS16 prevents targeting to lipid rafts and palmitoylation of an internal cysteine residue. *J. Biol. Chem.* **278**, 19301–19308.

Hollinger, S., and Hepler, J. R. (2004). Methods for measuring RGS protein phosphorylation by G protein-regulated kinases. *Methods Mol. Biol.* **237**, 205–219.

Hooks, S. B., and Haden, T. K. (2004). Purification and *in vitro* functional analysis of R7 subfamily RGS proteins in complex with Gβ5. *Methods Enzymol.* **390**, 163–177.

Iiri, T., Backlund, P. S., Jr., Jones, T. L. Z., Wedegaertner, P. B., and Bourne, H. R. (1996). Reciprocal regulation of Gsα by palmitate and the βγ subunit. *Proc. Natl. Acad. Sci. USA* **93,** 14592–14597.

Ishii, M., and Kurachi, Y. (2004). Assays of RGS protein modulation by phosphatidylinositides and calmodulin. *Methods Enzymol.* **389**(7), 105–118.

Johnson, E. N., Seasholtz, T. M., Waheed, A. A., Kreutz, B., Suzuki, N., Kozasa, T., Jones, T. L. Z., Brown, J. H., and Druey, K. M. (2003). RGS16 inhibits signalling through the Gα13 Rho axis. *Nature Cell Biol.* **5,** 1095–1103.

Kellett, E., Carr, I. C., and Milligan, G. (1999). Regulation of G protein activation and effector modulation by fusion proteins between the human 5 hydroxytryptamine (1A) receptor and the α subunit of G(il): Differences in receptor-constitutive activity imparted by single amino acid substitutions in G(il)α. *Mol. Pharmacol.* **56,** 684–692.

Krumins, A. M., and Gilman, A. G. (2002). Assay of RGS protein activity *in vitro* using purified components. *Methods Enzymol.* **344,** 673–685.

Linder, M. E., and Deschenes, R. J. (2003). New insights into the mechanisms of protein palmitoylation. *Biochemistry* **42,** 4311–4320.

Lobo, S., Greentree, W. K., Linder, M. E., and Deschenes, R. J. (2002). Identification of a Ras palmitoyltransferase in *Saccharomyces cerevisiae. J. Biol. Chem.* **277,** 41268–41273.

Martemyanov, K. A., and Arshavsky, V. Y. (2004). Kinetic approaches to study the function of RGS9 isoforms. *Methods Enzymol.* **390,** 196–209.

Miura, Y., Hanada, K., and Jones, T. L. Z. (2001). G(s) signaling is intact after disruption of lipid rafts. *Biochemistry* **40,** 15418–15423.

Moffett, S., Mouillac, B., Bonin, H., and Bouvier, M. (1993). Altered phosphorylation and desensitization patterns of a human β2 adrenergie receptor lacking the palmitoylated Cys341. *EMBO J.* **12,** 349–356.

Mouillac, B., Caron, M., Bonin, H., Dennis, M., and Bouvier, M. (1992). Agonist modulated palmitoylation of β2-adrenergic receptor in Sf9 cells. *J. Biol. Chem.* **267,** 21733–21737.

Nair, K. S., Balasubramanian, N., and Slepak, V. Z. (2002). Signal dependent translocation of transducin, RGS9-1-Gβ5L complex, and arrestin to detergent-resistant membrane rafts in photoreceptors. *Curr. Biol.* **12,** 421–425.

Osterhout, J. C., Waheed, A. A., Hiol, A., Ward, R. J., Davey, P. C., Nini, L., Wang, J., Milligan, G., Jones, T. L. Z., and Druey, K. M. (2003). Palmitoylation and RGS16 signaling II. Palmitoylation of a cysteine residue in the RGS box of RGS16 is critical for its GTPase activity and *in vivo* activity. *J. Biol. Chem.* **278,** 19309–19316.

Peitzsch, R. M., and McLaughlin, S. (1993). Binding of acylated peptides and fatty acids to phospholipid vesicles. pertinence to myristoylated proteins. *Biochemistry* **32,** 10436–10443.

Putilina, T., Wong, P., and Gentleman, S. (1999). The DHHC domain: A new highly conserved cysteine-rich motif. *Mol. Cell. Biochem.* **195,** 219–226.

Qanbar, R., and Bouvier, M. (2003). Role of palmitoylation/depalmitoylation reactions in G-protein coupled receptor function. *Pharmacol. Ther.* **97,** 1–33.

Rose, J. J., Taylor, J. B., Shi, J., Cockett, M. I., Jones, P. G., and Hepler, J. R. (2000). RGS7 is palmitoylated and exists as biochemically distinct forms. *J. Neurochem.* **75,** 2103–2112.

Ross, E. M. (2002). Quantitative assays for GTPase-activating proteins. *Methods Enzymol.* **344,** 601–617.

Roth, A. F., Feng, Y., Chen, L., and Davis, N. G. (2002). The yeast DHHC cysteinc-rich domain protein Akrlp is a palmitoyl transferase. *J. Cell Biol.* **159,** 23–28.

Simons, K., and Toomre, D. (2000). Lipid rafts and signal transduction. *Nature Rev. Mol. Cell Biol.* **1,** 31–39.

Srinivasa, S. P., Bernstein, L. S., Blumer, K. J., and Linder, M. E. (1998). Plasma membrane localization is required for RGS4 function in *Saccharomyces cerevisiae. Proc. Natl. Acad. Sci. USA* **95,** 5584–5589.

Tesmer, J. J., Berman, D. M., Gilman, A. G., and Sprang, S. R. (1997). Structure of RGS4 bound to AlF4–activated G($i\alpha$1): Stabilization of the transition state for GTP hydrolysis. *Cell* **89,** 251–261.

Tu, Y., Popov, S., Slaughter, C., and Ross, E. M. (1999). Palmitoylation of a conserved cysteine in the regulator of G protein signaling (RGS) domain modulates the GTPase-activating activity of RGS4 and RGS10. *J. Biol. Chem.* **274,** 38260–38267.

Tu, Y., Wang, J., and Ross, E. M. (1997). Inhibition of brain Gz GAP and other RGS proteins by palmitoylation of G protein α subunits. *Science* **278,** 1132–1135.

Tu, Y., Woodson, J., and Ross, E. M. (2001). Binding of regulator of G protein signaling (RGS) proteins to phospholipid bilayers. Contribution of location and/or orientation to Gtpase activating protein activity. *J. Biol. Chem.* **276,** 20160–20166.

Ugur, O., and Jones, T. L. Z. (2000). A proline rich region and nearby cysteine residues target XLαs to the Golgi complex region. *Mol. Biol. Cell* **11,** 1421–1432.

Ugur, O., Ongan, H. O., and Jones, T. L. Z. (2002). Partial rescue of functional interactions of a nonpalmitoylated mutant of the G-protein Gαs by fusion to the β adrenergic receptor. *Biochemistry* **42,** 2607–2615.

Waheed, A. A., and Jones, T. L. Z. (2002). Hsp90 interactions and acylation target the G protein Gα12, but not Gα13 to lipid rafts. *J. Biol. Chem.* **277,** 32409–32412.

Wang, A., and Dennis, E. A. (1999). Mammalian lysophospholipases. *Biochim. Biophys. Acta* **1439,** 1–16.

Willard, F. S., and Siderovski, D. P. (2004). Purification and *in vitro* functional analysis of the Arabidopsis thaliana regulator of G-protein signaling AtRGS1. *Methods Enzymol.* **389**(19), 56–71.

Subsection B

Assays of RGS Box
Activity and Allosteric Control

[4] Fluorescence-Based Assays for RGS
Box Function

By FRANCIS S. WILLARD, RANDALL J. KIMPLE, ADAM J. KIMPLE,
CHRISTOPHER A. JOHNSTON, and DAVID P. SIDEROVSKI

Abstract

Ligand-activated, seven transmembrane-spanning receptors interact with inactive G-protein heterotrimers ($G\alpha\beta\gamma$) to catalyze GTP loading and, consequently, activation of $G\alpha$ subunits and the liberation of $G\beta\gamma$. $G\alpha\cdot$GTP and $G\beta\gamma$ are then competent to regulate independent effector pathways. The duration of heterotrimeric G-protein signaling is determined by the lifetime of the $G\alpha$ subunit in the GTP-bound state. Signal termination is facilitated by the intrinsic guanosine triphosphatase (GTPase) activity of $G\alpha$ and subsequent reformation of the inactive heterotrimer. Regulators of G-protein signaling (RGS) proteins act enzymatically, via their hallmark "RGS box," as GTPase-accelerating proteins (GAPs) for $G\alpha$ subunits and thus function as negative regulators of G-protein signaling *in vitro* and *in vivo*. This article describes the use of fluorescence resonance energy transfer (FRET) to monitor the interaction between a $G\alpha$ subunit and an RGS box protein. Furthermore, this article describes optimization of this assay for high-throughput screening and the evaluation of mutant RGS box and $G\alpha$ proteins. Finally, this article describes the novel application of this FRET technique to measure the activity of RGS protein-derived GoLoco peptides that modulate $G\alpha$ activation by aluminum tetrafluoride.

Introduction

Seven transmembrane-spanning receptors couple to second messenger pathways via heterotrimeric, guanine nucleotide-binding proteins (G proteins), which are composed of three subunits termed $G\alpha$, $G\beta$, and $G\gamma$ (Gilman, 1987; Pierce *et al.*, 2002). The $G\alpha$ subunit is a molecular switch that is conformationally sensitive to guanine nucleotides. GDP-bound $G\alpha$

is inactive and tightly associated with $G\beta\gamma$. Activated receptors catalyze guanine nucleotide exchange on the GDP-bound heterotrimer. GTP binding to $G\alpha$ is thought to cause subunit dissociation and the activation of second messenger pathways by both $G\alpha$·GTP and $G\beta\gamma$. Signal termination is facilitated by the intrinsic guanosine triphosphatase (GTPase) activity of $G\alpha$. Thus, the duration of heterotrimeric G-protein signaling is determined by the lifetime of the $G\alpha$ subunit in the GTP-bound state.

Regulator of G-protein signaling (RGS) proteins, characterized by their hallmark, ~120 amino acid "RGS box," are GTPase-accelerating proteins (GAPs) for $G\alpha$ subunits and thus act as negative regulators of G-protein signaling *in vitro* and *in vivo* (Neubig and Siderovski, 2002). Given their dynamic spatiotemporal regulation and receptor selectivity, RGS proteins represent promising targets for developing new therapeutic interventions (Neubig and Siderovski, 2002). *In vitro* measurements of RGS box GAP activity generally encompass radiolabeled guanine nucleotide hydrolysis assays (Berman *et al.*, 1996a), coprecipitation assays (Druey and Kehrl, 1997), or surface plasmon resonance (Kimple *et al.*, 2001; Popov *et al.*, 1997; see also this volume, Willard and Siderovski, 2004). Measurement of the RGS box-catalyzed nucleotide cycle has also been performed using both $G\alpha$-subunit intrinsic fluorescence (Lan *et al.*, 2000) and quench flow kinetic methods (Mukhopadhyay and Ross, 1999). This article describes methods for the use of fluorescence resonance energy transfer (FRET) to monitor the archetypal $G\alpha$/RGS box pair, namely the $G\alpha_{i1}$/RGS4 interaction (Tesmer *et al.*, 1997), both kinetically and at end point using a high-throughput methodology. Finally, we examine the application of this FRET technique to assay GoLoco peptides derived from RGS proteins that inhibit $G\alpha_{i1}$ activation.

Materials

Unless otherwise specified, all chemicals are of the highest purity obtainable from Sigma (St. Louis, MO) or Fisher Scientific (Pittsburgh, PA).

Fluorescence Resonance Energy Transfer Assay
Design Considerations

Fluorescence resonance energy transfer is the nonradiative transfer of energy from a donor fluorophore to an acceptor fluorophore. For a comprehensive discussion of the physical principles and practicalities of FRET, the reader is directed to the excellent resources of dos Remedios and Lakowicz (1999) and Moens (1995). Essentially, according to the Förster equation, FRET will be determined by (1) donor fluorophore quantum yield, (2) donor/acceptor spectral overlap, (3) physical distance between donor and acceptor, and (4) orientation of the acceptor adsorption and donor emission dipoles. Derivatives of the enhanced *Aequorea victoria* green fluorescent protein (Prasher *et al.*, 1992), cyan fluorescent protein (ECFP) and yellow-fluorescent protein (EYFP), have been optimized for FRET given the wide spectral shift between the excitation wavelength required for ECFP and the emission wavelength of excited EYFP (Miyawaki and Tsien, 2000). At the present time, ECFP and EYFP, hereafter referred to as CFP and YFP, are the premium choice for genetically encoded acceptor and donor fluorophores, respectively.

To complement *in silico* computational approaches (Willard *et al.*, 2004) and facilitate high-throughput screening for small molecule inhibitors of RGS box/Gα interactions, we have developed a homogeneous FRET-based assay for the protein–protein interaction between RGS4 and $G\alpha_{i1}$ (Kimple *et al.*, 2003). This assay uses the exquisite sensitivity of RGS4 for transition-state (GDP·AlF$_4^-$) $G\alpha_{i1}$ versus inactivated (GDP) or pseudo-activated (GTPγS) state $G\alpha_{i1}$ (Berman *et al.*, 1996b). In theory, fusion of CFP to $G\alpha_{i1}$ and fusion of YFP to RGS4 provides a FRET pair that responds to aluminum tetrafluoride-induced activation of $G\alpha_{i1}$ by an attenuated emission of the CFP donor and increased emission of the YFP acceptor, as outlined in Fig. 1A.

Cloning and Purification of $G\alpha_{i1}$-CFP and YFP-RGS4

To create an expression construct for the YFP-RGS4 fusion protein, the citrine mutant of YFP [amino acids 1–237 of the pEYFP-C1 open reading frame (ORF); BD Biosciences Clontech, Palo Alto, CA] with a Q69M mutation (Heikal *et al.*, 2000) and the entire ORF of mouse RGS4 (amino acids 1–205; SwissProt RGS4_MOUSE) were cloned in-frame with the N-terminal hexahistidine tag (His$_6$) and tobacco-etch virus (TEV) protease cleavage site of the prokaryotic expression vector pProEXHTb (Invitrogen, Carlsbad, CA) using polymerase chain reaction and standard molecular biology techniques. In a similar fashion, the ORF of human $G\alpha_{i1}$ (amino acids 1–353; SwissProt GBI1_HUMAN) and the ORF of CFP (amino acids 1–239 of the pECFP-C1 ORF; BD Biosciences Clontech) were cloned in-frame with the His$_6$/TEV site of pProEXHTc to create the expression construct for the $G\alpha_{i1}$-CFP fusion protein.

Proteins were purified following overexpression in *Escherichia coli* using standard chromatographic procedures (Kimple *et al.*, 2001; see also this volume, Willard and Siderovski, 2004). Briefly, both $G\alpha_{i1}$-CFP and YFP-RGS4 fusion proteins were purified by the sequential application of Ni^{2+} affinity chromatography (HiTrap chelating HP; Amersham

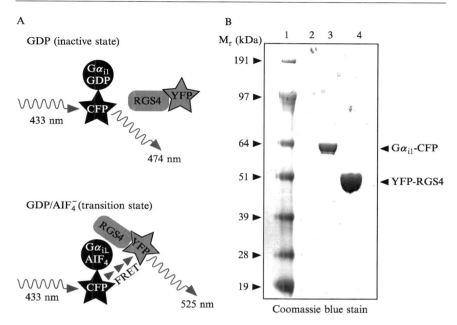

FIG. 1. (A) Schematic of fluorescence resonance energy transfer (FRET) assay for the $G\alpha_{i1}$–RGS4 interaction. In solution, $G\alpha_{i1}\cdot GDP$-CFP does not interact with YFP-RGS4 and thus excitation of CFP at 433 nm results in high donor fluorophore emission at 474 nm. Transition-state $G\alpha_{i1}\cdot GDP\cdot AlF_4^-$-CFP interacts with YFP-RGS4, bringing the CFP and YFP moieties into close proximity (<100 Å). Consequently, excitation energy of the donor fluorophore (CFP) is transferred to the acceptor fluorophore (YFP) and results in increased acceptor emission at 525 nm but reduced donor emission at 474 nm. (B) Electrophoretic analysis of purified $G\alpha_{i1}$-CFP and YFP-RGS4. Purified protein samples were resolved using 4–12% (w/v) SDS–polyacrylamide gel electrophoresis and Coomassie Blue staining. Lane 1, See Blue Plus 2 molecular weight standards (Invitrogen); lane 2, blank; lane 3, purified $G\alpha_{i1}$-CFP fusion protein; and lane 4, purified YFP-RGS4 fusion protein. Positions of molecular weight standards, $G\alpha_{i1}$-CFP, and YFP-RGS4 are denoted with arrowheads.

Bioscience, Uppsala, Sweden), anion-exchange chromatography (Source 15Q; Amersham), and gel-exclusion chromatography (HiPrep 26/60 Sephacryl S200 HR; Amersham). Figure 1B illustrates representative purifications of the His_6-$G\alpha_{i1}$-CFP and His_6-YFP-RGS4 fusion proteins, respectively. The predicted molecular masses and molar extinction coefficients (at 280 nm) are 71.5 kDa and 63,400 $M^{-1}cm^{-1}$ for His_6-$G\alpha_{i1}$-CFP and 53.5 kDa and 43,060 $M^{-1}cm^{-1}$ for YFP-RGS4. Protein quantitation is performed by denaturation with 8 M guanidine HCl (Sigma) and absorption measurements at 280 nm (see in this volume, Willard and Siderovski, 2004) to obviate any effects of fluorescence from the CFP or YFP moieties on other protein assay methodologies.

FRET-Based Assay of RGS4/Gα_{i1} Interaction

Measurements of CFP and YFP fluorescence are made in a LS55 luminescence spectrofluorimeter (Perkin Elmer, Boston, MA). The LS55 cuvette holder is water jacketed and thus can be connected to a circulating water bath to provide controlled temperature between 4 and 40°. Proteins are incubated in 2-ml quartz cuvettes (Fisher) with magnetic stirring bars (Fisher) containing 1 ml of buffer F (10 mM Tris–HCl, pH 8.0, 1 mM EDTA, 50 mM NaCl, 10 mM MgCl$_2$, 2 μM GDP). The water bath temperature is set to 20°, and magnetic stir bars are rotated using the interactive command menus in the LS55 control program "FL WinLab" (Perkin Elmer) and selecting STATUS, SAMPLER ACCESSORY, CELL CHANGER STIRRER HIGH. We have found the inclusion of low concentrations of NaCl to be essential for the long-term stability of protein complexes during FRET measurements. MgCl$_2$ and GDP are included in all solutions as they participate in the formation of the AlF$_4^-$-induced Gα transition state and their presence in all cuvettes obviates any background fluorescence changes due to these compounds.

Excitation and emission maxima for CFP, YFP, and FRET complexes can be determined using the SCAN utility and selecting PRESCAN, which provides a rapid evaluation of both excitation and emission spectra. Subsequently, the EMISSION and EXCITATION SCAN utilities can be used to obtain spectral data more accurately. Based on such analysis, the optimal conditions for detecting CFP/YFP FRET can be determined. To this end, CFP fluorescence is excited at 433 nm and sequential emission scans are recorded at 0.5-nm intervals between 450 and 600 nm, at a speed of 50 nm/min, with excitation and emission slit widths of 2.5 nm. Excitation and emission spectra are analyzed using Graph Pad Prism version 4.0 (Graph Pad Software, San Diego, CA). Spectra are corrected for buffer absorption by subtracting the values of a cuvette containing only buffer; data can then be expressed in relative fluorescence units and are scaled to give one specific spectrum, namely the YFP-RGS4 alone spectrum, a maximal emission of 100 (e.g., Fig. 2A).

The fluorescence emission spectrum ($\lambda_{ex} = 433$ nm) of 1 μM Gα_{i1}-CFP is illustrated in Fig. 2A. Gα_{i1}-CFP has an emission maximum at 474 nm within a broad spectrum characterized by a "shoulder" extending to a second inflexion point at 500 nm. One micromolar YFP-RGS4 has an emission maximum at 525 nm (Fig. 2A); however, at $\lambda_{ex} = 433$ nm, the relative intensity of YFP emission is low (as the excitation maximum of YFP is 516 nm). The fluorescence emission spectrum ($\lambda_{ex} = 433$ nm) of 1 μM Gα_{i1}-CFP and 1 μM YFP-RGS4 in the same cuvette is simply the sum of the individual Gα_{i1}-CFP and YFP-RGS4 spectra (Fig. 2A).

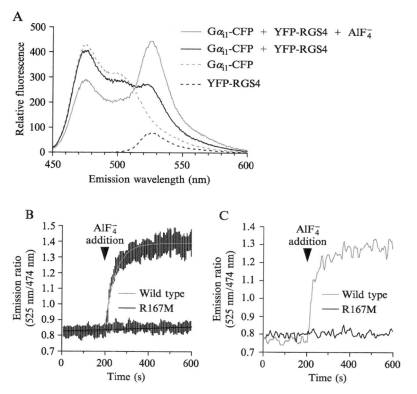

FIG. 2. Aluminum tetrafluoride-dependent FRET between CFP-Gα_{i1} and YFP-RGS4 fusion proteins. (A) Sequential emission scans between 450 and 600 nm were conducted using a LS55 spectrofluorimeter in the presence of various combinations of Gα_{i1}-CFP, YFP-RGS4, and/or AlF$_4^-$. The excitation wavelength for these emission spectral scans was 433 nm, with excitation and emission slit widths of 2.5 nm. Values are corrected by background subtraction of a buffer control and are normalized to make the peak fluorescence value for the YFP-RGS4 alone condition equal to 100. (B) Real-time measurement of AlF$_4^-$-induced FRET between Gα_{i1}-CFP and YFP-RGS4 proteins. Real-time emission ratio (525/474 nm) measurements were made from cuvettes containing 200 nM Gα_{i1}-CFP and 280 nM YFP-RGS4 (wild type) or 280 nM YFP-RGS4 point mutant (R167M) proteins. At 200 s, 20 mM NaF and 30 μM AlCl$_3$ were added to the cuvette to induce Gα_{i1} activation and binding to YFP-RGS4. Triplicate samples were fit to a single-exponential association function. Error bars denote the standard error of the mean. (C) One representative data set from the experiment conducted as described in B.

To measure the spectral dynamics of FRET induced by the binding of Gα_{i1} to RGS4, 40 μl of 0.5 M NaF and 1 μl of 30 mM AlCl$_3$ are added to the cuvette (for final concentrations of 20 mM and 30 μM, respectively). After a 5-min delay to allow full Gα_{i1} activation ($t_{1/2} = 33$ s at 20°;

assuming single exponential association as illustrated in Fig. 2B), the emission spectrum is read (Fig. 2A). Relative fluorescence values in the AlF_4^--containing sample are corrected for dilution by a factor of 1.041. (*Note*: This corrective step is only necessary in the analysis of spectral data, as emission ratio measurements are essentially independent of protein concentration in the micromolar range; see Fig. 3D.) In comparison to the emission spectrum of $G\alpha_{i1}$-CFP and YFP-RGS4 alone, the addition of AlF_4^- causes decreased emission at 474 nm (the normal emission peak for CFP) and a concomitant increase in emission at 525 nm (the normal emission peak for YFP) consistent with FRET (Fig. 2A). Thus, by tracking fluorescence emission at these two specific wavelengths (525 nm for YFP; 474 nm for CFP), one can calculate the ratio between the two emission intensities as a sensitive readout of RGS box/$G\alpha$ association. In this specific example (Fig. 2A), 525/474-nm emission ratio values were calculated as 0.66 (266.2/401.4) for control (no AlF_4^-) and 1.54 (443.3/287.4) in the presence of AlF_4^-, producing a 2.5-fold change in ratio by the interaction between $G\alpha_{i1}$-CFP and YFP-RGS4.

Real-time measurements of aluminum tetrafluoride-dependent $G\alpha$/ RGS box interactions have thus far only been described using surface plasmon resonance (Kimple *et al.*, 2001; Popov *et al.*, 1997; see also this volume, Willard and Siderovski, 2004). We have used the $G\alpha_{i1}$-CFP and YFP-RGS4 FRET system to evaluate the binding between $G\alpha_{i1}$ and RGS4 moieties in real time. AlF_4^--dependent $G\alpha$/RGS box interaction is measured in the LS55 spectrofluorimeter using 1 ml of buffer F containing 200 nM $G\alpha_{i1}$-CFP and 280 nM YFP-RGS4 at 20°. Using the TRUE RATIO mode, the 525/474-nm emission ratio is measured with a data interval of 5 s, encompassing 2s integrations at each wavelength. This is conducted using excitation at 433 nm and excitation and emission slit widths of 2.5 nm. After 200 s, the fluorescence emission ratio should have stabilized, and 41 μl of AlF_4^- (40 μl of 0.5 M NaF and 1 μl of 30 mM $AlCl_3$) is added to the cuvette to induce $G\alpha$ activation and FRET between the labeled RGS4 and $G\alpha_{i1}$ proteins. As illustrated in Fig. 2B and C, the addition of AlF_4^- increases the FRET emission ratio from 0.8 to 1.4. Increases are rapid and generally plateau after 200–400 s.

As a control to discount nonspecific fluorescence effects as the cause of emission ratio changes, mutant YFP-RGS4 proteins can be employed in this assay. For example, arginine-167 of RGS4 forms a salt bridge with two highly conserved $G\alpha_{i1}$ residues in the X-ray crystallographic structure of the transition state complex (Tesmer *et al.*, 1997). The R167M mutation inhibits GAP activity and the aluminum tetrafluoride-dependent interaction of RGS4 with $G\alpha_{i1}$ (Druey and Kehrl, 1997), thus R167M (or one of many other RGS box mutations described in the literature) serves as an appropriate negative control for these assays. Figures 2B and C illustrate the utility of YFP-RGS4(R167M) as a negative control for this FRET assay. Data from triplicate samples are averaged and fit to a single exponential association curve using Graph Pad Prism, as illustrated in Fig. 2B. The rate of AlF_4^--induced FRET increase is generally comparable to that of the AlF_4^--induced change in intrinsic tryptophan fluorescence that is indicative of $G\alpha$ activation (Higashijima et al., 1987). The half-time to maximal FRET is calculated, in this case, as 33.3 s. An example of raw data from an individual set of experiments is illustrated in Fig. 2C. These experiments were conducted with just 200–300 nM of protein and thus experimental "noise" is significant. Performing these assays with higher concentrations of protein (i.e., low micromolar amounts) essentially obviates this problem.

Development of a High-Throughput FRET Assay for $G\alpha_{i1}$/RGS4 Interaction

Aluminum tetrafluoride-dependent FRET between $G\alpha_{i1}$-CFP and YFP-RGS4 can be measured in a 96-well plate format using a SpectraMax Gemini fluorescence plate reader (Molecular Devices, Sunnyvale, CA). Samples are placed in Costar 96-well, flat-bottom, "special optics" black plates (Corning, Acton, MA). The SpectraMax Gemini is first optimized for CFP/YFP FRET using the SPECTRUM (FLUORESCENCE) function under the INSTRUMENT SETTINGS dialog box in the SOFTmax PRO v3.11 control software (Molecular Devices). This allows for excitation and emission scanning and optimization of cutoff filter settings, as recommended by the manufacturer. For experimental analysis of FRET, the ENDPOINT (FLUORESCENCE) function is used, specifying automatic photomultiplier tube adjustment (PMT AUTO), Costar plates, high sensitivity (30 reads per well), and thorough mixing of samples before readings (AUTOMIX 5 s). For dual ratio emission measurements, wavelength settings are as follows: channel 1, excitation 433 nm, emission 479 nm with a 455-nm cutoff filter and channel 2, excitation 433 nm, emission 529 nm with a 495-nm cutoff filter. FRET is measured using 190 μl buffer F per well containing 200 nM each of $G\alpha_{i1}$-CFP and YFP-RGS4 fusion proteins. The SpectraMax Gemini is temperature controlled and plates are equilibrated at 25° for 5 min. An initial reading is then taken to serve as the "preinteraction" measurement. The AlF_4^- ion is then formed by the addition of NaF and $AlCl_3$ to final concentrations of 20 mM and 30 μM to make up a final, per-well volume of 200 μl. The plate is then incubated for 5 min at 25°, a time period sufficient for complete $G\alpha_{i1}$ activation and $G\alpha_{i1}$/RGS4 complex formation (as determined in Figs. 2B and C). Readings are then made to serve as the "$+AlF_4^-$"

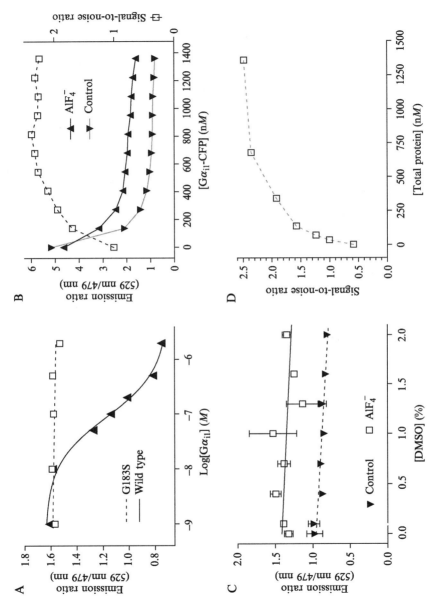

Fig. 3. Optimization of the CFP-$G\alpha_{i1}$/YFP-RGS4 FRET assay for high-throughput screening of RGS box inhibitors. FRET between CFP-$G\alpha_{i1}$ and YFP-RGS4 was measured using a SpectraMax Gemini 96-well plate fluorescence reader. All emission ratio scans were conducted using an excitation wavelength of 433 nm and dual emission wavelengths of 479 and 529 nm. Samples were

measurement. Initial data analysis (i.e., emission ratio calculation) is accomplished using SOFTmax PRO; subsequent data analysis is done using Graph Pad Prism. Readings from wells containing only buffer with or without AlF_4^- are subtracted from sample well readings to account for any buffer effects. We have not observed significant qualitative differences between data analyzed with or without background subtraction (Willard and Siderovski, unpublished observations).

To validate the utility of this assay for the discovery of inhibitors of the $G\alpha_{i1}$/RGS4 interaction, we investigated the disruption of the $G\alpha_{i1}$-CFP/YFP-RGS4 complex using nonfluorescent $G\alpha_{i1}$ proteins. Disruption of the interaction between $G\alpha_{i1}$-CFP and YFP-RGS4 is measured by first initiating the FRET interaction, as described earlier, with 200 nM $G\alpha_{i1}$-CFP, 200 nM YFP-RGS4, 20 mM NaF, and 30 μM $AlCl_3$. After a 5-min incubation at 25°, a broad range of concentrations (up to 2 μM) of His_6-$G\alpha_{i1}$ or RGS box-insensitive, glycine-183 to serine (G183S) His_6-$G\alpha_{i1}$ (DiBello et al., 1998; Lan et al., 1998) are added to individual wells. Following a 20-min incubation, FRET is quantitated and data are fit to a dose–response curve with Graph Pad Prism. Disruption of FRET by the wild-type unlabeled $G\alpha_{i1}$ protein is illustrated in Fig. 3A; this inhibitory effect occurs with an IC_{50} of 87 nM. The competitive and specific nature of this disruption is indicated by the failure of His_6-$G\alpha_{i1}$ G183S to disrupt FRET at a 20-fold molar excess; the glycine-183 to serine point mutation to $G\alpha_{i1}$ renders the protein unable to interact with RGS proteins (DiBello et al., 1998; Lan et al., 1998).

Further optimization of the $G\alpha_{i1}$/RGS4 FRET assay can be achieved by careful titration of the various assay components and reagents. For

---

equilibrated with all reaction components, save for AlF_4^-, for 5 min at 25° and then duplicate fluorescence readings were collected. The planar ion AlF_4^- was induced by the addition of NaF and $AlCl_3$ to final concentrations of 20 mM and 30 μM, respectively. Following a 5-min incubation, duplicate fluorescence readings were collected again. (A) Competitive inhibition of the CFP-$G\alpha_{i1}$/YFP-RGS4 interaction. This experiment was conducted in the presence of indicated concentrations of nonfluorescent His_6-$G\alpha_{i1}$ ("wild type") or RGS box-insensitive, point mutant His_6-$G\alpha_{i1}$ ("G183S"). Data are presented as emission ratio measurements following AlF_4^- addition. (B) Titration of CFP-$G\alpha_{i1}$ protein concentration required for effective FRET in the presence of 1400 nM YFP-RGS4. Data on the left ordinate are presented as emission ratio measurements before ("control") and after aluminium tetrafluoride addition ("AlF_4^-"). The right ordinate scale ("signal-to-noise" ratio) is defined as the ratio of AlF_4^--induced emission ratio to control emission ratio (as described in the text). (C) Tolerance of the FRET assay to dimethyl sulfoxide (DMSO). This experiment was conducted in the presence of indicated concentrations of DMSO. Data are presented as emission ratio measurements before ("control") and after aluminium tetrafluoride addition ("AlF_4^-"). (D) Dependence of the FRET assay on total fluorescent protein concentration. $G\alpha_{i1}$-CFP and YFP-RGS4 proteins were combined in an equimolar ratio and assayed for FRET in a dose-dependent manner up to a total protein concentration of 1.4 μM (1400 nM).

example, preliminary experiments achieved a signal-to-noise ratio of 1.3 (as defined later), whereas optimized assay conditions give a signal-to-noise ratio of over 3.0:

$$\text{signal-to-noise ratio} = \text{AlF}_4^- \left(\text{Em}_{529 \text{ nm}}/\text{Em}_{479 \text{ nm}}\right)/$$
$$\text{control} \left(\text{Em}_{529 \text{ nm}}/\text{Em}_{479 \text{ nm}}\right)$$

Figure 3B illustrates the signal-to-noise ratio using a titration of $G\alpha_{i1}$-CFP protein concentration against a constant concentration of 1400 nM YFP-RGS4. The 529/479-nm emission ratios stabilize as the $G\alpha_{i1}$-CFP concentration reaches that of YFP-RGS4; similarly, the signal-to-noise increases to above 2. Similar titrations should be performed with all buffer components to maximize the signal-to-noise ratio and facilitate sensitive detection of the $G\alpha_{i1}$/RGS4 interaction.

To create a high-throughput screen for small molecule inhibitors of RGS box proteins, it is necessary to validate the tolerance of the assay for the solvents in which the small molecule organic compounds are dissolved. Generally, the solvent of choice is dimethyl sulfoxide (DMSO) (Seethala and Fernandes, 2001). Testing of the DMSO tolerance of the $G\alpha_{i1}$/RGS4 FRET assay is conducted essentially as described earlier, but with the addition of DMSO [0–2% (v/v)] and the reading of FRET values before and after AlF_4^- addition. Figure 3C illustrates the DMSO sensitivity of this particular assay: the overall fluorescence emission ratio attenuation observed upon DMSO addition does not appear to affect signal-to-noise performance of this assay up to a concentration of 2% (v/v) DMSO. Clearly, for any compound library screening, it will be important to include solvent controls to eliminate false positives.

To examine the detection–response range of the $G\alpha_{i1}$/RGS4 FRET assay, $G\alpha_{i1}$-CFP and YFP-RGS4 fusion proteins are mixed in an equimolar ratio and a FRET experiment is conducted by titrating the total protein concentration (0–1400 nM) versus the signal-to-noise ratio (as defined earlier). Figure 3D illustrates that a plot of signal-to-noise ratio versus protein concentration is exponential and plateaus above 750 nM total protein. This is an important experimental consideration in assaying for RGS inhibitors, as the larger the molar ratio of inhibitor to RGS box protein that can be achieved, the higher the likelihood of observing inhibitor activity; *in vitro* inhibitors derived from compound libraries will probably have orders of magnitude less affinity to their RGS box targets than the affinity of the $G\alpha$ subunit/RGS box interaction. The intricacies and numerical considerations of high–throughput screening are well discussed in the literature (e.g., Copeland, 2003). Clearly, it is best to strive for a balance between assay sensitivity and high test compound concentration in such assays.

Use of RGS Box Proteins as Biosensors for G-Protein Activation

The FRET assay for the $G\alpha$/RGS box interaction can also be utilized to assay the effect of G-protein modulatory peptides on $G\alpha$ function. We have explored this approach using a synthetic peptide derived from the GoLoco motif of RGS12. GoLoco motifs are guanine nucleotide dissociation inhibitors (GDIs) for the adenylyl cyclase inhibitory class of G-protein α subunits. GoLoco motif-containing proteins have essential roles in metazoan cell division processes via their regulation of the heterotrimeric G-protein cycle (Kimple et al., 2002a; Willard et al., 2004b). Two of the more rigorously characterized GoLoco motifs are those present in RGS12 and RGS14 (Kimple et al., 2001, 2002b). The GoLoco motifs of RGS12 and RGS14 are potent GDIs for $G\alpha_i$ but not $G\alpha_o$ subunits. GoLoco motif peptides from RGS12, RGS14, and AGS3 have been found to stabilize the GDP-bound conformation of $G\alpha$ (De Vries et al., 2000; Kimple et al., 2001). This activity was measured as refractoriness of GoLoco-liganded $G\alpha$ subunits to AlF_4^--induced activation using intrinsic tryptophan fluorescence assays. The use of intrinsic tryptophan fluorescence can be limited by the weak fluorescence enhancement (60%) achieved upon G-protein activation (Higashijima et al., 1987) and the background fluorescence contributed by any protein cofactor added. We have thus abandoned the intrinsic tryptophan fluorescence assay for this new FRET assay system.

The modulatory effect of the RGS12 GoLoco motif on $G\alpha$ activation by AlF_4^- is measured as generally described in the previous section. $G\alpha_{i1}$-CFP (200 nM) and YFP-RGS4 (280 nM) are incubated at 25° in a 1-ml cuvette containing 10 mM Tris–HCl, pH 8.0, 200 mM NaCl, 1 mM EDTA, 10 mM $MgCl_2$, and 2 μM GDP. Test samples contain either a 2- or a 10-fold molar excess of a synthetic peptide encompassing the minimal GoLoco motif of rat RGS12 (RGS12-GoLoco; amino acids 1186–1221) or a vehicle (H_2O) control (Kimple et al., 2001). The RGS12-GoLoco peptide is dissolved in H_2O to a stock concentration of 400 μM; we have also found that purified recombinant GoLoco motif-containing proteins and DMSO-solubilized synthetic peptides are also compatible with this assay (see in this series: Kimple, Willard, and Siderovski, 2004). After incubation for 5 min at 25° to allow complex formation between $G\alpha$ and RGS12-GoLoco peptide, fluorescence measurements are initiated with excitation at 433 nm (2.5-nm slit width) and dual-emission wavelength detection at 474 nm (2.5-nm slit width) and 525 nm (2.5-nm slit width). (As the salient data obtained in this assay are kinetic, only one sample is measured using the LS55 spectrofluorimeter. However, for end point FRET studies, and those assays where kinetics are not imperative, multiple sample measurements can be conducted using a 96-well plate reader, as described earlier.)

After 150 s, the fluorescence emission ratio should have stabilized, and 41 μl of AlF_4^- (40 μl of 0.5 M NaF and 1 μl of 30 mM $AlCl_3$) is added to

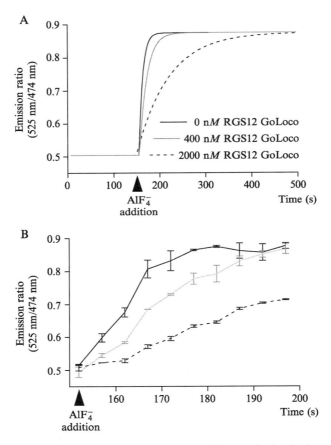

FIG. 4. Analysis of GoLoco motif peptide inhibition of $G\alpha_{i1}$ activation by the use of FRET between $G\alpha_{i1}$-CFP and YFP-RGS4. (A) FRET between 200 nM $G\alpha_{i1}$-CFP and 280 nM YFP-RGS4 was measured, in real time, as a ratio of emission intensity at 525 nm to an emission intensity at 474 nm. FRET pair samples contained RGS12 GoLoco peptide (400 or 2000 nM) or a vehicle control. At 150 s, planar ion AlF_4^- was created by the addition of NaF and $AlCl_3$ to final concentrations of 20 mM and 30 μM, respectively. Duplicate experiments were averaged and fit to a single exponential association function using Graph Pad Prism. (B) Raw data from the experiment conducted in A are presented over the period immediately following the addition of AlF_4^-, illustrating the dramatic attenuation of the G-protein activation rate in the presence of RGS12 GoLoco peptide. Data points are the mean ± SEM of duplicate experiments.

the cuvette to induce $G\alpha$ activation and FRET between $G\alpha_{i1}$-CFP and YFP-RGS4. In general, the addition of excess volume to reactions does not need to be accounted for with corrective calculations, as the emission ratio measurement for FRET does not depend on total protein concentration; rather, the relative protein concentration is the important factor (Figs. 3B and D). Figure 4A illustrates attenuation of $G\alpha_{i1}$ activation by the RGS12-GoLoco peptide, present at suprastoichiometric concentrations over the input amount of $G\alpha_{i1}$-CFP (200 nM). Figure 4A represents data from duplicate samples that were subsequently fit to a single exponential association model using Graph Pad Prism. This analysis yielded initial $G\alpha$ activation rates for samples containing 0 nM ($1.2 \times 10^{-1} \pm 7.3 \times 10^{-3}$ s^{-1}), 400 nM ($6.2 \times 10^{-2} \pm 1.6 \times 10^{-3}$ s^{-1}), and 2000 nM ($1.4 \times 10^{-2} \pm 2.7 \times 10^{-4}$ s^{-1}) RGS12-GoLoco peptide. Thus, a 2-fold molar excess of RGS12 GoLoco peptide (400 nM) to $G\alpha_{i1}$-CFP (200 nM) gave a 2-fold decrease in the $G\alpha_{i1}$ activation rate, whereas a 10-fold molar excess of RGS12-GoLoco peptide (2000 nM) gave a 12-fold decrease in activation rate. These findings suggest that this assay can accurately measure the attenuation of AlF_4^--mediated $G\alpha$ activation by GoLoco motif peptides and, presumably, other $G\alpha$-binding peptides/proteins. Raw data from the experiment in Fig. 4A over the time period immediately following AlF_4^- addition are presented in Fig. 4B to illustrate the dramatic attenuation of the G-protein activation rate in the presence of the RGS12-GoLoco peptide that can be observed in real time.

Concluding Remarks

We have described protocols for the accurate measurement of the interaction between $G\alpha_{i1}$ and RGS4 using FRET. This assay is effective in measuring the activities of both mutant $G\alpha_{i1}$ (i.e., G183S) and mutant RGS4 (i.e., R167M) proteins and should be amenable to any $G\alpha$/RGS box pair that can be purified as soluble recombinant proteins. There remains the potential for this assay to be modified and utilized *in vivo* to detect receptor-mediated heterotrimer activation and we are currently pursuing this approach. Furthermore, we have investigated the use of the $G\alpha_{i1}$-CFP and YFP-RGS4 protein pair to create a microtiter plate-formatted high-throughput assay for RGS-box/$G\alpha$ interaction inhibitors. Finally, we have described a novel assay for detecting the modulatory effects of other $G\alpha$ interactors (e.g., GoLoco motif peptides) on G-protein activation by aluminum tetrafluoride—an assay that is uniquely able to resolve the subtle kinetics of this activation process.

Acknowledgments

We thank Dr. Catherine Berlot for the provision of ECFP and EYFP cDNAs (Weis Centre for Research, Danville, PA). FSW is an American Heart Association Postdoctoral Fellow. RJK is supported by a predoctoral fellowship from the National Institutes of Mental Health (F30 MH64319). This work was funded by NIH Grants R01 GM062338 and P01 GM065533. Thanks to Dr. Benjamin Yerxa and the rest of Inspire Pharmaceuticals Inc. for support in the initial phase of this project. DPS is a recipient of the Burroughs Wellcome Fund New Investigator Award in the Basic Pharmacological Sciences and a Year 2000 Scholar of the EJLB Foundation.

References

Berman, D. M., Wilkie, T. M., and Gilman, A. G. (1996a). GAIP and RGS4 are GTPase activating proteins for the Gi subfamily of G protein alpha subunits. *Cell* **86**, 445–452.

Berman, D. M., Kozasa, T., and Gilman, A. G. (1996). The GTPase-activating protein RGS4 stabilizes the transition state for nucleotide hydrolysis. *J. Biol. Chem.* **271**, 27209–27212.

Copeland, R. A. (2003). Mechanistic considerations in high throughput screening. *Anal. Biochem.* **320**, 1–12.

De Vries, L., Fischer, T., Tronchere, H., Brothers, G. M., Strockbine, B., Siderovski, D. P., and Farquhar, M. G. (2000). Activator of G protein signaling 3 is a guanine dissociation inhibitor for G-alpha i subunits. *Proc. Natl. Acad. Sci. USA* **97**, 14364–14369.

DiBello, P. R., Garrison, T. R., Apanovitch, D. M., Hoffman, G., Shuey, D. J., Mason, K., Cockett, M. I., and Dohlman, H. G. (1998). Selective uncoupling of RGS action by a single point mutation in the G protein alpha-subunit. *J. Biol. Chem.* **273**, 5780–5784.

dos Remedios, C. G., and Moens, P. D. (1995). Fluorescence resonance energy transfer spectroscopy is a reliable "ruler" for measuring structural changes in proteins. Dispelling the problem of the unknown orientation factor. *J. Struct. Biol.* **115**, 175–185.

Druey, K. M., and Kehrl, J. H. (1997). Inhibition of regulator of G protein signaling function by two mutant RGS4 proteins. *Proc. Natl. Acad. Sci. USA* **94**, 12851–12856.

Gilman, A. G. (1987). G proteins: Transducers of receptor-generated signals. *Annu. Rev. Biochem.* **56**, 615–649.

Heikal, A. A., Hess, S. T., Baird, G. S., Tsien, R. Y., and Webb, W. W. (2000). Molecular spectroscopy and dynamics of intrinsically fluorescent proteins: Coral red (dsRed) and yellow (Citrine). *Proc. Natl. Acad. Sci. USA* **97**, 11996–12001.

Higashijima, T., Ferguson, K. M., Sternweis, P. C., Ross, E. M., Smigel, M. D., and Gilman, A. G. (1987). The effect of activating ligands on the intrinsic fluorescence of guanine nucleotide binding regulatory proteins. *J. Biol. Chem.* **262**, 752–756.

Kimple, R. J., Willard, F. S., and Siderovski, D. P. (2002a). The GoLoco Motif: Heralding a new tango between G protein signaling and cell division. *Mol. Intervent.* **2**, 88–100.

Kimple, R. J., Kimple, M. E., Betts, L., Sondek, J., and Siderovski, D. P. (2002b). Structural determinants for GoLoco induced inhibition of nucleotide release by G-alpha subunits. *Nature* **416**, 878–881.

Kimple, R. J., De Vries, L., Tronchere, H., Behe, C. I., Morris, R. A., Gist Farquhar, M., and Siderovski, D. P. (2001). RGS12 and RGS14 GoLoco motifs are G-alpha(i) interaction sites with guanine nucleotide dissociation inhibitor activity. *J. Biol. Chem.* **276**, 29275–29281.

Kimple, R. J., Jones, M. B., Shutes, A., Yerxa, B. R., Siderovski, D. P., and Willard, F. S. (2003). Established and emerging fluorescence based assays for G-protein function: Heterotrimeric G-protein-alpha subunits and regulator of G protein signaling (RGS) proteins. *Comb. Chem. High Throughput Screen.* **6**, 399–407.

Kimple, R. J., Willard, F. S., and Siderovski, D. P. (2004). Purification and *in vitro* functional analyses of R6S12 and RGS14 GoLoco-motif peptides. *Methods Enzymol.* **390**(26), 419–436.

Lakowicz, J. R. (1999). "Principles of Fluorescence Spectroscopy." Kluwer Academic/ Plenum, New York.

Lan, K. L., Sarvazyan, N. A., Taussig, R., Mackenzie, R. G., DiBello, P. R., Dohlman, H. G., and Neubig, R. R. (1998). A point mutation in Galphao and Galphail blocks interaction with regulator of G protein signaling proteins. *J. Biol. Chem.* **273**, 12794–12797.

Lan, K. L., Zhong, H., Nanamori, M., and Neubig, R. R. (2000). Rapid kinetics of regulator of G-protein signaling (RGS)-mediated Galphai and Galphao deactivation. Galpha specificity of RGS4 and RGS7. *J. Biol. Chem.* **275**, 33497–33503.

Miyawaki, A., and Tsien, R. Y. (2000). Monitoring protein conformations and interactions by fluorescence resonance energy transfer between mutants of green fluorescent protein. *Methods Enzymol.* **327**, 472–500.

Mukhopadhyay, S., and Ross, E. M. (1999). Rapid GTP binding and hydrolysis by G(q) promoted by receptor and GTPase-activating proteins. *Proc. Natl. Acad. Sci. USA* **96**, 9539–9544.

Neubig, R. R., and Siderovski, D. P. (2002). Regulators of G-protein signalling as new central nervous system drug targets. *Nature Rev. Drug Discov.* **1**, 187–197.

Pierce, K. L., Premont, R. T., and Lefkowitz, R. J. (2002). Seven-transmembrane receptors. *Nature Rev. Mol. Cell. Biol.* **3**, 639–650.

Popov, S., Yu, K., Kozasa, T., and Wilkie, T. M. (1997). The regulators of G protein signaling (RGS) domains of RGS4, RGS10, and GAIP retain GTPase activating protein activity *in vitro*. *Proc. Natl. Acad. Sci. USA* **94**, 7216–7220.

Prasher, D. C., Eckenrode, V. K., Ward, W. W., Prendergast, F. G., and Cormier, M. J. (1992). Primary structure of the Aequorea victoria green-fluorescent protein. *Gene* **111**, 229–233.

Seethala, R., and Fernandes, P. B. (2001). "Handbook of Drug Screening." Dekker, New York.

Tesmer, J. J., Berman, D. M., Gilman, A. G., and Sprang, S. R. (1997). Structure of RGS4 bound to AlF4–activated G(i alpha1): Stabilization of the transition state for GTP hydrolysis. *Cell* **89**, 251–261.

Willard, F. S., and Siderovski, D. P. (2004). *Methods Enzymol.* **389**(19), (this volume).

Willard, F. S., Kimple, A. J., Kimple, R. J., and Siderovski, D. P. (2004a). "Toward small molecule inhibitors of RGS proteins: Development of computational and *in vitro* fluorescence-based approaches." *FASEB J.* **18**(4), A 219 (abstract 165.4).

Willard, F. S., Kimple, R. J., and Siderovski, D. P. (2004b). Return of the GDI: The GoLoco Motif in cell division. *Annu. Rev. Biochem.* **73**, 925–951.

[5] Application of RGS Box Proteins to Evaluate G-Protein Selectivity in Receptor-Promoted Signaling

By Melinda D. Hains, David P. Siderovski, and T. Kendall Harden

Abstract

Regulator of G-protein signaling (RGS) domains bind directly to GTP-bound Gα subunits and accelerate their intrinsic GTPase activity by up to several thousandfold. The selectivity of RGS proteins for individual Gα subunits has been illustrated. Thus, the expression of RGS proteins can be used to inhibit signaling pathways activated by specific G protein-coupled receptors (GPCRs). This article describes the use of specific RGS domain constructs to discriminate among $G_{i/o}$, G_q-and $G_{12/13}$-mediated activation of phospholipase C (PLC) isozymes in COS-7 cells. Overexpression of the N terminus of GRK2 (amino acids 45–178) or p115 RhoGEF (amino acids 1–240) elicited selective inhibition of $G\alpha_q$- or $G\alpha_{12/13}$-mediated signaling to PLC activation, respectively. In contrast, RGS2 overexpression was found to inhibit PLC activation by both $G_{i/o}$- and G_q-coupled GPCRs. RGS4 exhibited dramatic receptor selectivity in its inhibitory actions; of the $G_{i/o}$- and G_q-coupled GPCRs tested (LPA$_1$, LPA$_2$, P2Y$_1$, S1P$_3$), only the G_q-coupled lysophosphatidic acid-activated LPA$_2$ receptor was found to be inhibited by RGS4 overexpression.

Introduction

Many extracellular stimuli, such as neurotransmitters, hormones, chemokines, inflammatory mediators, and odorants, exert their effects by activating phosphoinositide-hydrolyzing phospholipase C (PLC) isozymes (Berridge, 1987). Five classes of PLC isozymes underlie these signals: PLC-β, PLC-γ, PLC-δ, PLC-ε, and PLC-ζ (Rhee, 2001; Saunders et al., 2002). The G protein-coupled receptor (GPCR) superfamily activates PLC-β enzymes through activation of α subunits of the G_q family of G-protein heterotrimers or by $G\beta\gamma$ dimers released from activated G_i and potentially from other heterotrimeric G proteins. PLC-ε is regulated by Ras and Rho GTPases, as well as by $G\alpha_{12/13}$ and $G\beta\gamma$ subunits of heterotrimeric G proteins (Wing, 2003a) and therefore is also activated by GPCRs (Hains et al., 2004, Kelley et al., 2004).

Identification of the specific G proteins involved in GPCR-promoted signaling pathways is a critical step in understanding the mechanism by which receptors promote particular downstream signaling events. Tools such as pertussis toxin, $G\beta\gamma$ "sinks" such as $G\alpha$-transducin or the C-terminal fragment of the G protein-coupled receptor kinase 2 (GRK2), antisense oligonucleotides, RNA interference, or GoLoco motif peptides can be applied as reagents to delineate the G-protein subunits involved in the GPCR-mediated activation of PLC. However, such reagents are limited to specific G proteins (e.g., pertussis toxin and GoLoco motif peptides only uncouple GPCRs from heterotrimers of the $G_{i/o}$ family), may result in more widespread, nonspecific effects, or are expensive and time-consuming (e.g., antisense, RNAi).

A potentially important new means for the modification of GPCR signaling involves the application of regulator of G-protein signaling

(RGS) proteins. RGS proteins contain a conserved ~120 amino acid "RGS box" that accelerates $G\alpha$-mediated GTP hydrolysis and, therefore, inhibits GPCR-promoted signaling (Neubig and Siderovski, 2002). Although several investigators have applied RGS proteins to inhibit G-protein-dependent signaling, e.g., a decrease in D_2 dopamine receptor-stimulated signaling occurred in the striatum after ectopic expression of RGS9 (Rahman *et al.*, 2003), utilization of RGS proteins to control GPCR function has not been fully elucidated.

Greater than 30 RGS domain-containing proteins have been identified to date, and many of these are negative regulators of G-protein signaling. Given that certain RGS proteins exhibit selectivity for specific $G_{i/o}$, G_q, $G_{12/13}$, and G_s $G\alpha$ subunits (Neubig and Siderovski, 2002), these RGS proteins potentially can be utilized to delineate the G proteins involved in signaling pathways activated by GPCRs. This article describes the use of RGS proteins as antagonists of GPCR signaling and examines their effectiveness in delineating the G-protein-coupling profile of a number of GPCRs using the inositol lipid signaling pathway as an example of their application.

Methodology

Phospholipase C isozymes catalyze the hydrolysis of phosphatidylinositol 4,5-bisphosphate [PtdIns(4,5)P_2] to 1,2-diacylglycerol and inositol (1,4,5)-trisphosphate [Ins(1,4,5)P_3] in response to the activation of cell surface receptors. The production of Ins(1,4,5)P_3 is measured using a phosphoinositide hydrolysis assay that quantifies the amount of inositol phosphates produced as a result of PLC activation. Both PLC-β and PLC-ε are stimulated by GPCRs through the activation of G_q, $G_{i/o}$, and $G_{12/13}$ heterotrimers. Conversely, the activity of RGS proteins as negative regulators of GPCR signaling can be evaluated using inositol phosphate accumulation as a biochemical assay of receptor activation. The methodology described in this article uses COS-7 cells cotransfected with GPCRs, RGS proteins, and, in some cases, PLCs, and the phosphoinositide hydrolysis assay as a readout, to examine the specificity and capacity of specific RGS proteins to inhibit the GPCR activation of PLC.

Materials and Methods

RGS Constructs

A plethora of RGS proteins act on $G\alpha$ subunits of the $G_{i/o}$, G_q, $G_{12/13}$, and G_s families. We have used the following RGS proteins to discriminate among $G_{i/o}$-, G_q-, and $G_{12/13}$-promoted signaling: the RGS box of GRK2

Portion of RGS protein
used in assays:

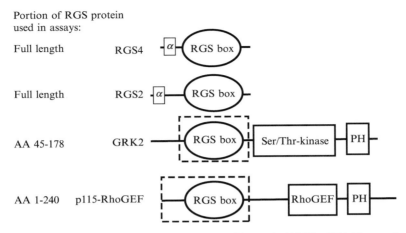

FIG. 1. Domain architecture of GRK2, RGS2, RGS4, and p115 RhoGEF. The portion of each RGS protein used in the analyses described in the text is described to the left of the structures. Dashed lines around GRK2 and p115 RhoGEF highlight the regions of these proteins subcloned into expression vectors for this study.

(GRK2RGS), RGS2, RGS4, and the RGS box of p115 RhoGEF (p115RGS). RGS2 and RGS4 are essentially limited to a core RGS domain structure with short N- and C-terminal polypeptide extensions, whereas GRK2 and p115 RhoGEF have an easily delineated multidomain architecture (Fig. 1). cDNA encoding full-length RGS4 with a triple hemagglutinin (3X HA) epitope tag was obtained from the Guthrie Research Institute (Sayre, PA; www.cdna.org), cDNA encoding full-length RGS2 was subcloned in frame with an N-terminal histidine (His_{10}) epitope tag into pcDNA3.1($-$) (Invitrogen Corp., Carlsbad, CA) (Cunningham *et al.*, 2001), cDNA encoding amino acids 45–178 of GRK2 in frame with an N-terminal HA epitope tag in pcDNA3 was kindly provided by Dr. Jeffrey Benovic (Thomas Jefferson University, Philadelphia, PA), and cDNA encoding the first 240 amino acids of human p115 RhoGEF was subcloned in frame with an N-terminal tandem HA epitope tag into a modified pcDNA3.1 vector (Snow *et al.*, 1998).

Other Constructs

cDNAs comprising the coding sequence of the human LPA_2 and $S1P_3$ receptor are inserted into the pCR3.1 vector. cDNA comprising the coding sequence of the mouse LPA_1 receptor is inserted into the pcDNA3 vector. The LPA_1, LPA_2, and $S1P_3$ receptor cDNAs were generously provided by Dr. Kevin R. Lynch (University of Virginia, Charlottesville, VA). The human purinergic P2Y$_1$ receptor, bearing an N-terminal HA epitope tag, was cloned as described previously (Schachter *et al.*, 1996). Rat PLC-ε in pCMV-script (Kelley *et al.*, 2001) was a kind gift of Dr. Grant Kelley (SUNY Syracuse, NY). cDNA of the constitutively active mutant of mouse Gα$_{12}$ [Gα$_{12}$ Q229L (Gα$_{12}$QL)] was kindly provided by Dr. Channing Der (University of North Carolina, Chapel Hill, NC). The constitutively active mutants of human Gα$_q$ [Gα$_q$ Q209L (Gα$_q$QL)] and human RhoB [RhoB G14V (RhoBGV)] were obtained from Guthrie Research Institute (Sayre, PA; www.cdna.org). The carboxyl-terminal domain of GRK2 (GRK2ct) was a kind gift of Dr. Robert J. Lefkowitz (Duke University, Durham, NC).

Cell Culture and Transfection of COS-7 Cells

COS-7 cells are maintained in Dulbecco's modified Eagle's medium (DMEM) supplemented with 10% fetal bovine serum, 4 mM L-glutamine, 200 U penicillin, and 0.2 mg/ml streptomycin (all obtained from Gibco, Invitrogen Corp.) at 37° in a 5% CO_2/95% air-humidified atmosphere. To plate COS-7 cells, remove medium from cells, wash cells once with 5 ml phosphate-buffered saline (PBS), remove PBS, and add 4 ml 0.25% trypsin/EDTA (all obtained from Gibco, Invitrogen Corp.). Following a 5-min incubation at 37°, suspend the cells in 10 ml of medium, remove 10 μl, and determine the cell number with a hemocytometer. A confluent 162-cm^2 flask of COS-7 cells will yield approximately 10 ml of cells at a concentration of 1×10^6 cells/ml. Bring cells to a density of 60,000–75,000 cells/ml with medium, plate 1 ml of cells/well in 12-well culture dishes (CoStar 3512), and incubate overnight (approximately 8 to 12 h). Following overnight incubation, prepare the transfection reagent in serum-free DMEM. Specifically, dilute 2 μl of FuGENE 6 transfection reagent (Roche Applied Science, Indianapolis, IN) directly into 41 μl serum-free DMEM. Make sure not to touch the side of the tube with the FuGENE 6 reagent, as this will significantly reduce transfection efficiency. Once diluted, flick the tube several times to mix (do not vortex) and incubate at room temperature for 5 min. While waiting, dilute all DNA plasmids to 100 ng/μl with sterile water. To each tube containing the 43 μl FuGENE 6/serum-free DMEM mix, add 700 ng total of the necessary DNA expression plasmids. For most experiments described in this article, 50 ng of GPCR, 100 ng of PLC-ε, and 550 ng of RGS protein DNA plasmids were added to the 43-μl mix (0.5 μl, 1 μl, and 5.5 μl of 100 ng/μl dilutions, respectively). When one of the aforementioned plasmids was not required, the total amount of DNA was brought up to 700 ng with empty pcDNA3 vector. Once the DNA is added to the FuGENE 6/serum-free mix, flick the tube several times to mix (do not vortex) and incubate at room temperature for 15–45 min to allow the DNA to complex with the FuGENE 6 transfection reagent. Add 50 μl of complexes to each well and return the culture dishes to the 37°, 5% CO_2/95% air atmosphere.

Phosphoinositide Hydrolysis Assay

Dilute 1 μCi (1 μl) of [*myo*-³H]inositol (American Radiolabeled Chemicals) into 400 μl serum-free, inositol-free DMEM (ICN Biomedicals, Inc., Aurora, OH) and mix by inverting the tube several times. Approximately 12–24 h after transfection, remove the medium from each well and replace with 400 μl of medium consisting of serum-free inositol-free DMEM containing [*myo*-³H]inositol (1 μCi/well). Return the dishes to the tissue culture incubator and incubate for 12–18 h at 37°. Phospholipase C activity is quantified 12–18 h after labeling by treating cells for 30–45 min in inositol-free DMEM containing 10 mM LiCl (to inhibit inositol phosphatases). Hormone-stimulated PLC activity was determined in the experiments described later by incubating cells in the absence or presence of lysophosphatidic acid [LPA; 1-oleoyl-sn-glycerol 3-phosphate sodium salt dissolved in water containing 1.0% fatty acid-free bovine serum albumin (BSA)], sphingosine 1-phosphate (S1P; dissolved in water containing 1.0% fatty acid-free BSA), or the purinergic receptor agonist 2-methylthioadenosine diphosphate (2-MeSADP) (LPA and S1P, Sigma, St. Louis, MO; 2-MeSADP, Research Biochemicals Inc., Natick, MA). The reaction is terminated after 30–45 min by aspirating the medium and adding 750 μl of ice-cold 50 mM formic acid (Fisher Scientific, formic acid 88%) for a minimum of 15 min at 4°. Neutralize cell extracts with 250 μl of 150 mM NH₄OH (Fisher Scientific, ammonium hydroxide) and add 1 ml neutralized sample to 1-ml packed volume columns of AG 1-X8 anion-exchange gel resin (Bio-Rad, formate form, 200–400 mesh size). Poly-Prep chromatography columns (Bio-Rad) are utilized because they contain a filter for stacking the resin and also have a large volume capacity to accommodate washes up to 10 ml. Prior to the addition of cell extracts, the columns should be prepared by washing with 5 ml of 2 M ammonium formate (Fisher Scientific, ammonium formate) and 0.1 M formic acid followed by two 10-ml washes of deionized water. After loading the cell extracts, wash the columns with 10 ml deionized water followed by 10 ml of 50 mM ammonium formate and elute the [³H]inositol phosphates into scintillation vials by adding 5 ml 1.2 M ammonium formate/0.1 M formic acid to the column. Add 10 ml scintillation fluid (Fisher Scientific, Scintisafe gel cocktail) to each vial and shake vigorously until the mixture becomes clear. If the mixture remains a "milky" color, add more scintillation fluid and repeat the shaking step. Quantify radioactivity by scintillation counting (Brown *et al.*, 1991).

Note: A convenient indicator of proper neutralization of the cell extracts, and an easy way to keep track of which columns have already received a sample, is as follows. Add 1–2 ml of inositol-free DMEM to the packed gel resin, which does not alter the binding capacity or specificity of the gel resin for the [^{3}H]inositol phosphates. The gel resin will appear purple–pink in color, and a pale purple color should remain after washing the columns twice with deionized water. The addition of 5 ml of 2 *M* ammonium formate/0.1 *M* formic acid changes this color from purple to yellow (presumably due to phenol red in the medium, a pH indicator dye). However, following two to three washes with 10 ml of deionized water, the columns return to a pale purple color. Upon the addition of neutralized sample, each column will turn yellow, providing indication of the proper neutralization of the sample and indication of the addition of the samples to each column.

Background

In choosing an RGS protein to use in interrogating GPCR coupling specificity, a number of properties of these proteins should be considered, including the specificity of the RGS boxes for each activated Gα subunit, the mechanism by which each RGS box inhibits signaling, and the multidomain structure of several of these proteins. The GPCR signaling regulator, GRK2, contains a conserved N-terminal RGS box that does not stimulate GTPase activity, but rather sequesters Gα subunits of the G$_q$ family (Carman *et al.*, 1999; see also in this series Day *et al.*, 2004; Sterne-Marr *et al.*, 2004). The advantages of using the RGS box of GRK2 are that it is (1) selective for Gα subunits of the G$_q$ family and (2) a potent effector antagonist for G$_q$-coupled GPCRs (as described later). However, a caveat of GRK2 is its multidomain architecture (Fig. 1). The C-terminal fragment of GRK2 binds G$\beta\gamma$ competitively via its pleckstrin homology (PH) domain (Koch *et al.*, 1994), and the kinase domain phosphorylates and desensitizes GPCRs (Oppermann *et al.*, 1996); thus, it is important to employ a construct of GRK2 restricted to its RGS box for discrimination of GPCR signaling through Gα_q.

RGS2, a founding member of the RGS protein family, stimulates the GTPase activity of Gα subunits of the G$_q$ family in biochemical assays (Heximer *et al.*, 1997; Ingi *et al.*, 1998). However, in receptor reconstitution and cellular assays, RGS2 acts as a negative regulator of both G$_{i/o}$- and G$_q$-coupled receptor signaling (Ingi *et al.*, 1998). These findings suggest that assay conditions alter the G-protein specificity of RGS2 from Gα_q to both

$G\alpha_q$ and $G\alpha_{i/o}$. Thus, when using RGS2 to discriminate GPCR signaling, pertussis toxin and/or expression of the $G\alpha_q$-specific GRK2RGS construct could be tested in parallel to support inhibitory actions observed with RGS2.

RGS4 stimulates the GTPase activity of $G\alpha$ subunits of the $G_{i/o}$ (Berman *et al.*, 1996; Popov *et al.*, 1997) and G_q family (Hepler *et al.*, 1997; Ingi *et al.*, 1998) via its RGS box. Similar to RGS2, RGS4 contains little sequence beyond the RGS box. However, RGS4 has been demonstrated to exert receptor-selective inhibitory activity via its amphipathic α-helical N terminus (Zeng *et al.*, 1998). In addition, the N terminus has also been shown to confer high potency inhibition of G_q-mediated receptor signaling *in vivo* (Zeng *et al.*, 1998). Thus, RGS4 should be applied as a tool to discriminate $G\alpha$ subunits in GPCR signaling with cognizance of the benefits and caveats associated with using either the full-length or the isolated RGS box construct. As described earlier, results can be supported by parallel experiments using pertussis toxin and expression of GRK2RGS.

The guanine nucleotide exchange factor (GEF) for Rho, p115 RhoGEF, contains an NH_2-terminal RGS box and acts as a GTPase-accelerating protein (GAP) for $G\alpha_{12}$ and $G\alpha_{13}$ (Kozasa *et al.*, 1998). This GAP activity requires sequences flanking the RGS box (Wells *et al.*, 2002) and, thus, it is important to use a construct of p115 RhoGEF that contains the N terminus of the protein in addition to the RGS box. p115 RhoGEF also contains a tandem Dbl homology domain (DH/RhoGEF) and PH domain C-terminal cassette (Fig. 1) that exhibits RhoA-specific GEF activity (Hart *et al.*, 1996) and could thus confound the use of p115 RhoGEF in discriminating $G_{12/13}$-coupled receptor signaling. Thus, full-length constructs containing the C-terminal DH and PH domains should be avoided in targeting $G\alpha_{12/13}$-dependent signaling with p115 RhoGEF overexpression.

Application

Specificity of RGS Box Constructs for Constitutively Active
   Mutants of $G\alpha_q$ and $G\alpha_{12}$ in COS-7 Cells

To demonstrate the selectivity of RGS box constructs as effector antagonists of specific $G\alpha$ subunits in COS-7 cells, we utilized constitutively active (GTPase-deficient) mutants of $G\alpha_q$, $G\alpha_{12}$, and RhoB. $G\alpha_qQL$ was expressed in COS-7 cells in the absence or presence of GRK2RGS, RGS2, RGS4, and p115RGS; marked activation of endogenous PLC-β was observed with the expression of $G\alpha_qQL$ alone (Fig. 2). GRK2RGS and RGS2 potently inhibited the activation of PLC-β by $G\alpha_qQL$ (Fig. 2). In contrast,

RGS4 only partially inhibited the activation of PLC-β, and p115RGS did not alter Gα_qQL-stimulated PLC activity. These results suggest that GRK2RGS and RGS2 are effective inhibitors of Gα_qQL-mediated activation of PLC-β and thus are useful tools for discriminating G$_q$-coupled receptor signaling in COS-7 cells. Full-length RGS4 and the RGS box of RGS4 have both been shown to inhibit Gα_q-promoted Ca^{2+} signaling in rat pancreatic acinar cells (Zeng *et al.*, 1998); however, RGS4 does not appear to be as effective as GRK2RGS or RGS2 for discriminating Gα_q signaling in COS-7 cells.

To examine the specificity of GRK2RGS, RGS2, RGS4, and p115RGS on Gα_{12}QL-mediated activation of PLC-ε, we cotransfected Gα_{12}QL and PLC-ε in the absence or presence of the RGS box constructs and examined the ability of each to attenuate inositol phosphate accumulation. Gα_{12}QL activated PLC-ε when both were coexpressed in COS-7 cells (Fig. 2).

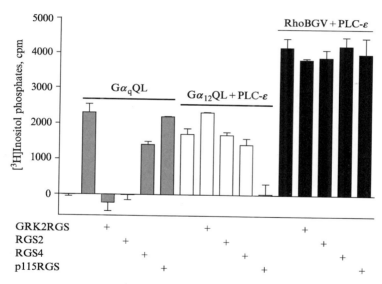

FIG. 2. Effect of GRK2RGS, RGS2, RGS4, and p115RGS expression on activation of PLC isozymes by constitutively active, GTPase-deficient mutants of Gα_q, Gα_{12}, and RhoB. COS-7 cells were transfected with the indicated DNA (supplemented to 700 ng total DNA with empty vector) and [^{3}H]inositol phosphate accumulation was measured 36 h later as described in the text. COS-7 cells were transfected with either 50 ng of empty pcDNA3 vector, Gα_qQL, Gα_{12}QL, or RhoBGV with or without PLC-ε (100 ng), and with or without GRK2RGS, RGS2, RGS4, or p115RGS (550 ng). [^{3}H]inositol phosphate accumulation was quantified by incubation for 45 min with 10 mM LiCl. Data shown are mean $\pm$ SD for duplicate samples in one experiment. Basal levels of [^{3}H]inositol phosphate production in the presence of empty pcDNA3 vector alone (approximately 3000 cpm) were subtracted from the values presented.

p115RGS, but not GRK2RGS, RGS2, nor RGS4, antagonized the activation of PLC-ε by Gα_{12}QL (Fig. 2). These results suggest that p115RGS selectively inhibits Gα subunits of the G$\alpha_{12/13}$ family, whereas GRK2RGS, RGS2, and RGS4 have no effect on this pathway. Thus, p115RGS is a useful tool for discriminating G$\alpha_{12/13}$-dependent receptor signaling in COS-7 cells.

A control experiment was carried out to demonstrate that the RGS box constructs do not indirectly affect the activation of PLC-ε by small GTPases of the Rho family. Wing and colleagues (2003b) showed that the Rho subfamily GTPases, RhoA, RhoB, and RhoC, each directly activate PLC-ε. Thus, expression constructs for PLC-ε and constitutively active (GTPase-deficient) RhoB (RhoBGV) were cotransfected in COS-7 cells in the absence or presence of GRK2RGS, RGS2, RGS4, and p115RGS. RhoBGV expression promoted PLC-ε-dependent increases in inositol phosphate accumulation in both the absence and the presence of the RGS box constructs (Fig. 2), suggesting that these RGS box constructs selectively inhibit Gα-promoted stimulation of PLC without affecting regulation by other activators.

RGS2 as a Useful Tool for Discriminating G$_{i/o}$-Mediated Signaling by the LPA$_1$ Receptor in COS-7 Cells

The LPA$_1$ receptor couples to G$_{i/o}$ heterotrimers in COS-7 cells to activate an endogenous PLC in response to LPA (Hains et al., 2004). This LPA-promoted response is pertussis toxin sensitive and is inhibited by coexpression of the G$\beta\gamma$-binding PH domain of GRK2 (GRKct) (Fig. 3B). Thus, to demonstrate the capacity of RGS box constructs to discriminate G$_{i/o}$-mediated signaling in COS-7 cells, we coexpressed the LPA$_1$ receptor with and without GRK2RGS, RGS2, RGS4, and p115RGS and quantified LPA-promoted inositol phosphate accumulation. RGS2, but not GRK2RGS, RGS4, nor p115RGS, blocked the effects of LPA (Fig. 3A). RGS4 is a potent GAP for G$\alpha_{i/o}$ subunits in vitro (Berman et al., 1996; Popov et al., 1997); thus, the lack of effect of RGS4 on LPA$_1$ receptor (G$_{i/o}$-coupled) signaling to PLC activation most likely reflects the receptor-selective nature of RGS4 activity (as observed and discussed later).

These results with LPA$_1$ receptor signaling demonstrate that RGS2 can inhibit G$_{i/o}$-mediated GPCR signaling in COS-7 cells, presumably by virtue of its GAP activity on G$\alpha_{i/o}$ subunits (Ingi et al., 1998). The ability of RGS2 to inhibit LPA$_1$ receptor signaling is consistent with previous reports of activity on G$_{i/o}$-coupled pathways; for example, RGS2 has been shown to inhibit M2 mAChR-mediated (G$_i$-coupled) MAP kinase activation in COS cells (Ingi et al., 1998). This ability of RGS2 to inhibit G$_{i/o}$-mediated signaling could be applied to other G$_{i/o}$-coupled receptors in COS-7 and

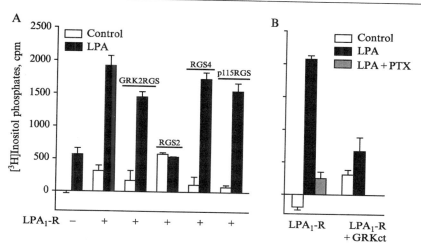

FIG. 3. $G_{i/o}$-dependent stimulation of an endogenous PLC by the LPA$_1$ receptor is inhibited by RGS2. (A) COS-7 cells were transfected with the indicated DNA (supplemented to 600 ng total DNA with empty vector), and [^{3}H]inositol phosphate accumulation was measured 36 h later as described in the text. COS-7 cells were transfected with either 50 ng of empty pcDNA3 vector or the LPA$_1$ receptor (LPA$_1$-R), and with or without GRK2RGS, RGS2, RGS4, and p115RGS (550 ng). Cells were then incubated for 45 min with 10 mM LiCl in the absence of agonist (Control) or presence of 10 μM lysophosphatidic acid (LPA). (B) A separate experiment (with the same y axis scale) in which 50 ng of empty vector or the LPA$_1$ receptor was cotransfected with or without 300 ng of GRKct in COS-7 cells. Cells were labeled with [^{3}H]inositol in the presence or absence of 100 ng/ml pertussis toxin (PTX; List Biochemicals, Campbell, CA) for 12 h, followed by incubation for 30 min with 10 mM LiCl in the absence of agonist (Control) or presence of 10 μM LPA. Data shown are mean $\pm$ SD for duplicate samples in one experiment. Basal levels of [^{3}H]inositol phosphate production in the presence of empty pcDNA3 vector alone (approximately 2500 cpm) were subtracted from the values presented.

other cells and should be kept in mind in light of reports that consider RGS2 as solely a negative regulator of $G\alpha_q$-mediated signaling (e.g., Heximer, 2004 in this series).

RGS2 and GRK2RGS Discriminate $G\alpha_q$-Mediated Signaling by P2Y$_1$ and S1P$_3$ Receptors in COS-7 Cells

Both the P2Y$_1$ receptor P2Y$_1$ and the S1P$_3$ receptor couple predominantly to $G\alpha$ subunits of the G_q family in COS-7 cells to stimulate endogenous PLC (Filtz *et al.*, 1994; Hains *et al.*, 2004). To examine the capacity of RGS box constructs to inhibit signaling by the P2Y$_1$ and S1P$_3$ receptors, we coexpressed each receptor in the absence or presence of GRK2RGS, RGS2, RGS4, and p115RGS and measured inositol phosphate accumulation. Both

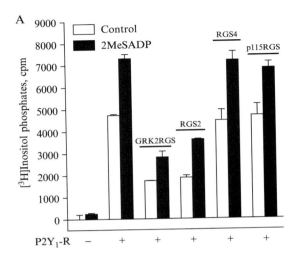

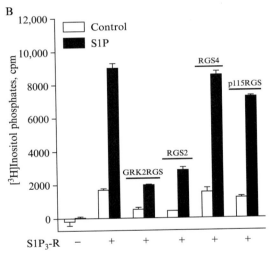

FIG. 4. $G\alpha_q$-dependent stimulation of an endogenous PLC by the P2Y$_1$ and S1P$_3$ receptors is inhibited by GRK2RGS and RGS2. COS-7 cells were transfected with the indicated DNA [supplemented to 400 ng (A) or 600 ng (B)], and [^{3}H]inositol phosphate accumulation was measured 36 h later as described in the text. (A) COS-7 cells were transfected with either 50 ng of empty pcDNA3 vector or the P2Y$_1$ receptor (P2Y$_1$-R), and with or without GRK2RGS, RGS2, RGS4, and p115RGS (350 ng). Cells were then incubated for 45 min with 10 mM LiCl in the absence of agonist (Control) or presence of 10 μM 2MeSADP. (Note that overexpression of the P2Y$_1$ receptor results in increased [^{3}H]inositol phosphate production in the absence of exogenous agonist application because of autocrine release by cells of the cognate adenine nucleotide agonist (Filtz *et al.*, 1994). (B) COS-7 cells were transfected with

GRK2RGS and RGS2 markedly inhibited agonist-promoted activation of PLC by the $P2Y_1$ and $S1P_3$ receptors, respectively (Fig. 4A and B). Conversely, RGS4 and p115RGS exhibited relatively little effect on the receptor-promoted activation of endogenous PLC. Thus, GRK2RGS and RGS2, but not RGS4 or p115RGS, are useful tools for discriminating $G\alpha_q$ signaling in COS-7 cells. RGS4 is a potent GAP for $G\alpha_q$ signaling *in vivo*; however, the N terminus of the protein can impose receptor-selective effects (Zeng *et al.*, 1998) as described earlier. The next section provides an example of a $G\alpha_q$-coupled receptor that is inhibited by GRK2RGS, RGS2, and RGS4.

RGS2, GRK2RGS, and RGS4 Inhibit $G\alpha_q$-Mediated Signaling by the LPA_2 Receptor

Thus far in this article, GRK2RGS and RGS2, but not RGS4, have been shown to be useful tools for inhibiting $G\alpha_q$-mediated GPCR signaling in COS-7 cells. However, in the case of the lysophosphatidic acid receptor LPA_2, RGS4 as well as GRK2RGS and RGS2 effectively inhibited LPA-promoted inositol phosphate accumulation (Fig. 5A). Similar to the results observed with the $P2Y_1$ and $S1P_3$ receptors, p115RGS did not inhibit PLC activation by the LPA_2 receptor. The expression level of HA-tagged GRK2RGS and RGS4 in the absence and presence of the LPA_2, $S1P_3$, or $P2Y_1$ receptor was examined by cell lysate immunoblotting using an anti-HA antibody. The expression level of GRK2RGS and RGS4 was not altered significantly in the presence of the receptors (Fig. 5B), suggesting that the lack of action of RGS4 in inhibiting the $S1P_3$ and $P2Y_1$ receptors is not due to the inhibition of RGS4 expression. These results suggest that GRK2RGS, RGS2, and RGS4 discriminate $G\alpha_q$-mediated signaling by the LPA_2 receptor. These data also demonstrate the receptor-selective effects of full-length RGS4 (i.e., inhibition of LPA_2, but not $S1P_3$ or $P2Y_1$, receptor signaling to PLC) and thus highlight the importance of using alternative reagents such as GRK2RGS or RGS2 to confirm observations of signaling inhibition (or lack thereof) using RGS4 in particular and $G\alpha_{i/o}/G\alpha_q$-specific RGS proteins in general.

---

either 50 ng of empty pcDNA3 vector or the $S1P_3$ receptor ($S1P_3$-R), and with or without GRK2RGS, RGS2, RGS4, and p115RGS (550 ng). Cells were then incubated for 45 min with 10 mM LiCl in the absence of agonist (Control) or presence of 10 μM sphingosine 1-phosphate (S1P). Data shown are mean $\pm$ SD for duplicate samples in one experiment. Basal levels of inositol phosphate production in the presence of empty pcDNA3 vector alone (approximately 2500 cpm) were subtracted from the values presented.

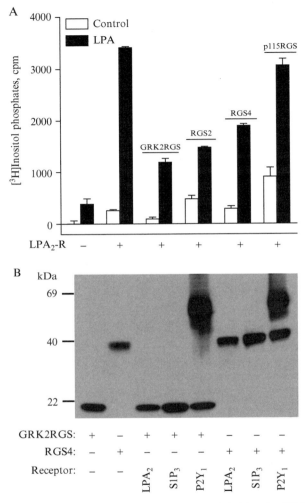

FIG. 5. $G\alpha_q$-dependent stimulation of an endogenous PLC by the LPA_2 receptor is inhibited by GRK2RGS, RGS2, and RGS4. COS-7 cells were transfected with the indicated DNA (supplemented to 400 ng), and [³H]inositol phosphate accumulation was measured 36 h later as described in the text. (A) COS-7 cells were transfected with either 50 ng of empty pcDNA3 vector or the LPA_2 receptor (LPA_2-R), and with or without GRK2RGS, RGS2, RGS4, and p115RGS (350 ng). Cells then were incubated for 45 min with 10 mM LiCl in the absence of agonist (Control) or presence of 10 μM LPA. Data shown are mean ± SD for duplicate samples in one experiment. Basal levels of [³H]inositol phosphate production in the presence of empty pcDNA3 vector alone (approximately 2000 cpm) were subtracted from the values presented. (B) COS-7 cells were transfected with 100 ng of empty pcDNA3 vector, the LPA_2 receptor, the $S1P_3$ receptor, or the $P2Y_1$ receptor with or without 550 ng of

p115RGS Discriminates $G\alpha_{12/13}$-Mediated Signaling by the LPA_1 Receptor in COS-7 Cells

The LPA_1 receptor couples to $G\alpha_{12/13}$ to activate PLC-ε in COS-7 cells (Hains *et al.*, 2004). To assess the action of RGS box constructs in discriminating GPCR-promoted activation of $G\alpha_{12/13}$, we coexpressed the LPA_1 receptor and PLC-ε in the absence and presence of the RGS proteins and examined the capacity of the constructs to inhibit PLC-ε activation by the LPA_1 receptor. In the absence of the RGS box constructs, the LPA_1 receptor promoted inositol phosphate accumulation when coexpressed with PLC-ε (Fig. 6). In contrast, in the presence of p115RGS, but not GRK2RGS, RGS2, or RGS4, LPA-mediated activation of PLC-ε by the LPA_1 receptor was inhibited markedly. These results suggest that p115RGS is a useful tool for discriminating $G\alpha_{12/13}$ GPCR signaling in COS-7 cells.

Concluding Remarks

RGS proteins specific for $G_{i/o}$, G_q, $G_{12/13}$, and G_s $G\alpha$ subunits have been described (Neubig and Siderovski, 2002), and these RGS proteins can be utilized to delineate the G proteins involved in signaling pathways activated by GPCRs. In addition to $G\alpha$ subunit selectivity, many RGS proteins also exhibit receptor selectivity, thus highlighting the importance of using additional RGS proteins or other inhibitors of $G\alpha$ signaling (e.g., GoLoco motif peptides: see in this series: Oxford & Webb, 2004), when employing RGS box constructs to delineate signaling pathways. For example, we have found that RGS4 demonstrates receptor selectivity in inhibiting $G\alpha_q$ signaling by the LPA_2, but not by the $P2Y_1$ or $S1P_3$ receptors. Thus, the lack of an effect of RGS4 overexpression on a particular GPCR signal transduction pathway should not be misconstrued as a lack of involvement of G_q class $G\alpha$ subunits in receptor/effector coupling.

RGS box-independent regulation of GPCR signaling also has been demonstrated and should be considered when choosing an RGS protein as a tool to probe GPCR/effector coupling. The N terminus of RGS2

GRK2RGS or RGS4. COS-7 cells were lysed in 500 μl of lysis buffer (150 mM NaCl, 20 mM Tris–HCl, pH 7.5, 2 mM EDTA, 1% Triton X-100, 0.1% protease inhibitors), sonicated for 5 min, and diluted 1:1 in 2.5× Laemmli sample buffer. Proteins were resolved by SDS–polyacrylamide gel electrophoresis, transferred onto nitrocellulose, and detected by a rat monoclonal anti-HA-peroxidase high-affinity antibody (clone 3F10; Roche Diagnostics Corporation, Indianapolis, IN) and enhanced chemiluminescence (Amersham Biosciences). Note that of the three receptors cotransfected with HA-tagged GRK2RGS or HA-tagged RGS4, the LPA_2 and $S1P_3$ receptors were not epitope tagged; however, the $P2Y_1$ receptor was N-terminally HA tagged and thus appears in the anti-HA immunoblot as a broad band centered about 60 kDa.

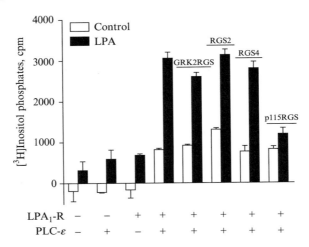

FIG. 6. Inhibition of LPA$_1$ receptor-mediated activation of PLC-ε by the RGS domain of p115 RhoGEF. COS-7 cells were transfected with the indicated DNA (supplemented to 700 ng total DNA with empty vector), and [^{3}H]inositol phosphate accumulation was measured 36 h later as described in the text. COS-7 cells were transfected with either 50 ng of empty pcDNA3 vector or the LPA$_1$ receptor (LPA$_1$-R), with or without PLC-ε (100 ng), and with or without individual RGS box expression vectors (550 ng) as indicated. Cells were then incubated for 45 min with 10 mM LiCl in the absence of agonist (Control) or presence of 10 μM LPA. Data shown are mean ± SD for duplicate samples in one experiment. Basal levels of inositol phosphate production in the presence of the empty pcDNA3 vector alone (approximately 3000 cpm) were subtracted from the values presented.

inhibits Gα_s Q227L and β_2-adrenergic receptor-stimulated cAMP accumulation in HEK293 cells expressing type V adenylyl cyclase (Salim *et al.*, 2003; see also in this series: Salim & Dessauer, 2004). Similarly, RGS16, a GAP for G$\alpha_{i/o}$ and Gα_q, binds to Gα_{13} and inhibits Gα_{13}-mediated signal transduction via its amino terminus (Johnson *et al.*, 2003).

This article demonstrated the application of full-length RGS proteins (RGS2, RGS4) and isolated RGS boxes (GRK2RGS, p115RGS) to discriminate G$\alpha_{i/o}$, Gα_q, and G$\alpha_{12/13}$ GPCR signaling in COS-7 cells. These constructs should prove useful for discerning the contributions of these three families of Gα subunits to GPCR signaling in a variety of systems when applied with the appropriate controls.

Acknowledgments

This work was supported by GM029536 (T.K.H.) and GM065533 (T.K.H. and D.P.S.) from the National Institute of General Medical Sciences (NIH). D.P.S. is a Year 2000 Neuroscience Scholar of the EJLB Foundation (Montréal, Canada).

References

Berman, D. M., Wilkie, T. M., and Gilman, A. G. (1996). GAIP and RGS4 are GTPase activating proteins for the Gi subfamily of G protein alpha subunits. *Cell* **86,** 445–452.

Berridge, M. J. (1987). Inositol trisphosphate and diacylglycerol: Two interacting second messengers. *Annu. Rev. Biochem.* **56,** 159–193.

Brown, H. A., Lazarowski, E. R., Boucher, R. C., and Harden, T. K. (1991). Evidence that UTP and ATP regulate phospholipase C through a common extracellular 5' nucleotide receptor in human airway epithelial cells. *Mol. Pharmacol.* **40,** 648–655.

Carman, C. V., Parent, J. L., Day, P. W., Pronin, A. N., Sternweis, P. M., Wedegaertner, P. B., Gilman, A. G., Benovic, J. L., and Kozasa, T. (1999). Selective regulation of Galpha(q/11) by an RGS domain in the G protein coupled receptor kinase, GRK2. *J. Biol. Chem.* **274,** 34483–34492.

Cunningham, M. L., Waldo, G. L., Hollinger, S., Hepler, J. R., and Harden, T. K. (2001). Protein kinase C phosphorylates RGS2 and modulates its capacity for negative regulation of Galpha 11 signaling. *J. Biol. Chem.* **276,** 5438–5444.

Day, P. W., Wedengartner, P. B., and Benovic, J. L. (2004). *Methods Enzymol.* **390,** 295–310.

Filtz, T. M., Li, Q., Boyer, J. L., Nicholas, R. A., and Harden, T. K. (1994). Expression of a cloned P2Y purinergic receptor that couples to phospholipase. *Mol. Pharmacol.* **46,** 8–14.

Hains, M. D., Wing, M. R., Rogan, S. C., Siderovski, D. P., Lynch, K. R., and Harden, T. K. (2004). Activation of phospholipase C-epsilon through Gα12/13-dependent stimulation of the LPA1 receptor. Submitted for publication.

Hart, M. J., Sharma, S., elMasry, N., Qiu, R. G., McCabe, P., Polakis, P., and Bollag, G. (1996). Identification of a novel guanine nucleotide exchange factor for the Rho GTPase. *J. Biol. Chem.* **271,** 25452–25458.

Hepler, J. R., Berman, D. M., Gilman, A. G., and Kozasa, T. (1997). RGS4 and GAIP are GTPase activating proteins for Gq alpha and block activation of phospholipase C beta by gamma thio GTP Gq alpha. *Proc. Natl. Acad. Sci. USA* **94,** 428–432.

Heximer, S. P., Watson, N., Linder, M. E., Blumer, K. J., and Hepler, J. R. (1997). RGS2/GOS8 is a selective inhibitor of Gqalpha function. *Proc. Natl. Acad. Sci. USA* **94,** 14389–14393.

Heximer, S. P. (2004). *Methods Enzymol.* **390,** 65–82.

Ingi, T., Krumins, A. M., Chidiac, P., Brothers, G. M., Chung, S., Snow, B. E., Barnes, C. A., Lanahan, A. A., Siderovski, D. P., Ross, E. M., Gilman, A. G., and Worley, P. F. (1998). Dynamic regulation of RGS2 suggests a novel mechanism in G-protein signaling and neuronal plasticity. *J. Neurosci.* **18,** 7178–7188.

Johnson, E. N., Seasholtz, T. M., Waheed, A. A., Kreutz, B., Suzuki, N., Kozasa, T., Jones, T. L., Brown, J. H., and Druey, K. M. (2003). RGS16 inhibits signalling through the G alpha 13-Rho axis. *Nature Cell Biol.* **5,** 1095–1103.

Kelley, G. G., Reks, S. E., Ondrako, J. M., and Smrcka, A. V. (2001). Phospholipase C(epsilon): A novel Ras effector. *EMBO J.* **20,** 743–754.

Kelley, G. G., Reks, S. E., and Smrcka, A. V. (2004). Hormonal regulation of phospholipase C epsilon through distinct and overlapping pathways involving G12 and Ras family G-proteins. *Biochem. J.* **378,** 129–139.

Koch, W. J., Hawes, B. E., Inglese, J., Luttrell, L. M., and Lefkowitz, R. J. (1994). Cellular expression of the carboxyl terminus of a G protein-coupled receptor kinase attenuates G beta gamma-mediated signaling. *J. Biol. Chem.* **269,** 6193–6197.

Kozasa, T., Jiang, X., Hart, M. J., Sternweis, P. M., Singer, W. D., Gilman, A. G., Bollag, G., and Sternweis, P. C. (1998). p115 RhoGEF, a GTPase activating protein for Galpha12 and Galpha13. *Science* **280,** 2109–2111.

Neubig, R. R., and Siderovski, D. P. (2002). Regulators of G-protein signalling as new central nervous system drug targets. *Nature Rev. Drug Discov.* **1**, 187–197.

Oppermann, M., Freedman, N. J., Alexander, R. W., and Lefkowitz, R. J. (1996). Phosphorylation of the type 1A angiotensin II receptor by G protein-coupled receptor kinases and protein kinase C. *J. Biol. Chem.* **271**, 13266–13272.

Oxford, G. S., and Webb, C. K. (2004). *Methods Enzymol.* **390**, 437–450.

Popov, S., Yu, K., Kozasa, T., and Wilkie, T. M. (1997). The regulators of G protein signaling (RGS) domains of RGS4, RGS10, and GAIP retain GTPase activating protein activity *in vitro*. *Proc. Natl. Acad. Sci. USA* **94**, 7216–7220.

Rahman, Z., Schwarz, J., Gold, S. J., Zachariou, V., Wein, M. N., Choi, K. H., Kovoor, A., Chen, C. K., DiLeone, R. J., Schwarz, S. C., Selley, D. E., Sim-Selley, L. J., Barrot, M., Luedtke, R. R., Self, D., Neve, R. L., Lester, H. A., Simon, M. I., and Nestler, E. J. (2003). RGS9 modulates dopamine signaling in the basal ganglia. *Neuron* **38**, 941–952.

Rhee, S. G. (2001). Regulation of phosphoinositide specific phospholipase C. *Annu. Rev. Biochem.* **70**, 281–312.

Salim, S., Sinnarajah, S., Kehrl, J. H., and Dessauer, C. W. (2003). Identification of RGS2 and type V adenylyl cyclase interaction sites. *J. Biol. Chem.* **278**, 15842–15849.

Salim, S., and Dessauer, C. W. (2004). *Methods Enzymol.* **390**, 83–99.

Saunders, C. M., Larman, M. G., Parrington, J., Cox, L. J., Royse, J., Blayney, L. M., Swann, K., and Lai, F. A. (2002). PLC zeta: A sperm-specific trigger of Ca(2+) oscillations in eggs and embryo development. *Development* **129**, 3533–3544.

Schachter, J. B., Li, Q., Boyer, J. L., Nicholas, R. A., and Harden, T. K. (1996). Second messenger cascade specificity and pharmacological selectivity of the human P2Y1 purinoceptor. *Br. J. Pharmacol.* **118**, 167–173.

Snow, B. E., Krumins, A. M., Brothers, G. M., Lee, S. F., Wall, M. A., Chung, S., Mangion, J., Arya, S., Gilman, A. G., and Siderovski, D. P. (1998). A G protein gamma subunit like domain shared between RGS11 and other RGS proteins specifies binding to Gbeta5 subunits. *Proc. Natl. Acad. Sci. USA* **95**, 13307–13312.

Sterne-Marr, R., Dhami, G. K., Tesmer, J. J. G., and Ferguson, S. G. (2004). *Methods Enzymol.* **390**, 310–336.

Wells, C. D., Liu, M. Y., Jackson, M., Gutowski, S., Sternweis, P. M., Rothstein, J. D., Kozasa, T., and Sternweis, P. C. (2002). Mechanisms for reversible regulation between G13 and Rho exchange factors. *J. Biol. Chem.* **277**, 1174–1181.

Wing, M., Bourdon, D. M., and Harden, T. K. (2003a). PLC epsilon: A shared effector protein in Ras-, Rho-, and G alpha beta gamma-mediated signaling. *Mol. Intervent.* **3**, 273–280.

Wing, M. R., Snyder, J. T., Sondek, J., and Harden, T. K. (2003b). Direct activation of phospholipase C-epsilon by Rho. *J. Biol. Chem.* **278**, 41253–41258.

Zeng, W., Xu, X., Popov, S., Mukhopadhyay, S., Chidiac, P., Swistok, J., Danho, W., Yagaloff, K. A., Fisher, S. L., Ross, E. M., Muallem, S., and Wilkie, T. M. (1998). The N terminal domain of RGS4 confers receptor-selective inhibition of G protein signaling. *J. Biol. Chem.* **273**, 34687–34690.

[6] Allosteric Regulation of GAP Activity by Phospholipids in Regulators of G-Protein Signaling

By YAPING TU and THOMAS M. WILKIE

Abstract

Regulators of G-protein signaling (RGS) proteins are GTPase-activating proteins (GAPs) for α subunits of the G_i and/or G_q class of heterotrimeric G proteins. RGS GAP activity is inhibited by phosphatidic acid (PA), lysophosphatidic acid (LPA), and phosphatidylinositol 3,4,5-trisphosphate (PIP_3) but not by other phospholipids, phosphoinositides, or diacylglycerol. Both PA and PIP_3 can inhibit RGS4 GAP activity and their inhibition is additive, suggesting that PA and PIP_3 interact with different domains of RGS4. The N terminus of RGS4 (1–57 amino acids) is required for PA binding and inhibition. Mutation at Lys20, far from the RGS domain of RGS4, decreases PA-mediated inhibition of RGS4 by more than 85%. Amino acid substitutions in helix 5 within the RGS domain of RGS4, opposite to the RGS/Gα protein contact face, reduce binding affinity and inhibition by PIP_3. Calmodulin binds all RGS proteins tested in a Ca^{2+}-dependent manner at two sites, one in the N-terminal 33 amino acids and another in the RGS domain. Ca^{2+}/calmodulin does not directly affect GAP activity of RGS4 but reverses PA and PIP_3-mediated inhibition. In summary, these results demonstrate that phospholipids such as PA and PIP_3 act as allosteric inhibitors of RGS proteins, and Ca^{2+}/calmodulin competition with PA and PIP_3 may provide an intracellular mechanism for feedback regulation of Ca^{2+} signaling evoked by G-protein-coupled agonists.

Introduction

Heterotrimeric G proteins transduce a wide variety of receptor-mediated signals across the plasma membrane (Gilman, 1987). Regulators of G-protein signaling (RGS proteins) are GTPase-activating proteins (GAPs) that stimulate the intrinsic GTPase activity of G-proteins α subunits. RGS proteins are one family of seven within the superfamily of RGS homology (RH) proteins (Ross and Wilkie, 2000). Mammals express four subfamilies of RGS proteins that are related by a conserved domain of approximately 130 amino acid residues, termed the RGS domain. The RGS domain of RGS4 and other RGS proteins bind α subunits of the G_i and/or G_q class of heterotrimeric G proteins. Residues in the RGS domain stabilize the three switch regions of $G\alpha$, thus promoting the formation of a transition state and GTP hydrolysis (Natochin *et al.*, 1998; Popov *et al.*, 1997; Posner *et al.*, 1999; Tesmer *et al.*, 1997). By accelerating GTP hydrolysis and inactivation of $G\alpha$ subunits, RGS proteins serve as negative regulators of G-protein-mediated signaling pathways and modulate G-protein-stimulated cellular responses (Ross and Wilkie, 2000).

The ability of RGS proteins to diminish the magnitude and duration of G-protein-dependent signaling mandates tight regulation of their GAP activity. However, mechanisms directly controlling GAP activity of RGS proteins are poorly understood. RGS proteins are subject to both transcriptional and posttranslational regulation (Ingi *et al.*, 1998; Popov *et al.*, 2000; Tu *et al.*, 1999). The GAP activity of RGS proteins can be modulated by phosphorylation or palmitoylation of either the RGS proteins (Derrien and Druey, 2001; Osterhout *et al.*, 2003; Tu *et al.*, 1999) or their $G\alpha$ substrates (Glick *et al.*, 1998; Tu *et al.*, 1997; Wang *et al.*, 1998). This article describes the allosteric regulation of RGS4 GAP activity by phosphatidic acid and phosphatidylinositol 3,4,5-triphosphate.

Reagents and Methods

Buffers

The following buffers are used for the preparation of RGS proteins and experiments.

Buffer A: 20 mM sodium phosphate, pH 7.8, 300 mM NaCl, 10 μg/ml leupeptin, 1 μg/ml aprotinin, 0.1 mM phenylmethylsulfonyl fluoride (PMSF)

Buffer B: 50 mM HEPES, pH 8.0, 1 mM dithiothreitol (DTT), 0.05% $C_{12}E_{10}$ (Merck), 5 mM EDTA

Buffer C: 10 mM Tris–HCl, pH 9.5, 100 mM NaCl, 5 mM MgCl$_2$

Buffer D: 10 mM HEPES, pH 8.0, 150 mM NaCl, 3 mM EDTA, 0.005% surfactant NP20

Buffer E: 10 mM HEPES, pH 7.4, 0.1 mM CaCl$_2$, 1 mM DTT

Buffer F: 25 mM HEPES, pH 7.5, 10 mM EDTA, 0.05% Triton X-100, 50 mM NaCl, 1 mM DTT

Buffer G: 25 mM HEPES, pH 7.5, 50 mM NaCl, 1 mM DTT, 1 mM EDTA, 2 mM MgCl$_2$

HMEN buffer: 50 mM HEPES, pH 7.5, 2 mM MgCl$_2$, 1 mM EGTA, 100 mM NaCl

Lipids and Detergents

Lysophosphatidic acid (LPA), phosphatidylserine (PS, bovine brain), phosphatidylinositol (egg yolk), cholesteryl hemisuccinate (CHS), imidazole, and Tween-20 are from Sigma. Phosphatidylcholine (PC, bovine brain), phosphatidic acid (PA, egg yolk), phosphatidylglycerol (PG, bovine brain), and phosphatidylethanolamine (PE, bovine liver) are products of Avanti Polar Lipids Inc. Dihexadecanoylphosphatidylinositol 3,4,5-trisphosphate (diC16-PIP$_3$) is synthesized as described by Popov et al., 2000.

Production and Purification of Recombinant Proteins

Gα_q, Gα_z, G$\beta_1\gamma_2$, and m1 muscarinic cholinergic receptor (m1AChR) are purified from Sf9 cells (Kozasa and Gilman, 1995; Parker et al., 1991). Myristoylated Gα_{i1} is purified from Escherichia coli (Lee et al., 1994).

The ΔN57 RGS4 deletion mutation is generated by polymerase chain reaction (PCR) with the primer 5'-GATCCATGGGCAAATGGGCT GAATCGCTGGAA and 5'-GCGTTCTGAACAAATCCAGAT. The product is cut with NcoI and BamHI and cloned into the corresponding sites of a modified pQE60. RGS4K20E is generated by the QuikChange mutagenesis kit (Stratagene) using the following primers (the mutated codons are in bold italics) (Ouyang et al., 2003): primer I, GCAAAGGA-TATG***GAA***CATCGGCTGGG; and primer 2, CCCAGCCGATG***TTC***CA TATCCTTTGC. The fragment coding for mutant RGS4 K112E/K113E is generated by PCR using the oligonucleotide containing the aforementioned mutation as a primer. All mutations are cloned into the pQE60 expression vector (Qiagen). RGS4 coding regions of all constructs are sequenced to verify that only the desired mutations had been introduced.

All recombinant RGS proteins are His$_6$ tagged at the amino terminus with the sequence MGH$_6$MG, and all cDNA clones are sequenced to verify that the correct recombinant RGS proteins are expressed in E. coli JM109. The expression of His$_6$-RGS4 and the K20E (Lys $\rightarrow$ Glu) mutant is induced by adding 1 mM isopropyl-β-D-thiogalactoside (IPTG) to midlog cultures (A_{600} = 0.7–0.8). Cultures are harvested after 3 h at 37°. To purify His$_6$-RGS4 and the K20E mutant, E. coli cells are allowed to lyse for 30 min at 0° in buffer A plus 1 mg/ml lysozyme and are then sonicated with four 30-s bursts using a microprobe. After centrifugation at 35,000g for 30 min, 2-mercaptoethanol is added to 10 mM and the supernatant is applied to Ni^{2+}-NTA agarose (Qiagen) that has been equilibrated with buffer A plus 10 mM 2-mercaptoethanol. The column is washed sequentially with 1 M NaCl, 1 M NaCl plus 10 mM imidazole, and 10 mM of imidazole alone in buffer A. RGS4 and mutants are then eluted with a 10–150 mM imidazole gradient in buffer A.

For expression of the RGS4 mutant with a deletion of the N-terminal 57 amino acid residues (ΔN57 RGS4), 1 liter of T7 medium with ampicillin (100 μg/ml) is inoculated with an overnight culture started from a single colony, IPTG (10 μM) induction is performed at $OD_{600} = 0.6$, cultures are shaken overnight, and cells are pelleted, lysed by freezing, and sonicated with buffer A plus 20 mM 2-mercaptoethanol. Lysozyme (0.2 mg/ml) and DNase I (5 μg/ml) are added to complete lysis and digest DNA. Soluble proteins are isolated from total lysate centrifuged at 12,000g (30 min at 4°). The supernatant is applied onto a 2-ml Ni^{2+}-NTA column (Qiagen, Chatsworth, CA) preequilibrated with buffer A, washed first with 20 ml buffer A and 0.2 M NaCl, and finally washed with 10 ml of buffer A with 10 mM imidazole (pH 8.0). Protein is eluted with 9 ml elution buffer (buffer A containing 150 mM imidazole, pH 8.0). All RGS proteins are concentrated with an Ultrafree 15 device (Millipore) in buffer B. Sodium dodecylsulfate–polyacrylamide gel electrophoresis (SDS–PAGE) analysis indicates that the protein is more than 90% pure by Coomassie Blue staining.

Binding of RGS4 or RGS4 Mutant to PA Assayed by Enzyme-Linked Immunosorbent Assay

All procedures are performed at room temperature. PA solutions (10 μg/well) diluted in methanol are allowed to bind onto the wells of a 96-well titer plate overnight. The wells are blocked with 3% bovine serum albumin in phosphate-buffered saline (PBS), pH 7.5, for 2 h. RGS4 proteins are loaded onto the wells and incubated for 1 h. The plate is then washed with PBS containing 0.1% Tween 20 for 5 min each, for a total of four washes. Bound proteins are detected using the anti-RGS4 C-terminal goat polyclonal antibody (1:500) followed by the donkey antigoat IgG-alkaline phosphatase conjugate (1:8000) and developed using *p*-nitrophenyl phosphate in buffer C. The resulting absorbancies are monitored at 405 nm using an ELISA plate reader (Bio-Rad). The magnitude of the absorbancies is related directly to the bound amount of RGS4 and/or its mutant proteins, and this measurement is used to determine the dissociation constant according to the one-site binding equation $y = a + b[x/(k + x)]$, where y is the measured binding of RGS4 or mutants, a is background binding, b is maximal phospholipid-stimulated binding, x is the concentration of PA, and k is the concentration of PA resulting in half-maximal binding.

Surface Plasmon Resonance

Surface plasmon resonance measurements are performed using the BIAcore 1000 instrument (BIAcore, Inc.) at 25°. The RGS protein is immobilized on the surface of the carboxymethylated dextran chip (CM5) using standard carbodiimide chemistry in accordance with the manufacturer's instructions. The lipid dissolved in running buffer D is injected over the chip surface with a flow rate of 5 μl/min. At least three different concentrations are used. Control injections are made with a blank chip without coupled RGS protein. The amount of coupled RGS protein is 2500 response units (RU), 1580 RU (RGS4 K112E/K113E), and 2900 RU (RGS16). Regeneration of the chip after each binding experiment is achieved by injecting 0.01% SDS in running buffer. Data are analyzed using the BIAevaluation 2.1 software (BIAcore, Inc.). Binding curves show a moderate heterogeneity of both association and dissociation phases. The minor components identified in the analysis of association and dissociation (less than 20% of total signal) are ignored. A relatively fast transition process evident at the beginning of the dissociation phase is not studied. The dissociation constant, K_d, is calculated for each sensorgram using kinetic constants k_a and k_d.

Ca^{2+}/Calmodulin Binding to RGS Proteins

The concentration of calmodulin is measured spectrophotometrically using the extinction coefficient 3060 $M^{-1}cm^{-1}$ at 278 nm in the presence of 1 mM EGTA (Erickson-Viitanen and DeGrado, 1987). RGS binding to calmodulin-agarose is detected by SDS–PAGE of supernatants as follows: 15 μl of wet calmodulin-agarose beads (26 μg of calmodulin) washed with buffer E is pelleted, the supernatant is removed, and the beads are mixed with RGS4 (1.5 nmol) equal in amount to coupled calmodulin in a final volume of 30 μl buffer. The suspension is incubated at room temperature with constant agitation. After 30 min, the beads are washed twice with 500 μl of buffer. The suspension is brought to the original volume and incubated for 10 min. EGTA (500 mM) and NaCl (5 M) are added to final concentrations 2 and 50 mM, respectively, and the suspension is incubated for an additional 10 min. Finally, 5 μl of SDS–PAGE loading buffer is added to the beads and incubated another 10 min. Supernatant aliquots of equal volume are withdrawn after each incubation step, followed by separation by SDS–PAGE and Coomassie Blue staining.

RGS4 GAP Assays: Single Turnover

$[\gamma^{-32}P]GTPase$ Assays. Single turnover assays using ~100 nM Gα_{i1}-$[\gamma^{-32}P]GTP$ are performed on ice as described previously (Wang et al., 1998). Purified Gα_{i1} (2 μM) is first bound to $[\gamma^{-32}P]GTP$ in buffer F. The reaction mixture is chilled on ice, and Gα_{i1}-$[\gamma^{-32}P]GTP$ is purified immediately by centrifugal gel filtration: Sephadex G-25 superfine resin (2.5 ml) is poured as a 75% (v/v) slurry into 1 × 3-cm polypropylene columns (Isolab QS-Q) and allowed to settle for at least 1 h at 4°. The column is centrifuged for 4 min at 2400 rpm (Beckman TJ-6) to remove excess buffer. The column is then placed in a 4-ml plastic scintillation vial (Beckman Omnivial), the sample ($\sim$100 μl) is applied gently to the center of the surface of the resin bed, and the column is centrifuged again for 4 min at 2800 rpm. Recovery of $G\alpha_{i1}$-$[\gamma$-^{32}P]GTP in the elute should be about 50% in approximately 120 μl. The rate of $G\alpha_{i1}$-$[\gamma$-^{32}P]GTP hydrolysis is measured at 4° in buffer G in the presence or absence of RGS4. RGS4 is preincubated with the indicated lipids for 30 min at 4° before initiation of the reaction. In this assay, GAP activity is defined as the increase in the first-order hydrolysis rate constant (k_{hydrol}) or is approximated as an increase in the initial rate of hydrolysis (Wang *et al.*, 1998).

RGS4 GAP Assays: Steady-State GTPase Assays

RGS4 GAP activity is determined by agonist-stimulated, steady-state GTPase assay using proteoliposomes reconstituted with receptor and heterotrimeric G proteins. m1AChR are reconstituted with recombinant $G\alpha_q$ and $\beta\gamma$ ($\beta\gamma$:α = 2.5) into unilamellar phospholipid vesicles (PE:CHS:PS) according to the method of Parker *et al.* (1991). Reconstitution is accomplished by gel filtration in detergent-free HMEN buffer on a 1-ml column of Sephacryl AcA34. Recovery of the m1 receptor in the vesicles after reconstitution, defined as the fraction of receptor that is applied initially to the column, is typically 20–30%, and the recovery of G proteins is around 10–20%. We keep PE/CHS/PS at a constant molar ratio of 60:10:20 in our reconstituted vesicles and add 10 mol% PC, PS, PG, LPA, or PA to evaluate the effect of different phospholipids on the GAP activity of RGS4. RGS4 is added to the vesicles and incubated for 30 min at 30° prior to assay. Steady-state assays are carried out at 30° for 10 min. Data are given as increases in steady-state GTPase activity (Wang *et al.*, 1998).

Regulation of RGS Domain GAP Activity

Lysophosphatidic Acid and Phosphatidic Acid Inhibit RGS4 GAP Activity

It was reported previously that RGS4 binds to anionic phospholipid liposomes (Bernstein *et al.*, 2000). However, relatively little is known about what modulates RGS4 activity on signaling pathways when bound to the membrane. Figure 1 shows the effects of different anionic phospholipid vesicles on GAP activity of RGS4 in a single-turnover assay. After preincubation with 50 nM RGS4, LPA showed the greatest inhibitory activity, with maximal inhibition exceeding 90% and an IC$_{50}$ of $\sim$1.5 μM (Fig. 1A); PA also strongly inhibited RGS4 GAP activity with an IC$_{50}$ of 5 μM,

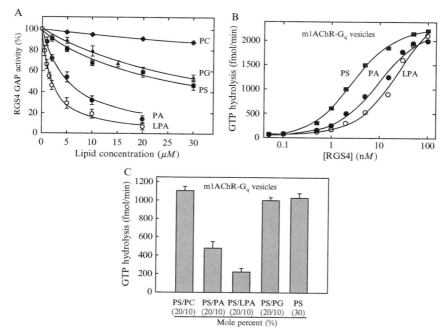

FIG. 1. LPA and PA inhibit RGS4 GAP activity in both single turnover assay (A) and steady-state GTPase assays (B and C). (A) RGS4 was incubated with different phospholipids for 30 min at 4°. The final RGS4 concentration in the assay was 0.08 μM and the phospholipid concentration varied from 0.5 to 30 μM. Two-microliter aliquots were withdrawn and measured for GAP activity using $G\alpha_{i1}[\gamma^{-32}P]GTP$ as the substrate. The first-order hydrolysis rate constant (k_{hydrol}) was determined in the presence and absence of RGS4 and the difference was defined as GAP activity. One hundred percent GAP activity was equal to 1.2–1.5 min^{-1} in different experiments. These values are averages of three experiments with less than 10% standard error. (B) m1AChR-G_q vesicles (1.4 nM G_q and 0.4 nM m1AChR reconstituted in PE/CHS/PS (60:10:30, ■), 1.3 nM G_q and 0.3 nM m1AChR reconstituted in PE/CHS/PS/PA (60:10:20:10, ●), or 1.6 nM G_q and 0.4 nM m1AChR reconstituted in PE/CHS/PS/PA (60:10:20:10, ○; 6 μM final lipid concentration) were mixed with RGS4 (0.05–100 nM) and the mixture was preincubated at 30° for 30 min prior to assay. Carbachol-stimulated GTPase activity was assayed as described in the text. For reference, basal GTPase activities, in the absence of RGS4, of the three proteoliposome vesicle formulations were about 60 fmol/min. (C) RGS4 (2 nM) was incubated with 6 μM m1AChR-G_q at 30° for 30 min prior to assay. PE/CHS/PS/PG (60:10:20:10) vesicles contained 1.3 nM G_q and 0.3 nM m1AChR. PE/CHS/PS/PC (60:10:20:10) vesicles contained 1.3 nM G_q and 0.4 nM m1AChR and the other vesicles were as described in B.

consistent with a previous report (Ouyang *et al.*, 2003). Other anionic phospholipids, PS and PG, could only weakly inhibit GAP activity of RGS4 (Fig. 1A), with maximal inhibition of less than 50% and a much lower potency (IC$_{50}$ > 30 μM). Even at concentrations of up to 100 μM, neutral phospholipids (PC and PE) never inhibited GAP activity of RGS4 by more than 10%, thus they are considered essentially inactive. None of the phospholipids tested affected the intrinsic GTPase activity of $G\alpha_{i1}$ (data not shown). The GAP assays shown in Fig. 1A used $G\alpha_{i1}$-GTP as a substrate, but similar results were obtained in experiments using $G\alpha_z$-GTP and $G\alpha_o$-GTP (data not shown). These data show that although RGS4 can bind to anionic phospholipids and insert into membranes without great preference (Ouyang et al., 2003), LPA and PA may play distinctive and important roles in regulating RGS4 GAP function.

To further characterize the effect of phospholipids on RGS4 activity, we performed receptor-mediated, steady-state GTP hydrolysis assays. This assay is presumably more physiologically relevant because it monitors RGS4-mediated enhancement of agonist-stimulated, steady-state GTPase activity in proteoliposomes reconstituted with receptor and heterotrimeric G proteins. A lipid mixture of PE/CHS/PS (60:10:20) was used for our reconstituted proteoliposomes because the binding of RGS4 through anionic phospholipids appears to be necessary for efficient GAP activity (Tu et al., 2001). To evaluate the effects of different phospholipids on RGS4 GAP activity, the remaining 10 mol% was PC, PS, PG, PA, or LPA. The agonist-stimulated GTPase activity of unilamellar phospholipid vesicles containing m1AChR and heterotrimeric G_q protein was measured in the presence of increasing concentrations of RGS4 (Fig. 1B). When an agonist-bound receptor drives GDP/GTP exchange in these vesicles, the hydrolysis of G_q-bound GTP becomes rate limiting and a GAP increases steady-state hydrolysis until the overall reaction again approaches the rate of receptor-catalyzed GDP/GTP exchange (Mukhopadhyay and Ross, 1999). In m1AChR-G_q vesicles (PE/CHS/PS, 60:10:30), 50 nM RGS4 increases agonist-stimulated GTPase activity about 40-fold with an EC_{50} ~ 2.5 nM (Fig. 1B). Partial replacement of PS by LPA (PE/CHS/PS/LPA, 60:10:20: 10) inhibited RGS4 GAP activity by approximately 10-fold (EC_{50} ~ 25 nM). An inhibitory effect was also observed in the presence of 10 mol% of PA (EC_{50} ~ 10 nM). In contrast, 10 mol% of PG (PE/CHS/PS/PG, 60:10:20:10) or PC (PE/CHS/PS/PC, 60:10:20:10) exhibited no effect or slightly increased RGS4 GAP activity (Fig. 1C). A further decrease of PS ($<15\%$) dramatically decreased both the efficiency of reconstitution and the receptor-stimulated GTPγS binding of G_q, presumably because PS is essential for both the association of G proteins with membranes and the coupling between receptor and G_q (Parker et al., 1991). It should be noted that RGS4 binding to all four proteoliposomes was virtually identical and almost 100% under the conditions used for the GAP assay. The maximal GTPase activity was also similar for each vesicle composition, presumably because GDP/GTP exchange became rate limiting (Mukhopadhyay and

Ross, 1999). These results further support the idea that LPA and PA may play important roles in the regulation of RGS4–Gα interaction.

Lysophosphatidic acid is a serum phospholipid with growth factor-like activities for many cell types. It acts through specific G-protein-coupled receptors on the cell surface. Although PA only constitutes a minor portion of the total phospholipid pool in resting cells, there has been intense interest in the role of PA as a phospholipid second messenger. Our data indicate that both PA and LPA are potent inhibitors of RGS4 GAP activity in *in vitro* assays. While the physiological importance of LPA-mediated inhibition of RGS activity is still unclear, PA may act as a positive regulator of Gα-mediated signaling pathways by inhibiting RGS protein GAP activity.

The N Terminus of RGS4 Is Required for PA Binding and Inhibition

The first 57 amino acids of RGS4 are essential for its localization to membranes (Bernstein *et al.*, 2000; Tu *et al.*, 2001). To determine the importance of this region for interaction with PA, we generated both deletion (missing amino acids 1–57) and single point mutants of RGS4. All the RGS4 mutants used in our study reacted with the anti-RGS4 C-terminal goat polyclonal antibody with relatively equal affinity in an ELISA (Ouyang *et al.*, 2003). Therefore, any differences that may be observed between proteins in the phospholipid-binding assay would be due to their differential binding to phospholipids and not because of variation in their binding to the primary antibody. The RGS4 N-terminal truncation mutant ΔN57 RGS4 did not bind to PA (Fig. 2A), and removing the first 57 amino acids resulted in the loss of the PA-conferred inhibitory effect on RGS4 GAP activity in a solution-based, single-turnover GTP hydrolysis assay (Fig. 2B). These results suggest that residues between 1 and 57 play an important role in both association of RGS4 with PA and PA-conferred inhibition of RGS4 GAP activity. Furthermore, a mutation (K20E) at Lys20 (Lys → Glu) decreased PA-mediated inhibition of RGS4 by more than 85% (Fig. 2B). Protein–lipid-binding assays indicated that this mutation did not have any significant effect on the initial binding of RGS4 to PA vesicles (Fig. 2A). Both the wild-type RGS4 and the K20E mutant have similar dissociation constants for PA vesicles ($K_d \sim 25$ nM). K20E was also as effective as wild-type RGS4 in the single-turnover GAP assay in the absence of PA vesicles (Fig. 2B). Although the RGS domain alone can function as a GAP, the N-terminal domain of RGS4 is required for efficient interaction with Gα proteins (Tu *et al.*, 2001; Zeng *et al.*, 1998). Our data strengthen the argument for allosteric interactions between the N terminus and the RGS domain of RGS4 because Lys20 lies far from the RGS domain in the primary structure.

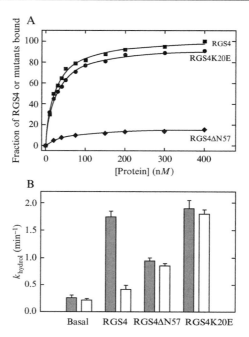

FIG. 2. The RGS4 N terminus is required for PA-mediated inhibition of GAP activity. (A) Binding of wild-type RGS4 or its mutants to PA as determined by an ELISA. Each well of an ELISA plate was coated with 10 μg of PA and blocked using bovine serum albumin. Increasing amounts of RGS4 or its mutants were aliquoted to individual wells as indicated. Bound proteins were analyzed using an antibody to the C terminus of RGS4. Mutation at Lys[20] of RGS4 did not affect the PA–RGS4 interaction. (B) Values of k_{hydrol} (min^{-1}) of RGS4 and its mutants in the presence and absence of PA. RGS4 (2 μM) or its mutants were preincubated with 200 μM PA vesicles (open bars) or buffer as a control (solid bars) for 30 min at 4° before initiation of the reaction. The molar ratios of PA vesicles to proteins were 100:1. GAP activity was assayed as in Fig. 1. Values shown are averages of three experiments ± SD. Mutation at Lys[20] of RGS4 diminished PA-mediated inhibition of GAP activity.

Phosphatidylinositol 3,4,5-Trisphosphate-Dependent Inhibition of RGS GAP Activity

Brief dialysis of recombinant RGS4 into patch clamped pancreatic acinar cells potently inhibits Ca^{2+} signaling evoked by G_i-and G_q-coupled receptor agonists (see also in this volume: Luo *et al.*, 2004; Zeng *et al.*, 1998). This suggests the possibility that endogenous RGS proteins might be relatively inactive prior to agonist stimulation of Ca^{2+} signaling and that recombinant RGS proteins escape this inhibition. To identify inhibitors of RGS4 GAP activity on $G\alpha_{i1}$ in the single turnover assay, we tested

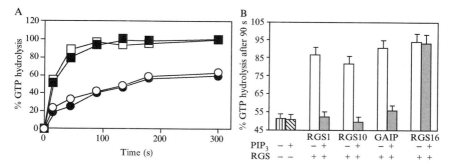

Fig. 3. DiC16-PIP$_3$ inhibits RGS4 GAP activity. (A) Single turnover assay with RGS4 and Gα_{i1}-GTP in the presence of PC lipid vesicles contained 20% diC16-PIP$_3$ (○) or 20% PIP$_2$ (■) RGS4 (0.72 μM) was preincubated with 10 μl of vesicles on ice in 11 μl volume for 20 min. Control reactions contained no lipid, either with (□) or without RGS4 (●). (B) GAP activity of RGS proteins in the presence of diC16-PIP$_3$. RGS1 (2 μM), RGS10 (0.5 μM), GAIP (0.24 μM), and RGS16 (2 μM) were preincubated with (solid bars) or without (open bars) 200 μM PIP$_3$ on ice for 20 min. GAP activity was assayed as in A.

various compounds related to and including either the substrate or products of PLCβ.

Figure 3A shows that PIP$_3$ inhibits RGS4 GAP activity following pre-incubation with phosphatidylcholine vesicles containing 20% PIP$_3$. The kinetic curve of GTP hydrolysis in the presence of PIP$_3$ closely approximated the basal activity of Gα_{i1} without RGS4. Inhibition of RGS4 GAP activity was dependent on the concentration of PIP$_3$ (Popov *et al.*, 2000). In control experiments, the intrinsic GTPase activity of Gα_{i1} was unaffected by PIP$_3$. The GAP activity of each RGS protein that was tested in the single turnover assay, except RGS16, was inhibited by PIP$_3$ (Fig. 3B).

In contrast to PIP$_3$ inhibition of RGS GAP activity, no effect was observed following the incubation of RGS4 with 400 μM PIP$_2$ from bovine brain (Fig. 3A). As summarized in Table I, an analog of PIP$_3$, diC16-PIP$_3$, was the most potent inhibitor of RGS4 GAP activity. Interestingly, dioctanoylphosphatidylinositol 3,4,5-trisphosphate (diC8-PIP$_3$), which only differed from PIP$_3$ (diC16-PIP$_3$) in the length of the fatty acid chains, did not inhibit RGS4 GAP activity. Highly charged head group derivatives of PIP$_3$ lacking the fatty acyl moieties, including inositol 1,3,4,5-tetrakisphosphate (IP$_4$), also show no effect. No inhibitory activity was detected using PI and several synthetic PIP$_2$ lipids such as diC16-3,5-PIP$_2$. RGS4 GAP activity was also not affected by diacylglycerol (DAG) and IP$_3$, both being the reaction products of phospholipase Cβ. Thus, diC16-PIP$_3$ is the only phosphatidylinositide lipid that inhibits RGS4 GAP activity, and its activity appears to require both the long chain fatty acid moiety and the highly charged head group.

TABLE I
PIP$_3$ CHARGE DENSITY AND FATTY ACID CHAIN
LENGTH EFFECTS ON RGS4 GAP ACTIVITY[a]

Lipid	RGS4 binding	RGS4 GAP activity
diC16-3,4,5-PIP$_3$	+	−
4,5-PIP$_2$ (brain)	−	+/−[b]
4-PIP (brain)	NT	+/−[c]
PI (brain)	−	+
diC8-3,4,5-PIP$_3$	−	+
diC8-3,4-PIP$_2$	−	+
diC8-3,5-PIP$_2$	−	+
diC8-4,5-PIP$_2$	−	+
Gro-3,4,5-IP$_3$	NT	+
1,3,4,5-IP$_4$	NT	+
DAG	NT	+

[a] +, binding detected or RGS4 GAP activity is normal; −, binding or GAP activity not detected; NT, not tested.
[b] Inhibition detectable above 1 mM 4,5-PIP$_2$.
[c] Marginal effect at 7 mM 4-PIP.

RGS4 Is a PIP$_3$-Binding Protein

Protein–lipid-binding affinities were estimated by surface plasmon resonance measurements on the BIAcore (Biacore, Inc). RGS4 coupled to the carboxymethylated dextran surface of the BIAcore chip bound diC16-PIP$_3$ (Fig. 4A) but not diC8-PIP$_3$, PIP$_2$, or other phosphoinositides, consistent with the observation that RGS GAP activity was most sensitive to inhibition by diC16-PIP$_3$. The association phase of diC16-PIP$_3$ binding was typically complete within several minutes, whereas dissociation was slow (Fig. 4A). No diC16-PIP$_3$ binding was detected on a blank chip. The K_d value of diC16-PIP$_3$ binding to RGS4 (44 ± 19 nM) was calculated from the on and off rates extracted from the binding curves (Fig. 4A, and data not shown). The positively charged patch on the surface of helixes 4 and 5 in the RGS domain of RGS4 (residues 99–113) appeared to be a good candidate for binding to PIP$_3$. Glutamates were substituted for the wild-type lysine residues at positions 112 and 113 in helix 5 of RGS4. The mutant bound PIP$_3$ with almost 10-fold lower affinity than did wild-type RGS4. Similarly, low binding affinity was observed with RGS16 (Fig. 4A). The comparatively weak binding of these proteins to PIP$_3$ correlated with their relative insensitivity to PIP$_3$ inhibition of GAP activity (Figs. 3C and 4B). Helixes 4 and 5 are conserved in many RGS domains (Table II). PIP$_3$ binding to the RGS

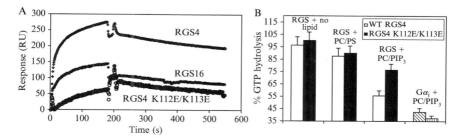

Fig. 4. RGS4 binds PIP$_3$. BIAcore sensorgrams of diC16-PIP$_3$ (4 μM) binding to RGS proteins immobilized on the chip (background binding subtracted). (A) Binding curves for RGS16 and RGS4 K112E/K113E were normalized to 2500 RU of coupled RGS4. (B) Single turnover assay with Gα_{i1}-GTP. The GAP activity of mutant RGS4 K112E/K113E (solid bars) after 90 s is less sensitive than wild-type RGS4 (white bars) to inhibition by PC/PIP$_3$ vesicles (8:2).

TABLE II
PUTATIVE CALMODULIN-BINDING REGIONS IN RGS
DOMAIN PROTEIN [a]

Protein		Sequence	
Consensus		$-$h++h+XXX+hXX+h++h	
rRGS4	(97)	EYKKIKSPSKLSPKAKKI	(114)
hGAIP	(125)	ELKAEANQHVVDEKARLI	(142)
hRGS1	(107)	DYKKTES.DLLPCKAEEI	(123)
hRGS2	(118)	DFKKTKSPQKLSSKARKI	(135)
hRGS3	(429)	DFKKVKSQSKMASKAKKI	(446)
hRGS10	(68)	DFKKMQDKTQMQEKAKEI	(85)
mRGS16	(99)	EFKKIRSATKLASRAHHI	(116)

[a] The amino acid residues used are $-$, negative; $+$, positive; h, hydrophobic; X, any; (.), gap in alignment.

domain at charged residues in helix 5, opposite to the RGS/Gα contact face, further supports the idea that the RGS/Gα interaction can be allosterically modulated to regulate RGS protein activity in cells.

PA and PIP$_3$ appear to act at different sites to regulate RGS4 GAP activity in an allosteric fashion. While PA-mediated inhibition depends on the N-terminal domain of RGS4, mutations in helix 5 of the RGS domain of RGS4 decrease both the affinity of RGS4 to PIP$_3$ and the inhibition of GAP activity by PIP$_3$. As shown in Fig. 5A, both PA and PIP$_3$ (10 mole% in PC vesicles) inhibit RGS4 GAP activity and their inhibition is additive, which further argues that PA and PIP$_3$ interact with different domains of RGS4.

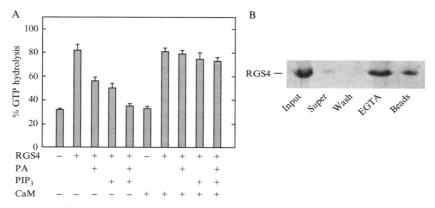

FIG. 5. Ca^{2+}/calmodulin binds RGS4 and antagonizes PA and PIP$_3$ inhibition of RGS4 GAP activity. (A) Single turnover GAP assay with PA or PIP$_3$ lipid vesicles, as in Fig. 3. Before initiating the assay, RGS4 (1 μM) was incubated for 20 min on ice with 50 μM PA or PIP$_3$ vesicles in the presence or absence of 100 μM Ca^{2+}/CaM. Two-microliter aliquots were withdrawn and measured for GAP using $G\alpha_{i1}$-GTP as the substrate. Values are shown as percentage of $G\alpha_{i1}$-bound GTP hydrolyzed at 30 s and are averages of three experiments $\pm$ SD. (B) RGS4 binding to CaM is Ca^{2+} dependent. RGS4 binding to CaM-agarose beads is dissociated by EGTA. Proteins were visualized on Coomassie-stained PAGE gels.

Ca^{2+}/Calmodulin Binds RGS4 and Antagonizes PA and PIP$_3$ Inhibition of RGS4 GAP Activity

As shown in Fig. 5A, 1.6 μM Ca^{2+}/calmodulin preincubated with RGS4 did not alter its GAP activity toward $G\alpha_{i1}$ in a single turnover assay. However, both PA- and PIP$_3$-mediated inhibition of RGS4 GAP activity was reversed by coincubation of RGS4 with Ca^{2+}/calmodulin.

There are two potential calmodulin-binding regions in RGS4: one in the N-terminal 33 amino acids and another between residues 99 and 113, in helixes 4 and 5 of the RGS domain. These regions contain amphiphathic sequences with bulky hydrophobic residues at certain positions and clusters of positively charged amino acids similar to calmodulin-binding sites in other proteins (Table II). Figure 5B indicates that RGS4 binds calmodulin-agarose beads in a Ca^{2+}-dependent manner. The binding was stable to buffer washes, but bound RGS4 could be eluted from beads with either EGTA or SDS (Fig. 5B). An RGS4 N-terminal 33 amino acid peptide (P$_{1-33}$), which conveys high affinity and receptor-selective regulation of G$_q$ signaling (Zeng et al., 1998), bound Ca^{2+}/calmodulin and competed with RGS4 for binding to calmodulin-agarose beads (Popov et al., 2000). In contrast, a scrambled sequence composed of the same amino acids

(ΨP_{1-33}) did not bind to calmodulin beads and did not compete with RGS4 binding (Popov *et al.*, 2000). A previously characterized amphipathic calmodulin-binding peptide from CaM kinase II (Payne *et al.*, 1988) also competed with RGS4 binding to calmodulin-agarose beads. Interestingly, RGS4 apparently forms a heterotrimeric complex with $G\alpha_{i1}$–GDP–AlF_4^- and Ca^{2+}/calmodulin (Popov *et al.*, 2000), consistent with their predicted distinct binding sites on RGS4.

Conclusions

RGS proteins can serve as intracellular sensors and thereby as feedback regulators of G-protein signaling. Regulation of RGS GAP activity can provide a mechanism for controlling the duration and/or frequency of oscillations in intracellular signaling. Using *in vitro* assays, the phospholipids PA, LPA, and PIP_3 were discovered to regulate allosterically RGS GAP activity. However, it remains to be determined if these lipids are *in vivo* regulators of RGS GAP activity and, thereby, G-protein signaling.

Acknowledgments

TMW is supported by NIH GM61395, PAR-98-057, and the Welch Foundation (I-1382). YT is supported by National Scientist Development Grant From American Heart Association (0230338N) and Health Future Foundation (HFF) at Creighton University (200365-823720-120). YT thanks Dr. Elliott Ross for his mentorship during his postdoctoral training. YT also thanks his collaborator Dr. Fuyu Yang (Institute of Biophysics, Chinese Academy of Science) for contributions toward parts of the research reviewed here.

References

Bernstein, L. S., Grillo, A. A., Loranger, S. S., and Linder, M. E. (2000). RGS4 binds to membranes through an amphipathic alpha helix. *J. Biol. Chem.* **275**, 18520–18526.

Derrien, A., and Druey, K. M. (2001). RGS16 function is regulated by epidermal growth factor receptor-mediated tyrosine phosphorylation. *J. Biol. Chem.* **276**, 48532–48538.

Erickson-Viitanen, S., and DeGrado, W. F. (1987). Recognition and characterization of calmodulin-binding sequences in peptides and proteins. *Methods Enzymol.* **139**, 455–478.

Gilman, A. G. (1987). G proteins: transducers of receptor-generated signals. *Annu. Rev. Biochem.* **56**, 615–649.

Glick, J. L., Meigs, T. E., Miron, A., and Casey, P. J. (1998). RGSZ1, a G_z-selective regulator of G-protein signaling whose action is sensitive to the phosphorylation state of G_z-alpha. *J. Biol. Chem.* **273**, 26008–26013.

Ingi, T., Krumins, A. M., Chidiac, P., Brothers, G. M., Chung, S., Snow, B. E., Barnes, C. A., Lanahan, A. A., Siderovski, D. P., Ross, E. M., Gilman, A. G., and Worley, P. F. (1998).

Dynamic regulation of RGS2 suggests a novel mechanism in G-protein signaling and neuronal plasticity. *J. Neurosci.* **18,** 7178–7188.

Kozasa, T., and Gilman, A. G. (1995). Purification of recombinant G-proteins from Sf9 cells by hexahistidine tagging of associated subunits. Characterization of alpha 12 and inhibition of adenylyl cyclase by alpha z. *J. Biol. Chem.* **270,** 1734–1741.

Lee, E., Linder, M. E., and Gilman, A. G. (1994). Expression of G-protein alpha subunits in *Escherichia coli. Methods Enzymol.* **237,** 146–164.

Luo, X., Ahn, W., Muallem, S., and Zeng, W. (2004). *Methods Enzymol.* **389**(8), 119–130.

Mukhopadhyay, S., and Ross, E. M. (1999). Rapid GTP binding and hydrolysis by Gq promoted by receptor and GTPase activating proteins. *Proc. Natl. Acad. Sci. USA* **96,** 9539–9544.

Natochin, M., McEntaffer, R. L., and Artemyev, N. O. (1998). Mutational analysis of the Asn residue essential for RGS protein binding to G proteins. *J. Biol. Chem.* **273,** 6731–6735.

Osterhout, J. L., Waheed, A. A., Hiol, A., Ward, R. J., Davey, P. C., Nini, L., Wang, J., Milligan, G., Jones, T. L., and Druey, K. M. (2003). Palmitoylation regulates regulator of G-protein signaling (RGS) 16 function. II. Palmitoylation of a cysteine residue in the RGS box is critical for RGS16 GTPase accelerating activity and regulation of Gi-coupled signaling. *J. Biol. Chem.* **278,** 19309–19316.

Ouyang, Y. S., Tu, Y., Barker, S. A., and Yang, F. (2003). Regulators of G-protein signaling (RGS)4, insertion into model membranes and inhibition of activity by phosphatidic acid. *J. Biol. Chem.* **278,** 11115–11122.

Parker, E. M., Kameyama, K., Higashijima, T., and Ross, E. M. (1991). Reconstitutively active G-protein coupled receptors purified from baculovirus infected insect cells. *J. Biol. Chem.* **266,** 519–527.

Payne, M. E., Fong, Y. L., Ono, T., Colbran, R. J., Kemp, B. E., Soderling, T. R., and Means, A. R. (1988). Calcium/calmodulin dependent protein kinase II. Characterization of distinct calmodulin binding and inhibitory domains. *J. Biol. Chem.* **263,** 7190–7195.

Popov, S., Yu, K., Kozasa, T., and Wilkie, T. M. (1997). The regulators of G protein signaling (RGS) domains of RGS4, RGS10, and GAIP retain GTPase activating protein activity *in vitro. Proc. Natl. Acad. Sci. USA* **94,** 7216–7220.

Popov, S. G., Krishna, U. M., Falck, J. R., and Wilkie, T. M. (2000). Ca^{2+}/Calmodulin reverses phosphatidylinositol 3,4,5-trisphosphate-dependent inhibition of regulators of G protein signaling GTPase-activating protein activity. *J. Biol. Chem.* **275,** 18962–18968.

Posner, B. A., Mukhopadhyay, S., Tesmer, J. J., Gilman, A. G., and Ross, E. M. (1999). Modulation of the affinity and selectivity of RGS protein interaction with G-alpha subunits by a conserved asparagine/serine residue. *Biochemistry* **38,** 7773–7779.

Ross, E. M., and Wilkie, T. M. (2000). GTPase-activating proteins for heterotrimeric G proteins. Regulators of G protein signaling (RGS) and RGS-like proteins. *Annu. Rev. Biochem.* **69,** 795–827.

Tesmer, J. J., Berman, D. M., Gilman, A. G., and Sprang, S. R. (1997). Structure of RGS4 bound to AlF4 activated $G\alpha_{il}$: Stabilization of the transition state for GTP hydrolysis. *Cell* **89,** 251–261.

Tu, Y., Wang, J., and Ross, E. M. (1997). Inhibition of brain G_z GAP and other RGS proteins by palmitoylation of G protein α subunits. *Science* **278,** 1132–1135.

Tu, Y., Popov, S., Slaughter, C., and Ross, E. M. (1999). Palmitoylation of a conserved cysteine in the regulator of G protein signaling (RGS) domain modulates the GTPase-activating activity of RGS4 and RGS10. *J. Biol. Chem.* **274,** 38260–38267.

Tu, Y., Woodson, J., and Ross, E. M. (2001). Binding of regulator of G-protein signaling (RGS) proteins to phospholipid bilayers. Contribution of location and/or orientation to GTPase-activating protein activity. *J. Biol. Chem.* **276,** 20160–20166.

Wang, J., Tu, Y., Mukhopadhyay, S., Chidiac, P., Biddlecome, G. H., and Ross, E. M. (1998). GTPase Activating Proteins (GAPs) for Heterotrimoric G-Proteins. *In* "G Proteins: Techniques of Analysis" (D. R. Manning, ed.), pp. 123–151. CRC Press, Boca Raton, FL.

Zeng, W., Xu, X., Popov, S., Mukhopadhyay, S., Chidiac, P., Swistok, J., Danho, W., Yagaloff, K. A., Fisher, S. L., Ross, E. M., Muallem, S., and Wilkie, T. M. (1998). The N-terminal domain of RGS4 confers receptor selective inhibition of G-protein signaling. *J. Biol. Chem.* **273**, 34687–34690.

[7] Assays of RGS Protein Modulation by Phosphatidylinositides and Calmodulin

By MASARU ISHII and YOSHIHISA KURACHI

Abstract

Regulator of G-protein signaling (RGS) proteins are a family of proteins that accelerate intrinsic GTP hydrolysis on α subunits of trimeric G proteins. They play crucial roles in the physiological regulation of G-protein-mediated cell signaling. If RGS proteins were active unrestrictedly, they would completely suppress G-protein-mediated signaling. Therefore, it is important to understand how the actions of RGS proteins are regulated under different physiological conditions. We have discovered a physiological mode of regulation of a RGS protein in cardiac myocytes. The voltage-dependent formation of Ca^{2+}/calmodulin (CaM) facilitated the GTPase activity of RGS proteins by removing intrinsic inhibition mediated by the phospholipid phosphatidylinositol-3,4,5-trisphosphate. This modulation of RGS protein action underlies the characteristic "relaxation" behavior of G-protein-gated K^+ channels in native cardiac myocytes. This article describes briefly the discovery of this novel mode of RGS protein modulation in native cardiac myocytes and then gives details of the biochemical and electrophysiological assays used for the functional investigation of this modulation. These assays would be useful for dissecting the physiological modes of action of RGS proteins in controlling G-protein-mediated signaling machinery.

Introduction

Once stimulated by hormones or neurotransmitters, heptahelical G-protein-coupled receptors accelerate the exchange of the guanine nucleotide GDP for GTP on heterotrimeric $G\alpha$ subunits, which leads to the dissociation of Gα and G$\beta\gamma$ subunits (Gilman, 1987). Both GTP-bound Gα subunits and free G$\beta\gamma$ subunits can modulate the function of different target molecules. GTP bound to a Gα subunit is hydrolyzed to GDP by the intrinsic GTPase activity of Gα, and GDP–Gα reassociates with $\beta\gamma$ subunits to terminate the activating process. The time for which Gα and G$\beta\gamma$ subunits remain separate and active is usually short. It depends on how quickly the Gα subunit hydrolyzes its bound GTP. Gα subunit intrinsic GTPase activity is not sufficient, however, as it takes several minutes to hydrolyze GTP. GTPase activity of Gα is enhanced greatly by additional proteins, which can be either its target protein (e.g., phospholipase Cβ1) or a specific modulator known as a regulator of G-protein signaling (RGS).

RGS proteins act as GTPase-activating proteins (GAP) specific to the α subunit of trimeric G proteins and they play a crucial role in shutting off G-protein-mediated cell responses in eukaryotes (reviewed in Hollinger and Hepler, 2002; Ishii and Kurachi, 2003; Ross and Wilkie, 2000). To date, more than 30 kinds of mammalian RGS proteins have been identified. These proteins share a conserved structure of $\sim$120 amino acids, designated the "RGS domain," which is responsible for their GAP activity. The problem with our understanding of the role of RGS proteins is that if they are constitutively active, they would completely block the G-protein cycle and abolish G-protein-coupled receptor signaling. This has been demonstrated using *in vitro* reconstitution protein expression systems. It is therefore clear that *in vivo* and in native cells, RGS proteins must be subject to some form of regulation. Regulation can be achieved through the control of either the protein function and/or the subcellular localization.

This article highlights a physiological mode of regulation of RGS protein function, which operates in native cardiac myocytes (Fujita *et al.*, 1999; Inanobe *et al.*, 2000; Ishii *et al.*, 2001, 2002; Kurachi and Ishii, 2004). The action of the RGS protein is inhibited by the binding of a phospholipid, phosphatidylinositol-3,4,5,-trisphosphate [PtdIns(3,4,5)P$_3$], which is recovered by the binding of Ca^{2+}/calmodulin (CaM) (Ishii *et al.*, 2001, 2002; Popov *et al.*, 2000). It turns out that this reciprocal regulation of RGS protein action underlies a characteristic behavior of cardiac G-protein-gated K$^+$ channel current known as "relaxation." This article (1) outlines the background of the fortuitous discovery of this function of RGS protein in native cardiac myocytes and then (2) describes the detailed methods of the different assays used to dissect the function of RGS proteins. These assays enabled us to obtain insight into the function and kinetics of RGS proteins under physiological conditions.

Background

Agonist Concentration-Dependent Relaxation Behavior of G-Protein-Gated K^+ Channels

G-protein-gated inward rectifier K^+ (K_G) channels, which are activated directly by G-protein $\beta\gamma$ subunits released from pertussis toxin-sensitive G proteins, are responsible for the acetylcholine (ACh)-induced deceleration of heartbeat and neurotransmitter-evoked slow inhibitory postsynaptic potentials in different neurons (reviewed in Kurachi and Ishii, 2004; Yamada *et al.*, 1998). The K_G channel belongs to the inward rectifier K^+ (Kir) channel family, and the cardiac K_G channel is a heterotetramer composed of two kinds of Kir channel subunit, GIRK1/Kir3.1 and GIRK4/Kir3.4. One of the characteristics of native cardiac K_G channel current is the slow time-dependent increase during hyperpolarized potentials, which is known as "relaxation" (Fig. 1A). Upon hyperpolarization to -100 mV, the inward K_G current immediately jumps to one level (I_{ins}) and then slowly increases to a steady level (I_{max}). The immediate increase in current reflects the rapid relief from the blockade of outward K_G current by intracellular Mg^{2+} and/or polyamines, which is a general feature of Kir channels. The slow increase is a unique characteristic of the K_G current and reflects a time-dependent release from inhibition of K_G channel gating that is associated with depolarization. Because Kir channels lack endogenous voltage-sensing units, it is surprising that the K_G channel exhibits such an apparently voltage-dependent behavior.

More surprisingly, this relaxation behavior is dependent on the concentration of agonist (e.g., ACh) for the G-protein-coupled receptor (Fig. 1B). Relaxation is more obvious when activated by a low dose of ACh (0.1 μM) than by a high dose of ACh (1 μM). This observation also contradicts the notion that relaxation represents a genuinely voltage-dependent channel characteristic. Since its first description in sinoatrial node cells, the molecular mechanism underlying this characteristic property of the K_G current has remained an enigma. Because agonist concentration-dependent relaxation behavior is not observed in the K_G current reconstituted by expressing the Kir subunits and m_2-muscarinic receptors in *Xenopus laevis* oocytes, it has been suggested that some exogenous elements should be involved.

RGS Protein Confers Agonist Concentration-Dependent Relaxation Behavior to the K_G Channel

We showed that coexpression of an RGS protein, RGS4, in Kir3.1/Kir3.4/m_2R-expressing oocytes reconstitutes the agonist concentration-dependent relaxation behavior to the K_G current (Fujita *et al.*, 2000).

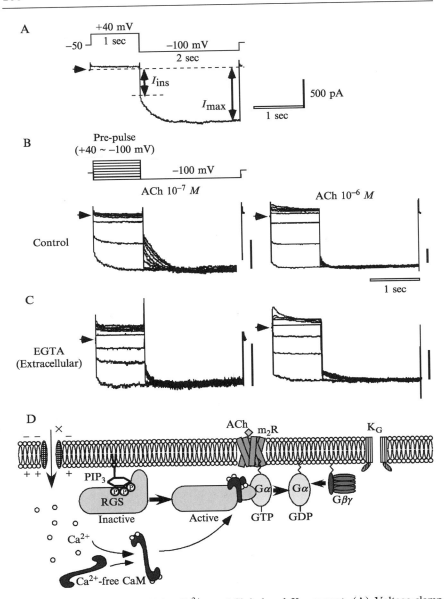

FIG. 1. Effects of extracellular Ca^{2+} on ACh-induced K_G current. (A) Voltage-clamp protocol (top) and a typical ACh-induced K_G current in an isolated atrial myocyte (bottom). Inward current upon stepping voltage to -100 mV changes first instantaneously (I_{ins}) and then increases slowly to a steady state (I_{max}). (B) K_G current evoked by 10^{-7} M (left) or 10^{-6} M (right) of ACh in control conditions. Currents at -100 mV were recorded following

Further studies using site-directed mutagenesis on RGS4 revealed that the interaction between the RGS domain of RGS4 and the $G\alpha$ subunit was essential for reconstituting such characteristic behavior (Inanobe et al., 2001). These results led us to wonder how the RGS protein does so.

The next major step was the finding that the removal of extracellular Ca^{2+} ions abolished agonist concentration-dependent relaxation behavior in native cardiac myocytes (Fig. 1C). Because RGS4 binds to CaM in the presence of Ca^{2+} and because a CaM inhibitor peptide abrogated relaxation, we proposed the underlying mechanism shown in Fig. 1D.

During depolarization, the CaM complex is formed upon Ca^{2+} influx across the plasma membrane. The Ca^{2+}/CaM complex binds to RGS protein. This facilitates its action, which accelerates the G-protein cycle and thus decreases the active K_G channel number. The time-dependent slow increase in K_G current upon sudden hyperpolarization then reflects the reverse reaction, that is, less Ca^{2+} influx, less Ca^{2+}/CaM, lesser activity of RGS, and more available $\beta\gamma$ subunits (Ishii et al.. 2001). The question then arose as to what was inhibiting RGS protein action in the absence of CaM. Subsequent studies revealed that in resting conditions the RGS protein was inhibited by a phospholipid, $PtdIns(3,4,5)P_3$. Ca^{2+}/CaM relieved $PtdIns(3,4,5)P_3$-mediated inhibition and restored GAP activity of the RGS protein (Ishii et al., 2002). Therefore, relaxation behavior has turned out to reflect the membrane potential-dependent control of the G-protein cycle via RGS proteins and is not a reflection of voltage-dependent K_G channel gating. This mechanism of regulation of a RGS protein may not be restricted to the K_G channel system but it should also be applicable to the control of adenylyl cyclase and phospholipase C and thus have a general importance in the control of different G-protein-mediated signaling systems.

---

prepulses between -100 and 40 mV in 20-mV steps as indicated by the schematic diagram above the traces. With 10^{-7} M ACh, voltage-dependent relaxation behavior was prominent. (C) K_G current evoked by 10^{-7} M ACh when extracellular-free Ca^{2+} was chelated by EGTA. The agonist concentration-dependent alteration of the relaxation component is abolished. In each current trace arrowheads indicate the zero current level, and vertical scale bars represent 500 pA. (D) Schematic representation of voltage-dependent relaxation of the K_G current resulting from Ca^{2+}/CaM-dependent facilitation of the action of RGS proteins. At hyperpolarized voltages when $[Ca^{2+}]_i$ is low, the action of RGS is inhibited by $PtdIns(3,4,5)P_3$. Once intracellular Ca^{2+} concentration is increased, e.g., upon depolarization, Ca^{2+}/CaM binds to RGS proteins, removes the inhibition due to $PtdIns(3,4,5)P_3$, and evokes the negative regulation of the G-protein cycle. When the Ca^{2+} concentration decreases to the basal level, CaM dissociates from the RGS proteins and their action is once again inhibited by $PtdIns(3,4,5)P_3$. Reproduced with permission from Ishii et al. (2001).

Assays

This section introduces some of the essential assays utilized in our studies (Fujita *et al.*, 2000; Inanobe *et al.*, 2001; Ishii *et al.*, 2001, 2002). First of all, we describe the method to obtain large amounts of recombinant RGS protein necessary for studying RGS protein function. Second, we focus on *in vitro* assays detecting the interaction between RGS proteins and phospholipids, which are important in demonstrating the functional regulation of RGS proteins. Finally, we concisely describe the electrophysiological methods used to investigate RGS protein action. The K_G channel, which is activated by G-protein $\beta\gamma$ subunits, is essentially free from complicated time-dependent channel gating, and the activity of K_G channels reflects the state of the G-protein cycle. Therefore, by measuring the K_G channel current, we can monitor the trimeric G-protein cycle with fairly high temporal resolution. This system also illustrates the temporal behavior of RGS protein action.

Expression and Purification of RGS Proteins

Construction of the plasmids used for expressing glutathione S-transferase (GST) fusioned RGS proteins used the methods described by Sambrook *et al.* (1989). A prokaryotic expression vector pGEX-2T is available commercially from Amersham Biosciences (Uppsala, Sweden). This vector is designed to fuse GST and the recombinant protein in its N-terminal region. The open reading frame (ORF) of rat RGS4 cDNA (kindly provided by Dr. C. Doupnik), tagged with *Bam*HI and *Eco*RI sites at the 5′ and 3′ ends, respectively, by polymerase chain reaction (PCR), is subcloned into the multicloning site (MCS) of pGEX-2T vector. The pGEX-2T/RGS4 plasmid is then transformed into *Escherichia coli* strain BL21/DE3. This strain lacks some of the proteases that can degrade proteins during purification (Grodberg and Dunn, 1988). The overnight culture of BL21/DE3 cells harboring pGEX-2T/RGS4 is then diluted 1000-fold in 200 ml of LB medium containing 50 μg/ml ampicillin and is cultured. When the OD_{600} reaches 0.5, 1 mM isopropyl-β-D-thiogalactopyranoside (IPTG) is added and cultured for 2 h. Cell cultures are maintained at 37° with shaking at 200 rpm until harvesting. Bacteria are collected by centrifugation at 4° for 30 min at 8000g. The pellet is then resuspended with icy cold lysis buffer containing (in mM) 140 NaCl, 5 KCl, 20 HEPES–NaOH (pH 7.5), 0.5 EDTA, 0.5 phenylmethylsulfonyl fluoride, 0.5 dithiothreitol (DTT), a cocktail of protease inhibitors, 2% (v/v) Triton X-100, and 1% (w/v) CHAPS. The lysate is sonicated for 1 min and is then ultracentrifugated at 100,000g for 1 h. The supernatant fraction containing the extracted GST-RGS4 protein is incubated with 1 ml glutathione–Sepharose

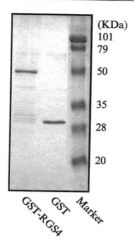

FIG. 2. Sodium dodecyl sulfate–polyacrylamide gel electrophoresis (SDS–PAGE) of affinity-purified GST-RGS4 and GST proteins. Samples (1 μg each) are resolved by SDS–PAGE in an 12% polyacrylamide gel, and the proteins are visualized by staining with Coomassie brilliant blue.

4B resin (Amersham Biosciences) for 30 min at 4° with occasional stirring. The slurry is collected in a chromatography column (Bio-Rad Laboratories, Hercules, CA) and is then washed with 50 ml of wash buffer containing (in mM) 500 NaCl, 20 Tris–HCl (pH 8.0), 0.5 EDTA, 0.5 DTT, and 3% (w/v) CHAPS. The GST-RGS4 protein is then eluted with 5 ml of elution buffer containing (in mM) 150 NaCl, 20 Tris–HCl (pH 8.0), 10 glutathione, and 0.5% (w/v) CHAPS. This method normally yields 1–5 mg/ml of the recombinant GST-RGS4 protein with >90% purity (Fig. 2).

In Vitro Binding Assays between RGS Protein and Phospholipids

We and other groups have demonstrated that the RGS protein interacts specifically with phosphatidylinositol-3,4,5,-trisphosphate (Ishii et al., 2002; Popov et al., 2000). In order to examine the binding between RGS protein and phospholipids in vitro, we have utilized two methods.

Overlay Assay. A protein–lipid overlay assay is a simple and easy method for checking protein–lipid binding (Cheever et al., 2001; Ishii et al., 2002). Different lipids are spotted on a nitrocellulose membrane. This membrane (PIP-Strips) can be purchased from Echelon Research Laboratories (Salt Lake City, UT). The membrane is blocked with 3% (w/v) fatty acid-free bovine serum albumin (Sigma-Aldrich) in TBST [150 mM NaCl, 10 mM Tris–HCl (pH 8.0), and 0.1% (v/v) Tween 20] for

1 h at 4°. The blocked membrane is incubated for 2 h at 4° with 100 ng/ml GST-RGS4 in TBST with shaking. The membrane is then washed five times for 5 min each with TBST. After washing, the membrane is incubated with anti-GST goat antibody (Amersham Biosciences) for 1 h at 1:2000 dilution, followed by additional washing and incubation with the horseradish peroxidase (HRP)-conjugated antigoat IgG rabbit antibody. After final washing, enhanced chemiluminescence is used to visualize the binding of GST fusions to phospholipids. With this assay we detected potential interactions between the RGS4 protein and all of the tested phosphoinositides, as well as lysophosphatidyl acid (LPA) (Fig. 3).

Cosedimentation assay. A protein–lipid cosedimentation assay helps us detect more specific interactions in a quantitative manner (Itoh et al., 2001). Phospholipids are dissolved with a mixture of chloroform and methanol (chloroform/methanol = 2/1). Dissolved lipids are evaporated under a stream of nitrogen gas and are resuspended by vortexing in a sonication buffer containing (in mM) 150 NaCl, 10 HEPES–NaOH (pH 7.4). Small unilamellar liposomes are generated by sonication of the lipid suspension in a water bath at room temperature for 20 min. Alternatively, unilamillar liposomes can be generated easily using a commercially available extrusion apparatus (Avestin Inc., Ottawa, ON, Canada) (MacDonald *et al.*, 1991). Liposomes [phosphatidylethanolamine (PE)/phosphatidylcholine (PC) = 4/1; total 100 μg] containing 0 to 1% (w/w) of phosphoinositides are mixed

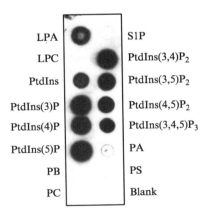

FIG. 3. Interactions between different phospholipids and RGS4. A protein–lipid overlay assay of GST-RGS4. GST-RGS4 binds specifically to PI, PIPs, and LPA. Spotted phospholipids are identified on either side of the assays. LPA, lysophosphatidic acid; LPC, lysophosphocholine; S1P, sphingosine-1-phosphate; PA, phosphatidic acid; PE, phosphatidylethanolamine; PC, phosphatidylcholine; PS, phosphatidylserine. Reproduced with permission from Ishii *et al.* (2002).

with 5 μg of GST-RGS4 and are incubated for 30 min at 4°. Then the mixture is ultracentrifuged at 100,000g for 1 h. Any protein bound to the liposome is cosedimented with the precipitate. Proteins in supernatant (s) and precipitate (p) are separated by SDS–PAGE and transferred to polyvinylidene difluoride (PVDF) membranes. The PVDF membranes are incubated with anti-GST polyclonal antibodies (Amersham Biosciences) diluted to 1:2000 in "buffer A" containing 5% (w/v) skimmed milk and 0.2% (w/v) Lubrol PX in 50 mM Tris–HCl (pH 8.0) and 80 mM NaCl. After being washed three times with buffer A for 10 min each, the membrane is incubated with the HRP-conjugated antirabbit IgG antibody (Wako Pure Chemical, Osaka, Japan) diluted to 1:2000 (v/v) in buffer A for 2 h at room temperature, followed by three washes with 2% Lubrol PX and 0.2% (w/v) SDS in 50 mM Tris–HCl (pH 8.0) and 150 mM NaCl. Immunoreactive bands are developed with the SuperSignal chemiluminescence kit (Pierce, Rockford, IL). This assay should detect more specific and physiological protein–lipid interaction than the overlay assay (Fig. 3), and detailed analyses using this method are now under examination.

Measurement of RGS Protein Action by Recording K_G Channel Activity in Inside-Out Patch Membranes

The G-protein-gated K$^+$ channel is activated directly by G-protein $\beta\gamma$ subunits without the mandatory participation of cytosolic second messengers (Kurachi *et al.*, 1989; Logothetis *et al.*, 1987). This type of G-protein regulation of ion channels has been defined as "membrane delimited." This concept was established by a series of studies showing that the K_G channel is activated in cell-free inside-out patches of atrial myocyte membranes by GTP (in the presence of agonists in the patch pipette), nonhydrolyzable GTP analogs or purified/recombinant G-protein subunits (Kurachi *et al.*, 1986, 1989; Logothetis *et al.*, 1987). In recording K_G channel activity we can visualize with high-time resolution the real-time state of G-protein signaling on the excised patch membrane. This technique is also helpful for understanding the modulation of the G-protein cycle by regulators such as RGS proteins.

Isolation of Rat Atrial Myocyte. Single rat atrial myocytes are isolated enzymatically from hearts removed from adult male Wister–Kyoto rats. Rats are anesthetized deeply by an intraperitoneal injection of pentobarbital. A glass capillary (1.5–1.8 diameter) is inserted into the aorta, and the heart is perfused in a retrograde manner through the coronary arteries. After perfusing nominally Ca^{2+}-free Tyrode's solution containing (in mM) 136 NaCl, 5.4 KCl, 0.6 MgCl$_2$, and 5 HEPES–NaOH (pH 7.4), the heart is digested by perfusing "enzyme solution" at 37° for 10 min. The enzyme

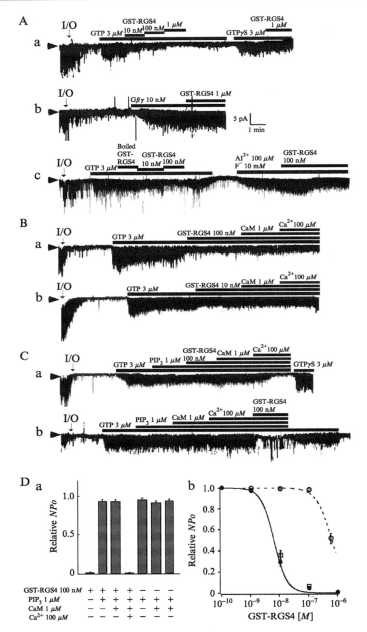

FIG. 4. Effects of RGS4, PtdIns(3,4,5)P₃, and Ca²⁺/CaM on single channel K_G currents in patches from atrial myocytes. Experiments were performed on excised inside-out membrane patches in symmetrical 150 mM K⁺ solutions with 0.3 μM ACh in the pipette. *Arrowheads*

solution consists of nominally Ca^{2+}-free Tyrode's solution, which also contains 0.08% (w/v) collagenase (Collagenase S-1; Nitta Geratin, Osaka, Japan), 0.12% (w/v) hyaluronidase (Sigma-Aldrich, St. Louis, MO), 0.05% (w/v) bovine serum albumin (BSA) (Sigma), and 0.015% (w/v) protease (Sigma). Digested cardiac tissues are removed and then suspended in "stop solution" of nominally Ca^{2+}-free Tyrode's solution containing 0.1%(w/v) BSA. In the stop solution, myocytes are dissociated by shaking for 5 min at room temperature. Dissociated myocytes are seeded on glass coverslips (15 mm diameter) coated with poly-D-lysine (Sigma) kept in a humidified environment of 0.5% CO_2 at 37° and are fed with medium M199 (PAA Laboratories) containing gentamycin and kanamycin (25 $\mu g/ml$, each).

Patch-Clamp Techniques. The patch-clamp technique and, more specifically, its inside-out variant, which allows free access to the cytosolic side of the plasma membrane, is employed for electrophysiological recordings (Hamill *et al.*, 1981). Patch electrodes are made by pulling, in two steps, capillary borosilicate glass tubes (1.5–1.8 × 100 mm, Kimax51, Kimble Products, Toredo, OH) with the aid of a vertical puller (PC-10, Narishige, Tokyo, Japan). The tip of the electrode is coated with Sylgard (Dow Corning, Midland, MI) to reduce random noise and is fire polished. The tip resistance is 5–8 MΩ when filled with the pipette solution. All recordings from inside-out patches are made at a holding potential of -70 mV. The single-channel current is recorded using a patch-clamp amplifier (Axon 200A, Axon Instruments, Foster City, CA) and is stored on videotape through a PCM converter system (VR-10B; Instrutech Corp., Great Neck, NY). For analysis, data are replayed from tape, low pass filtered at 2 kHz (-3 dB) by an eightpole Bessel filter, digitized by an AD converter (ITC-16;

---

indicate the zero current level. Vertical bars represent 5 pA, and horizontal bars represent 1 min. I/O indicates the excision of the patch from the myocyte. Compounds are applied in bath solution to the internal surface of patches for the periods indicated by the bars above the current records. (A) Effects of GST-RGS4 on channel activity. GST-RGS4 inhibits the GTP-activated K_G channel in a dose-dependent manner (a) but does not inhibit (b) G$\beta\gamma$- and (c) GDP-AlF$_4^-$-activated K_G channels. (B) The effect of CaM on the inhibitory action of GST-RGS4. The addition of 1 μM CaM and 100 μM Ca^{2+} had no effect on the inhibition of K_G channels by 100 nM (a) or 10 nM (b) GST-RGS4. (C) The effect of PtdIns(3,4,5)P$_3$. One micromolar PtdIns(3,4,5)P$_3$ severely reduced the inhibitory effect of GST-RGS4, an effect that is reversed by Ca^{2+}/CaM (a). In adding PtdIns(3,4,5)P$_3$ and Ca^{2+}/CaM first, the effect of GST-RGS4 was not inhibited (b). (D, a) The NPo of K_G channel currents recorded under different conditions (indicated below) relative to that seen in 3 μM GTP. N represents the number of K_G channels in a patch, and Po represents the open probability of each channel. Symbols and bars indicate the mean $\pm$ SEM. $n = 10$ for each. (b) Dose-dependent inhibition by GST-RGS4 of K_G channel currents in the absence ($\bullet$) or presence ($\bigcirc$) of 1 μM PtdIns(3,4,5)P$_3$, and in the presence of 1 μM PtdIns(3,4,5)P$_3$ and Ca^{2+}/CaM ($\square$). Symbols and bars indicate the mean $\pm$ SEM. $n = 6$ for each. Reproduced with permission from Ishii *et al.* (2002).

Instrutech. Corp.), acquired continuously by a computer (Macintosh G4; Apple Computer, Inc., Cupertino, CA) with Pulse software (HEKA Elektronik, Lambrecht, Germany), and analyzed off-line with standard software.

For single-channel recordings, the pipette solution that bathes the external surface of the cell membrane contains (in mM) 150 KCl, 1 CaCl$_2$, 1 MgCl$_2$, 5 HEPES–KOH (pH 7.4), and 0.3 μM acetylcholine chloride (Sigma-Aldrich). During inside-out patch-clamp recordings, the bath is perfused with an "internal" solution composed of (in mM), 140 KCl, 2 MgCl$_2$, 5 EGTA, and 5 HEPES–KOH (pH 7.3). Guanosine 5'-triphosphate (GTP, Na salt, Sigma-Aldrich) and its nonhydrolyzable analog, guanosine 5'-O-(3-thiophophate) (GTPγS, Li salt, Boehringer Mannheim, Mannheim, Germany), are added to the internal solution as required. When AlF$_4^-$ is to be applied, a cocktail of 100 μM AlCl$_3$ (Sigma-Aldrich) and 10 mM NaF (Sigma-Aldrich) is added to the solution. Proteins such as GST-RGS4, CaM (Sigma-Aldrich), and purified G$\beta\gamma$ subunits from bovine brain (Kobayashi *et al.*, 1990) are diluted in internal solution just before use.

Detection of RGS Protein Action in Inside-Out Patch Measurement of K_G *Channel Activity.* To examine the function of the RGS4 protein, the purified GST-RGS4 protein is applied to inside-out membrane patches excised from atrial myocytes. K_G channel activity is stimulated by including ACh in the pipette solution bathing the external surface of the cell membrane and superfusing GTP across the exposed internal membrane surface of the patch (Fig. 4). The fusion protein GST-RGS4 applied to the internal surface of patches inhibits K_G channel activity in a dose-dependent manner (Fig. 4Aa). The GST construct alone has no effect (not shown), along with the boiled RGS4 protein (Fig. 4Ac) or mutant RGS4 (N128H), which lacks GAP activity (not shown). Inhibition of K_G channel activity by 1 μM GST-RGS4 is essentially irreversible upon wash out of the protein (Fig. 4Aa), although channel activity can be recovered by the application of GTPγS (Fig. 4Aa), following which GST-RGS4 is without effect. GST-RGS4 is also without effect when K_G channels are irreversibly activated by the application of either G$\beta\gamma$ subunits (Fig. 4Ab) or AlF$_4^-$ (Fig. 4Ac). These results show that GTP hydrolysis is required for GST-RGS4 to reduce K_G channel activity and that GST-RGS4 therefore accelerates the intrinsic GTPase activity of the Gα subunit.

The effect of RGS4 on K_G channels in intact atrial myocytes is facilitated by Ca^{2+}/CaM (see earlier discussion). However, in excised membrane patches, the action of RGS4 is not facilitated by Ca^{2+}/CaM (Fig. 4B). In the presence of PtdIns(3,4,5)P$_3$, the inhibitory effect of GST-RGS4 is suppressed and it then requires the addition of CaM and Ca^{2+} to reestablish the inhibition of K_G channels by GST-RGS4 (Fig. 4Ca). CaM and Ca^{2+} themselves have no effect on channel activity in the presence of

PtdIns(3,4,5)P$_3$ and required the addition of GST-RGS4 to inhibit channel opening (Fig. 4Cb).

These results can be summarized as follows (Fig. 4D). The purified GST-RGS4 protein inhibits K$_G$ channel activity (IC$_{50}$ ~80 nM, Hill coefficient ~1.4). Ca^{2+}/CaM has no effect on this action in excised membrane patches. However, the phospholipid PtdIns(3,4,5)P$_3$ blocks the effect of GST-RGS4. The effect of the phospholipid is reversed by Ca^{2+}/CaM. These data suggest that modulation of the effect of RGS4 on K$_G$ channel activity by Ca^{2+}/CaM occurs not at the level of the channel, but via the interaction of RGS4 with PtdIns(3,4,5)P$_3$. This assay can directly detect the modulation of RGS protein action by calmodulin and phospholipids.

Summary

This article outlined studies that reveal a novel mode of physiological regulation of RGS proteins by calmodulin and phospholipid. Then it described the biochemical and electrophysiological assays found to be useful for studying such modulation of RGS proteins. The combination of these techniques helps us glimpse the whole picture of the physiological control of RGS protein action.

Acknowledgments

We thank Dr. Ian Findlay (Université de Tours, Tours, France) for critically reading this manuscript. This work was supported by the Grant-in-Aid for Specific Research on Priority Area (12144207) (to Y.K), by the Grant-in-Aid for Scientific Research (A) (15209008) (to Y.K), by the Grant-in-Aid for Encouragement of Young Scientists (13770044, 15790133) (to M.I) from the Ministry of Education, Science, Sports and Culture of Japan, by the Japan Heart Foundation Dr. Hiroshi Irisawa Commemorative Research Grant (to M.I), and by the Grant-in-Aid from Ichiro Kanehara Foundation (to M.I).

References

Cheever, M. L., Sato, T. K., de Beer, T., Kutateladze, T. G., Emr, S. D., and Overdulin, M. (2001). Phox domain interaction with PtdIns(3)P targets the Vam7 t-SNARE to vacuole membranes. *Nature Cell Biol.* **3,** 613–618.

Fujita, S., Inanobe, A., Chachin, M., Aizawa, Y., and Kurachi, Y. (2000). A regulator of G protein signaling (RGS) protein confers agonist-dependent relaxation gating to a G protein-gated K$^+$ channel. *J. Physiol.* **526,** 341–347.

Gilman, A. G. (1987). G proteins: transducers of receptor-generated signals. *Annu. Rev. Biochem.* **56,** 615–649.

Grodberg, J., and Dunn, J. J. (1988). OmpT encodes the *Escherichia coli* outer membrane protease that cleaves T7 RNA polymerase during purification. *J. Bacteriol.* **170,** 1245–1253.

Hamill, O. P., Marty, A., Neher, E., Sakmann, B., and Sigworth, F. J. (1981). Improved patch-clamp techniques for high-resolution current recording from cells and cell free membrane patches. *Pflüg. Arch.* **391,** 85–100.

Hollinger, S., and Hepler, J. R. (2002). Cellular regulation of RGS proteins: modulators and integrators of G protein signaling. *Pharmacol. Rev.* **54,** 527–559.

Inanobe, A., Fujita, S., Makino, Y., Matsushita, K., Ishii, M., Chachin, M., and Kurachi, Y. (2001). Interaction between the RGS domain of RGS4 with G protein α subunits mediates the voltage-dependent relaxation of the G protein-gated potassium channel. *J. Physiol.* **535,** 133–143.

Ishii, M., Inanobe, A., Fujita, S., Makino, Y., Hosoya, Y., and Kurachi, Y. (2001). Ca^{2+} elevation evoked by membrane-depolarization regulates G protein-cycle via RGS proteins in the heart. *Circ. Res.* **89,** 1045–1050.

Ishii, M., Inanobe, A., and Kurachi, Y. (2002). PIP3 inhibition of RGS protein and its reversal by Ca^{2+}/calmodulin mediate voltage-dependent control of G protein cycle in a cardiac K^+ channel. *Proc. Natl. Acad. Sci. USA* **99,** 4325–4330.

Ishii, M., and Kurachi, Y. (2003). Physiological actions of regulators of G-protein signaling (RGS) proteins. *Life Sci.* **74,** 163–171.

Itoh, T., Koshiba, S., Kigawa, T., Kikuchi, A., Yokoyama, S., and Takenawa, T. (2001). Role of the ENTH domain in phosphatidylinositol 4,5-bisphosphate binding and endocytosis. *Science* **291,** 1047–1051.

Kobayashi, I., Shibasaki, H., Takahashi, K., Tohyama, K., Kurachi, Y., Ito, H., Ui, M., and Katada, T. (1990). Purification and characterization of five different alpha subunits of guanine-nucleotide-binding proteins in bovine brain membranes. Their physiological properties concerning the activities of adenylate cyclase and atrial muscarinic K^+ channels. *Eur. J. Biochem.* **191,** 499–506.

Kurachi, Y., and Ishii, M. (2004). Cell signal control of the G-protein gated potassium channel and its subcellular localization. *J. Physiol.* **554,** 285–294.

Kurachi, Y., Ito, H., Sugimoto, T., Katada, T., and Ui, M. (1989). Activation of atrial muscarinic K^+ channels by low concentrations of $\beta\gamma$ subunits of rat brain G protein. *Pflüg. Arch.* **413,** 325–327.

Kurachi, Y., Nakajima, T., and Sugimoto, T. (1986). On the mechanism of activation of muscarinic K^+ channels by adenosine in isolated atrial cells: Involvement of GTP-binding proteins. *Pflüg. Arch.* **407,** 264–274.

Logothetis, D. E., Kurachi, Y., Galper, J., Neer, E. J., and Clapham, D. E. (1987). The $\beta\gamma$ subunits of GTP binding proteins activate the muscarinic K^+ channel in heart. *Nature* **325,** 321–326.

MacDonald, R. C., MacDonald, R. I., Menco, B. M., Takeshita, K., Subbarao, N. K., and Hu, L. (1991). Small-volume extrusion apparatus for preparation of large, unilamellar vesicles. *Biochem. Biophys. Acta* **1061,** 297–303.

Popov, S. G., Krishna, U. M., Falck, J. R., and Wilkie, T. M. (2000). Ca^{2+}/calmodulin reverse phosphatidylinositol 3,4,5-triphosphate-dependent inhibition of regulators of G protein signaling GTPase activating protein activity. *J. Biol. Chem.* **275,** 18962–18968.

Ross, E. M., and Wilkie, T. M. (2000). GTPase activating proteins for heterotrimeric G proteins: Regulators of G protein signaling (RGS) and RGS-like proteins. *Annu. Rev. Biochem.* **69,** 795–827.

Sambrook, J., Fritsch, E. F., and Maniatis, T. (1989). "Molecular Cloning: A Laboratory Manual," 2nd Ed., Cold Spring Harbor Laboratory Press, Cold Spring Harbor, NY.

Yamada, M., Inanobe, A., and Kurachi, Y. (1998). G protein regulation of potassium ion channels. *Pharmacol. Rev.* **50,** 723–760.

Subsection C

Electrophysiological Methods and
RGS-Insensitive Gα Subunits

[8]   Analyses of RGS Protein Control
of Agonist-Evoked Ca^{2+} Signaling

By Xiang Luo, Wooin Ahn, Shmuel Muallem, and Weizhong Zeng

Abstract

Analysis of the function of regulator of G-protein signaling (RGS) protein function and their selectivity of action *in vivo* is complicated by the expression of multiple RGS proteins in a single cell and requires precise control of cytosolic RGS protein concentration. This article describes two experimental systems using pancreatic acinar cells suitable for such analyses. The first is pancreatic acini permeabilized with streptolysin O, which retains agonist responsiveness while allowing RGS proteins and molecules with molecular masses of up to 25–30 kDa access to the cytosol. The second is a whole cell recording of the Ca^{2+}-activated Cl$^-$ current of single pancreatic acinar cells as a reporter of $[Ca^{2+}]_i$. This system can be used to introduce to the cytosol any protein of interest, including recombinant RGS proteins and RGS protein-scavenging antibodies. The use of these systems to study the specificity of RGS proteins action, the function of their domains, and the role of RGS proteins in controlling Ca^{2+} oscillations is discussed.

Introduction

The off reaction in the G-protein turnover cycle is the hydrolysis of GTP by Gα and reassociation of Gα with Gβγ. This reaction is slow *in vitro* and is not compatible with fast termination of the GPCR-dependent response *in vivo*. The Gα GTPase activity is catalyzed by PLCβ (Berstein *et al.*, 1992) and regulator of G-protein signaling (RGS) proteins (Ross and Wilkie, 2000). The role of PLCβ GTPase-activating proteins (GAP) *in vivo* is not well understood, largely because it is difficult to separate the catalytic (PIP$_2$ hydrolysis) and GAP activities of PLCβ *in vivo*. The activity and role of the family of RGS proteins are understood better. The RGS protein family includes about 23 members that accelerate the GTPase activities of Gα$_q$, Gα$_i$, Gα$_o$, and Gα$_z$. Another member, RGS-PX1, acts on Gα$_s$ (Zheng

METHODS IN ENZYMOLOGY, VOL. 389

et al., 2001). RGS proteins are typified by a conserved sequence of 120 amino acids termed the RGS domain or RGS box, flanked by divergent N and C termini.

Two features make it difficult to understand the specificity of RGS protein action using *in vitro* systems. First, RGS proteins show limited specificity toward different Gα subunits *in vitro* (Ross and Wilkie, 2000). RGS2 has limited preference to Gα_q (Heximer *et al.*, 1997), RGSz prefers Gα_z (Wang *et al.*, 1998), and RGS-PX1 specifically catalyzes the GTPase activity of Gα_s (Zheng *et al.*, 2001), but otherwise RGS proteins similarly catalyze the GTPase activity of several Gα subunits (Kiselyov *et al.*, 2003; Ross and Wilkie, 2000; Wieland and Mittmann, 2003). Second, many cell types express multiple RGS proteins that can act on several Gα subunits (Wieland and Mittmann, 2003). This is also illustrated in Fig. 1 for pancreatic acini, and similar results were observed with submandibular acinar and duct cells, although there were clear differences in the level of expression of the different RGS proteins.

Therefore, it is necessary to study RGS protein function *in vivo*. Three approaches are available to study RGS protein function *in vivo*. The first is overexpression of selective RGS proteins. This approach is not very useful, as the expression of several RGS proteins inhibits signaling evoked by several classes of Gα subunits (Ross and Wilkie, 2000). A second approach is selective deletion of RGS protein genes in mice or by RNAi. This was achieved for RGS9 (Chen *et al.*, 2000) and RGS2 (Oliveira-dos-Santos *et al.*, 2000) in mice and will likely be used extensively for other RGS proteins with RNAi. However, in addition to the expense and time involved in such an approach, it is not suitable for analysis of the function of specific domains of each RGS protein. The third approach is the introduction of recombinant RGS proteins or their domains into permeabilized cells or the infusion of recombinant proteins into single cells and measuring

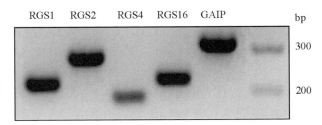

FIG. 1. RT-PCR analysis of RGS proteins in a single pancreatic acinar cell. Single acinar cell were collected individually with patch pipettes and used to extract mRNA for the RT-PCR analysis. The presence of mRNA only for the indicated RGS proteins was probed.

their effect on a well-defined cell-signaling function. Among the advantages of this approach is the ability to control the exact concentration of RGS proteins and their domains to determine concentration dependence curves so that specificity and potency of action can be studied. The goal of this section is to describe how to use such experimental systems for the analysis of the role of RGS proteins in Ca^{2+} signaling.

Selection of the Experimental System

The permeabilized cell system is much less technically demanding than the single cell system. In addition, it can be used to measure the effect of RGS proteins on several signaling pathways, such as Ca^{2+}, cAMP, and MAPK activation, in the same cells and to perform combined functional and biochemical assays. A prerequisite for using this system is that the permeabilized cells will (a) retain agonist responsiveness and (b) that the cells will become permeable to the RGS protein or construct of interest. This system was used extensively in pancreatic acinar and salivary gland cells. Thus, the experimental conditions for use of the system are readily available and the experimental system is highly reproducible. Furthermore, it is relatively easy to prepare well responsive cells, cells obtained from the pancreas of one rat are sufficient for at least 15 assays, and the assay requires only a fluorometer to monitor Fluo-3 fluorescence. However, note that the assay can be used reliably only with proteins of up to 25–30 kDa, which excludes many RGS proteins and inhibitory antibodies. Additionally, the permeabilized cell system is not suitable for studying functions that can be resolved only in single cells, such as Ca^{2+} oscillations and Ca^{2+} waves. The use of single cells requires techniques to introduce the proteins into the cells and monitor a signaling function on the single cell level. Injection of RGS and other proteins into small cells is difficult to control. Infusion through a patch pipette is much more reliable. Dyes or monitoring probes can also be introduced into the cell through the pipette, and the technique can be used to monitor current that is activated by receptor stimulation.

Assay of RGS Protein Function in Permeabilized
   Pancreatic Acinar Cells

Materials

Collagenase P (Boehringer Mannheim), Fluo-3 (Teff Labs), streptolysin O (SLO; Becton-Dickinson and Co.), 74-μm nylon mesh (Small Parts Inc.), and Chelex-100 (Sigma). All other chemicals are from Sigma.

Solutions

> *Solution A*: 140 mM NaCl, 5 mM KCl, 1 mM MgCl$_2$, 1 mM CaCl$_2$, 10 mM glucose, 10 mM HEPES (pH 7.4, with NaOH), measure and adjust osmolarity to 310 mOsm with concentrated (2 M) NaCl. Osmolarity can be measured with a reliable osmometer such as Osmette A from Precision Systems, Inc.
>
> *PSA*. 100 ml solution A supplemented with 100 mg bovine serum albumin (BSA, fraction V), 100 mg sodium pyruvate, and 15 mg soybean trypsin inhibitor (Type II, Sigma)
>
> *PSB*: 8 ml (for rat pancreas) or 15 ml (for mouse pancreas) PSA and 2.5 mg collagenase P
>
> *HK*: 145 mM KCl, 20 mM NaCl, and 20 mM HEPES (pH 7.4 with KOH)
>
> *HK-Ca^{2+}*: HK solution treated with Chelex-100. Wash 100 g of resin three times with 1 liter double-distilled H$_2$O, decant all H$_2$O, and add 1 liter HK solution and stir gently overnight. Let resin settle down and use solution. Resin can be used to reduce Ca^{2+} concentration of up to 5 liters HK to about 600 nM.
>
> *Permeabilization medium*: To the HK-Ca^{2+} solution add 0.02% soybean trypsin inhibitor, 3 mM ATP, 5 mM MgCl$_2$, 10 mM creatine phosphate, 5 units/ml creatine phosphokinase, 10 μM antimycin A, 10 μM oligomycin, 1 μM Fluo-3, and 0.4 unit (3 mg)/ml of SLO

Preparation of Pancreatic Cells

The pancreas of a 100- to 150-g rat or of a 18- to 25-g mouse is removed to a paraffin slab and stretched and anchored to paraffin. The large pancreatic blood vessels and the pancreatic envelope are injected with 10 ml of PSA solution using a 27- or 30-gauge needle, and all visible fat, connective tissues, and blood vessels are removed. Extra fluid is then blotted. The tissue is minced finely (the finer, the better) with a dissection scissor, transferred to a 50-ml plastic centrifuge tube filled with PSA solution, and centrifuged at 1000 rpm ($\sim$200g) for 10 s. The supernatant fluid is decanted; the tissue is suspended in an oxygenated PSB solution and incubated at 37° for 9–10 min (rat) or 5–6 min (mouse) in a shaking water bath at a speed of 200 rpm. At 2-min intervals, the suspension is vigorously shaken manually for about 10 s. Once acini are released to the digestion medium, the digestion is stopped by two washes with 30 ml PSA. The pellet is suspended in 15 ml PSA and filtered through a 74-μm nylon mesh. Acini are collected by a 10-s centrifugation at 200g, resuspended in about 4 ml (mouse) or 15 ml (rat) fresh PSA, and kept on ice until use.

Measurement of RGS Protein Effects on GPCR-Evoked Ca²⁺ Uptake and Release in Permeabilized Cells

1. Fluo-3 fluorescence is monitored at excitation and emission wavelengths of 488 and 530 nm, respectively. Permeabilization medium (0.45 ml) is placed in cuvettes and prewarmed to 37°. RGS proteins or constructs are added to the medium before or shortly after the addition of cells to the permeabilization medium. Recombinant RGS proteins are prepared as His- or GST-tagged fusions as detailed elsewhere (Xu et al., 1999) and dialyzed against an HK solution to ensure the removal of all detergents, divalent ions, and their chelators, such as EDTA and EGTA, all of which affect the assay markedly.

2. About 1 ml of the acinar cell suspension is transferred to a 15-ml centrifuge tube and the cells are washed once with the HK solution and once with the HK-Ca²⁺ solution by a 10-s centrifugation at 3000 rpm. The final pellet is resuspended in 50 μl of the HK-Ca²⁺ solution and transferred to the prewarmed permeabilization medium in the cuvette. The suspension is stirred with a pedal stirrer (Spectrocell Inc.) that fits into the cuvette, and the recording of Fluo-3 fluorescence is initiated on the addition of cells.

3. The Ca²⁺ concentration of this suspension is about 500–600 nM. At the SLO concentration used, acini are permeabilized almost instantly to Ca²⁺ so that Ca²⁺ uptake into the ER begins as soon as recording starts. Ca²⁺ uptake continues until the medium Ca²⁺ concentration is reduced to about 50–75 nM. The slope of this curve reflects the rate of Ca²⁺ uptake into the ER by the ER SERCA pump. For details on the analysis of Ca²⁺ pumping with this technique see Ahn et al., 2003.

4. After stabilization of the medium Ca²⁺ concentration, Ca²⁺ release can be initiated by the addition of Ca²⁺ releasing second messengers such as IP₃ and cyclic ADP ribose (cADPR) or the cells can be stimulated with agonists that activate specific receptors. Pancreatic acini express the Gq-coupled muscarinic M3, CCK, and bombesin receptors.

5. At the end of each experiment, 2 μM IP₃ is added to the medium to completely discharge the IP₃-mobilizable pool so that the extent of Ca²⁺ mobilization can be determined.

6. The fluorescence signals are calibrated by the addition of 2 mM CaCl₂ to the medium to obtain the Ca²⁺-saturated fluorescence (F_{max}) and then with 10 mM EGTA and 40 mM NaOH to obtain the Ca²⁺-free fluorescence (F_{min}). The Ca²⁺ concentration is calculated using the equation:

$$[Ca^{2+}] = 370 \times (F - F_{min})/(F_{max} - F)$$

Figure 2 illustrates an example of such measurements and the type of information that can be obtained.

Similar experiments, summarized in Xu *et al.*, 1999 and Zeng *et al.*, 1998, led to the conclusion that RGS proteins have remarkable selectivity toward GPCRs and that the N-terminal domain of RGS proteins plays an important role in receptor recognition. For example, RGS2 inhibited Ca^{2+} signaling evoked by the M3 and CCK receptors in pancreatic acinar cells equally well, but RGS16 showed a 1000-fold preference toward the M3 receptor (Xu *et al.*, 1999). Further work showed that RGS proteins catalyze the α2A-AR-activated GTPase activity of $G\alpha_o$, but not of $G\alpha_{i1}$, $G\alpha_{i2}$, or $G\alpha_{i3}$ and act in the rank order RGS16 > RGS1 > RGS-GAIP (Hoffmann

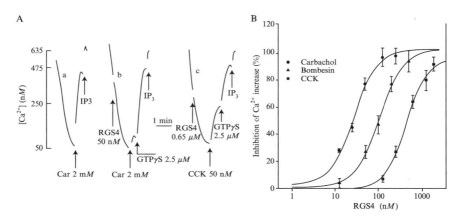

FIG. 2. Effect of RGS4 on stimulation of Ca^{2+} release by M3, bombesin, and CCK receptors. (A) Pancreatic acini were added to permeabilization medium in a fluorometer cuvette and Fluo-3 fluorescence was monitored. Note that Fluo-3 fluorescence declined rapidly as a result of Ca^{2+} uptake into the endoplasmic reticulum (ER). The slope of this curve reports the rate of Ca^{2+} uptake into the ER by the ER Ca^{2+} ATPase pump (trace a). When medium Ca^{2+} began to stabilize the cells were stimulated with a supramaximal concentration of 2 mM carbachol (Car), which resulted in a rapid increase in fluorescence, indicating Ca^{2+} release from the ER. Finally, residual ER Ca^{2+} was discharged by adding 2 μM IP_3 to the incubation medium. To test the effect of RGS4 on Ca^{2+} signaling, RGS4 was added to the incubation medium (b, c) after the start of permeabilization, as indicated by the arrows. Trace b shows that 50 nM RGS4 in the medium inhibited the response to stimulation of the M3 receptors by 90%, whereas trace c shows that 0.65 μM RGS4 inhibited the response to stimulation of the CCK receptor by 60%. The specificity of inhibition is shown by its reversal by the nonhydrolyzable GTP analog GTPγS. GTPγS locks $G\alpha$ subunits in an active conformation (Ross and Wilkie, 2000). (B) Summary of the dose responses for inhibition by RGS4 of Ca^{2+} release triggered by the stimulation of M3, bombesin, and CCK receptors. Reproduced from Xu *et al.*, with permission.

et al., 2001). Hence, regulation by RGS proteins emerged as an important mechanism in obtaining receptor-specific Ca^{2+} signaling.

Assay of RGS Protein Effects on GPCR-Evoked Ca^{2+} Signaling in Single Cells

Assaying RGS protein function in single cells is achieved using the whole cell configuration of the patch-clamp technique. Establishing whole cell configuration requires forming a tight seal on the membrane. In addition, the current has to be measured in single cells so that the current is recorded only from the cell infused with the protein of interest. A modified cell preparation technique is used to prepare single cells suitable for current recording.

Preparation of Single Pancreatic Acinar Cells for Current Recording

Solutions. The composition of *solution A* and *PSA* are as described earlier.

PSB: PSA containing collagenase type IV (CLSP, Worthington), 160 units/ml

PSC: Ca^{2+}- and Mg^{2+}-free solution containing 140 mM NaCl, 5 mM KCl, 0.5 mM EGTA, 10 mM HEPES (pH 7.4, with NaOH), and 10 mM glucose

PSD: PSC containing 0.025% trypsin (Sigma)

PSE: PSA containing collagenase type IV (CLSP, Worthington), 80 units/ml

Pipette solution: 140 mM KCl, 1 mM $MgCl_2$, 0.1 mM EGTA, 5 mM Na_2ATP, and 10 mM HEPES (pH 7.3 with KOH). The free Ca^{2+} concentration of this solution is approximately 50 nM.

Single pancreatic acinar cells are prepared by a six-step method.

1. Rats (100–150 g) or mice (18–25 g) are sacrificed. The pancreas is removed rapidly, injected with PSA, and cleaned from fat, mesentery, and connective tissue as described earlier. The pancreatic tissue is minced finely and washed once with PSA by centrifugation at 1000 rpm for 1 min.

2. The pellet is suspended in 6 ml PSB, and the tissue is digested for 6 min at 37° in a shaking water bath at 200 times/min.

3. The partially digested tissue is washed twice in Ca^{2+}- and Mg^{2+}-free solution PSC by centrifugation at 1000 rpm for 1 min; the pellet is resuspended in 10 ml of trypsin-containing solution PSD and is incubated for 5 min at 37° by shaking at 100 times/min.

4. The tissue is washed twice with solution PSA by a 1-min centrifugation at 1000 rpm to remove the trypsin.

5. The washed pellet is resuspended and digested in 7 ml of solution PSE containing 80 units/ml collagenase type IV by a 3-min incubation at 37° in a shaking water bath at 200 times/min.

6. The liberated single acinar cell suspension is filtered through a nylon mesh with a pore size of 100 μm and is washed three times with solution PSA by centrifugation at 1000 rpm for 1 min, and the cells are resuspended in solution PSA and kept on ice until use.

About 50% of the cells obtained with this procedure are single cells, and better than 90% of the single cells are suitable for patch-clamp experiments. The cells can be used for up to 8 h without loss of responsiveness.

Whole Cell Current Recording

The whole cell current recording procedure of the patch-clamp technique (Hamill *et al.*, 1981) is used to measure the Ca^{2+}-activated chloride current. This current faithfully reports the Ca^{2+} concentration in acinar cells (Thorn and Petersen, 1993). The acinar cells are placed in a 35-mm petri dish, and the dish is placed on a stage of an inverted microscope. After about a 5-min incubation to allow attachment of cells to the dish, the cells are perfused with about 20 ml of solution A to wash away the unattached cells. Patch pipettes are fabricated from borosilicate glass (Warner Instrument Corp. o.d. 1.2 mm, i.d. 0.69 mm) using a two-step glass micropipette puller (such as a PP-830, Narishige, Japan) and are heat polished, leading to a tip diameter of about 1–2 μm. The pipette resistance should be 2–5 MΩ when it is filled with pipette solution. Giga ohm seals (>5 GΩ) are obtained within 20 s after attachment of the pipette to the cell membrane and release of the positive pressure applied to the pipette interior by gentle suction. After obtaining the giga ohm seal the holding potential is changed from 0 to −40 mV. The whole cell configuration is established by gentle suction or a voltage pulse of 0.5 V for 0.3–1 ms. Formation of a successful whole cell configuration is indicated by a sudden increase in capacitive transient current and a decrease in resistance to about 1 GΩ. The current output from the patch-clamp amplifier (Axopatch-200B, Axon Instruments) is usually filtered at 20 Hz, sampled at 1 kHz, digitized, and stored in a computer with the aid of a Digi-Data 1200 interface and appropriate software, such as the pClamp software from Axon Instruments. A stable recording can last for longer than 1 h without a change in response to agonists such as carbachol or CCK. This is important, as the infusion of proteins with high molecular weights, such as antibodies, requires long dialysis before stimulation (see later). Experiments are commonly carried out at room temperature.

When attempting to introduce proteins into cells at a precise concentration, such as RGS proteins or scavenging antibodies, it is necessary to allow time for the equilibration of proteins between the pipette solution and the cytosol. A diffusional equilibrium between the pipette solution and the cytosol is reached within seconds for small ions such as Na^+ and K^+, whereas it takes many minutes for larger molecules such as proteins, depending on their molecular weight. The equation for determining the diffusional time constant was derived empirically in chromaffin cells (Pusch and Neher, 1988) and is applicable to other cell types. The diffusional time constant of compounds with different molecular weights obeys the equation:

$$\tau(s) = (0.6 \pm 0.17) \times AR \times M^{1/3}$$

where τ is the diffusion time constant in seconds, AR is access resistance of the pipette in $M\Omega$ for a whole cell recording configuration, and M is the molecular mass in daltons. For example, the molecular mass of recombinant RGS4 is about 25 kDa. The average AR for pancreatic acinar cells is about 15 $M\Omega$. Pancreatic acinar cells have a diameter of about 17 μm. Therefore, the estimated diffusion time constant for RGS4 to reach equilibrium between the patch pipette and the pancreatic acinar cell cytosol is between 4 and 5 min.

The single cell recording system is very economical, requiring only small amounts of recombinant proteins. The activity of multiple receptors can be probed in one cell simply by stimulating with one agonist, washing out the agonist by perfusion, and, after a recovery period, stimulating with a second agonist. The mechanism of action of RGS proteins and their constructs can be determined accurately, particularly if the effect is subtle. For example, the mechanism of action of RGS4 mutants that interfere with Ca^{2+} signaling could be determined and used to study the role of RGS proteins in Ca^{2+} oscillations. Positively charged and hydrophobic amino acids on helixes 4 and 5 of the RGS domain of RGS4 participate in the binding of calmodulin and phosphatidylinositol 3,4,5-trisphosphate (PIP$_3$) to RGS4. Mutation of these charges altered the effects of calmodulin and PIP$_3$ on RGS4 GAP activity (Popov *et al*, 2000). One such mutant, RGS4 (K99Q, K100Q), is very useful in dissecting RGS protein action. Introduction of RGS4 (K99Q, K100Q) into permeabilized cells stimulated PLC activity to generate IP$_3$ and release Ca^{2+} from the IP$_3$-mobilizable intracellular Ca^{2+} pool. Furthermore, the RGS4 (K99Q, K100Q) mutant evoked Ca^{2+} oscillations (Luo *et al.*, 2001). These results indicate that

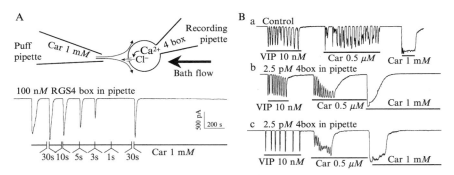

FIG. 3. (A) Stimulation-dependent action of the RGS4 box *in vivo*. (B) Control of Ca^{2+} oscillations by RGS proteins. (A) The current recording pipette contained 100 nM of the 4box. A second pipette containing solution A and 1 mM carbachol was advanced close to the cell. The flow direction of bath perfusion is indicated by the thick arrow. Stimulation was applied by puffing the solution containing carbachol onto the cell for the desired period of time and stimulation was terminated by stopping the puffing and washing away the carbachol by the bath flow. (B) Cells were infused with a low concentration of the 4box (2.5 pM, two traces from two different cells with similar behavior). The extent of inhibition by the 4box was assessed from the inhibition of Ca^{2+} signaling evoked by the stimulation of G_i with VIP. Cells were then stimulated with a low concentration of carbachol (0.5 μM) to evoke G_q-dependent Ca^{2+} oscillations. Reproduced from Luo *et al.*, with permission.

in vivo RGS4 (K99Q, K100Q) antagonized the action of native RGS proteins to activate Gq in an agonist-independent manner.

Another probe useful for dissecting the role of RGS proteins in Ca^{2+} oscillations is the RGS box of RGS4 (4box). Figure 3 shows how the 4box functions as a dominant negative of native RGS proteins *in vivo* and how this tool was used to show the central role of RGS proteins in Ca^{2+} oscillations.

Inhibition of Ca^{2+} signaling by the 4box required prior stimulation of the receptors. Thus, in Fig. 3A, infusion of as much as 100 nM of the 4box into the cell did not prevent the initial stimulus that was similar in control cells and cells infused with the 4box. However, soon after attending the maximum stimulated state, the signal was terminated and returned to the baseline level rapidly. Importantly, the inhibited state was maintained as long as the cell was stimulated and it was sufficient to remove the agonist for only 10 s to restore a maximal stimulated state upon restimulation. Furthermore, even in the presence of a very high concentration of the 4box, repetitive cell stimulation always generated a maximal signal, indicating that the 4box had no effect on Ca^{2+} signaling prior to cell stimulation. This behavior indicates that the 4box, which lacks a receptor recognition domain, does not recognize the receptor in the Ca^{2+} signaling complex of resting cells. Once the cell is stimulated to generate $G\alpha \cdot GTP$, the 4box binds to $G\alpha \cdot GTP$ and is recruited to the receptor complex. In the complex the 4box rapidly and maximally terminates stimulation due to the high concentration of the 4box used, which provides sufficient GAP activity to terminate the stimulated state. Inhibition of the stimulated state is maintained for as long as the cells are stimulated, as the 4box remains trapped in the Ca^{2+} signaling complex. During this time the internal Ca^{2+} store reloads with Ca^{2+}. Washing away the agonist terminates the production of $G\alpha \cdot GTP$, resulting in dissociation of the 4box from the signaling complex to allow restimulation and generation of the maximally stimulated state. Hence, recruitment and dissociation of the 4box depends on the stimulated state of the receptor.

In Fig. 3B, cells were infused with a low concentration of the 4box (2.5 pM) and were stimulated with VIP and then a low concentration of carbachol (0.5 μM). The extent of inhibition by the 4box was assessed from the inhibition of Ca^{2+} signaling evoked by the stimulation of G_i with VIP. The differential effect of the 4box on G_q (stimulated with carbachol)- and G_i (stimulated with VIP)-mediated Ca^{2+} signaling suggests a possible differential role of RGS proteins in the two forms of Ca^{2+} signaling. In the case of G_q-evoked Ca^{2+} signaling, a low concentration of the 4box initially augmented rather than inhibited the Ca^{2+} signal and converted Ca^{2+} oscillations to a sustained response. Thus, in the absence of the 4box during each Ca^{2+} spike, Ca^{2+} returned to the basal level. However, in the presence of a low concentration of the 4box, Ca^{2+} remained progressively higher (Fig. 3B, trace b) or remained elevated for a long period before the stimulated state was eventually terminated and Ca^{2+} returned to the basal level. The significance of these findings is as follows. In this case the weak agonist stimulation $G\alpha \cdot GTP$ is generated only periodically. In the absence of the 4box, $G\alpha \cdot GTP$ is recognized by the native RGS protein(s) (Tesmer et al., 1997) that terminates the stimulated state to return Ca^{2+} to the basal level. Once the native RGS protein dissociates from the Ca^{2+}-signaling complexes, another Ca^{2+} spike is generated and so on. In the presence of a low concentration of the 4box, a fraction of the $G\alpha \cdot GTP$ binds to the 4box in each cycle, which prevents its binding to the native RGS proteins. Since the 4box is not as good a GAP as full-length RGS4 (Zeng et al., 1998) and probably the native RGS proteins present in acinar cells (see Fig. 1), it does not terminate but rather stabilizes the stimulated state by preventing binding of the native RGS proteins to $G\alpha \cdot GTP$. Binding of the 4box to increasing amounts of $G\alpha \cdot GTP$ in each cycle of Ca^{2+} spiking increases the concentration of active $G\alpha \cdot GTP$ and the stimulated state gradually until converting the Ca^{2+} oscillations to a stable and sustained response. Eventually, sufficient 4box accumulates in the Ca^{2+}-signaling complex to terminate the stimulated state. This and other experiments with RGS protein mutants and RGS protein-scavenging antibodies led to the conclusion that RGS proteins play a central role in Ca^{2+} signaling by providing a biochemical control of $[Ca^{2+}]_i$ oscillations (Luo *et al.*, 2001).

Conclusions

The methods described here provide useful techniques to assay RGS protein function accurately in Ca^{2+} signaling. They are particularly suitable when information on receptor and $G\alpha$ specificity and the mechanism of action *in vivo* is required.

References

Ahn, W., Lee, M. G., Kim, K. H., and Muallem, S. (2003). *J. Biol. Chem.* **278,** 20795–20801.

Berstein, G., Blank, J. L., Jhon, D. Y., Exton, J. H., Rhee, S. G., and Ross, E. M. (1992). *Cell* **70,** 411–418.

Chen, C. K., Burns, M. E., He, W., Wensel, T. G., Baylor, D. A., and Simon, M. I. (2000). *Nature* **403,** 557–560.

Hamill, O. P., Marty, A., Neher, E., Sakmann, B., and Sigworth, F. J. (1981). *Pflüg. Arch.* **391,** 85.

Heximer, S. P., Watson, N., Linder, M. E., Blumer, K. J., and Hepler, J. R. (1997). *Proc. Natl. Acad. Sci. USA* **94,** 14389–14393.

Hoffmann, H., Ward, R. J., Cavalli, A., Carr, I. C., and Milligan, G. (2001). *J. Neurochem.* **78,** 797–806.

Kiselyov, K., Shin, D. M., and Muallem, S. (2003). *Cell Signal* **15,** 243–253.

Luo, X., Popov, S., Bera, A. K., Wilkie, T. M., and Muallem, S. (2001). *Mol. Cell* **7,** 651–660.

Oliveira-dos-Santos, A. J., Matsumoto, G., Snow, B. E., Bai, D., Houston, F. P., Whishaw, I. Q., Mariathasan, S., Sasaki, T., Wakeham, A., Ohashi, P. S., Roder, J. C., Barnes, C. A., Siderovski, D. P., and Penniger, J. M. (2000). *Proc. Natl. Acad. Sci. USA* **97,** 12272–12277.

Popov, S. G., Krishna, U. M., Falck, J. R., and Wilkie, T. M. (2000). *J. Biol. Chem.* **275,** 18962–18968.

Pusch, M., and Neher, E. (1988). *Pflüg. Arch.* **11,** 204.

Ross, E. M., and Wilkie, T. M. (2000). *Annu. Rev. Biochem.* **69,** 795–827.

Tesmer, J. J., Berman, D. M., Gilman, A. G., and Sprang, S. R. (1997). *Cell* **89,** 251–261.

Thorn, P., and Petersen, O. H. (1993). *J. Biol. Chem.* **268,** 23219.

Wang, A., Ducret, A., Tu, Y., Kozasa, T., Aebersold, R., and Ross, E. M. (1998). *J. Biol. Chem.* **273,** 26014–26025.

Wieland, T., and Mittmann, C. (2003). *Pharmacol. Ther.* **97,** 95–115.

Xu, X., Zeng, W., Popov, S., Berman, D. M., Davignon, I., Yu, K., Yowe, D., Offermanns, S., Muallem, S., and Wilkie, T. M. (1999). *J. Biol. Chem.* **274,** 3549–3556.

Zeng, W., Xu, X., Popov, S., Mukhopadhyay, S., Chidiac, P., Swistok, J., Danho, W., Yagaloff, K. A., Fisher, S. L., Ross, E. M., Muallem, S., and Wilkie, T. M. (1998). *J. Biol. Chem.* **273,** 34687–34690.

Zeng, W., Xu, X., and Muallem, S. (1996). *J. Biol. Chem.* **271,** 18520.

Zheng, Y. C., Ma, Y. C., Ostrom, R. S., Lavoie, C., Gill, G. N., Insel, P. A., Huang, X. Y., and Farquhar, M. G. (2001). *Science* **294,** 1939–1942.

[9] Measuring the Modulatory Effects of RGS Proteins on GIRK Channels

By CRAIG A. DOUPNIK, CRISTINA JAÉN, and QINGLI ZHANG

Abstract

Discovery of "regulators of G-protein signaling" (RGS) as GTPase-activating proteins for heterotrimeric G proteins has provided a highly sought "missing link," reconciling past discrepancies between the *in vitro* GTPase activity of purified G proteins and the kinetics of physiological responses mediated by G-protein signaling *in vivo*. With the number of RGS genes in the mammalian genome at more than 30, associating specific RGS proteins to specific G-protein-coupled receptor (GPCR) signaling events has become a focus of RGS investigators. The ubiquitous expression of multiple RGS proteins has complicated this effort, yet the outlook has been encouraged with the identification of RGS9 as the determinant mediating rapid recovery of the transducin-dependent photoresponse. G-protein-gated inwardly rectifying potassium (GIRK) channels that mediate inhibitory synaptic transmission via GPCR activation of pertussis toxin-sensitive G proteins are similarly accelerated by RGS proteins when reconstituted in heterologous cell expression systems and fully reproduce the gating properties of native GIRK channels in neurons and cardiomyocytes. The endogenous neuronal and cardiac RGS protein(s) that accelerate GPCR → GIRK channel-gating kinetics are currently not known. This article describes methods used to measure the receptor-dependent GIRK channel-gating parameters reconstituted in Chinese hamster ovary (CHO-K1) cells and *Xenopus* oocytes, as well as rat atrial myocytes and rat cerebellar granule neurons as model cells with native GPCR → GIRK channel signaling. Applications of these methods for structure–function-based studies of RGS proteins, G proteins, and GPCRs are discussed. We also describe single cell reverse transcriptase polymerase chain reaction methods developed to profile atrial myocyte and neuronal RGS expression to identify specific RGS proteins for targeted knockdown or knockout.

Introduction

G-protein-gated inwardly rectifying potassium (GIRK) channels are K^+-selective ion channels "opened" by a direct interaction with $G\beta\gamma$ subunits (Logothetis *et al.*, 1987). Physiologically, GIRK channels play an instrumental role in suppressing membrane excitability during the activation of G-protein-coupled receptors (GPCRs) in neurons, cardiomyocytes, and endocrine cells (Stanfield *et al.*, 2002; Yamada *et al.*, 1998). Receptor-independent "basal" GIRK channel activity may also contribute toward establishing the cell resting membrane potential. The discovery of genes encoding inward rectifier K^+ (Kir) channel subunits in 1993 led to the identification of the Kir channel gene family, which includes a subfamily (Kir3.0) encoding four GIRK channel subunits (Doupnik *et al.*, 1995a). Functional GIRK channels in mammals are now known to be heterotetramers composed of Kir3.1, Kir3.2, Kir3.3, and Kir3.4 subunits (Stanfield *et al.*, 2002; Yamada *et al.*, 1998). The cardiac GIRK channel consists primarily of Kir3.1 and Kir3.4 subunits, whereas neuronal GIRK channels are more diverse, having an overlapping expression of Kir3.1, Kir3.2, Kir3.3, and Kir3.4 subunits in different regions of the brain.

Role of RGS Proteins in GIRK Channel Gating

Confounding early studies of cloned GIRK channels was a recognized discrepancy in the kinetics of receptor-dependent channel activation and deactivation associated with agonist application and washout, respectively (Doupnik *et al.*, 1995b, 1996). The receptor-dependent gating kinetics of cloned GIRK channels were significantly slower than the kinetics of native G-protein-gated K^+ channels studied in isolated cardiomyocytes and neurons (Inomata *et al.*, 1989; Sodickson and Bean, 1996). Breitwieser and Szabo (1988) had demonstrated earlier that the GTP/GDP exchange rate and subsequent GTP hydrolysis were the rate-limiting determinants for muscarinic receptor-dependent gating of the cardiac GIRK channel. These steps in the G-protein cycle effectively rate limit the production of free $G\beta\gamma$ dimers via receptor-catalyzed GDP release and the clearance of $G\beta\gamma$ dimers via reassociation with $G\alpha(GDP)$ following GTP hydrolysis. Thus the limiting rates of $G\alpha$ GDP release and GTP hydrolysis are now generally thought to be reflected in the time course for GIRK activation and deactivation associated with agonist application and washout, respectively. Breitwieser and Szabo (1988) found that the rate constant for GIRK deactivation was nearly two orders of magnitude faster than the reported GTPase activity of purified $G\alpha_o$ subunits measured *in vitro*, postulating that "interaction with the effector (e.g., the ion channel) or some as yet unidentified regulatory component" enhances the GTPase activity of the G protein (Szabo and Otero, 1990).

The discovery of "regulators of G-protein signaling" (RGS proteins) as GTPase-activating proteins (GAPs) for heterotrimeric G proteins (Berman *et al.*, 1996; Chen *et al.*, 1996; Hunt *et al.*, 1996; Watson *et al.*,

1996) provoked reconstitution studies to examine whether RGS proteins could accelerate the receptor-dependent gating kinetics of cloned GIRK channels (Doupnik *et al.*, 1997; Saitoh *et al.*, 1997). These studies indeed reveal that coexpression of certain RGS proteins fully reconstitutes the GIRK activation and deactivation kinetics observed in native cells and that RGS proteins are likely to be integral components of cardiac and neuronal GPCR–G-protein–GIRK channel-signaling complexes. These studies also highlight how electrophysiological approaches measuring the modulatory effects of RGS proteins on GPCR → GIRK signaling provide a high-resolution cellular assay for measuring the kinetic actions of RGS proteins on GPCR signaling *in vivo*. Electrophysiological approaches have similarly been utilized to establish the role of RGS9 in accelerating the kinetics of the photoresponse in photoreceptor cells (Chen *et al.*, 2000).

This article describes methods used currently to study the modulatory actions of RGS proteins on GIRK channel-gating properties. These include conventional electrophysiological methods applied to Chinese hamster ovary (CHO-K1) cells and *Xenopus* oocytes expressing cloned genes, and also toward rat atrial myocytes and rat cerebellar granule (CG) neurons for comparisons with native GIRK-expressing cells. We found that both the CHO-K1 and the *Xenopus* oocyte expression system offer complementary approaches for mechanistic and structure–function studies of GPCR → GIRK channel signaling. We also describe the application of single cell reverse transcriptase polymerase chain reaction (RT-PCR) analysis to profile endogenous RGS expression in native GIRK-expressing cells, with the long-term goal of identifying endogenous RGS proteins that modulate cardiac and neuronal ion channels by specific GPCRs.

Measuring RGS Modulation of GIRK Channels in CHO-K1 Cells

Considerations in selecting a mammalian cell system for heterologous expression importantly include what is known about the endogenous expression of the various proteins under investigation. CHO-K1 cells were initially selected for GPCR-GIRK coexpression studies primarily because of their small size and round geometry, which make them well suited for whole cell patch-clamp recordings (Doupnik *et al.*, 1997; Ehrengruber *et al.*, 1998). They also display a high cotransfection efficiency using cationic lipid-based transfection methods, a critical attribute for reconstituting expression of multiple proteins within a single cell (Ehrengruber *et al.*, 1998). CHO-K1 cells do not express endogenous GIRK channel subunits, yet they do express various GPCRs and RGS proteins. The endogenous expression of RGS mRNA in CHO-K1 cells has been partially characterized (RGS1, RGS2, RGS3, RGS4, RGS10, RGS16, and RGS19) with

RGS2 being significantly expressed, RGS4 not expressed, and the others being expressed at moderate to low levels based on RT-PCR analysis (Boutet-Robinet *et al.*, 2003; Takesono *et al.*, 1999).

Our typical CHO-K1 cotransfection procedure involves introducing five different DNA vectors into single cells. Two of the vectors encode GIRK channel subunits, either Kir3.1 and Kir3.4 to emulate cardiac heteromeric GIRK channels or Kir3.1 and Kir3.2a to emulate neuronal GIRK channels. We then express either the human muscarinic m2 receptor or the human serotonin 1A receptor for agonist activation of the coexpressed GIRK channels (Jaén and Doupnik, 2002). In general, any GPCR that couples to pertussis toxin (PTX)-sensitive G proteins appears to activate coexpressed GIRK channels (Leaney and Tinker, 2000). To examine RGS modulation, a DNA vector encoding a specific RGS protein is included. For RGS negative controls (−RGS), an empty pcDNA3.1 vector is used in place of the RGS vector. All of the cDNAs we use are cloned in the pcDNA3.1 vector (Invitrogen) that contains the human cytomegalovirus (CMV) promoter/enhancer for high-level expression in mammalian cells. To identify positively transfected cells for electrophysiological recording, a DNA vector encoding enhanced green fluorescent protein (EGFP) is also included (pGreenLantern-1, GIBCO).

A couple of points regarding cDNA construction and expression are worth noting. First, for comparative studies of different RGS proteins or different GPCRs, uniformity in cDNA design is necessary to minimize potential differences in protein expression that may affect functional parameters and lead to erroneous structure/activity interpretations. The cDNA design features should include removal of the 5′- and 3′-untranslated regions (UTRs) that often accompany cDNA clones obtained from a variety of sources and can differentially affect RNA stability and thus protein expression levels. Also, the starting ATG codon should contain the same upstream consensus Kozak sequence (i.e., GCCACC) for equivalent translation initiation. Both of these manipulations can be accomplished by PCR amplification of the cDNA coding region, with the Kozak sequence and restriction sequences for subcloning designed in the PCR primers. The final coding sequence should then be confirmed by automated DNA sequencing. Although these steps help minimize potential sources causing differential protein expression, they do not impact any differences attributed to the coding sequence that can potentially affect, for example, protein degradation rates. Measures of steady-state protein expression levels by immunoprecipitation provide an independent and direct means of verifying cellular protein levels.

In an effort to reduce the number of DNA vectors used in our cotransfection protocol, we subcloned both Kir3.1 and Kir3.2a into the pBudCE4.1

vector (Invitrogen), which enables expression of two genes via two different high-level expression promoters, the CMV promoter and the human elongation factor 1α subunit (EF-1α) promoter. Similarly, we subcloned EGFP behind the EF-1α promoter to serve as a reporter vector and then subsequently subcloned different RGS proteins behind the CMV promoter to create a vector expressing both RGS protein and EGFP. These constructs together reduce the cotransfection protocol to three DNA vectors (Kir3.1/Kir3.2a-pBudCE4.1, RGS/EGFP-pBudCE4.1, and GPCR-pcDNA3.1) with incorporated epitope tags for future coimmunoprecipitation experiments.

CHO-K1 Cotransfection Protocol

CHO-K1 cells obtained from ATCC (American Type Culture Collection) are cultured in α-modified Eagle's medium (α-MEM) containing 5% fetal bovine serum and 0.1 mg/ml streptomycin and are maintained in a humidified 5% CO_2 incubator at 37°. One day after low-density plating in 35-mm Corning cell culture dishes, cells 30–50% confluent are transfected with DNA–cationic lipid complexes composed of lipofectamine (Invitrogen) and the five different DNA vectors following the manufacturer's protocol. The lipofectamine (μg) to total DNA (μg) ratio is kept constant at 5:1 when preforming the complexes in serum-free OPTI-MEM medium. The amount of each DNA vector per dish is as follows: Kir3.1 (0.2 μg), Kir3.2a (0.2 μg), GPCR (0.2 μg), EGFP (0.1 μg), and either RGS or the empty pcDNA3.1 vector (1.0 μg). The cells are incubated overnight in serum-free OPTI-MEM medium, and 24 to 36 h posttransfection are used for electrophysiological recordings. The transfection efficiency can approach 50% as evidenced by the percentage of GFP-positive cells, which are identified readily with an epifluorescence-equipped inverted microscope (Fig. 1A).

Whole Cell Patch-Clamp Recording of Receptor-Activated GIRK Currents

Critical to resolving RGS-modulated GIRK current kinetics in mammalian cells is establishing an electrophysiology setup capable of rapid solution changes for agonist application and washout during whole cell voltage-clamp recording. We currently use the SF-77B Fast-Step perfusion system (Warner Instruments), which consists of a three-barrel array made of 700-μm^2 capillary tubes, delivering a gravity-driven flow of three independent solutions in parallel (Fig. 1B). Each barrel can receive input via a manifold connecting up to six different solution reservoirs to expand the solution testing capability. The movement and position of the barrel array are computer controlled, having a limiting step speed of ~20 ms according to the manufacturer. In practice, the time constant for solution exchange determined by K^+ concentration jumps at the surface of an attached CHO-K1 cell

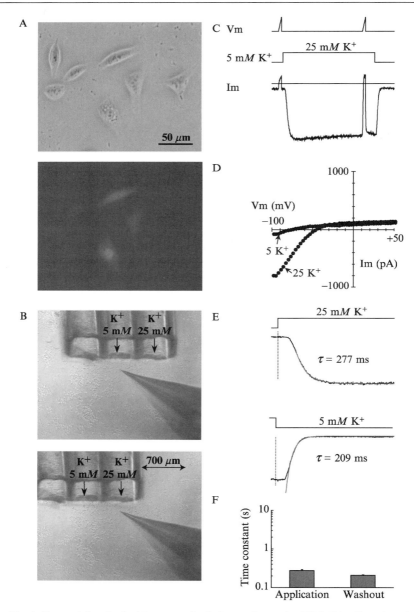

FIG. 1. Determining the limiting rates of solution exchange for CHO-K1 cells under whole cell patch-clamp recording. (A) CHO-K1 cells 24 h after DNA transfection visualized under phase-contrast (*top*) and epifluorescence (*bottom*) microscopy. Cells were transfected with EGFP, GIRK channel subunits (Kir3.1/Kir3.2a), and the muscarinic m2 receptor.

under whole cell recording is ~240 ms (Fig. 1), although it can be as fast as 120–140 ms at the highest flow rates generated (~75 cm column height). These solution exchange rates are comparable to some (Breitwieser and Szabo, 1988; Bunemann et al., 1996), although they are somewhat slower than the 10- to 50-ms time constants reported by others using similar configurations (Karschin et al., 1991; Sodickson and Bean, 1996). Nonetheless, they are sufficient to temporally resolve the GIRK current kinetics observed at room temperature (22–24°) with and without RGS coexpression.

Single GFP-positive CHO-K1 cells are selected for electrophysiological recordings using standard whole cell tight-seal patch clamp methods (Hamill et al., 1981). Cells are initially washed and placed in an external solution that consists of (in mM) NaCl 145, KCl 5, CaCl$_2$ 2, MgCl$_2$ 1; glucose 10, and HEPES 5 (pH 7.4). After gigaseal formation and breaking into the cell for whole cell recording, 2 min are allowed to permit equilibration of the intracellular solution. The composition of the internal pipette solution is (in mM) KCl 120, NaCl 10, MgCl$_2$ 5, EGTA 1, HEPES 5, ATP 5, and GTP 0.2 (pH 7.2). First after breaking into the cell, the membrane capacitance (a direct measure of cell surface area) is determined via amplifier compensation and is later used to express the maximal current amplitude as a current density (pA/pF) for cell-to-cell comparisons.

Agonist-evoked inward K$^+$ currents are recorded from a holding potential of −100 mV, which is sufficiently negative to the experimentally set K$^+$ equilibrium potential (E$_K$ = −40 mV). Thus after establishing

---

(B) Alignment and movement of perfusion barrels for rapid solution exchange. (*Top*): The three-barrel array positioned with the patch-clamped cell (see patch electrode) being superfused with 5 mM K$^+$ solution. Flow through both the middle and the right barrels is continuous and gravity driven, whereas the left barrel is typically not used in these experiments. (*Bottom*): The position of the barrels following computer-controlled movement (700 μm), where the 25 mM K$^+$ barrel is now aligned with the recorded cell. (C) Electrophysiological recording of a CHO-K1 cell voltage clamped at −100 mV (Vm) during superfusion with the 5 mM K$^+$ solution and during the triggered application (15 s) and washout of the 25 mM K$^+$ solution. Voltage ramps (−100 to 50 mV, 500 ms) were evoked before and during 25 mM K$^+$ solution application to evaluate the voltage dependence of receptor-independent currents. Changes in the whole cell current (I$_m$) in response to the voltage ramps and solution change are shown; the horizontal line indicates the zero current level. (D) Current–voltage plots derived from the voltage ramps evoked before (5 mM K$^+$) and during the 25 mM K$^+$ solution application. The reversal potential shift is consistent with a K-selective conductance having steep inward rectification; both are signature features of GIRK currents. (E) The time course for changes in I$_m$ with 25 mM K$^+$ solution application and washout was fit with a single exponential function (red lines) to derive time constants (τ) for solution exchange. Note the delay in the onset, which is attributed to the distance of the perfusion barrels from the recorded cells and the associated volume dead space. (F) Plots of the solution exchange time constant obtained from applications and washouts of 25 mM K$^+$ are nearly equivalent. Data are means $\pm$ SEM ($n = 3$). (See color insert.)

whole cell recording and clamping the membrane potential to -100 mV, the cell is initially superfused with a high K^+ solution composed of (in mM) NaCl 125, KCl 25, CaCl$_2$ 2, MgCl$_2$ 1; glucose 10, and HEPES 5 (pH 7.4). The solution is applied via one of the 700-μm^2 capillary tubes positioned next to the cell (Fig. 1B) and connected to a 20-ml syringe reservoir where the flow rate is gravity controlled by adjusting the syringe height. After a stable baseline holding current is established, the agonist is applied (in the high K^+ solution) via the step movement of the barrel array so that agonist flow via the adjacent barrel is positioned in line with the recorded cell. We typically apply the agonist for 15 s to minimize receptor desensitization, followed by agonist washout with the step movement back to the high K^+ solution barrel. The agonist barrel is connected via PE-10 tubing to a manifold that connects to six different 20-ml syringes via PE-50 tubing, each containing a different agonist concentration controlled by a two-way valve. By closing one syringe and opening another, the agonist concentration flowing out of the barrel can be changed after a brief delay due to tube dead space. We wait 1 min for each manifold solution change when testing six different agonist concentrations. Using this procedure, full agonist dose–response relations can be obtained from single cells to derive EC$_{50}$ values from fitted Hill functions (Fig. 2A and C).

Our voltage-clamp recordings are performed using an Axoclamp 1D amplifier (Axon Instruments). Current signals are sampled and digitized via a Digidata 1200B A/D board that also synchronizes digital output signals to the SF-77B Fast-step controller. Axon pCLAMP8.0 software is used to trigger the perfusion barrel movements along with 500-ms voltage ramps (-100 to 50 mV) evoked before and during agonist application to assess the voltage-dependent properties of the agonist-evoked currents. Characteristic features of GIRK currents include steep inward rectification and K^+ selectivity (i.e., a reversal potential near the E_K). The analog current signals are low pass filtered with the integrated four-pole Bessel filter of the amplifier at a corner frequency of 50 Hz and is then sampled digitally at 100 Hz.

The time constants for GIRK current activation (τ_{act}) and deactivation (τ_{deact}) are derived by fitting a single exponential function to the rising or decaying portion of the current (Fig. 2D) using nonlinear least-squares curve-fitting software (Clampfit 8.0). The derived τ_{act} values are dependent on agonist concentration, whereas τ_{deact} is independent of agonist concentration (Fig. 2D). The latency or delay that precedes the exponential time course can also be quantified by using the initial triggering signal as a reference point and the inflection point for current onset. Note that RGS expression not only accelerates the activation and deactivation time course, but also shortens the latency period (Fig. 2D) and may reflect the

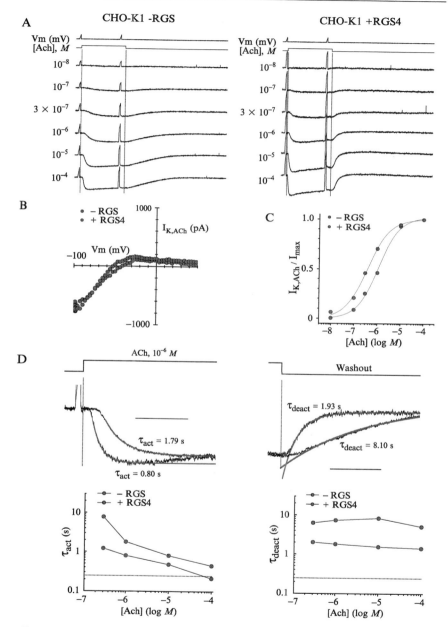

FIG. 2. Quantitative analysis of RGS-modulated GIRK current kinetics in CHO-K1 cells. (A) Acetylcholine (ACh)-activated whole cell GIRK currents recorded from CHO-K1 cells cotransfected with the muscarinic m2 receptor, Kir3.1 and Kir3.2 subunits, and EGFP, with

RGS-facilitated coupling of GPCRs and GIRK channels independent of RGS GAP activity (Jeong and Ikeda, 2001).

Measuring Native GIRK Channel-Gating Properties in Cardiomyocytes and Neurons

The electrophysiological methods and conditions used to measure GIRK channel properties in rat atrial myocytes and CG neurons are as described for CHO-K1 cells, thus enabling a direct comparison among the different GIRK-expressing cells. Neonatal rat atrial myocytes and rat CG neurons both exhibit receptor-evoked GIRK currents that are sustained in primary culture and can be characterized electrophysiologically (Fig. 3). Muscarinic m2 receptor-activated GIRK currents in rat atrial myocytes and $GABA_B$ receptor-activated GIRK currents in rat CG neurons both display rapid activation and deactivation kinetics, suggesting modulation by endogenous RGS proteins (Fig. 3). These features have been described previously, as atrial GIRK channels have a rich history of investigation (Noma and Trautwein, 1978; Yamada *et al.*, 1998). Neuronal GIRK channels have been studied only recently in CG neurons, prompted largely from the discovery of the mouse *weaver* gene that contains a point mutation in the Kir3.2 subunit that disrupts K^+ selectivity, causing CG cell death and

---

(*right*) or without (*left*) rat RGS4. Families of currents from each cell are from different ACh concentrations applied for 15 s from a holding potential of -100 mV as described in Fig. 1. Vertical dotted reference lines indicate triggering times for the movement of the perfusion barrels for agonist application and washout. (B) Current–voltage plots for $I_{K,ACh}$ derived from RGS4 (green circles) and non-RGS-transfected CHO-K1 cells in A. Plots were obtained by subtracting the ramp-evoked current prior to ACh application from the ramp-evoked current obtained during 1 μM ACh application. Both I–V plots display steep inward rectification and similar reversal potential near the K^+ equilibrium potential (-40 mV), indicative of receptor-evoked GIRK currents. (C) Steady-state dose–response relations for ACh-evoked GIRK currents ($I_{K,ACh}$) in the absence (red circles) and presence of coexpressed RGS4 (green circles). The peak $I_{K,ACh}$ amplitude for each ACh concentration was normalized to the maximal amplitude elicited by 100 μM ACh from each cell. The normalized $I_{K,ACh}$ amplitudes were then fit with a Hill function ($I_{K,ACh}/I_{Max} = 1/\{1+[ACh]/(EC_{50})^{[nH]}\}$) to derive the effective ACh concentration evoking a half-maximal response (EC_{50} value) and the Hill coefficient value (nH). RGS4 expression causes a small rightward shift of the ACh dose–response curve (EC_{50} values: $-$RGS, 0.82 ± 0.16 μM ACh, $n = 10$; +RGS4, 1.39 ± 0.30 μM ACh, $n = 6$), contrary to earlier observations (Doupnik *et al.*, 1997). (D) Time constants for ACh-evoked GIRK current activation (τ_{act}) and deactivation (τ_{deact}) were derived from single exponential fits to the currents. The fitted exponential functions are superimposed on current data ($-$RGS, red; RGS4, green) from GIRK currents elicited by 1 μM ACh. The current amplitudes have been normalized for comparison. The horizontal line in each panel represents a 5-s scale bar. (*Bottom*) Plots of τ_{act} and τ_{deact} derived from currents elicited by the different ACh concentrations, illustrating RGS4 effects on both kinetic parameters. (See color insert.)

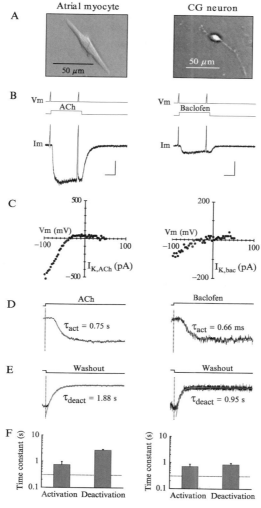

FIG. 3. Quantitative analysis of native GIRK currents recorded from rat atrial myocytes and rat cerebellar granule (CG) neurons. (A) Phase-contrast images of a typical neonatal rat atrial myocyte and a rat CG neuron maintained in primary culture and selected for electrophysiological recordings. (B) ACh (10 μM) and the GABA$_B$ receptor agonist baclofen (100 μM) evoke characteristic GIRK currents from atrial myocytes and CG neurons, respectively. Both display steep inward rectification (C) and rapid activation (D, F) and deactivation kinetics (E, F), with deactivation of baclofen-activated GIRK currents being slightly faster than ACh-activated GIRK currents. Data are means ± SEM. Dashed lines in F refer to time constants for solution exchange from Fig. 1 and represent the limit of resolving kinetic events any faster using the described methods. (See color insert.)

ataxia (Kofuji *et al.*, 1996; Patil *et al.*, 1995; Slesinger *et al.*, 1996; Surmeier *et al.*, 1996). Both of these GIRK-expressing cells have very different physiological roles and provide suitable cell systems to further explore endogenous RGS modulation in a native cell environment.

Atrial myocytes and CG neurons from neonatal rat pups are both amiable to enzymatic isolation and primary culture for experimental manipulations. Issues of experimental concern include changes in the gene expression profile of the cells with development and during primary culture and should be considered when extrapolating findings to adult and intact tissues. The methods used for isolating and culturing atrial myocytes from Sprague–Dawley rat pups (postnatal days 4–6) are described in detail elsewhere (Doupnik *et al.*, 2001). For postnatal day 4–6 rat CG neurons, we use a protocol modified from Slesinger and Lansman (1991). Following ip injection of sodium pentobarbital (4 mg/100 g body weight) to induce deep anesthesia, rat pup cerebella (2–4) are removed rapidly and placed in a 35-mm culture dish containing ice-cold calcium and magnesium-free (CMF) Tyrode's solution (in mM): 136.9 NaCl, 5.4 KCl, 6.0 NaHCO$_3$, 0.33 Na$_2$HPO$_4$, 5.5 D-glucose, 5.0 HEPES, at pH 7.4 (NaOH), containing 100 U/ml penicillin and 0.1 mg/ml streptomycin. The tissue is minced, washed with CMF Tyrode's solution, and then digested with 0.5 ml of trypsin/EDTA solution (GIBCO 25300-054) for 10 min at room temperature. The digestion is stopped by placing on ice and adding "isolation medium" that consists of modified Eagle's medium with Earle's salts (GIBCO 11095-080) supplemented with 10% heat-inactivated horse serum, 25 mM KCl, 6 mg/ml D-glucose, 2 mM glutamine, 0.5 U/ml DNase I, 0.5 U/ml penicillin, and 0.5 μg/ml streptomycin. The digested tissue is then triturated using a 1-ml sterile pipette, and the dispersed cells are plated at low density on poly-L-lysine-coated 35-mm Corning cell culture dishes. The cells are incubated for 5 h at 37° in a 5% CO$_2$ atmosphere, and then the culture medium is changed to serum-free Neurobasal-A medium (GIBCO 10888-022) with B-27 supplement and 25 mM KCl, 2 mM glutamine, 0.5 U/ml penicillin, and 0.5 μg/ml streptomycin. The cerebellar cell cultures are then maintained in a humidified incubator at 37° with a 5% CO$_2$ atmosphere for 24–48 h before extensive neurite outgrowth occurs. All procedures for the use and handling of rats are approved by our institutional animal care and use committee in accordance with NIH guidelines.

Single Cell RT-PCR Analysis of Endogenous RGS Expression

Since RGS overexpression in CHO-K1 cells effectively accelerates the kinetics of GIRK channel gating to rates similar to those observed in rat atrial myocytes and CG neurons (compare Figs. 2 and 3), determining which of the 19 possible RGS genes (representing the four major RGS subfamilies (Ross and Wilkie, 2000)) are expressed in these cell types and responsible for accelerating GIRK channel gating kinetics becomes a daunting challenge.

To begin to address which RGS proteins are expressed endogenously in rat atrial myocytes, we developed single cell RT-PCR methods for detecting and profiling RGS expression in spontaneously beating atrial myocytes maintained in primary culture (Doupnik et al., 2001). The selective harvesting of easily distinguished single atrial myocytes limits the assay to atrial myocyte mRNA and effectively excludes mRNA from the various other cells within the atrial myocardium that are present in the cultures. We have also applied this approach to profiling RGS expression in rat CG neurons, which are abundant and can similarly be distinguished in culture by their relative small size and simple bipolar morphology compared to other cerebellar cells present in the culture dish (Fig. 3). Positive Kir3.4 expression in atrial myocytes and Kir3.2 expression in CG neurons confirm GIRK channel expression in these cells (Fig. 4). Results indicate that rat atrial myocytes express 7 different RGS genes, whereas rat CG neurons express at least 13 (Fig. 4). All of the RGS genes expressed in atrial myocytes were also expressed in CG neurons, and each RGS subfamily (R4, R7, R12, and RZ) is represented in both expression profiles. Although a profile of protein expression has not been correlated with mRNA data and relative RGS protein levels are unknown, the single cell RT-PCR results clearly indicate that numerous RGS proteins are likely to be present in these native GIRK-expressing cells.

Designing Intron-Spanning Gene-Specific Primers

The RT-PCR approach utilizes gene-specific primers that selectively amplify mRNA transcripts from a specific RGS gene from a single cell. Because all mammalian RGS genes are poly-intronic (Sierra et al., 2002), intron-spanning primers were designed to distinguish mRNA-derived PCR products from genomic DNA-derived products (Doupnik et al., 2001). At the time of our original study, sequence information for rat RGS genes was limiting so mouse and human RGS sequences were used as alternatives for primer design. The effectiveness of each RGS primer set was confirmed by positive controls using samples of rat brain poly(A) mRNA. A full list of the RGS primer sequences is provided elsewhere (Doupnik et al., 2001), with the exception of the intron-spanning GIRK2 primers that were designed from the mouse GIRK2 sequence (5' to 3'): forward primer CCTGCCGGGGCTGATGTGA; reverse primer TTGGT CCTGTCTCGGCTGATGTGT; annealing temperature 58°.

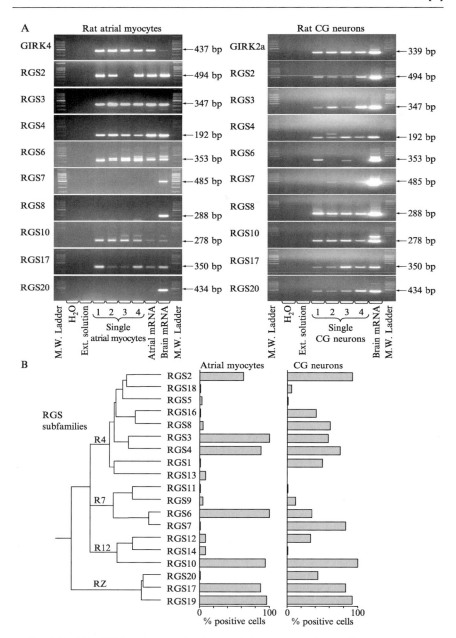

FIG. 4. Profiling RGS gene expression in rat atrial myocytes and cerebellar granule neurons by single cell RT-PCR analysis. (A) Separation of RT-PCR products by agarose gel electrophoresis. RT-PCR was performed on single rat atrial myocytes (*left*) and rat CG

Single Cell Harvesting

Single rat atrial myocytes or rat CG neurons are harvested from culture dishes using a micropipette (15–30 μm tip diameter) fabricated from borosilicate glass tubes (1.5 mm outside diameter, 0.86 mm inside diameter, GC150F-10, Warner Instruments) by a programmable microelectrode puller (P-97, Sutter Instruments). Latex gloves are worn throughout the handling and harvesting procedure to minimize potential sources of contamination. The micropipette is attached to a microelectrode holder used for patch-clamp recordings, allowing application of negative or positive pressure via an attached syringe. The culture dishes are first washed with a solution consisting of (mM) 145 NaCl, 5 KCl, 2 CaCl$_2$, 1 MgCl$_2$, 10 D-glucose, 5 HEPES, at pH 7.4 (NaOH) at room temperature ($\sim$23°). A single atrial myocyte having a spindle morphology and spontaneous contractile activity or a single CG neuron having bipolar morphology (Fig. 3) is then drawn into the micropipette by negative pressure. The micropipettes are not filled with solution initially, but contain $\sim$5 μl of the external solution after the cell has been harvested. The contents of the micropipette are then expelled into a PCR tube by positive pressure and the tube is placed on ice.

For each experiment, four to six single cells are harvested and tested in parallel with negative and positive controls. Two negative control samples are a 5-μl sample of RNase-free H$_2$O and a 5-μl sample of the external solution. Positive controls include poly(A)$^+$ mRNA from neonatal rat atria (0.5–1.0 ng) and neonatal rat whole brain (0.5–5.0 ng). Experiments are generally repeated three times ($\geq$12 myocytes) from separate dissections/cultures for each RGS examined to account for animal, culture, and cell variability.

RT-PCR Analysis

One-step RT-PCR is carried out using the intron-spanning gene-specific primers according to the manufacturer's protocol (OneStep RT-PCR, Qiagen Inc.). For each 5-μl sample, 20 μl of a RT-PCR master mix is added

---

neurons (*right*) as described in the text. Results using selected intron-spanning RGS-selective primer sets (Doupnik *et al.*, 2001), as well as primer sets for GIRK4 (Kir3.4) for atrial myocytes and GIRK2 (Kir3.2) for CG neurons, are shown. Negative controls included water and external solution (5 μl), and positive controls included postnatal poly(A) + mRNA of rat atria and brain (0.5–5.0 ng). The predicted molecular size for each RGS PCR product is indicated on the right of each gel. (B) Profile of RGS expression in rat atrial myocytes and rat CG neurons. The percentage of cells sampled and positive for RGS expression by RT-PCR analysis is shown for each RGS within the R4, R7, R12, and RZ subfamilies examined. The number of cells tested for each RGS ranges from 8 to 24 and is from at least two separate cultures. Atrial myocyte data have been reproduced from Doupnik *et al.* (2001).

and contains the following: forward and reverse primers at 0.6 μM each (final concentration), dNTPs at 400 μM each (final concentration), Omniscript and Sensiscript reverse transcriptases, HotStar *Taq* DNA polymerase, RNase inhibitor, and a buffer solution containing Tris–Cl, KCl, $(NH_4)_2SO_4$, $MgCl_2$, and dithiothreitol. Concentrations of enzymes and buffer components are as recommended by the manufacturer (1X concentration, Qiagen, Inc.) and include the 1X "Q solution," which effectively reduces nonspecific bands produced by mispriming events.

Each 25-μl sample is then placed in a PCR thermocycler (GeneAmp 2400, PE Biosystems, Inc.) for the following temperature protocol: 50° for 30 min (reverse transcription), 95° for 15 min (activation of HotStar *Taq* polymerase), 45 cycles of 94° for 30 s (melt), 3–4° below the primer annealing temperature for 30 s, and 72° for 60–90 s (extension). At the end of the cycling period, samples are held at 72° for 10 min (final extension) and the reaction is stopped by cooling to 4°. According to the manufacturer (Qiagen Inc.), this PCR cycling protocol (45–50 cycles) is expected to allow detection of mRNA transcripts in the general range of 10 to 100 copies per cell. The PCR samples are then analyzed by 2% agarose gel electrophoresis, and the products are visualized by ethidium bromide staining and UV illumination. Gel images are captured using a digital gel-imaging system (BioImager, Genomic Solutions Inc.) and are scored for positive or negative expression based on visual detection of the expected gel band.

Measuring RGS Modulation of GIRK Channels in *Xenopus* Oocytes

Heterologous expression in CHO-K1 cells offers the advantage of a mammalian cell "background" for closer comparisons with mammalian cardiomyocytes and neurons, in addition to the high performance of whole cell patch-clamp recording. Disadvantages, however, include a relatively low throughput in obtaining high-quality data and a general lack of control over the protein expression levels produced by transient DNA transfection methods. The *Xenopus* oocyte system offers a relatively high-throughput expression system where quantitative injections of different concentrations and proportions of *in vitro*-synthesized cRNAs provide a means to examine a wide range of expression conditions that can be advantageous for mechanistic and structure–function studies (Zhang *et al.*, 2002, 2004). Disadvantages of the oocyte system include the nonmammalian "background," the reduced performance of two-electrode voltage clamping, and the slower solution exchange rates associated with the larger oocytes. *Xenopus* oocytes do express an endogenous Kir3.4 homolog (XIR) that can coassemble with mammalian Kir3.1 to form functional GIRK channels (Hedin *et al.*, 1996). We typically express both rat Kir3.1 and mouse Kir3.2a in oocytes to emulate neuronal GIRK channels, although rat Kir3.1/XIR channels may also contribute to a small portion of the GIRK current amplitude measured. Endogenous RGS expression in oocytes has not been studied, although the slower receptor-dependent GIRK kinetics in the absence of heterologous RGS expression suggests that endogenous oocyte RGS expression may be less than that found in CHO-K1 cells (Doupnik et al., 1997).

To investigate specificity in the RGS modulation of GIRK channel-gating properties, we examined the effects of two distinct RGS proteins (RGS4 and RGS7) on GIRK channels activated by muscarinic m2 receptors coupled selectively to individual $G\alpha_i$ and $G\alpha_o$ subunits (Zhang et al., 2002). To accomplish this, we individually expressed five different PTX-insensitive $G\alpha_i$ and $G\alpha_o$ mutant subunits (Wise et al., 1997) together with the catalytically active PTX-S1 subunit to block endogenous heterotrimeric $G_{i/o}$ protein coupling (Vivaudou et al., 1997). Utilizing this approach, GIRK channel-gating parameters were analyzed and compared with and without RGS coexpression for each PTX-insensitive $G\alpha_i$ and $G\alpha_o$ subunit to fully characterize GIRK–receptor coupling properties. The gating parameters analyzed were the (1) maximal agonist-evoked GIRK current amplitude, (2) receptor-independent "basal" GIRK current amplitude, (3) agonist EC_{50} for GIRK activation, (4) GIRK activation time constant (τ_{act}), and (5) GIRK deactivation time constant (τ_{deact}) (Fig. 5).

Receptor activation via specific PTX-insensitive $G\alpha$ subunits provides an additional level of control in reconstituting GPCR → GIRK channel signaling and has led to several new observations (Zhang et al., 2002). First, receptor-dependent GIRK channel-gating kinetics differ for $G\alpha_i$ versus $G\alpha_o$-coupled receptors, with $G\alpha_i$ activation being significantly slower than $G\alpha_o$. Second, RGS7 selectively modulates $G\alpha_o$-coupled receptors over $G\alpha_i$-coupled receptors and is disrupted by association with its binding partner $G\beta5$. Third, the level of $G\alpha$ expression and receptor precoupling affects the ability of RGS proteins to suppress maximal GIRK current amplitudes in conjunction with accelerated channel-gating kinetics.

Isolation of Xenopus Oocytes

Oocytes from *Xenopus laevis* (Xenopus Express, Plant City, FL) are dissociated enzymatically from ovarian tissue with 1.8% collagenase A (Boehringer Mannheim) in Ca^{2+}-free oocyte Ringer's (OR) solution during a 1.5- to 2-h digestion period at room temperature. The OR solution is composed of (in mM) 82.5 NaCl, 2.5 KCl, 1.0 $CaCl_2$, 1.0 $MgCl_2$, 1.0 $NaHPO_4$, and 5.0 HEPES, at pH 7.5 (NaOH). Stage V–VI oocytes (see Ferrell, 1999) are then selected and maintained in oocyte culture medium (OCM) in 35-mm dishes on an orbital shaker at 19°. OCM is composed of OR solution

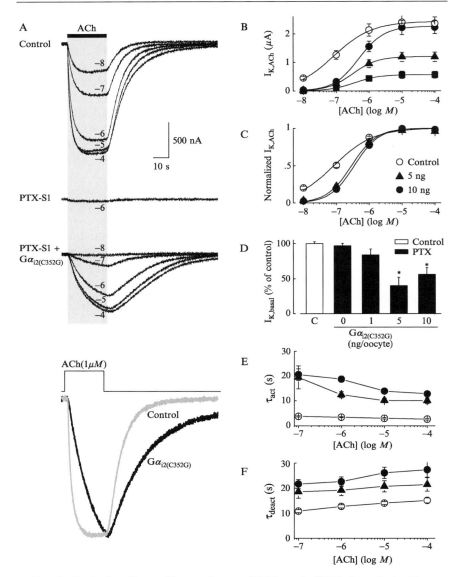

FIG. 5. Conferring Gα-specific coupling to GPCRs and GIRK channels in *Xenopus* oocytes. (A) Typical ACh-evoked GIRK currents ($I_{K,ACh}$) recorded from different oocytes from three separate experimental groups. (*Upper traces*) $I_{K,ACh}$ from a "Control" oocyte expressing muscarinic m2 receptors and Kir3.1/Kir3.2a channel subunits, utilizing endogenous $G\alpha_{i/o}$ proteins for receptor activation. (*Middle traces*) Coexpression of PTX-S1 (1 ng cRNA/oocyte) ADP-ribosylates endogenous $G\alpha_{i/o}$ proteins and decouples m2 receptor-GIRK activation. (*Lower traces*) Expression of a PTX-insensitive $G\alpha_{i2(C352G)}$ subunits (5 ng cRNA)

containing 2.5 mM sodium pyruvate and 5% heat-inactivated horse serum and is changed at least once a day during culture.

Heterologous Expression in Xenopus Oocytes

Linearized cDNA-containing vectors for rat Kir3.1, mouse Kir3.2a, human m2 receptor, PTX-S1, PTX-insensitive $G\alpha_{i/o}$ subunits, and rat RGS4 or bovine RGS7 were used to synthesize cRNA in vitro using the appropriate RNA polymerase (T7 or T3) as described by the manufacturer (mMessage mMachine, Ambion). Concentrations and quality of the cRNAs are checked by absorbance at 260 nm and formaldehyde-denatured agarose gel electrophoresis. The day after isolation, oocytes are injected with a mixture of cRNAs dissolved in diethylpyrocarbonate-treated H_2O at a final injection volume of 50 nl using a positive-displacement nanoliter injector (Nanoliter2000, World Precision Instruments). A series of up to six different cRNA mixtures are typically assembled and tested in parallel to provide internal controls for PTX inactivation efficiency and intraoocyte batch comparisons among the different experimental groups tested. In general, all oocytes are injected with cRNAs for Kir3.1 (0.5 ng), Kir3.2a (0.5 ng), and the m2 receptor (0.5 ng). The amounts of other cRNAs, including PTX-insensitive $G\alpha$ subunits and RGSs, are varied (1, 5, or 10 ng/oocyte) to examine the effects of expression level on the GIRK channel-gating kinetics. Rat or mouse PTX-insensitive $G\alpha$ cDNAs were constructed by PCR (kindly provided by Stephen Ikeda, NIH) and have a common 5' Kozak sequence followed by the $G\alpha$ coding region, which contains the C-terminal $C \rightarrow G$ mutation that renders the subunit insensitive to PTX-mediated ADP ribosylation. The 5' and 3'-untranslated regions of the $G\alpha$ cDNAs have been removed and the expression of each PTX-insensitive $G\alpha$ protein

---

with PTX-S1 rescues m2 receptor-coupled GIRK currents. All GIRK currents were elicited by a 25-s application of different concentrations of ACh as indicated. (*Bottom traces*) Superimposed $I_{K,ACh}$ elicited by 1 μM ACh from the control oocyte (gray trace) and the $G\alpha_{i2(C352G)}$-coupled oocyte (black trace). Peak amplitudes are normalized to illustrate the kinetic differences in the activation and deactivation time courses. (B) ACh dose–response relations for GIRK activation via m2 receptors coupled to oocyte $G\alpha_{i/o}$ subunits (○, control) and $G\alpha_{i2(C352G)}$ at different levels of expression (1 ■, 5 ▲, and 10 ● ng cRNA/oocyte). (C) ACh dose–response curves from B normalized to maximal $I_{K,ACh}$. (D) Comparison of receptor-independent basal GIRK currents ($I_{K,basal}$) with varying levels of $G\alpha_{i2(C352G)}$ expression. $I_{K,basal}$ is expressed as the percentage change in the "control group" mean value determined for each batch of oocytes. (E) Activation time constants (τ_{act}) and (F) deactivation time constants (τ_{deact}) for GIRK currents coupled to varying levels of $G\alpha_{i2(C352G)}$ expression and different concentrations of ACh. ○, control; ▲, 5 ng; and ●, 10 ng of $G\alpha_{i2(C352G)}$ cRNA/oocyte. Data in B–F represent the mean ± SEM from at least three batches of oocytes with the number of oocytes indicated. Asterisk denotes a $P < 0.05$. Reproduced with permission from Zhang et al. (2002).

is nearly equivalent (Zhang *et al.*, 2002). The catalytically active PTX-S1 subunit included in the cRNA mixture (1 ng cRNA/oocyte) is >95% effective at uncoupling endogenous $G\alpha_{i/o}$ proteins (Vivaudou *et al.*, 1997; Zhang *et al.*, 2002).

Electrophysiological Recordings

GIRK currents are measured by standard two-electrode voltage-clamp recording techniques at room temperature (21–23°)(Stuhmer and Parekh, 1995). Electrodes filled with 3 M KCl and having tip resistances of 0.8–1.0 $M\Omega$ are connected to a voltage clamp amplifier (GeneClamp 500, Axon Instruments) and impaled into the oocyte. Inward GIRK currents are recorded from oocytes voltage clamped at a holding potential of −80 mV, analogous to the approach described for mammalian cell recordings.

Oocytes are placed in a recording chamber perfused continuously with a minimal Ringer's solution composed of 98 mM NaCl, 1 mM MgCl$_2$, and 5 mM HEPES at pH 7.5 (NaOH). After electrode impalement and clamping the membrane potential to −80 mV, the perfusion solution is changed to a high K$^+$ solution composed of 20 mM KCl, 78 mM NaCl, 1 mM MgCl$_2$, and 5 mM HEPES at pH7.5 (NaOH). The resulting increase in inward current represents a "basal" K$^+$ current ($I_{K,basal}$) that is due primarily to receptor-independent GIRK channel activity (Dascal *et al.*, 1993). Rapid agonist application and washout are performed similar to that described for the mammalian cell recording system, where the SF-77B Fast-step system (Warner Instruments) is used to switch the position of two perfusion barrels positioned next to the oocyte. For oocyte superfusion, two PE-160 tubes (1.14 mm inside diameter) are secured to each other and positioned next to the oocyte with a micromanipulator. The perfusion barrels contain a high K$^+$ solution (barrel A) and a high K$^+$ solution plus receptor agonist (barrel B). For barrel B, a range of agonist concentrations can be tested for each oocyte via a manifold that connects multiple 60-ml syringe reservoirs containing different agonist concentrations. Flow through the perfusion barrels is gravity driven, and the time constant for solution exchange is <1 s (2-mm barrel movement) as determined by the time course change in receptor-independent GIRK current with switching between 20 and 40 mM external K$^+$.

Axon Instruments software (pCLAMP 8) is used to acquire data, trigger agonist applications via SF-77B barrel movements, and analyze the time-dependent GIRK current kinetics (τ_{act} and τ_{deact}). On a practical note, we find the activation and deactivation kinetics of GIRK currents in oocytes to sometimes have multiple kinetic components that are not always reproduced within a particular expression group, and therefore may reflect nonuniform solution exchanges or other experimental artifacts. This can complicate the data analysis of the current traces, so we either exclude these responses or examine the single dominant kinetic component that is observed reproducibly within a given experimental group.

Future Applications

The methodological approaches outlined here have several potential applications to further investigate the actions of RGS proteins on G-protein signaling. Heterologous expression of cloned genes in CHO-K1 cells and *Xenopus* oocytes provides an excellent system for structure–function-based studies, with the GIRK channel effector providing unparalleled temporal resolution for kinetic analysis of early signaling events. Currently we are utilizing these methods to further study the structural determinants of $G\alpha$ subunits (Doupnik *et al.*, 2004; Zhang *et al.*, 2004) and RGS proteins (Jaén and Doupnik, 2002) in GPCR → GIRK channel-signaling kinetics. Investigations of RGS effects on GPCR modulation of other heterologously expressed ion channels have also been explored by other investigators (Melliti *et al.*, 1999).

Incorporating methods of RGS knockout (Chen *et al.*, 2000) and endogenous RGS knockdown via RGS-specific antibodies (Diverse-Pierluissi *et al.*, 1999), antisense nucleic acids, (Garzon *et al.*, 2001), RGS-specific ribozymes (Wang *et al.*, 2002), or RNA interference (Dykxhoorn *et al.*, 2003) will ultimately address questions of RGS specificity in the GPCR activation of cardiac and neuronal GIRK channels. Given the large number of RGS proteins expressed within single CG neurons and cardiomyocytes and the potential for RGS redundancy in GPCR signaling, coupling via RGS-resistant $G\alpha_{i/o}$ mutant subunits (Lan *et al.*, 1998) will also aid in establishing the degree of RGS influence on the native signaling kinetics as demonstrated by Jeong and Ikeda (2000, 2001).

Finally, our single cell RT-PCR sampling method to profile RGS expression in atrial myocytes and CG neurons demonstrates the feasibility of this approach for profiling expression patterns of similar size gene families (i.e., $G\alpha$, $G\beta$, and $G\gamma$ gene families) in a small laboratory setting. Expression profiling with cDNA microarray chips would expand the gene expression profile, but this approach typically requires RNA pooled from a large number of cells that are typically heterogeneous in origin. However, advances incorporating T7 RNA amplification may make single cell gene profiling with cDNA microarrays more feasible (Kamme *et al.*, 2003).

Acknowledgment

This work was made possible through the generous support of the American Heart Association–Florida and Puerto Rico Affiliate.

References

Berman, D. M., Wilkie, T. M., and Gilman, A. G. (1996). GAIP and RGS4 are GTPase-activating proteins for the Gi subfamily of G protein alpha subunits. *Cell* **86**, 445–452.

Boutet-Robinet, E. A., Finana, F., Wurch, T., Pauwels, P. J., and De Vries, L. (2003). Endogenous RGS proteins facilitate dopamine D(2S) receptor coupling to Gαo proteins and Ca^{2+} responses in CHO-K1 cells. *FEBS Lett.* **533**, 67–71.

Breitwieser, G. E., and Szabo, G. (1988). Mechanism of muscarinic receptor-induced K^+ channel activation as revealed by hydrolysis-resistant GTP analogues. *J. Gen. Physiol.* **91**, 469–494.

Bunemann, M., Brandts, B., and Pott, L. (1996). Downregulation of muscarinic M2 receptors linked to K^+ current in cultured guinea-pig atrial myocytes. *J. Physiol.* **494**, 351–362.

Chen, C. K., Burns, M. E., He, W., Wensel, T. G., Baylor, D. A., and Simon, M. I. (2000). Slowed recovery of rod photoresponse in mice lacking the GTPase accelerating protein RGS9-1. *Nature* **403**, 557–560.

Chen, C.-K., Wieland, T., and Simon, M. I. (1996). RGS-r, a retinal specific RGS protein, binds an intermediate conformation of transducin and enhances recycling. *Proc. Natl. Acad. Sci. USA* **93**, 12885–12889.

Dascal, N., Schreibmayer, W., Lim, N. F., Wang, W., Chavkin, C., DiMagno, L., Labarca, C., Kieffer, B. L., Gaveriaux-Ruff, C., Trollinger, D., Lester, H. A., and Davidson, N. (1993). Atrial G protein-activated K^+ channel: Expression cloning and molecular properties. *Proc. Natl. Acad. Sci. USA* **90**, 10235–10239.

Diverse-Pierluissi, M. A., Fischer, T., Jordan, J. D., Schiff, M., Ortiz, D. F., Farquhar, M. G., and De Vries, L. (1999). Regulators of G protein signaling proteins as determinants of the rate of desensitization of presynaptic calcium channels. *J. Biol. Chem.* **274**, 14490–14494.

Doupnik, C. A., Davidson, N., and Lester, H. A. (1995a). The inward rectifier potassium channel family. *Curr. Opin. Neurobiol.* **5**, 268–277.

Doupnik, C. A., Davidson, N., Lester, H. A., and Kofuji, P. (1997). RGS proteins reconstitute the rapid gating kinetics of Gβγ activated inwardly rectifying K^+ channels. *Proc. Natl. Acad. Sci. USA* **94**, 10461–10466.

Doupnik, C. A., Dessauer, C. W., Slepak, V. Z., Gilman, A. G., Davidson, N., and Lester, H. A. (1996). Time resolved kinetics of direct $G_{\beta 1\gamma 2}$ interactions with the carboxyl terminus of Kir3.4 inward rectifier K^+ channel subunits. *Neuropharmacology* **35**, 923–931.

Doupnik, C. A., Lim, N. F., Kofuji, P., Davidson, N., and Lester, H. A. (1995b). Intrinsic gating properties of a cloned G protein-activated inward rectifier K^+ channel. *J. Gen. Physiol.* **106**, 1–23.

Doupnik, C. A., Xu, T., and Shinaman, J. M. (2001). Profile of RGS expression in single rat atrial myocytes. *Biochim. Biophys. Acta* **1522**, 97–107.

Doupnik, C. A., Zhang, Q., and Dickson, A. (2004). Gα-specific interdomain interactions effecting GDP release and GIRK activation kinetics. *Biophys. J. Abstr.* **86**, 438a.

Dykxhoorn, D. M., Novina, C. D., and Sharp, P. A. (2003). Killing the messenger: Short RNAs that silence gene expression. *Nature Rev. Mol. Cell. Biol.* **4**, 457–467.

Ehrengruber, M. U., Lanzrein, M., Xu, Y., Jasek, M. C., Kantor, D. B., Schuman, E. M., Lester, H. A., and Davidson, N. (1998). Recombinant adenovirus-mediated expression in nervous system of genes coding for ion channels and other molecules involved in synaptic function. *Methods Enzymol.* **293**, 483–503.

Ferrell, J. E. (1999). *Xenopus* oocyte maturation: New lessons from a good egg. *Bioessays* **21**, 833–842.

Garzon, J., Rodriguez-Diaz, M., Lopez-Fando, A., and Sanchez-Blazquez, P. (2001). RGS9 proteins facilitate acute tolerance to mu-opioid effects. *Eur. J. Neurosci.* **13**, 801–811.

Hamill, O. P., Marty, A., Neher, E., Sakmann, B., and Sigworth, F. J. (1981). Improved patch-clamp techniques for high resolution current recording from cells and cell-free membrane patches. *Pflüg. Arch.* **391,** 85–100.

Hedin, K. E., Lim, N. F., and Clapham, D. E. (1996). Cloning of a *Xenopus laevis* inwardly rectifying K$^+$ channel subunit that permits GIRK1 expression of I$_{KACh}$ currents in oocytes. *Neuron* **16,** 423–429.

Hunt, T. W., Fields, T. A., Casey, P. J., and Peralta, E. G. (1996). RGS10 is a selective activator of Gα_i GTPase activity. *Nature* **383,** 175–177.

Inomata, N., Ishihara, T., and Akaike, N. (1989). Activation kinetics of the acetylcholine-gated potassium current in isolated atrial cells. *Am. J. Physiol.* **257,** C646–C650.

Jaén, C., and Doupnik, C. A. (2002). Effects of a novel short isoform of RGS3 on GIRK currents evoked by serotonin 1A and muscarinic m2 receptors. Program No. 438.11. Abstract Viewer/Itinerary Planner. Society for Neuroscience, Washington, DC.

Jeong, S. W., and Ikeda, S. R. (2000). Endogenous regulator of G-protein signaling proteins modify N-type calcium channel modulation in rat sympathetic neurons. *J. Neurosci.* **20,** 4489–4496.

Jeong, S. W., and Ikeda, S. R. (2001). Differential regulation of G protein-gated inwardly rectifying K(+) channel kinetics by distinct domains of RGS8. *J. Physiol.* **535,** 335–347.

Kamme, F., Salunga, R., Yu, J., Tran, D. T., Zhu, J., Luo, L., Bittner, A., Guo, H. Q., Miller, N., Wan, J., and Erlander, M. (2003). Single-cell microarray analysis in hippocampus CA1: Demonstration and validation of cellular heterogeneity. *J. Neurosci.* **23,** 3607–3615.

Karschin, A., Ho, B., Labarca, C., Elroy-Stein, O., Moss, B., Davidson, N., and Lester, H. A. (1991). Heterologously expressed serotonin 1A receptors couple to muscarinic K$^+$ channels in heart. *Proc. Natl. Acad. Sci. USA* **88,** 5694–5698.

Kofuji, P., Hofer, M., Millen, K. J., Millonig, J. H., Davidson, N., Lester, H. A., and Hatten, M. E. (1996). Functional analysis of the weaver mutant GIRK2 K$^+$ channel and rescue of weaver granule cells. *Neuron* **16,** 941–952.

Lan, K. L., Sarvazyan, N. A., Taussig, R., Mackenzie, R. G., DiBello, P. R., Dohlman, H. G., and Neubig, R. R. (1998). A point mutation in Gαo and Gαi1 blocks interaction with regulator of G protein signaling proteins. *J. Biol. Chem.* **273,** 12794–12797.

Leaney, J. L., and Tinker, A. (2000). The role of members of the pertussis toxin-sensitive family of G proteins in coupling receptors to the activation of the G protein-gated inwardly rectifying potassium channel. *Proc. Natl. Acad. Sci. USA* **97,** 5651–5656.

Logothetis, D. E., Kurachi, Y., Golper, J., Neer, E. J., and Clapham, D. (1987). The $\beta\gamma$-subunit of GTP-binding proteins activate the muscarinic K$^+$ channel in heart. *Nature* **325,** 321–326.

Melliti, K., Meza, U., Fisher, R., and Adams, B. (1999). Regulators of G protein signaling attenuate the G protein-mediated inhibition of N-type Ca channels. *J. Gen. Physiol.* **113,** 97–110.

Noma, A., and Trautwein, W. (1978). Relaxation of the ACh-induced potassium current in the rabbit sinoatrial node cell. *Pflüg. Arch.* **377,** 193–200.

Patil, N., Cox, D. R., Bhat, D., Faham, M., Myers, R. M., and Peterson, A. S. (1995). A potassium channel mutation in weaver mice implicates membrane excitability in granule cell differentiation. *Nature Genet.* **11,** 126–129.

Ross, E. M., and Wilkie, T. M. (2000). GTPase-activating proteins for heterotrimeric G proteins: Regulators of G protein signaling (RGS) and RGS-like proteins. *Annu. Rev. Biochem.* **69,** 795–827.

Saitoh, O., Kubo, Y., Miyatani, Y., Asano, T., and Nakata, H. (1997). RGS8 accelerates G-protein-mediated modulation of K$^+$ currents. *Nature* **390,** 525–529.

Sierra, D. A., Gilbert, D. J., Householder, D., Grishin, N. V., Yu, K., Ukidwe, P., Barker, S. A., He, W., Wensel, T. G., Otero, G., Brown, G., Copeland, N. G., Jenkins, N. A., and Wilkie, T. M. (2002). Evolution of the regulators of G-protein signaling multigene family in mouse and human. *Genomics* **79,** 177–185.

Slesinger, P. A., and Lansman, J. B. (1991). Inactivation of calcium currents in granule cells cultured from mouse cerebellum. *J. Physiol.* **435,** 101–121.

Slesinger, P. A., Patil, N., Liso, Y. J., Jan, Y. N., Jan, L. Y., and Cox, D. R. (1996). Functional effects of the mouse weaver mutation on G protein-gated inwardly rectifying K^+ channels. *Neuron* **16,** 321–331.

Sodickson, D. L., and Bean, B. P. (1996). $GABA_B$ receptor-activated inwardly rectifying potassium current in dissociated hippocampal CA3 neurons. *J. Neurosci.* **16,** 6374–6385.

Stanfield, P. R., Nakajima, S., and Nakajima, Y. (2002). Constitutively active and G-protein coupled inward rectifier K^+ channels: Kir2.0 and Kir3.0. *Rev. Physiol. Biochem. Pharmacol.* **145,** 47–179.

Stuhmer, W., and Parekh, A. B. (1995). Electrophysiological recordings from *Xenopus* oocytes. *In* Single-Channel Recording (B. Sakmann and E. Neher, eds.), pp. 341–356. Plenum Press, New York.

Surmeier, D. J., Mermelstein, P. G., and Goldowitz, D. (1996). The weaver mutation of GIRK2 results in a loss of inwardly rectifying K+ current in cerebellar granule cells. *Proc. Natl. Acad. Sci. USA* **93,** 11191–11195.

Szabo, G., and Otero, A. S. (1990). G protein mediated regulation of K^+ channels in heart. *Annu. Rev. Physiol.* **52,** 293–305.

Takesono, A., Zahner, J., Blumer, K. J., Nagao, T., and Kurose, H. (1999). Negative regulation of $\alpha 2$-adrenergic receptor-mediated Gi signalling by a novel pathway. *Biochem. J.* **343**(Pt. 1), 77–85.

Vivaudou, M., Chan, K. W., Sui, J. L., Jan, L. Y., Reuveny, E., and Logothetis, D. E. (1997). Probing the G-protein regulation of GIRK1 and GIRK4, the two subunits of the KACh channel, using functional homomeric mutants. *J. Biol. Chem.* **272,** 31553–31560.

Wang, Q., Liu, M., Mullah, B., Siderovski, D. P., and Neubig, R. R. (2002). Receptor-selective effects of endogenous RGS3 and RGS5 to regulate MAP kinase activation in rat vascular smooth muscle cells. *J. Biol. Chem.* **277,** 24949–24958.

Watson, N., Linder, M. E., Druey, K. M., Kehrl, J. H., and Blumer, K. J. (1996). RGS family members: GTPase-activating proteins for heterotrimeric G-protein α–subunits. *Nature* **383,** 172–177.

Wise, A., Watson-Koken, M. A., Rees, S., Lee, M., and Milligan, G. (1997). Interactions of the $\alpha 2A$-adrenoceptor with multiple Gi-family G-proteins: Studies with pertussis toxin-resistant G-protein mutants. *Biochem. J.* **321,** 721–728.

Yamada, M., Inanobe, A., and Kurachi, Y. (1998). G protein regulation of potassium ion channels. *Pharmacol. Rev.* **50,** 723–760.

Zhang, Q., Pacheco, M. A., and Doupnik, C. A. (2002). Gating properties of GIRK channels activated by $G\alpha o$ and $G\alpha i$-coupled muscarinic m2 receptors in *Xenopus* oocytes: The role of receptor precoupling in RGS modulation. *J. Physiol.* **545**(2), 355–373.

Zhang, Q., Dickson, A., and Doupnik, C. A. (2004). $G\beta\gamma$-activated inwardly rectifying K^+ (GIRK) channel activation kinetics via $G\alpha i$ and $G\alpha o$-coupled receptors are determined by $G\alpha$-specific interdomain interactions that affect GDP release rates. *J. Biol. Chem.* **279,** 29787–29796.

[10] Assays for G-Protein-Coupled Receptor Signaling Using RGS-Insensitive Gα Subunits

By MARY J. CLARK and JOHN R. TRAYNOR

Abstract

Regulators of G-protein signaling (RGS) proteins, by their action on $G\alpha_{i/o}$ proteins, may enhance receptor–effector signaling by physical or kinetic scaffolding mechanisms. However, more than 30 mammalian proteins with RGS activity have been identified so it is difficult to determine which RGS protein is most relevant to a particular receptor system and in any particular cell. To avoid this problem, one approach is to examine agonist-stimulated second messenger signaling in cells expressing Gα proteins that are insensitive to the GTPase accelerating property of all RGS proteins. This article describes protocols for the preparation and analysis of C6 rat glioma cells stably expressing RGS- and pertussis toxin-insensitive Gα subunits; pertussis toxin treatment uncouples endogenous $G\alpha_{i/o}$ proteins and allows for the determination of the expressed RGS-insensitive Gα activity. Methods to determine signaling at the level of adenylyl cyclase, the extracellular signal-regulated kinase (ERK1/2) mitogen-activated protein kinase pathway, and intracellular Ca^{2+} levels are described. As a typical G-protein-coupled receptor, we have used the μ-opioid receptor expressed in C6 cells together with RGS-insensitive $G\alpha_o$. In these cells, agonist inhibition of adenylyl cyclase and stimulation of ERK1/2 phosphorylation were enhanced markedly. In contrast, increases in intracellular calcium were less affected. The altered signaling in cells expressing RGS-insensitive $G\alpha_o$ subunits allows for determination of the role of endogenous RGS proteins to limit and/or direct signaling.

Introduction

Agonist occupation of G-protein-coupled receptors (GPCRs) leads to activation of heterotrimeric G proteins by exchange of GDP for GTP on Gα and dissociation of the Gα-GTP and Gβγ subunits. The dissociated Gα-GTP and Gβγ subunits interact with a wide variety of downstream signaling proteins, including adenylyl cyclase, voltage-operated calcium channels, inwardly rectifying K^+ channels, intracellular calcium stores, and the extracellular signal-regulated (ERK) mitogen-activated protein kinase

(MAP kinase) pathway. Intrinsic GTPase activity of $G\alpha$ causes $G\alpha$-bound GTP to be hydrolyzed to GDP, which allows for reassociation of the inactive $G\alpha$-GDP with $G\beta\gamma$ subunits, thereby terminating signaling through both $G\alpha$ and $G\beta\gamma$. G-protein signaling is controlled negatively by a family of more than 30 proteins known as regulators of G-protein signaling (RGS) proteins (Hollinger and Hepler, 2002; Neubig and Siderovski, 2002). These proteins act as GTPase-accelerating proteins (GAPs) for $G\alpha$ and speed up hydrolysis of the $G\alpha$-bound GTP, thus directly reducing the steady-state levels of $G\alpha$-GTP and indirectly reducing the levels of $G\beta\gamma$ and so inhibiting signaling.

The importance of RGS proteins in the dynamic control of signaling is supported by changes in mRNA for these proteins under a variety of conditions (reviewed in Neubig and Siderovski, 2002). However, the physiological consequences of RGS protein activity are not well defined. Different RGS proteins interact with varying preference with members of the $G_{i/o}$ and/or G_q families of G proteins to reduce signaling (Neubig and Siderovski, 2002). In addition to the RGS domain, RGS proteins have a variety of domains for protein–protein interaction (Hollinger and Hepler, 2002; Neubig and Siderovski, 2002) and so selectivity for activation of particular pathways may be obtained by scaffolding mechanisms linking RGS proteins to particular $G\alpha$ proteins and/or receptor proteins and signaling pathways. Given the complexity of GPCR signaling, an attractive hypothesis is that RGS proteins regulate pathways differentially and have a controlling influence on signaling, which would be cell specific and depend on the type of receptor, signaling pathways, G proteins, and RGS proteins expressed.

The RGS family of proteins is defined by the RGS domain or RGS box. Although highly conserved, it is possible that this region could provide control of signaling by a kinetic scaffolding mechanism (Zhong *et al.*, 2003). Under such a mechanism, the GAP activity of RGS proteins stimulates the rapid hydrolysis of $G\alpha$-bound GTP to permit rapid recoupling of receptor and G protein. This maintains active $G\alpha$-GTP and $G\beta\gamma$ proteins close to the receptor so they can interact with nearby signaling proteins. Thus RGS protein action maintains receptor-mediated G-protein signaling even though it deactivates $G\alpha$ protein. This explains how, in some cases, RGS proteins can speed the kinetics of responses without compromising steady-state signaling strength (e.g., Doupnik *et al.*, 2004) and so may increase the efficiency of signaling. In contrast, spillover of $G\alpha$-GTP and $G\beta\gamma$ to more distant effectors is prevented by the GAP activity of an RGS protein and so signaling selectivity is obtained.

RGS domain-containing proteins have a wide variety of non-RGS domains that, when the RGS protein binds to $G\alpha$, can act as a physical scaffold to bring together several different proteins involved in signaling cascades and so provide receptor or pathway selectivity. For example, in rat vascular smooth muscle cells, RGS3 is a negative modulator of muscarinic m_3 receptor signaling, whereas RGS5 is a negative modulator of angiotensin AT1a receptor signaling through $G_{q/11}$ (Wang et al., 2002). In some cases, the selectivity of interaction of an RGS protein has been suggested to involve direct interaction with receptors. RGS12 binds to the carboxy terminus of the interleukin-8 receptor (Snow et al., 1998) and RGS-mediated inhibition of Ca^{2+} signaling in rat pancreatic acinar cells is selective for muscarinic receptors, possibly due to interaction of the N-terminal domain of RGS4 with the receptors (Zeng et al., 1998).

The presence of many different subtypes of G proteins, RGS proteins, and the number of signaling pathways that can be activated makes the question of whether RGS proteins contribute to intracellular signaling strength and specificity a difficult one to address. As more than 30 mammalian proteins with RGS activity have been identified to date (Hollinger and Hepler, 2002; Neubig and Siderovski, 2002), the choice of which RGS protein is most relevant to a particular receptor system and in any particular cell is extremely difficult. Moreover, because of the variety of Gα and RGS proteins, there may be redundancy, such that several Gα and RGS protein pairs may interact to perform the same function. To avoid this problem, our approach has been to examine agonist-stimulated second messenger responses in cells expressing Gα proteins that are insensitive to the GAP activity of all RGS proteins. This article describes protocols for the preparation of cells expressing RGS-insensitive Gα subunits, the determination of Gα activity (guanine nucleotide binding and hydrolysis), and methods to determine signaling through several typical pathways (adenylyl cyclase, MAP kinase, and intracellular Ca^{2+} levels). As a typical GPCR, we have used the μ-opioid receptor expressed in C6 rat glioma cells together with RGS-insensitive $G\alpha_o$ (Clark et al., 2003).

Overview of Methods and Rationale

The interaction of RGS proteins with Gα subunits occurs between the conformationally flexible "switch" regions of the Gα and components of the RGS consensus domain of the RGS proteins. Within the "switch 1" region, Gly[183] in $G\alpha_{i1}$ (Gly[184] in $G\alpha_o$) is a major contributor to the contact surface area between the Gα subunit and the RGS protein and there is direct contact of this Gly with Ser[85] and Tyr[84] of RGS4 (Tesmer et al., 1997). Consequently, mutation of this Gly to Ser, thereby introducing a hydroxymethyl side chain, interferes with these interactions and also disrupts adjacent amino acids from their interaction with the RGS domain.

Therefore, $G\alpha_o$ (G184S) does not interact with RGS proteins and so renders the $G\alpha_o$ insensitive to the GAP activity of all members of the RGS family (RGS-i) without affecting intrinsic GTPase activity or the kinetics of GDP release (Lan et al., 1998). Importantly, this Gly is highly conserved in the $G\alpha_{i/o}$ family and the $G\alpha_{i/o}$–RGS interaction interface is within the conserved RGS domain such that all members of the $G\alpha_{i/o}$ family are amenable to this methodology.

The interaction of adenylyl cyclase inhibitory $G\alpha$ proteins (other than $G\alpha_z$) with GPCRs is prevented by treatment with pertussis toxin (PTX). This causes ADP-ribosylation of a Cys residue in the C terminus of the $G\alpha$ subunit. In order to study the activity of RGS-i $G\alpha_{i/o}$ without interference from endogenous $G\alpha$ proteins, $G\alpha_{i/o}$ is also mutated at this C-terminal Cys (C351) to Gly, which then confers insensitivity to PTX (PTX-i) (Milligan, 1998). The RGS/PTX-i mutant $G\alpha_{i/o}$ is expressed in cells that are then treated with PTX to prevent receptor coupling through endogenously expressed PTX-sensitive G proteins. Because the degree of agonist-induced response is controlled by the level of receptor expression and the concentration of interacting $G\alpha$ (Kenakin, 2002), it is essential that levels of RGS-i $G\alpha$ and receptor are monitored closely. Consequently, we have compared stable clones that express similar concentrations of the $G\alpha$ under study and have similar receptor expression, such that effects on signaling pathways are easier to measure and differences easier to define.

Here we describe the use of RGS-i $G\alpha_o$ to investigate signaling through the μ-opioid receptor, a typical GPCR that signals through $G_{i/o}$ heterotrimers to a multitude of second messengers and cellular effectors, including adenylyl cyclase (Law et al., 2000), intracellular calcium stores (Smart and Lambert, 1996), and the extracellular signal-regulated (ERK) mitogen-activated protein kinase (MAP) pathway (Gutstein et al., 1997). The presence of the G184S mutation prevents the ability of endogenous RGS proteins expressed in these cells to stimulate GTPase activity; this results in an increase in signaling. However, this increase appears to show some selectivity, as not all signaling pathways are altered to the same extent. For these studies, we use C6 rat glioma cells because these cells do not express $G\alpha_o$ endogenously (Charpentier et al., 1993) and so the presence of this protein can be demonstrated readily by Western blotting. Because the μ-opioid receptor is expressed in C6 cells using a geneticin-resistant vector, we used zeocin resistance to select for the mutant $G\alpha_o$. Note that in many of the studies, we use the opioid agonist morphine to stimulate second messengers. This drug is controlled and, in the United States, a DEA license is needed to obtain the compound. However, many peptides that act at the μ-opioid receptor (e.g., [D-Ala[2], MePhe[4], Gly-ol[5]]enkephalin; DAMGO) are available without restriction.

Generation and Characterization of Stable PTX-i and
  RGS/PTX-i Gα Expressing Cells

PTX-i and RGS/PTX-i Gα Mutants

Our initial studies used mouse PTX-i Gα$_o$ in the pCI vector, a gift from Steve Ikeda (NIH/NIAAA, Bethesda, MD). Human Gα proteins containing a PTX-insensitive mutation (C351G) in the pcDNA3.1 expression vector with CMV promoter and ampicillin resistance (Invitrogen Life Technologies, Carlsbad, CA) are now available commercially from the Guthrie cDNA Resource Center (Sayre, PA). The RGS-i mutation G184S is introduced into human PTX-i Gα$_o$ using the QuikChange site-directed mutagenesis kit (Stratagene, La Jolla, CA) according to the manufacturer's instructions with the primer 5′ GGGTCAAAACCACT*TCC*ATCGTAG AAAC 3′, followed by DNA sequencing to verify the mutation. Similarly, the mutation can be introduced into human Gα$_i$ proteins using the following primers:

  hGαi_1: 5′GAGTGAAAACTACA*TCA*ATTGTTGAAACCC 3′
  hGαi_2: 5′ GTAAAGACCACG*TCG*ATCGTGGAG 3′
  hGαi_3: 5′CGAGAGTGAAGACCACA*TCC*ATTGTAGAAAC 3′

The Gα$_o$ protein cDNA is inserted into the Zeocin-resistant vector, pcDNA3.1zeo-(Invitrogen). Excise PTX-i or RGS/PTX-i mouse Gα$_o$ DNA from the plasmid vector (pCI Vector) by a 2-h incubation at 37° with *Not*I and *Nhe*I restriction enzymes (Promega, Madison, WI); alternately, cut PTX-i or RGS/PTX-i human Gα$_o$ DNA from the plasmid vector (pcDNA3.1) with *Nhe*I and *Xba*I restriction enzymes (Promega). Both are done in MultiCore buffer as per the manufacturer's instructions. The pcDNA3.1zeo- recipient vector is cut with the same restriction enzymes. Separate the DNA (75 V for 40–60 min) on a 1% agarose gel prepared in 0.4 *M* Tris–Acetate buffer, pH 8.0, with 1 m*M* EDTA and 0.5 μg/ml ethidium bromide (Boehringer-Mannheim, Indianapolis, IN) for visualization and compare with 1 kb plus DNA ladder (Invitrogen). Remove the 1.1-kb Gα$_o$ DNA and the ~5-kb pcDNA3.1zeo- cut vector from the gel with a razor blade and transfer to separate sterile microcentrifuge tubes. Extract the DNA from the gels using the QIAquick gel extraction kit (Qiagen, Inc., Valencia, CA) following the instructions provided by the manufacturer. Ligate the vector and insert with T4 DNA ligase (Invitrogen) with at least a 3:1 insert-to-vector molar ratio, following the manufacturer's instructions. Transform JM109 supercompetent *Escherichia coli* bacteria (Stratagene, La Jolla, CA) with the ligation product and grow overnight on LB agar (Invitrogen) plates with 0.1 mg/ml ampicillin (Sigma Chemical Co., St. Louis, MO) to select for plasmid vector-containing cells. Select isolated colonies and grow in 5 ml LB broth (Invitrogen) with ampicillin overnight with shaking. Isolate DNA from the bacterial cells with a QIAprep spin miniprep kit (Qiagen Inc.). The presence of the $G\alpha_o$ insert in the plasmid vector may be confirmed by cutting with appropriate restriction enzymes and running on a 1% agarose gel with ethidium bromide.

Cell Culture and Transfection

C6 glioma cells stably expressing the rat μ-opioid receptor (C6μ) (obtained from Huda Akil, University of Michigan; Lee *et al.*, 1999) are maintained in Dulbecco's modified Eagles medium (DMEM with high glucose, L-glutamine, and pyridoxine HCl, without sodium pyruvate; Invitrogen) containing 10% fetal bovine serum (FBS, Invitrogen) under 5% CO_2 in the presence of 0.25 mg/ml geneticin (G418; Invitrogen, as a 50-mg/ml solution) to maintain expression of the μ-opioid receptor. One day before transfection, subculture cells into the same medium so as to reach 60–80% confluency at transfection. Transfect plasmid DNA (e.g., $G\alpha_o$ in pc-DNA3.1zeo– vector) into the cells using Lipofectamine plus reagent (Invitrogen) at a 1:4:5 $\mu g/\mu l/\mu l$ DNA/lipofectamine/plus reagent ratio. After 48 h, subculture the cells to varying densities and grow in the aforementioned medium with the addition of 0.4 mg/ml zeocin (Invitrogen, as a 100-mg/ml solution). Isolate colonies using cloning cylinders, transfer to 24-well plates, and grow up sufficient cells in the aforementioned medium to screen for expression of $G\alpha_o$ by Western blotting (see later). Maintain chosen clones under the same conditions and typically subculture at a ratio of 1:20 to 1:30 with partial replacement of the medium on day 4 and the day before subculturing or harvesting at day 7.

Membrane Preparation

Treat confluent C6μ cells expressing the PTX-i or RGS/PTX-i $G\alpha_o$ with 100 ng/ml PTX (List Biological Laboratories, Inc., Campbell, CA) overnight. Wash cells two times with ice-cold phosphate-buffered saline (0.9% NaCl, 0.61 mM Na_2HPO_4, 0.38 mM KH_2PO_4, pH 7.4) and then detach from plates by incubation in harvesting buffer (20 mM HEPES, pH 7.4, 150 mM NaCl, 0.68 mM EDTA) at room temperature and pellet by centrifugation in 50-ml conical sterile polypropylene tubes at 200 g for 3 min. Suspend the cell pellet in ice-cold 50 mM Tris–HCl buffer, pH 7.4, and homogenize with a Tissue Tearor (7-mm probe diameter; Biospec Products, Bartlesville, OK) for 20 s at setting 4. Centrifuge this homogenate at 20,000g for 20 min at 4° and then resuspend the pellet in 50 mM Tris–HCl, pH 7.4, using the Tissue Tearor for 10 s at setting 2, followed by recentrifugation as described earlier. Finally, dilute the cell pellet in 50 mM

Tris–HCl buffer, pH 7.4, to approximately 1.0 mg protein/ml and freeze in aliquots at $-80°$. To determine protein concentration, dissolve membrane samples with 1 N NaOH for 30 min at room temperature, neutralize with 1 M acetic acid, and assay by the method of Bradford (1976) using a bovine serum albumin (BSA) standard.

Determination of Gα Expression

It is essential when comparing cells that levels of the Gα protein under investigation are the same. Western blot analysis is used to compare expression levels. For example, for Gα$_o$: Dilute membrane proteins (20 μg), 10–40 ng His$_6$-Gα$_o$ standards (obtained from Richard Neubig, University of Michigan; prepared as described by Lan et al., 1998), or molecular mass markers (23–132 kDa; Santa Cruz Biotechnology, Inc., Santa Cruz, CA) with sample buffer (62.5 mM Tris–HCl, pH 6.8, 10% glycerol, 2% SDS, 5% 2-mercaptoethanol, 0.05% bromophenol blue), boil for 3 min, and apply to a 12% SDS–PAGE gel (Protogel, National Diagnostics, Atlanta, GA) for 30 min at 70 V, followed by 150 V for 60 min or until the dye front has run off the gel using a Mini-Protean II electrophoresis cell (Bio-Rad, Hercules, CA). Equilibrate the gel, nitrocellulose membrane (45 μm, Osmonics, Minnetonka, MN), Whatman filter papers, and sponges in transfer buffer [25 mM Tris base, pH 8.3 (do not adjust), 192 mM glycine, 1% SDS, 20% methanol] for 20 min before transfer at 100 V for 60 min at room temperature (Mini Trans-Blot electrophoretic transfer cell, Bio-Rad, Hercules, CA). Block membrane in 5% BSA in TBS (20 mM Tris–HCl, pH 7.4, 500 mM NaCl) overnight at 4° and then probe with a 1:200 dilution (in TBS with 0.1% Tween 20 and 5% BSA) of anti-Gα$_o$ antibody (K-20, Santa Cruz) for 60 min at room temperature. Wash membrane three times for 5 min in TBS–Tween and then treat with a 1:12,000 dilution (in TBS with 0.1% Tween 20 and 3% BSA) of goat anti-rabbit IgG-HRP (Santa Cruz) for 60 min at room temperature. Wash membrane three times for 5 min each with TBS–Tween followed by one wash for 5 min in TBS. Visualize bands by enhanced chemiluminescence using a 5-min treatment with SuperSignal West Pico chemiluminescent substrate from Pierce (Rockford, IL) followed by immediately exposing the membrane to Kodak BioMax MR film or recording and quantifying with a Kodak Image Station 440.

Receptor Binding Assay

Since receptor expression level can influence agonist activity, it is necessary to confirm that receptor numbers are the same in clones chosen for further study. Incubate membranes (10–20 μg protein) in 5 ml polypropylene tubes with 50 mM Tris–HCl, pH 7.4, and 0.2–28 nM of the μ-receptor ligand [^{3}H]DAMGO (~42 Ci/mmol; Perkin Elmer Life Sciences, Boston, MA) with or without the opioid antagonist naloxone (50 μM) in a total volume of 0.2 ml for 60 min in a shaking water bath at 25°. Filter the tube contents through glass fiber filters (Schleicher and Schuell No. 32, Keene, NH) mounted in a Brandel cell harvester (Gaithersburg, MD) and quickly rinse three times with 3 ml ice-cold 50 mM Tris–HCl, pH 7.4. Place each filter in a scintillation vial, add 4 ml EcoLume scintillation cocktail (ICN, Aurora, OH), allow to stand for at least 6 h so that radioactivity is distributed evenly in the cocktail, and then count in a liquid scintillation counter. Obtain specific binding by subtraction of radioactive counts in the presence of naloxone from total binding. Radioactive counts are converted to picomoles bound per milligram protein and fit to a one-site binding hyperbola using GraphPad Prism (GraphPad, San Diego, CA) to determine B_{max} and affinity K_D of the radioligand. Naloxone and all other chemicals used in this assay are available from Sigma.

Assays for Gα Protein Activity

[^{35}S]GTPγS Binding Assay

This assay measures agonist-stimulated G-protein activation by the binding of [^{35}S]GTPγS, a nonhydrolyzable analog of GTP, which displaces GDP on Gα on receptor activation. Cellular membranes, prepared as described earlier, from PTX-treated cells expressing PTX-i or RGS/PTX-i Gα proteins are used so agonist activation of endogenous Gα is eliminated. Bound [^{35}S]GTPγS is separated from unbound by rapid filtration. Cell membrane preparations are used, as [^{35}S]GTPγS does not cross the plasma membrane, although in several instances permeabilized cells have been employed (e.g., Alt et al., 2001). The degree of agonist-stimulated [^{35}S]GTPγS binding is not expected to differ between PTX-i and PTX/ RGS-i Gα containing cells as the RGS activity is downstream of guanine nucleotide binding to the Gα protein. However, the assay confirms the functional coupling of the receptor to particular Gα subunits and is another control on the level of receptors and Gα protein expressed.

To cell membrane homogenates (10–20 μg protein) in a mixture of 20 mM Tris–HCl, pH 7.4, 5 mM MgCl$_2$, 100 mM NaCl, 0.1 mM dithiothreitol (freshly prepared) in 5-ml polypropylene tubes, add 30 μM GDP and keep on ice. To this mixture, add 0.001–10 μM agonist or vehicle and finally 0.1 nM [^{35}S]GTPγS (approximately 1250 Ci/mmol; Perkin Elmer Life Sciences, Boston, MA). The GDP is present to reduce basal binding of

[^{35}S]GTPγS and will need to be optimized with different cell preparations as discussed in Harrison and Traynor, 2003. Incubate samples for 60 min in a shaking water bath at 25°. Stop the reaction by rapid filtration through glass fiber filters (Schleicher and Schuell No. 32) mounted in a Brandel cell harvester and rinse three times with 3 ml ice-cold wash buffer (50 mM Tris–HCl, pH 7.4, 5 mM MgCl$_2$, and 100 mM NaCl). Determine the level of bound [^{35}S]GTPγS (retained on the filters) by liquid scintillation counting as described for the receptor-binding assay. Data are presented as percentage increase over basal ([^{35}S]GTPγS bound in the absence of agonist) or as picomoles [^{35}S]GTPγS bound per milligram protein. To determine the rate of [^{35}S]GTPγS binding, incubate membranes (approximately 20 μg) for 10 min at 25° under the same conditions. Add a maximal concentration of ligand (usually 10 μM) and allow the mixture to equilibrate for a further 10 min before the addition of 0.1 nM [^{35}S]GTPγS to start the reaction. At various times (3 min to 2 h) separate bound and free [^{35}S]GTPγS and quantify as described earlier. GDP, dithiothreitol, and other reagents are obtained from Sigma.

GTPase Assay

This assay measures the GTP hydrolysis step of the G-protein activation/deactivation cycle, which is stimulated by agonist in a dose-dependent manner. This assay is necessary to confirm that the expressed Gα$_o$G184S mutant is insensitive to RGS protein action.

Prewarm cell membranes (14–20 μg protein) for approximately 10 min at 30° in 10 mM Tris, pH 7.6, 2 mM MgCl$_2$, 20 mM NaCl, 0.2 mM EDTA, 0.1 mM dithiothreitol (fresh solution), ATP-regenerating system (0.2 mM ATP, 0.2 mM AppNHp, 50 units/ml creatine phosphokinase, and 5 mM phosphocreatine), and 0.001–10 μM agonist or vehicle with or without 1 μM GST-RGS8 (obtained from Richard Neubig, University of Michigan; prepared as described by Lan et al., 1998). Initiate the reaction by the addition of 0.1 μM [γ-^{32}P]GTP (Perkin Elmer Life Sciences) prewarmed to 30°, for a final volume of 0.1 ml. Stop the reaction after various times (15–120 s) by the addition of ice-cold 15% charcoal with 20 mM phosphoric acid in 0.1% gelatin. After at least 30 min on ice, centrifuge samples at 4000g for 20 min at 4° and take a 0.3-ml aliquot from the supernatant for liquid scintillation counting using 4-ml EcoLume scintillation cocktail. Use samples without membranes as blanks, subtract from each value, and convert data to picomoles GTP hydrolyzed per milligram per minute. ATP, AppNHp, and other biochemicals are obtained from Sigma.

Second Messenger Assays

These assays, downstream of G protein, assess the consequences for signaling of the presence of RGS-insensitive $G\alpha$ proteins. Three different signaling pathways are described, namely adenylyl cyclase, the MAP kinase pathway, and increases in intracellular calcium.

Adenylyl Cyclase

The μ-opioid receptor couples to inhibitory G proteins to inhibit adenylyl cyclase stimulated by forskolin or $G\alpha_s$, measured as a reduction in cyclic AMP accumulation in the presence of a phosphodiesterase inhibitor. The assay is also used to measure adenylyl cyclase supersensitivity that is seen as an increase in forskolin or $G\alpha_s$-stimulated cAMP accumulation following removal of a chronically applied agonist acting at a receptor coupled to $G\alpha_{i/o}$ proteins (Watts, 2002).

Plate cells to confluency in 24-well plates the day before the assay and treat overnight with 100 ng/ml PTX to uncouple endogenous $G\alpha_{i/o}$ proteins. To measure acute effects, rinse cells with serum-free medium and then incubate with serum-free medium containing 30 μM forskolin, 1 mM iso-butyl-1-methylxanthine (IBMX, a phosphodiesterase inhibitor), and 0.001–10 μM agonist or vehicle for 30 min at 37°. To measure supersensitization, plate cells to confluency in 24-well plates 2 days before the assay and treat with 100 ng/ml PTX to inactivate endogenous $G\alpha_{i/o}$. Add an appropriate concentration of agonist the following day and leave in the medium overnight. To start the assay, replace medium with serum-free medium containing 3–30 μM forskolin (this may be adjusted to optimize the level of supersensitization), 1 mM IBMX, and a maximal concentration of opioid antagonist (10 μM naloxone) to displace the agonist for 5 min at 37°. Alternately, the agonist can be removed by two washes with serum-free medium at 37° before adding medium with forskolin and IBMX.

To quantify cAMP produced in these assays, stop the reaction by replacing the medium with ice-cold 3% perchloric acid. After at least 30 min at 4°, remove a 0.4-ml sample from each well, neutralize with 0.08 ml 2.5 M KHCO$_3$, vortex, and centrifuge at 15,000g for 1 min. Quantify accumulated cAMP in 10 μl of supernatant from each sample using a radioimmunoassay kit available from Diagnostic Products Corp (Los Angeles, CA). Detailed instructions are provided with the kit. Inhibition of cyclic AMP formation is determined as a percentage of forskolin-stimulated cyclic AMP accumulation in the absence of opioid agonist or as picomoles cAMP accumulated per milligram protein. Forskolin, IBMX, and other reagents for the adenylyl cyclase assay may be obtained from Sigma.

p44/42 MAP Kinase Phosphorylation

This intact cell assay measures the agonist-stimulated phosphorylation of p44/42 MAP kinase (ERK1, ERK2), which modulates cytoskeleton proteins, phospholipase A_2, and several transcription factors. This assay utilizes an antibody for phosphorylated ERK1 and ERK2 in a quantitative Western blot assay of samples from intact cells treated for various times or with various doses of stimulating agonist.

Plate cells expressing PTX-insensitive $Gα_o$ into 6-well plates with 2 ml medium (DMEM with 10% FBS) the day before the assay to reach 70–90% confluency on the day of the assay and treat overnight with 100 ng/ml PTX. Replace the medium with serum-free medium for 2 h before the start of the assay to reduce basal phosphorylation levels. To start the assay, add vehicle (usually water), agonist, or 10% FBS as a positive control and stop after 1–20 min by rinsing the cells twice with ice-cold PBS and then add 0.1 ml ice-cold SDS sample buffer (62.5 mM Tris–HCl, pH 6.8, 2% SDS, 10% glycerol, 50 mM DTT, and 0.01% bromophenol blue). Remove samples from the wells by mixing the sample buffer over the well surface with a pipette tip before transferring to a 1.5-ml microcentrifuge tube and place in a water bath sonicator for 10–15 sec, boil for 5 min, and then centrifuge for 5 min. [Protein is determined in parallel wells. Rinse wells two times with ice-cold PBS, remove cells with a 5-min treatment with 1 ml lifting buffer (5.6 mM glucose, 5 mM KCl, 5 mM HEPES, 137 mM NaCl, 1 mM EGTA, pH 7.4) freeze to break cells, and determine protein using a BCA protein assay kit (Pierce, Rockford, IL).] Load the supernatant samples (120 $μ$g protein) and molecular weight markers (23–132 kDa; Santa Cruz) onto a 12% SDS–PAGE gel and run at 70 V for 30 min followed by 150 V for 60 min (Mini-Protean II electrophoresis cell; Bio-Rad) or until the dye front runs off the gel. Equilibrate the gel, nitrocellulose membrane (45 $μ$m, Osmonics), filter papers, and sponges in transfer buffer [25 mM Tris base, pH 8.3 (do not adjust), 192 mM glycine, 1% SDS, 20% methanol] for 20 min before transfer at 100 V for 60 min at room temperature (Mini Trans-Blot electrophoretic transfer cell). Wash the membrane for 5 min in TBS (20 mM Tris–HCl, pH 7.4, 500 mM NaCl) and then replace the TBS with 5% nonfat dry milk in TBS for 60 min at room temperature to block the membrane. Wash three times for 5 min with TBS with 0.1% Tween 20. Incubate the membrane overnight at 4° with a 1:2000 dilution (in 5% nonfat dry milk in TBS) of anti-phospho-p44/42 MAP kinase antibody (ERK 1/2, Cell Signaling Technology, Inc., Beverly, MA), followed by three washes for 5 min with TBS–Tween. Then incubate with a 1:2000 dilution (in 5% nonfat dry milk in TBS) of anti-mouse IgG-HRP (Santa Cruz) for 60 min at room temperature, followed by three washes for 5 min each with TBS–Tween. Visualize the bands using Super Signal WestPico (Pierce) enhanced chemiluminescence, quantify using a Kodak Image Station 440 as sum intensity (pixels), and give as percentage basal levels (in the absence of agonist).

To ensure equal loading, it is necessary to measure total ERK levels. Strip the membranes of antibody by incubating four times for 5 min each in 0.2 N NaCl with 1% SDS, followed by 1 wash for 5 min in TBS. Block the membrane in 5% nonfat dry milk in TBS for 60 min at room temperature and wash three times for 5 min each with TBS–Tween. Incubate the membrane overnight at 4° with a 1:1000 dilution (in 5% nonfat dry milk in TBS) of anti-p44/42 MAP kinase antibody (ERK 1/2, Cell Signaling Technology), wash three times for 5 min each in TBS–Tween, and then incubate with a 1:2000 dilution (in 5% nonfat dry milk in TBS) of antirabbit IgG-HRP (Santa Cruz) for 60 min at room temperature. After three washes with TBS–Tween for 5 min each, visualize and quantify the bands as described earlier.

Release of Intracellular Calcium

Treat confluent cells in a 75-cm² flask overnight with 100 ng/ml PTX and 5 μM forskolin. The next day, harvest cells as described under membrane preparation and centrifuge at 200g for 3 min. Wash the cell suspension twice by resuspension and centrifugation (200g) in 50-ml conical polypropylene tubes with 30 ml Krebs–HEPES buffer (143.3 mM NaCl, 4.7 mM KCl, 2.5 mM CaCl₂, 1.2 mM MgSO₄, 1.2 mM KH₂PO₄, 11.7 mM glucose, and 10 mM HEPES, pH 7.4). After the second spin, resuspend the cells in 3 ml of Krebs–HEPES buffer and load with 3 μM fura-2-acetoxymethyl ester (fura-2; 10 μl of a 1 mM stock made up in dimethyl sulfoxide, Sigma) for 30 min in a 37° shaking water bath in the dark. Centrifuge cells at 200 g, resuspend the cell pellet in 30 ml Krebs–HEPES, and then leave in the dark at room temperature for 20 min to allow for de-esterification of the acetoxymethyl ester group of the fura-2. Collect cells by centrifugation, wash in 30 ml Krebs–HEPES buffer, resuspend in 4 ml of Krebs–HEPES buffer, and store in the dark until use, which must occur within 30 min. To assay for the increase in intracellular Ca²⁺ levels, put cells suspension (1 ml) into a glass cuvette containing a magnetic stirrer and place in a Shimazdu RF5000 spectrofluorophotometer (Columbia, MD) at 37°. Set the excitation wavelength at 340/380 nm and emission wavelength at 510 nm. Read emission for at least 60 s or until a steady baseline is achieved. Prepare a solution of morphine (or other agonist) at 100 times the final required concentration in water and add 10 μl to the cuvette using a Hamilton syringe. Emission is read until a steady level is achieved. Calculate the change in intracellular

Ca^{2+} as (peak 340/380 ratio) − (basal 340/380 ratio). To test for a role of extracellular Ca^{2+} entry, use nominally Ca^{2+}-free buffer containing 0.1 mM EGTA and include in the final resuspension only.

Data Analysis

Concentration–response data from adenylyl cyclase, MAP kinase, and Ca^{2+} assays are fitted to sigmoidal concentration response curves using GraphPad Prism to determine EC_{50} and maximal effect.

Example of Experimental Findings

In the system described earlier, the μ-opioid agonist [D-Ala^2, MePhe^4, Gly-ol^5]enkephalin (DAMGO) shows increased potency and efficacy of signaling to adenylyl cyclase in cells expressing RGS-insensitive (RGS/PTX-i) Gα_o compared with RGS-sensitive (PTX-i) Gα_o (Fig. 1). Signaling

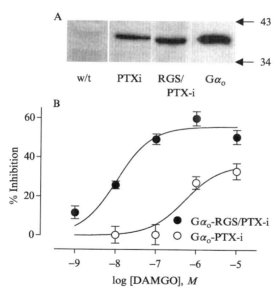

FIG. 1. (A) Western blot of Gα_o in wild-type (w/t) C6μ cells and C6μ cells transfected stably with Gα_oPTX-i or Gα_oRGS/PTX-i compared with recombinant His$_6$-Gα_o as a Western blot standard. (B) Inhibition of forskolin (30 μM)-stimulated cyclic AMP accumulation by the μ-opioid agonist DAMGO in PTX-treated C6μ Gα_oPTX-i or C6μ Gα_oRGS/PTX-i cells. Values are expressed as inhibition of forskolin-stimulated cAMP, which was the same in Gα_oPTX-i (6.6 ± 0.5 pmol/mg) as Gα_oRGS/PTX-i (6.8 ± 0.9 pmol/mg) expressing cells in the absence of DAMGO treatment. Adapted with permission from Clark *et al.* (2003).

through the MAP kinase pathway also shows an increased potency with the full agonist DAMGO, but not an increased maximal effect, although the maximal effect of the partial agonist morphine is enhanced significantly (Clark *et al.*, 2003). In contrast, the ability of DAMGO or morphine to stimulate the release of calcium from intracellular stores is altered to a much lesser extent in cells expressing RGS-insensitive $G\alpha_o$ compared with RGS-sensitive $G\alpha_o$ (Clark *et al.*, 2003). These results confirm that RGS proteins can modulate effector signaling by a single G protein and may play an important role in directing effector responses to μ-opioid receptor signaling.

Concluding Remarks

The use of RGS-i mutant $G\alpha$ proteins to study the signaling consequences of blocking RGS protein GAP activity is widely applicable to the study of all $G\alpha$ proteins and receptors that interact with these proteins and the analysis of these effects on a wide variety of intracellular signaling pathways. Complications arise from endogenous $G\alpha$ proteins expressed in the cell under study. For PTX-sensitive proteins, this can be overcome by use of the C351G mutants, allowing for elimination of activation of endogenous G proteins. Where this cannot be done, a viable option would be the use of Sf9 cells (*Spodoptera frugiperda*), which express very low levels of endogenous G proteins as described by Windh and Manning (2002).

A limitation of the use of RGS-insensitive $G\alpha$ proteins is that they do not identify which particular RGS protein (or proteins) is responsible for a particular action. The method is further restricted to effects that require the RGS-domain binding to a $G\alpha$ subunit and does not allow for study of signaling activity that may be independent of the conserved RGS domain. Nonetheless, the methodology does provide a starting point to view RGS protein actions at particular $G\alpha$ and receptor combinations and to allow identification of the signaling consequences for the cell of RGS protein activity.

References

Alt, A., McFadyen, I. J., Fan, C. D., Woods, J. H., and Traynor, J. R. (2001). Stimulation of guanosine-5'-O-(3-[^{35}S]thio)triphosphate binding in digitonin-permeabilized C6 rat glioma cells: Evidence for an organized association of μ-opioid receptors and G protein. *J. Pharmacol. Exp. Ther.* **298,** 116–121.

Bradford, M. M. (1976). A rapid and sensitive method for the quantitation of microgram quantities of protein using the principle of protein-dye binding. *Anal. Biochem.* **72,** 248–254.

Charpentier, N., Prezeau, L., Carrette, J., Betorelli, R., Le Cam, G., Manzoni, D., Bockaert, J., and Homburger, V. (1993). Transfected $Go_1\alpha$ inhibits the calcium dependence of β-adrenergic stimulated cAMP accumulation in C6 glioma cells. *J. Biol. Chem.* **268,** 8980–8989.

Clark, M. J., Harrison, C., Zhong, H., Neubig, R. R., and Traynor, J. R. (2003). Endogenous RGS protein action modulates μ-opioid signaling through Gα_o: Effects on adenylyl cyclase, extracellular signal-regulated kinases and intracellular calcium pathways. *J. Biol. Chem.* **278**, 9418–9425.

Doupnik, C. A., Jaén, C., and Zhang, Q. (2004). Measuring the modulating effects of RGS proteins on GIRK channel gating. *Methods Enzymol.* **389**(9), 131–154.

Gutstein, H. B., Rubie, E. A., Mansour, A., Akil, H., and Woodgett, J. R. (1997). Opioid effects on mitogen-activated protein kinase signaling cascades. *Anesthesiology* **87**, 1118–1126.

Harrison, C., and Traynor, J. R. (2003). The [^{35}S]GTPγS binding assay: Approaches and applications in pharmacology. *Life Sci.* **74**, 489–508.

Hollinger, S., and Hepler, J. R. (2002). Cellular regulation of RGS proteins: Modulators and integrators of G protein signaling. *Pharmacol. Rev.* **54**, 527–559.

Kenakin, T. (2002). Drug efficacy at G protein-coupled receptors. *Annu. Rev. Pharmacol. Toxicol.* **42**, 349–379.

Lan, K. L., Sarvazyan, N. A., Taussig, R., MacKenzie, R. G., DiBello, P. R., Dohlman, H. G., and Neubig, R. R. (1998). A point mutation in Gα_o and Gα_{i1} blocks interaction with regulator of G protein signaling proteins. *J. Biol. Chem.* **21**, 12794–12797.

Law, P. Y., Wong, Y. H., and Loh, H. H. (2000). Molecular mechanisms and regulation of opioid receptor signaling. *Annu. Rev. Pharmacol. Toxicol.* **40**, 389–430.

Lee, K. O., Akil, H., Woods, J. H., and Traynor, J. R. (1999). Differential binding properties of oripavines at cloned μ- and δ-opioid receptors. *Eur. J. Pharmacol.* **378**, 323–330.

Milligan, G. (1998). Techniques used in the identification and analysis of function of pertussis toxin-sensitive guanine nucleotide binding proteins. *Biochem. J.* **255**, 1–13.

Neubig, R. R., and Siderovski, D. P. (2002). Regulators of G-protein signaling as new central nervous system drug targets. *Nature Rev. Drug Disc.* **1**, 187–197.

Smart, D., and Lambert, D. G. (1996). The stimulatory effects of opioids and their possible role in the development of tolerance. *Trends Pharmacol. Sci.* **17**, 264–269.

Snow, B. E., Hall, R. A., Krumins, A. M., Brothers, G. M., Bouchard, D., Brothers, C. A., Chung, S., Mangion, J., Gilman, A. G., Lefkowitz, R. J., and Siderovski, D. P. (1998). GTPase activating specificity of RGS12 and binding specificity of an alternatively spliced PDZ (PSD-95/Dlg/ZO-1) domain. *J. Biol. Chem.* **10**, 17749–17755.

Tesmer, J. J. G., Berman, D. M., Gilman, A. G., and Sprang, S. R. (1997). Structure of RGS4 bound to AlF$_4^-$-activated G$_{i\alpha 1}$: Stabilization of the transition state for GTP hydrolysis. *Cell* **89**, 251–261.

Wang, Q., Liu, M., Bashar, M., Siderovski, D. P., and Neubig, R. R. (2002). Receptor-selective effects of endogenous RGS3 and RGS5 to regulate mitogen-activated protein kinase activation in rat vascular smooth muscle cells. *J. Biol. Chem.* **277**, 24949–24958.

Watts, V. J. (2002). Molecular mechanisms for heterologous sensitization of adenylate cyclase. *J. Pharmacol. Exp. Ther.* **302**, 1–7.

Windh, R. T., and Manning, D. R. (2002). Analysis of G protein activation in Sf9 and mammalian cells by agonist-promoted [^{35}S]GTPγS binding. *Methods Enzymol.* **344**, 3–14.

Zeng, W., Xu, X., Popov, S., Mukhopadhyay, S., Chidiac, P., Swistok, P., Danho, W., Yagaloff, K. A., Fisher, S. L., Ross, E. M., Muallem, S., and Wilkie, T. M. (1998). The N-terminal domain of RGS4 confers receptor-selective inhibition of G protein signaling. *J. Biol. Chem.* **273**, 34687–34690.

Zhong, H., Wade, S. M., Woolf, P. J., Linderman, J. J., Traynor, J. R., and Neubig, R. R. (2003). A spatial focusing model for G protein signals: Regulators of G protein signaling (RGS) protein mediated kinetic scaffolding. *J. Biol. Chem.* **278**, 7278–7284.

[11]   Use of RGS-Insensitive Gα Subunits to
Study Endogenous RGS Protein Action on
G-Protein Modulation of N-Type Calcium
Channels in Sympathetic Neurons

By Stephen R. Ikeda and Seong-Woo Jeong

Abstract

Regulators of G-protein signaling (RGS) proteins are a large family of signaling proteins that control both the magnitude and temporal characteristics of heterotrimeric G-protein-mediated signaling. A current challenge is to define how endogenous RGS protein function impacts G-protein modulation of ionic channels in mammalian neurons. The experimental strategy described here utilizes distinct mutations in Gα subunits that confer *Bordetella pertussis* toxin (PTX) and RGS protein insensitivity. The native signaling pathway in rat sympathetic neurons that mediates voltage-dependent modulation of N-type Ca^{2+} channels is ablated by PTX treatment and the signaling is reconstituted by expressing a PTX/RGS-insensitive Gα mutant along with Gβ and Gγ subunits. As neurons are resistant to conventional transfection modalities, heterologous expression is accomplished by the direct microinjection of plasmids into the nucleus of the neuron. An advantage of this approach is that knowledge of the specific RGS subtypes participating in the pathway is not required. From the resulting alterations in the kinetics and pharmacology of G-protein-coupled receptor modulation of N-type Ca^{2+} channels, we can infer the role endogenous RGS proteins play in the signaling pathway.

Introduction

Regulators of G-protein signaling (RGS) proteins are a large family of signaling proteins that play a key role in controlling both the magnitude and temporal characteristics of heterotrimeric G-protein-mediated signaling (Hollinger and Hepler, 2002; Neubig and Siderovski, 2002). Although the intricacies of this regulation remain to be investigated, the canonical pathway involves binding of RGS proteins to activated (GTP-bound) Gα subunits following ligand binding to G-protein-coupled receptors (GPCR). The interaction with Gα subunits has been demonstrated to have two distinct effects. First, the intrinsic GTPase activity of the Gα subunit is accelerated, thereby decreasing the time course of G-protein activation.

Second, binding of the RGS proteins can mask interaction of the Gα subunit with effector molecules, resulting in an antagonist effect. At present, the majority of what we know about RGS proteins derives from *in vitro* biochemical studies or overexpression studies carried out in clonal cell lines. A challenge, then, is to design an experimental paradigm in which the role of endogenous RGS protein function can be elucidated.

This article describes methods used to study the physiological roles of endogenous RGS proteins on ion channel modulation by GPCRs. The basic idea is to reconstitute the signaling pathway in primary cultures of sympathetic neurons following disruption of the native G-protein signaling pathway with *Bordetella pertussis* toxin (PTX). Using this strategy, we can introduce Gα subunits with two key mutations: (1) a mutation that renders the Gα subunit resistant to the actions of PTX, thereby allowing replacement of native Gα subunits, and (2) a mutation that abolishes interaction between Gα subunits and the RGS core domain of RGS proteins. An advantage of this strategy, when compared with RGS knockdown techniques (either at the cellular or organism level), is that presumably all RGS protein interactions with the mutated Gα will be disrupted, thus alleviating concerns that redundancy might mask the effects of a single protein knockdown. As a readout of RGS protein activity, we examined the time course of voltage-dependent (VD) N-type Ca^{2+} channel inhibition following agonist application and removal. VD modulation is believed to involve direct binding of the Gβγ subunit to the ion channel without the participation of intervening enzymatic steps (Dolphin, 2003; Ikeda and Dunlap, 1999). Hence, the time course of Ca^{2+} channel modulation should closely parallel the "free" intracellular Gβγ concentration. By comparing the time course of agonist action in the reconstituted signaling pathway with the native system, we can infer what role endogenous RGS proteins play in influencing the kinetics and magnitude of GPCR-mediated modulation of ion channels.

Gα Mutations Conferring Pertussis Toxin and RGS Insensitivity

Since PTX pretreatment will be used to uncouple endogenous G proteins, the heterologously introduced Gα subunits must first be rendered insensitive to PTX. Within cells, the S1 subunit of PTX catalyzes the ADP-ribosylation of a cysteine residue—four positions removed from the C termini (Fig. 1) of susceptible Gα subunits (i.e., $Gα_{oA}$, $Gα_{oB}$, $Gα_{i1}$, $Gα_{i2}$, $Gα_{i3}$, $Gα_t$, and $Gα_{gust}$). Therefore, mutation of this cysteine to any other residue should confer PTX resistance to the Gα subunit. The choice of the residue with which to replace the cysteine is not entirely straightforward. Recent studies have used a Cys-to-Gly substitution with good results. However, a comprehensive study of $Gα_{i1}$ coupling to $α_{2a}$-adrenoceptors

(α_2-AR) in which Cys[351] was mutated to all other possible amino acids indicated that the hydrophobicity of the substituted residue greatly influenced the extent of Gα activation by the receptor (Bahia *et al.*, 1999; Jackson *et al.*, 1999). Of the three amino acids commonly substituted for Cys, the rank order based on the extent of receptor activation of Gα (as measured by GTPγS binding) was Ile > Ser > Gly. Based on this result, one would predict that Cys to Ile would be the preferred substitution. Our own studies, however, indicated that Gly mutation in Gα_{oA} (C351G) reconstituted α_2-AR-mediated N-type Ca^{2+} channel modulation to a greater extent than the mutations incorporating either Ile or Ser (Jeong and Ikeda, 2000a). Therefore, we utilized the Cys-to-Gly mutation as a basis for our combined PTX/RGS–insensitive (PTX/RGS-i) Gα mutants. However, it seems prudent to examine the three commonly used cysteine substitutions (Ile, Ser, and Gly), which cover a wide hydrophobic range, to determine which works best in a particular system.

The mutations to render the Gα subunit insensitive to the actions of the RGS core domain (Fig. 1) are based on the elegant studies of the Dohlman, Neubig, and Artemyev laboratories. DiBello *et al.* (1998) discovered a single residue mutation in the yeast Gα subunit, Gpa1, that conferred a

```
mGNAoA    1 MGCTLSAEERAALERSKAIEKNLKEDGISAAKDVKLLLLGAGESGKSTIVKQMKIIHEDGFSGEDVKQYKPVVYSNTIQS   80
mGNAoB    1 MGCTLSAEERAALERSKAIEKNLKEDGISAAKDVKLLLLGAGESGKSTIVKQMKIIHEDGFSGEDVKQYKPVVYSNTIQS   80
rGNAi1    1 MGCTLSAEDKAAVERSKMIDRNLREDGEKAAREVKLLLLGAGESGKSTIVKQMKIIHEAGYSEEECKQYKAVVYSNTIQS   80
rGNAi2    1 MGCTVSAEDKAAAERSKMIDKNLREDGEKAAREVKLLLLGAGESGKSTIVKQMKIIHEDGYSEEECRQYRAVVYSNTIQS   80
rGNAi3    1 MGCTLSAEDKAAVERSKMIDRNLREDGEKAAKEVKLLLLGAGESGKSTIVKQMKIIHEDGYSEDECKQYKVVVYSNTIQS   80
            ****:***::** **** *:::**:*** **:::******************** *:* :: :**: *********

mGNAoA   81 LAAIVRAMDTLGVEYGDKERKTDSKMVCDVVSRMEDTEPFSAELLSAMMRLWGDSGIQECFNRSREYQLNDSAKYYLDSL  160
mGNAoB   81 LAAIVRAMDTLGVEYGDKERKTDSKMVCDVVSRMEDTEPFSAELLSAMMRLWGDSGIQECFNRSREYQLNDSAKYYLDSL  160
rGNAi1   81 IIAIIRAMGRLKIDFGDAARADDARQLFVLAGAAEEG-FMTAELAGVIKRLWKDSGVQACFNRSREYQLNDSAAYYLNDL  159
rGNAi2   81 IMAIVKAMGNLQIDFADPQRADDARQLFALSCAAEEQGMLPEDLSGVIRRLWADHGVQACFSRSREYQLNDSAAYYLNDL  160
rGNAi3   81 IIAIIRAMGRLKIDFGEAARADDARQLFVLAGSAEEG-VMTSELAGVIKRLWRDGGVQACFSRSREYQLNDSASYYLNDL  159
            : **::** * ::: : *   *::: : : *:   :*  : *** * *:* ** ********** ** ***
```

```
mGNAoA  161 DRIGAGDYQPTEQDILRTRVKTTGIVETHFTFKNLHFRLFDVGGQRSERKKWIHCFEDVTAIIFCVALSGYDQVLHEDET  240
mGNAoB  161 DRIGAGDYQPTEQDILRTRVKTTGIVETHFTFKNLHFRLFDVGGQRSERKKWIHCFEDVTAIIFCVALSGYDQVLHEDET  240
rGNAi1  160 DRIAQPNYIPTQQDVLRTRVKTTGIVETHFTFKDLHFKMFDVGGQRSERKKWIHCFEGVTAIIFCVALSDYDLVLAEDEE  239
rGNAi2  161 ERIAQSDYIPTQQDVLRTRVKTTGIVETHFTFKDLHFKMFDVGGQRSERKKWIHCFEGVTAIIFCVALSAYDLVLAEDEE  240
rGNAi3  160 DRISQTNYIPTQQDVLRTRVKTTGIVETHFTFKELYFKMFDVGGQRSERKKWIHCFEGVTAIIFCVALSDYDLVLAEDEE  239
            :** * **:**:******** ** *:; :****************** ********** ** ** ***
```

```
mGNAoA  241 TNRMHESLMLFDSICNNKFFIDTSIILFLNKKDLFGEKIKKSPLTICFPEYPGSNTYEDAAAYIQTQFESKNR-SPNKEI  319
mGNAoB  241 TNRMHESLKLFDSICNNKWFTDTSIILFLNKKDIFEEKIKKSPLTICFPEYTGPSAFTEAVAHIGGQYESKNK-SAHKEV  319
rGNAi1  240 MNRMHESMKLFDSICNNKWFTDTSIILFLNKKDLFEEKIKKSPLTICYPEYAGSNTYEEAAAYIQCQFEDLNKRKDTKEI  319
rGNAi2  241 MNRMHESMKLFDSICNNKWFTDTSIILFLNKKDLFEEKITQSPLTICFPPEYTGANKYDEAASYIQSKFEDLNKRKDTKEI  320
rGNAi3  240 MNRMHESMKLFDSICNNKWFTDTSIILFLNKKDLFEEKIKRSPLTICYPEYTGSNTYEEAAAYIQCQFEDLNRRKDTKEV  319
            ******: ********* * ***********:* *** ;*****:*** * : :* : ** ::* *:   **:
```

```
mGNAoA  320 YCHMTCATDTNNIQVVFDAVTDIIIANNLRGCGLY*  355
mGNAoB  320 YSHVTCATDTNNIQFVFDAVTDVIIAKNLRGCGLY*  355
rGNAi1  320 YTHFTCATDTKNVQFVFDAVTDVIIKNNLKDCGLF*  355
rGNAi2  321 YTHFTCATDTKNVQFVFDAVTDVIIKNNLKDCGLF*  356
rGNAi3  320 YTHFTCATDTKNVQFVFDAVTDVIIKNNLKECGLY*  355
            *:* ***** *:* *******:** **: ***:*
```

Fig. 1. Mutations conferring *Bordetella pertussis* toxin (PTX) and RGS protein insensitivity to mammalian G-protein α subunits. Clustal W alignment of rodent Gα subunits of the G$_{i/o}$ class. Amino acid identity is indicated by an asterisk and conserved substitution by a colon. Residues conferring RGS insensitivity when mutated are outlined in green and PTX insensitivity in yellow. Genbank accession numbers: mouse Gα_{oA} (M36777), mouse Gα_{oB} (M36778), rat Gα_{i1} (NM_013145), rat Gα_{i2} (NM_031035), and rat Gα_{i3} (NM_013106). (See color insert.)

phenotype similar to the loss of *SST2*, a yeast RGS homolog. The Gly-to-Ser substitution at position 302 of Gpa1 corresponded to a highly conserved residue in the switch region 1 of mammalian Gα and greatly decreased the ability of Sst2 to bind to and accelerate the GTPase activity of Gpa1. Based on these results, Lan *et al.* (1998) demonstrated that an equivalent mutation introduced into mammalian $G\alpha_o$ (G184S) and $G\alpha_{i1}$ (G183S) disrupted both biochemical actions and binding of RGS4 to the mutant Gα subunits. Importantly, this study also showed that the time course of GTPγS binding and GTP hydrolysis, in the absence of RGS proteins, was similar for both wild-type and mutant Gα subunits. Using a completely independent approach, Natochin and Artemyev (1998) identified another mutation, S202D in $G\alpha_t$, that produced a similar phenotype to that mentioned earlier. The mutation was identified by comparing the RGS domain contact residues (as predicted from structural studies) in $G\alpha_s$ and $G\alpha_t$ that are divergent. We have used both mutations—separately and together—within the context of $G\alpha_{oA}$ with similar, although not identical, results (Jeong and Ikeda, 2000b). As the Gly→Ser substitution is better characterized, we currently prefer this substitution when constructing RGS-insensitive mutant Gα subunits.

The point mutations conferring PTX and RGS insensitivity (Fig. 1) can be introduced into the Gα subunits using a variety of standard techniques and thus detailed protocols will not be provided here. The PTX-insensitive mutations are introduced easily by amplifying the entire Gα reading frame using polymerase chain reaction and a high fidelity polymerase (e.g., *Pfu*). The mutation is introduced with a mutagenic reverse primer (e.g., see Jeong and Ikeda, 2000a) and the product is subcloned into the vector of choice. We typically employ the mammalian expression vector pCI (Promega, Madison, WI) or pcDNA3.1(+) (Invitrogen Life Technologies, Carlsbad, CA). Both vectors incorporate a CMV promoter for high-level expression in mammalian cells. Alternatively, one can purchase sequence-verified PTX-insensitive (C→I, C→S, C→G) human Gα subunits in the pcDNA3.1(+) vector from the Guthrie cDNA Resource Center, (Sayre, PA) (www.cdna.org) for a moderate charge. Using the PTX–insensitive Gα clones as a template, a mutation conferring RGS insensitivity is introduced using commercial site-directed mutagenesis kits. Initially, we favored using the GeneEditor kit (Promega) because only a single mutagenic primer was required. However, the cost of synthetic oligonucleotide primers is no longer a dominant factor in the construction of mutants; hence, we currently use the QuikChange method (Stratagene, La Jolla, CA). The advantage of the latter technique is that it does not rely on mutating the vector antibiotic sensitivity as a selection strategy; hence, further rounds of mutagensis can be performed without moving the insert to a new vector. Following mutagenesis, the entire insert should be sequenced to confirm that the desired mutation was introduced and that spurious mutations did not occur during amplification. Plasmid DNA is prepared with Qiagen (Chatworth, CA) miniprep kits and is stored at $-20°$ at a concentration of 1 $\mu g/\mu l$.

Expression of G–Protein Subunits in Rat Sympathetic Neurons

Sympathetic neurons isolated from the adult rat superior cervical ganglion (SCG) represent a convenient and well-established model system for studying the GPCR-mediated modulation of voltage-gated ion channels in a physiologically relevant setting. Some advantages of using the system include (1) a relatively homogeneous population of neurons with regard to GPCR and voltage-gated ion channel content; (2) numerous well-characterized interactions between GPCRs and N-type Ca^{2+} channels; (3) good spatial voltage-clamp control of membrane potential during short term (4–24 h) tissue culture; and (4) large neuronal somata (approximately 20–30 μm diameter) with a clearly visible nucleus when visualized with phase-contrast microscopy. Of course, to be useful in the current context, the ability to express heterologously precise combinations of proteins is a necessity. Unlike clonal cell lines, primary cultures of neurons are resistant to typical transfection modalities, such as calcium phosphate, lipofection, and electroporation. The reasons for low transfection efficiencies in neurons are not entirely clear, but it seems likely that entry of plasmid DNA into the nucleus comprises the rate-limiting step. In dividing cells, the nuclear envelope breaks down during cell division, thereby allowing plasmid DNA to enter the nucleus where transcription occurs—an event absent in postmitotic neurons. Thus, even direct introduction of cDNA constructs into the cytoplasm of neurons results in minimal expression (unpublished observation). However, high levels of heterologous expression in postmitotic neurons have been obtained by the direct microinjection of cDNA into the nucleus.

We have published fairly detailed protocols on the dissection, enzymatic isolation, and short-term tissue culture of adult rat sympathetic neurons (Ikeda, 2004). Hence, the following includes only an abbreviated description of the procedures. Further details are available in the aforementioned reference.

Enzymatic Isolation of Adult Rat Superior Cervical Ganglion Neurons

1. The detailed anatomy of the rat superior cervical ganglion is described by Hedger and Webber (1976) and should be consulted if one is unfamiliar with the dissection. It is best to "practice" the dissection a few times before attempting the full procedure.

2. The following materials should be available prior to starting the procedure: (i) chilled (4°) Hank's balanced salt solution (HBSS), (ii) modified Earle's balanced salt solution (mEBSS)—see later; (iii) minimal essential medium (MEM) with Earle's salts supplemented with 1% glutamine, 1% penicillin–streptomycin solution, and 10% fetal calf serum (Invitrogen Life Technologies); (iv) poly-L-lysine-coated 35-mm tissue culture dishes (six per preparation); and (v) autoclaved glass 10 × 10-mm cloning cylinders (Bellco Glass Inc., Vineland, NJ).

3. Modified EBSS (100 ml) is prepared by adding 10 ml of 10× concentrated liquid EBSS (Sigma), 1.0 ml of 1 M HEPES solution (Sigma-Aldrich, St. Louis, MO), and 0.36 g of glucose to 70 ml of deionized water. The pH of the solution is adjusted to 7.4 with NaOH (1 N) followed by the addition of 0.220 g NaHCO$_3$. After the volume is adjusted to 100 ml in a volumetric flask, the solution is sterilized with a 0.2-μm filter flask and stored in 10-ml aliquots (15-ml centrifuge tubes) at 4°.

4. Adult male Wistar rats, 250–350 g, are anesthetized with CO$_2$ and decapitated using a laboratory guillotine. The head is placed immediately in a beaker containing chilled HBSS to cool for a few minutes.

5. A ventral approach is used to remove both carotid artery bifurcations (i.e., where the common carotid artery branches into the internal and external carotid arteries) from the head with the aid of a good binocular dissection microscope and fiber-optic light source. The SCG are located deep within the carotid bifurcation and can be visualized as fleshy fusiform-shaped (~1 × 3-mm) objects attached to the medial aspect of the bifurcation. The SCG results from the fusion of several cervical sympathetic chain ganglia during development. Hence, the gross shape of the ganglia varies from nearly spherical to bilobed (i.e., "hourglass-shaped") in different animals.

6. The Y-shaped bifurcation is slightly stretched out (medial side up) on the bottom of a dissecting dish using insect pins. The preparation is covered immediately with chilled HBSS. The dissecting dish is constructed from a 60-mm polystyrene tissue culture dish filled to a depth of about 5 mm with a silicon elastomer (Sylgard 184, World Precision Instruments, Sarasota, FL).

7. Ganglia are removed from the surrounding connective tissue sheath with fine forceps (#5 Dumont) and iridectomy scissors (Fine Science Tools, Foster City, CA) and are placed in a 35-mm tissue culture dish containing HBSS. Several transverse cuts (perpendicular to the long axes) are made in each ganglia with iridectomy scissors to allow access of the enzymatic solution. These procedures are done under a dissecting microscope with fiber-optic illumination. Adjusting the angle of illumination (e.g., dark field) can aid greatly in visualization of the process.

8. The enzymatic dissociation solution is prepared at this point by dissolving 6 mg collegenase D (Roche Applied Science, Indianapolis, IN), 4 mg trypsin (type TRL, Worthington Biochemical Corp., Lakewood, NJ), and 1 mg DNase I, type II (Sigma-Aldrich) in 10 ml of sterile room temperature mEBSS. Enzyme concentrations must be adjusted for each lot of enzymes.

9. Filter 6 ml of enzyme solution through a 0.2-μm syringe filter (Millex-GV, Millipore, Bedford, MA) into a 25-cm^2 tissue culture flask with a plug seal cap (e.g., Falcon #3013). Transfer ganglia to the flask using a fire-polished Pasteur pipette, flush the flask with 95% O_2/5% CO_2 for about a minute, and then seal the flask cap tightly.

10. Incubate the flask in a shaking ($\sim$240 strokes/min) water bath at 37° for 1 h.

11. Disperse the neurons by grasping the neck of the flask between the thumb and first finger and shaking the flask vigorously for 10 s. Following the shake, there should be no (or at most a few) visible pieces of ganglia remaining.

12. From this point on, procedures should be carried out in a laminar flow tissue culture hood when possible. Add 5 ml of MEM (preheated to 37°) to the flask and transfer the contents to a 15-ml polypropylene (not polystyrene) centrifuge tube. Centrifuge at 50g in a swinging bucket centrifuge (Eppendorf 5804R, Brinkmann Instruments, Westbury, NY) for 6 min. Remove as much solution as possible, resuspend the (small) pellet in 10 ml fresh MEM, and repeat the centrifugation. It is important to use a centrifuge that correctly maintains a relatively slow speed ($\sim$500 rpm). In our experience, centrifugation at greater than 50g damages the neurons and increases the number of nonneuronal elements in the preparation.

13. Carefully remove the supernatant and resuspend the small cell pellet in 1.2 ml of MEM. The cell suspension is triturated gently 15–20 $\times$ through the opening of a 1-ml (blue) filtered ("aerosol-resistant") pipette tip to disperse any remaining cell clumps.

14. To each of six poly-L-lysine-treated 35-mm culture dishes, pipette 200 μl of cell suspension into a glass cloning cylinder that has been centered in the dishes. Gently add 2 ml of MEM to the culture dish, taking care not to dislodge the cloning cylinder.

15. Incubate the dishes for about 1 h in a tissue culture incubator (37°, humidified atmosphere of 5% CO_2 in air) to allow the neurons to attach to the substrate. The cloning rings can then be removed gently using sterile forceps and dishes returned to the incubator. After about 3–4 h, the neurons are attached firmly to the substrate.

16. Isolated SCG neuronal somata are roughly spherical with a diameter of 20–35 μm. A few short stumps of dendritic process are present shortly after dissociation, but these are reabsorbed over the next few hours. The nucleus, containing one to three prominent nucleoli, should be clearly visible with a phase-contrast microscope (200 × magnification). Each rat SCG ganglia contains approximately 25,000 neurons—a yield of 10% (from adult rats) is adequate for our experiments.

Intranuclear Injection: General Principles

Direct microinjection of cDNA-containing vectors into the nuclei of neurons has numerous advantages over other conventional transfection methodologies, especially within the context of reconstituting heterotrimeric G-protein pathways. Intranuclear injection allows for the convenient and reliable expression of multiple constructs, in this case cDNAs for Gα, Gβ, Gγ, and green fluorescent protein (GFP), simply by mixing the appropriate plasmid constructs prior to injection. Perhaps most importantly, the stoichiometry of the expressed G-protein subunits can be adjusted by altering the ratio of the injected plasmids—something that would be difficult using viral approaches. Achieving the proper "balance" of Gα:Gβγ expression is critical for reconstitution, as addressed later. There are, of course, disadvantages in using microinjection techniques to achieve heterologous expression. In general, the equipment is expensive and the total number of expressing cells obtained is small. The latter factor is not a great disadvantage for single cell assays such as electrophysiology or optical methodologies but precludes the use of microinjection for most biochemical assays. Finally, microinjection can be time-consuming and many find it tedious.

Microinjection is a "tactile" skill and thus is easiest to learn by observing the process in a laboratory that uses the technique on a regular basis. The following description attempts to capture some of the more important aspects of microinjection that we have garnered from using the technique on a daily basis for several years (for additional details, see Ikeda, 2004). Our microinjection workstation (Fig. 2) consists of an Eppendorf FemtoJet 5247 microinjection unit and 5171 micromanipulator (Brinkmann Instruments); a Nikon Diaphot TMD-inverted microscope with 20× 0.4 NA phase-contrast objective; a remote head CCD camera (Model 6410, Cohu Inc., San Diego, CA), and a 12-inch black and white video monitor (Model PVM-122, Sony Corp., Japan). The microscope and manipulator are attached to a vibration isolation table (Technical Manufacturing Corporation, Peabody, MA) using hardware from Newport Corporation (Irvine,

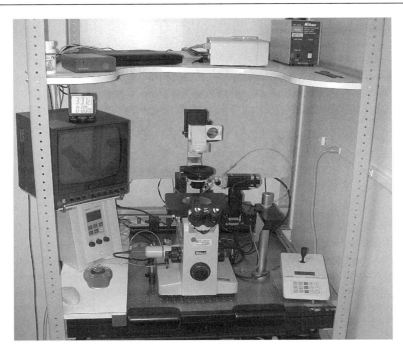

FIG. 2. Workstation for microinjection of isolated sympathetic neurons. Major equipment consists of an inverted microscope (Nikon Diaphot TMD), a micromanipulator (Eppendorf 5171), an injection pressure apparatus (Eppendorf FemtoJet 5247), a remote head CCD video camera (Cohu Model 6410), and a 12-inch B&W video monitor (Sony Model PVM-122). The microscope and manipulator head are mounted on a vibration isolation table (TMC Corporation) using optical mounting hardware (Newport Corporation). The 5171 micromanipulator and injection FemtoJet 5247 are linked to a Powerbook G3 laptop computer (Apple Computer) via USB-serial converters (Keyspan Corporation). Further automation of microinjection sequences is programmed using Igor Pro (Wavemetrics) using a video-editing controller (ShuttlePro, Contour A/V Designs) as an input device (lower left of the photograph, to the right of the computer mouse).

CA). It is useful to remove the rubber feet that come attached to most microscopes if the microscope is mounted on a vibration isolation table. The video monitor should be mounted on a mechanical arm for adjustment to eye level. The final degree of magnification is influenced by both the video coupler and the size of the CCD chip in the video head—in our case, a 20× objective produces an ~900× image on the video monitor. The general layout of our microinjection workstation is illustrated in Fig. 2. Ideally, the workstation should be located in a small separate room that is quiet and out of the main laboratory traffic.

Visualization of the injection process is critical for successful nuclear injections. One must be able to clearly see the nucleus and the slight change in refractive index that occurs during a successful injection (see Fig. 3). If the microscope objective has a correction collar, this should be adjusted to compensate for the thickness of the plastic culture dishes. Phase contrast should be optimized by aligning the phase rings properly. Unfortunately, the entry of the injection pipette into the culture medium forms a meniscus that greatly degrades phase ring alignment and hence contrast. Moreover, the degree of distortion is variable depending on the depth of the medium and the position of the pipette in the dish. One can rapidly reestablish acceptable alignment under these conditions by slightly rotating the phase turret out of the detent position while observing the image on the video monitor for maximal contrast. We have also observed that some newer phase objectives (e.g., Nikon CFI_{60} objectives) seem to produce less contrast than older objectives from the same manufacturer. Nikon has introduced a line of "apodized" phase objectives that claim to reduce the phase "halo" effect; however, we have no first-hand experience with this technology. Another determinant of contrast is the setup of the video system. By adjusting the degree of illumination, contrast can sometimes be improved by saturating parts of the image. If the video system has automatic gain control (AGC) or adjustable gamma, one can try altering these parameters to see if contrast improves. Finally, solutions with low ionic strength (e.g., 10 mM Tris buffer) are visualized better with phase contrast than isotonic solutions (e.g., 140 mM KCl) when injected into the nucleus (presumably due to refractive index differences).

The characteristics of the microinjection pipette have a major impact on the percentage of successfully injected neurons. Eppendorf manufactures injection pipettes (Femtotips, Brinkmann Instruments) that are sterile, RNase free, and mounted in holders compatible with their injection system. However, the pipettes are expensive and of fixed geometry. We currently make injection pipettes by pulling thin-walled (o.d. 1.2 mm, i.d. 0.9 mm, 100 mm length) filament-containing borosilicate tubing (TW120F-4, World Precision Instruments) using a Flaming/Brown type pipette puller (P-97; Sutter Instrument Co., Novato, CA) equipped with a 3 × 3-mm platinum–iridium filament (part# FB330B, Sutter Instrument Co.). The pipettes are pulled in two stages with the following settings (heat, pull, velocity, time): (1) 560, 115, 12, 250 and (2) 580, 130, 65, 250. Finding appropriate settings is a time-consuming trial-and-error process, as the tip opening cannot be visualized with light microscopy. One can start by examining an Eppendorf Femtotip under a microscope and attempt to emulate the general geometry. Miller *et al.* (2002) provide a detailed description on the effects of pipette puller (Sutter P-97) settings on pipette

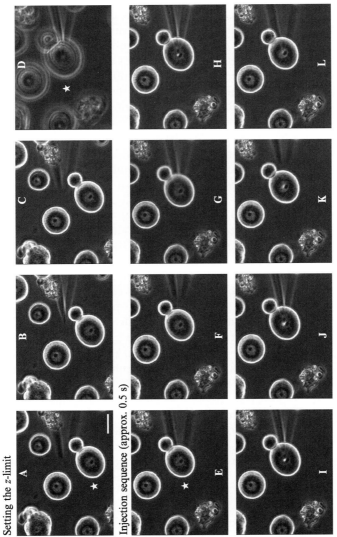

FIG. 3. Injection sequence. Video frames were taken by routing the video signal through a digital video camcorder and downloading to a computer. The isolated neurons are mouse primary sensory neurons (dorsal root ganglion). The horizontal calibration bar in A represents 25 μm. Neurons were visualized with a $40\times$ phase-contrast objective (note: we routinely use a $20\times$ objective for injections). (A) Alignment of the neuron to be injected (star) just below the injection pipette (which remains centered and stationary in the XY plane throughout). (B) The z axis limit is set by lowering the pipette to the level of the nucleolus. (C) The pipette is raised approximately 30 μm to clear the neuron.

Setting the z-limit

Injection sequence (approx. 0.5 s)

geometry. Sutter also has information on their website (www.sutter.com) that may be helpful for finding initial settings. In the end, one must test the pipettes by injecting neurons with a reporter construct (e.g., pEGFP-N1 encoding green fluorescent protein; BD Biosciences Clontech, Palo Alto, CA) and testing for functional expression. With experience, the visual appearance (Fig. 3) of the injection is adequate for judging the suitability of a given pipette geometry.

Intranuclear Injection: Protocol

1. Neurons are prepared the morning of injection as described earlier. It usually requires a minimum of 3 h at 37° for the neurons to adhere tightly enough to the substrate to allow injection. Insufficient attachment is indicated by large lateral movement or detachment of the neuron from the dish during injection.

2. Centrifuge the DNA solution to be injected (composition dependent on experiment, see later) at >10,000g for 60 min to remove any particles that might clog the injection pipette—this step is critical as plugging of the injection pipette represents a major problem. We typically place 10–20 μl of DNA solution into microtubes made from plain, that is, nonheparinized, hematocrit glass (cut to 1-inch length and fire polished shut on one end). The microtubes are inserted into 1.5-ml microcentrifuge tubes and centrifuged in a microcentrifuge (e.g., Eppendorf 5417C, Brinkmann Instruments) equipped with a swinging bucket rotor.

3. Pull 10–20 injection pipettes (see earlier discussion) while the DNA solutions are being centrifuged.

4. When the injection solution centrifugation is complete, remove a 35-mm dish containing neurons from the incubator and place it on the rectangular stage of the inverted microscope. Although the injections are done in a "nonsterile" environment, we seldom experience contamination during short-term culture (24–48 h).

5. Transfer about 1.0–1.5 μl of DNA solution into an injection pipette using a Eppendorf microloader (Brinkmann Instruments) pipette tip. Be careful not to touch the bottom of the microtube with the pipette tip while removing the solution as this will disturb any sediment that has collected there. The glass filament in the injection pipette will wick the solution (by capillary action) to the tip within a minute or so.

(D) The nucleus is centered under the pipette by moving the neuron with the rectangular stage of the inverted microscope. (E and F) The automated axial injection sequence (about 0.5 s) initiated by the 5171 manipulator. In E, the focal plane is set to the nucleus so that the injection can be monitored. The injection time, depicted in H–J, was set to 0.3 s. Note the expanding phase bright "cloud" signifying a successful nuclear injection.

6. Mount the injection pipette in the holder and adjust the pressure and time settings on the FemtoJet unit. Typically, injection pressure (p_i) is set between 120–200 hPa (hectopascal $\cong$ 0.015 psi compensation pressure (p_c) to 45 hPa (this value is not critical), and injection duration (t_i) to 0.3 s. The 5171 manipulator is set for axial injection, 45° injection angle, and velocity between 300 and 700 μm/s.

7. Locate the neurons in the dish. The cloning cylinder should have confined the neurons to a circular area in the center of the dish. Decide on a pattern to traverse through the field of cells while injecting, for example, start at the "bottom" of the circle and advance toward the "top" using a horizontal raster pattern, and then lower the pipette into the medium at the starting point.

8. Use the "clean" function (produces maximum pressure, around 6000 hPa) on the FemtoJet to ensure that solution is exiting the pipette. A large movement of the pipette during this procedure usually indicates a clogged pipette.

9. The sequence for adjusting the injection z limit (depth) is illustrated in Fig. 3A–E. The neurons illustrated are mouse dorsal root ganglion neurons, not rat SCG neurons, but the appearance and procedures are similar.

10. The pipette is kept in the center of the field about 30–40 μm above the surface of the dish (to avoid hitting the neurons). The neuron of interest (marked with the star) is moved into proximity of the pipette by moving the microscope stage (Fig. 3A).

11. Focus on the *nucleolus* and lower the pipette tip until it is in the same plane (Fig. 3B). Set the z limit of the 5171 manipulator to this depth.

12. Raise the pipette about 30 μm to clear the neuron surface (Fig. 3C).

13. Adjust the focus to the pipette tip and move the stage so that the pipette tip is centered over the nucleus (Fig. 3D).

14. Return the focal plane to the *nucleolus* (Fig. 3E). At this point, the injection sequence can be initiated.

15. The injection sequence is captured from the video system using a digital video camcorder. Figure 3E–L represent single video frames extracted from the automated injection sequence lasting about 0.5 s.

   a. The pipette is positioned laterally in preparation for the 45° axial injection (Fig. 3F).
   b. Axial movement of the pipette toward the target (Fig. 3G).
   c. Penetration of the nucleus and phase change accompanying injection of solution (Fig. 3H).
   d. Expanding cloud of injected solution as pipette remains in the nucleus for 0.3 s (Fig. 3I and J).

e. Withdrawal of the pipette from neuron (Fig. 3K). Note that the phase change is already diminishing.

f. Return of the pipette to preinjection position. The appearance of the neuron has returned to the preinjection state.

16. Several neurons can usually be injected sequentially without resetting the z limit. However, if the neurons vary greatly in size or movement of the stage produces a shift of focus (due to unevenness in the culture dish and/or the dish bottom not being parallel with the x/y plane), the entire z limit procedure should be repeated.

17. The injection z limit can be "tweaked" down a bit (we usually use 1-μm increments, although smaller steps are possible) if the pipette appears to be just "denting" the nucleus (usually indicated by a small phase bright dot but no obvious "puff" of injected material).

18. We usually inject 50–100 neurons per dish. The success rate (as determined by EGFP expression) ranges from 10 to 30%.

19. *Note.* Much of the tedium of resetting the z limit can be automated by sending a series of commands to the 5171 manipulator and FemtoJet via the RS232 serial port from a personal computer. We currently use a programmable video-editing controller (ShuttlePro, Contour A/V Designs, Windham, NH) connected to a Powerbook G3 laptop computer (Apple Computer, Cupertino, CA) running a custom program written for Igor Pro (version 4.05A, WaveMetrics, Lake Oswego, OR). Each key push or rotation of the shuttle ring initiates a subroutine that sends a predetermined series of keystrokes (as ASCII characters via the VDT XOP included with Igor Pro) to either the 5171 manipulator or the FemtoJet via a USB four-port serial adapter (Keyspan, Richmond, CA). Details of the programming are available from the one of the authors (S.R. Ikeda). Unfortunately, the latest Eppendorf injection manipulator (InjectMan, Brinkmann Instruments) does not appear to support the same functionality.

Reconstituting G-Protein Modulation of N-Type Ca²⁺ Channel with PTX/RGS–i Gα Mutants

Rat sympathetic neurons have a well-defined signaling pathway that starts with the activation of α_2-adrenergic receptors (α_2-AR) and results in a characteristic voltage-dependent (VD) modulation of N-type Ca^{2+} channels (Ikeda, 1996). Modulation is blocked almost completely by PTX pretreatment, implicating a Gα$_{i/o}$–containing heterotrimer. Available evidence suggests that the Gα$_o$ may be the primary Gα subunit involved in coupling to the receptor (Delmas *et al.*, 1999). It is the "freed" Gβγ subunit, however, that is believed to interact directly with the channel to produce modulation. Studies have indicated that expression of all three

G-protein subunits, Gα, Gβ, and Gγ, is required to reconstitute modulation following PTX treatment (Jeong and Ikeda, 2000a). This is in contrast to other studies in which only expression of the Gα subunit was required (Chen and Lambert, 2000; Straiker et al., 2002). We believe that the discrepancy arises from the short period of expression (ca. 18–24 h) used in our studies. During this interval, expression of Gα alone appears to result in an excess of Gα subunits in the GDP state, that is, the neuron does not upregulate expression of Gβ and Gγ sufficiently to reestablish a balanced G$\alpha\beta\gamma$ stoichiometry. Consequently, the result of Gα (alone) expression is a block of the signaling pathway, presumably via a G$\beta\gamma$ "buffering" or "sequestering" effect of excess Gα-GDP (Jeong and Ikeda, 1999). An advantage of using a short period of expression is that the neurons do not have time to extend long neurites and are thus spatially compact—an important consideration when studying voltage-gated channels with voltage-clamp methodology.

The necessity of expressing Gα, Gβ, and Gγ, however, requires that expression levels be "balanced" to achieve a meaningful result. Fortunately, the biophysical characteristics of VD N-type calcium channel modulation can be used as an functional assay to assess G-protein subunit stoichiometry. A hallmark of VD modulation is the relief of inhibition at very depolarized membrane potentials (Bean, 1989). The phenomenon is believed to arise from a decreased affinity of the calcium channel α_1 pore-forming subunit for G$\beta\gamma$ at depolarized voltages—and thus a relief of channel inhibition. Using a voltage protocol introduced by Elmslie et al. (1990) consisting of two identical test pulses (where the current amplitude is measured) separated by a depolarizing conditioning pulse (Fig. 4A), a parameter known as facilitation ratio (FR) is generated. The FR is calculated as the ratio of the postpulse (i.e., after the condition pulse) to prepulse current amplitude. The FR determined in the absence of overt G-protein activation is known as the basal facilitation ratio (BFR) and is an indication of the amount of "free" G$\beta\gamma$ available to modulate the channel. The BFR in control SCG neurons (measured under the conditions we employ) is usually 1.2–1.3, suggesting a tonic level of "free" G$\beta\gamma$ (Ikeda, 1991). If G$\beta\gamma$ is expressed in great excess over Gα, the BFR is $\gg 1$, mimicking the effect of GPCR-mediated G-protein activation and heterotrimer separation. Conversely, a BFR < 1.2 (usually about 0.8–1.0) is indicative of excess Gα expression and "buffering" of the tonic level of G$\beta\gamma$ seen in this system. In this situation, the BFR appears to decrease below unity because a small degree of voltage-dependent channel inactivation is unmasked by the absence of tonic G$\beta\gamma$ modulation. For successful G-protein reconstitution experiments, the ratio of G-protein subunit plasmids to be injected must be determined empirically based on the BFR criteria mentioned earlier (Jeong and Ikeda, 2000a; Kammermeier et al., 2003).

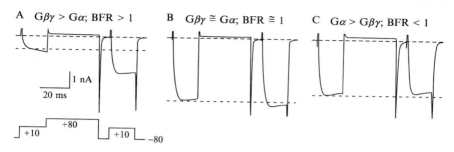

Fig. 4. Functional assay for the "balance" of expressed G-protein stoichiometry. Whole cell patch-clamp recording of calcium channel currents from rat superior cervical ganglion neurons injected previously with plasmid cDNA encoding Gα, Gβ, and Gγ subunits. The external solution contained 10 mM Ca²⁺ as the charge carrier. Currents were evoked with the triple-pulse protocol illustrated below A. The basal facilitation ratio (BFR) is defined as the ratio of the current amplitude evoked by depolarization to 10 mV following ("postpulse current") the conditional pulse (80 mV) to that preceding it ("prepulse current"), all obtained in the absence of agonist application. (A) Excess Gβγ expression. The prepulse current shows slowing of current activation and there is a large relief of block in the postpulse current, resulting in a BFR ≫1. (B) Balanced expression of Gβγ and Gα. The current amplitudes in each test pulse are approximately equal. (C) Excess Gα expression. The current amplitude of the prepulse current exceeds that of the postpulse current. The modest decrease in postpulse current likely represents voltage-dependent inactivation.

Calcium Channel Modulation Using PTX/RGS-i Gα Subunits: Protocol

1. Day 1. In the morning, prepare six 35-mm dishes of SCG neurons from two ganglia (one rat). By the afternoon (>3 h of incubation), the neurons should be sufficiently adherent to allow injection.

2. Prepare two sets of DNA for injection. The first, the control set (PTX-i), consists of $Gα_{oA}(C351G):Gβ_1:Gγ_2$ at a ratio of 2.8–3.0:10:10 (ng/μl). The second set (RGS/PTX-i) consists of $Gα_{oA}(G184S:C351G): Gβ_1:Gγ_2$ at a ratio of 4.2–4.4:10:10 (ng/μl). pEGFP-N1 (BD Biosciences Clontech) is added to both sets of DNA as a coinjection reporter at a final concentration of 5 ng/μl. All cDNA are cloned as the open reading frame (i.e., without 5'- or 3'-untranslated regions) in the mammalian vector pCI. DNA is made up in TE (10 mM Tris, 1 mM EDTA, pH 8). As mentioned previously, G-protein cDNA ratios must be determined empirically. Be sure to mix the DNA mixtures well (e.g., pipette up and down 10 times) and centrifuge (see earlier discussion) prior to injection.

3. Prepare injection pipettes while the DNA is centrifuging.

4. Inject half the dishes, approximately 50–100 neurons per dish with each mixture. Return each dish to the incubator after injecting.

5. Add PTX (holotoxin, List Biologicals, Campbell, CA) to achieve a final concentration of 500 ng/ml and return the dishes to the incubator for overnight incubation. Initial experiments should include a dish or two of cells without PTX.

6. Prepare calcium channel patch-clamp recording solutions if not already prepared.

7. External recording solution (1 liter): Add (slowly) 9.084 ml methanesulfonic acid (MS) 99% (Aldrich) to about 700 ml distilled water while stirring. Slowly mix in 50 ml tetraethylammonium hydroxide (TEA-OH) solution (35 wt%, Sigma-Aldrich). Add 2.38 g HEPES (free acid), 2.7 g glucose, and 1.47 g $CaCl_2 \cdot 2H_2O$ to the mixture. Titrate the solution to a pH of 7.4 with TEA-OH and bring the volume up to 1 liter in a volumetric flask. The osmolality should be approximately 320 mOsm/kg. Filter the solution through a 0.2-μm filter and refrigerate. The solution composition is 145 mM TEA-MS, 10 mM HEPES, 10 mM $CaCl_2$, and 15 mM glucose. Add tetrodotoxin citrate (EMD Biosciences, Inc., San Diego, CA) from a 1 mM stock solution (made up with H_2O) to a final concentration of 0.1–0.3 μM prior to use.

8. Pipette recording solution (100 ml batch). To about 50 ml of deionzied H_2O (with stirring), add 2 ml 1 N HCl, 2.343 g N-methyl-D-glucamine (NMG, Sigma-Aldrich), 0.823 ml TEA-OH (35 wt% solution), 0.418 g EGTA (free acid), 0.238 g HEPES (free acid), 0.342 g sucrose, and 0.1 ml of 1 M $CaCl_2$ (1 M stock in H_2O). Bring the pH to about 8.5 with MS solution 99%. Prior to adding MS, the solution will be basic (around 12) because of the NMG. Since this is titration of a strong base with a strong acid, the pH does not change much initially but then changes rapidly once the solution begins to neutralize—it is easy to overshoot the desired pH. Add 0.203 g MgATP, 0.017 g Na_2GTP, and 0.635 g Tris creatine phosphate (all from Sigma-Aldrich). The solution pH will decrease following addition as these compounds are acidic. Bring the final pH to 7.2 with MS (you may want to dilute the MS about 1:10 for more control over the titration as the pH nears 7.2). Adjust the final volume to 100 ml in a volumetric flask. The osmolality should be approximately 300 mOsm/kg. Filter the solution through a 0.2-μm filter and store in 3- to 5-ml aliquots at $-80°$. The solution composition is 120 mM NMG-MS, 20 mM TEA-MS, 11 mM EGTA, 1 mM $CaCl_2$, 10 mM HEPES, 4 mM MgATP, 0.3 mM Na_2GTP, and 14 mM Tris creatine phosphate. The solution contains 20 mM Cl^-.

9. Day 2. Identify successfully injected neurons for EGFP fluorescence on the inverted microscope used for patch recording. With a typical 100-W Hg lamp epifluorescence attachment and appropriate filter cube,

EGFP-expressing neurons are visualized easily with a 20× 0.4 NA objective. There should be 10–20 fluorescent neurons per dish.

10. The neurons are patch clamped using standard procedures—details are beyond the scope of this chapter. We currently use Axopatch 200B (Axon Instruments, Foster City, CA) amplifiers and pull recording electrodes of about 2 MΩ (when filled with pipette solution). Calcium currents are generally about 2 nA when evoked by a test pulse to 10 mV from a holding potential of −80 mV. Series resistance compensation (>80%) should be employed to improve voltage-clamp quality. Currents may be filtered at 2–5 kHz.

11. Agonists (e.g., 10 μM norepinephrine HCl) are applied locally to each neuron with a homemade multibarrel perfusion device similar to those available commercially (ALA Scientific Instruments, Westbury, NY).

12. The magnitude and time course of current inhibition produced by agonist can be assessed by evoking currents at 0.1–0.2 Hz. A few "unlabeled" neurons should be tested to ensure that the PTX treatment was successful (current inhibition by 10 μM norepinephrine should be about 10% when compared with 50–60% in untreated or successfully reconstituted neurons).

13. Even with optimization of the G-protein DNA concentrations used for injection, there is still considerable variability in the apparent expressed Gα:Gβγ stoichiometry. We typically use a BFR of 1.0–1.3 as a good indicator of appropriate "balance." When the magnitude of current inhibition is to be determined, it is important to establish a BFR range to use as a criteria for the inclusion of recordings in a given data set.

Conclusions

Using reconstitution of signaling pathways with PTX/RGS-i Gα mutants, we have been able to infer how endogenous RGS proteins impact GPCR-mediated calcium channel modulation in sympathetic neurons. In the presence of PTX/RGS-i Gα heterotrimers, agonist washout was slow to recover and the potency of agonist effects was increased (Jeong and Ikeda, 2000b). These effects are consistent with the ability of RGS proteins to accelerate the GTPase rate of wild-type Gα *in vitro* and suggest that endogenous RGS proteins have a similar function in neurons. Somewhat surprisingly, agonist efficacy was unaltered when modulation occurred through PTX/RGS-i Gα containing heterotrimers. Hence, RGS proteins appear to augment the temporal fidelity of agonist responses without altering the maximal response greatly. The described system has the advantage that the identities of the endogenous RGS protein(s) mediating these effects are not required for interpretation of the results, although clearly it will be important to determine the role that specific RGS protein subtypes subserve in channel modulation.

An additional advantage to this system is that experiments can be carried out in mature mammalian neurons with other components of the signaling system, that is, GPCRs and Ca^{2+} channels, expressed at native concentrations and locations. There are, however, some significant disadvantages that impact the methodology. First, the PTX/RGS-i $G\alpha$ used in these studies contains two mutations—the consequences of which we may not have completely understand. Bahia *et al.* (1998) and Jackson *et al.* (1999) have documented alterations in receptor coupling induced by the amino acid substitution conferring PTX insensitivity. In addition, we have seen evidence that elements outside the core RGS box of RGS proteins may interact in some way with PTX/RGS-i $G\alpha$ subunits (Jeong and Ikeda, 2001). Second, the concentration of expressed G-protein subunits is unknown and may influence the outcome of the experiments. Finally, the procedures detailed earlier are somewhat tedious and time-consuming. These problems aside, the method has given us a first approximation of the role that endogenous RGS proteins play in neuronal Ca^{2+} channel modulation.

References

Bahia, D. S., Wise, A., Fanelli, F., Lee, M., Rees, S., and Milligan, G. (1998). Hydrophobicity of residue 351 of the G protein $G_{i1}\alpha$ determines the extent of activation by the α_{2a}-adrenoceptor. *Biochemistry* **37**, 11555–11562.

Bean, B. P. (1989). Neurotransmitter inhibition of neuronal calcium currents by changes in channel voltage dependence. *Nature* **340**, 153–156.

Chen, H., and Lambert, N. A. (2000). Endogenous regulators of G protein signaling proteins regulate presynaptic inhibition at rat hippocampal synapses. *Proc. Natl. Acad. Sci. USA* **97**, 12810–12815.

Delmas, P., Abogadie, F. C., Milligan, G., Buckley, N. J., and Brown, D. A. (1999). $\beta\gamma$ dimers derived from Go and Gi proteins contribute different components of adrenergic inhibition of Ca^{2+} channels in rat sympathetic neurons. *J. Physiol.* **518**, 23–26.

DiBello, P. R., Garrison, T. R., Apanovitch, D. M., Hoffman, G., Shuey, D. J., Mason, K., Cockett, M. I., and Dohlman, H. G. (1998). Selective uncoupling of RGS action by a single point mutation in the G protein α-subunit. *J. Biol. Chem.* **273**, 5780–5784.

Dolphin, A. C. (2003). G protein modulation of voltage-gated calcium channels. *Pharmacol. Rev.* **55**, 607–627.

Elmslie, K. S., Zhou, W., and Jones, S. W. (1990). LHRH and GTP-γ-S modify calcium current activation in bullfrog sympathetic neurons. *Neuron* **5**, 75–80.

Hedger, J. H., and Webber, R. H. (1976). Anatomical study of the cervical sympathetic trunk and ganglia in the albino rat (*Mus norvegicus albinos*). *Acta Anat.* **96**, 206–217.

Hollinger, S., and Hepler, J. R. (2002). Cellular regulation of RGS proteins: Modulators and integrators of G protein signaling. *Pharmacol. Rev.* **54**, 527–559.

Ikeda, S. R. (1991). Double-pulse calcium channel current facilitation in adult rat sympathetic neurons. *J. Physiol.* **439**, 181–214.

Ikeda, S. R. (1996). Voltage-dependent modulation of N-type calcium channels by G-protein βγ subunits. *Nature* **380**, 255–258.

Ikeda, S. R. (2004). Expression of G-protein signaling components in adult mammalian neurons by microinjection. *Methods Mol. Biol.* **259**, 167–181.

Ikeda, S. R., and Dunlap, K. (1999). Voltage-dependent modulation of N-type calcium channels: Role of G protein subunits. *Adv. Second Messenger Phosphoprotein Res.* **33**, 131–151.

Jackson, V. N., Bahia, D. S., and Milligan, G. (1999). Modulation of relative intrinsic activity of agonists at the alpha-2A adrenoceptor by mutation of residue 351 of G protein $G_{i1\alpha}$. *Mol. Pharmacol.* **55**, 195–201.

Jeong, S. W., and Ikeda, S. R. (1999). Sequestration of G-protein βγ subunits by different G-protein α subunits blocks voltage-dependent modulation of Ca^{2+} channels in rat sympathetic neurons. *J. Neurosci.* **19**, 4755–4761.

Jeong, S. W., and Ikeda, S. R. (2000a). Effect of G protein heterotrimer composition on coupling of neurotransmitter receptors to N-type Ca^{2+} channel modulation in sympathetic neurons. *Proc. Natl. Acad. Sci. USA* **97**, 907–912.

Jeong, S. W., and Ikeda, S. R. (2000b). Endogenous regulator of G-protein signaling proteins modify N-type calcium channel modulation in rat sympathetic neurons. *J. Neurosci.* **20**, 4489–4496.

Jeong, S. W., and Ikeda, S. R. (2001). Differential regulation of G protein-gated inwardly rectifying K^+ channel kinetics by distinct domains of RGS8. *J. Physiol.* **535**, 335–347.

Kammermeier, P. J., Davis, M. I., and Ikeda, S. R. (2003). Specificity of metabotropic glutamate receptor 2 coupling to G proteins. *Mol. Pharmacol.* **63**, 183–191.

Lan, K. L., Sarvazyan, N. A., Taussig, R., Mackenzie, R. G., DiBello, P. R., Dohlman, H. G., and Neubig, R. R. (1998). A point mutation in $G\alpha_o$ and $G\alpha_{i1}$ blocks interaction with regulator of G protein signaling proteins. *J. Biol. Chem.* **273**, 12794–12797.

Miller, D. F. B., Holtzman, S. L., and Kaufman, T. C. (2002). Customized microinjection glass capillary needles for P-element transformations in *Drosophila melanogaster*. *BioTechniques* **33**, 366–375.

Natochin, M., and Artemyev, N. O. (1998). Substitution of transducin Ser202 by Asp abolishes G-protein/RGS interaction. *J. Biol. Chem.* **273**, 4300–4303.

Neubig, R. R., and Siderovski, D. P. (2002). Regulators of G-protein signalling as new central nervous system drug targets. *Nature Rev. Drug Discov.* **1**, 187–197.

Straiker, A. J., Borden, C. R., and Sullivan, J. M. (2002). G-protein alpha subunit isoforms couple differentially to receptors that mediate presynaptic inhibition at rat hippocampal synapses. *J. Neurosci.* **22**, 2460–2468.

[12] Endogenous RGS Proteins Regulate Presynaptic and Postsynaptic Function: Functional Expression of RGS-Insensitive Gα Subunits in Central Nervous System Neurons

By Huanmian Chen, Michael A. Clark, and Nevin A. Lambert

Abstract

Regulators of G-protein signaling (RGS)-insensitive (RGSi) G-protein α subunits can be used to indirectly determine the function of endogenous RGS proteins in native cells. This article describes the application of RGSi Gα subunits to the study of endogenous RGS function in central nervous system (CNS) neurons. Presynaptic inhibition of neurotransmitter release was reconstituted in primary neurons using RGSi $G\alpha_{i/o}$ subunits, whereas postsynaptic regulation of potassium channels was reconstituted using RGSi chimeras of $G\alpha_q$ and $G\alpha_i$. These studies have shown that endogenous RGS proteins are essential for the rapid termination of some G-protein-mediated signals in CNS neurons, whereas these proteins are much less important for the regulation of other signals. Together, these techniques have helped reveal the complexity of RGS regulation of CNS function.

Introduction

Regulators of G-protein signaling (RGS proteins) have been studied extensively as GTPase-accelerating proteins (GAPs) for heterotrimeric G proteins. Their ability to bind to GTP-bound Gα subunits also allows them to act as effector antagonists, and additional functional roles for these proteins have been suggested (Neubig and Siderovski, 2002). Most studies of RGS function have used model cell systems or purified proteins. These experiments have provided a sound basis for informed predictions regarding the role of RGS proteins *in vivo*. Unfortunately, these predictions have been tested in relatively few native tissues. Genetic deletion studies have confirmed that individual RGS proteins are essential for some physiological processes. However, most cells express several RGS proteins, and these are likely to be at least somewhat functionally redundant. Thus genetic deletion is interpreted most easily in the few tissues (e.g., the retina) or organisms (e.g., *C. elegans*) where relatively few RGS proteins are expressed (Chen *et al.*, 2000; Dong *et al.*, 2000). Dominant-negative strategies are hampered by the mechanism of GAP activity, which appears to require only RGS binding (Tesmer *et al.*, 1997). Thus GAP-defective RGS mutants also bind Gα subunits poorly, rendering them ineffective as dominant negatives.

An alternative approach to determine the function of native RGS proteins is to render Gα subunits insensitive to RGS proteins (RGSi). Chemical mutagenesis of the yeast Gα_i homolog GPA1 revealed a mutation (G302S) that greatly decreased the sensitivity of this subunit to the yeast RGS protein SST2 (DiBello *et al.*, 1998). The analogous mutation has a similar effect on mammalian Gα subunits, making them insensitive to RGS binding or GAP activity (Lan *et al.*, 1998; Natochin and Artemyev, 1998). It was suggested that these mutations would be useful for determining the functional roles of endogenous RGS proteins in intact cells and organisms. We have made use of RGSi Gα subunits to examine endogenous RGS function in primary central nervous system (CNS) neurons. We have focused on presynaptic inhibition of synaptic transmission and post-synaptic modulation of ion channels and used electrophysiological techniques to assess G-protein-coupled receptor (GPCR) signaling. These responses are mediated by G-protein heterotrimers containing G$\alpha_{i/o}$ and G$\alpha_{q/11}$ family subunits, respectively. The rapid kinetics of GPCR-mediated synaptic responses compared to the intrinsic GTP hydrolysis rates of these Gα subunits suggests that RGS GAP activity could be essential for normal signaling. In addition to mutations conferring the RGSi phenotype, we have adopted strategies to separate signals mediated by heterologously expressed RGSi subunits from those mediated by native (RGS-sensitive) Gα subunits. These strategies provide RGSi subunits with a unique ability to couple to GPCRs not shared by native Gα subunits. Finally, in order to integrate RGSi subunits into native signaling pathways in CNS neurons, we utilize both viral and nonviral gene transfer techniques. Here we provide methodological details for using RGSi proteins in CNS neurons, an account of the advantages and disadvantages of this approach, and the caveats associated with interpreting results. RGSi Gα subunits were first used to study endogenous RGS function in native neurons by Jeong and Ikeda (2000) (see also Ikeda and Jeong, 2004).

Presynaptic Inhibition: Replacing Endogenous G$\alpha_{i/o}$ Subunits with RGSi Variants

Activation of several different GPCRs can inhibit neurotransmitter release from central and peripheral presynaptic nerve terminals. The primary mechanism of this inhibition is thought to be direct binding of G$\beta\gamma$ subunits to the voltage-gated calcium channels that mediate neurotransmission (Wu and Saggau, 1997). Under physiological conditions, presynaptic inhibition can rise to a maximum within a few hundred milliseconds and can decay over the course of just a few seconds. Although the precise identities of the G-protein subunits that mediate presynaptic inhibition are unknown, the most widely studied examples are completely abolished by pertussis toxin (PTX), implicating heterotrimers containing $G\alpha$ subunits of the $G\alpha_{i/o}$ family ($G\alpha_{i1-3}$, $G\alpha_o$).

Pertussis toxin functionally inactivates $G\alpha_{i/o}$ proteins by ADP ribosylating a cysteine residue near the carboxy terminus (in the -4 position). Ribosylated subunits are unable to interact productively with GPCRs. Mutation of this residue to an alternative amino acid renders the resulting $G\alpha$ subunit PTX insensitive (PTXi), and the expression of PTXi $G\alpha$ subunits restores signaling in cells after native $G\alpha_{i/o}$ proteins are inactivated with PTX. We expressed PTXi $G\alpha_{i/o}$ subunits in hippocampal neurons to rescue presynaptic inhibition mediated by adenosine A_1 receptors (A1Rs) after treatment with PTX. Expression of double-mutant RGSi/PTXi $G\alpha$ subunits allowed us to infer the functional role of endogenous RGS proteins in CNS nerve terminals (Chen and Lambert, 2000).

Hippocampal Microisland Cultures

In order to study the function of heterologously expressed G proteins in CNS nerve terminals, we sought a simplified preparation that was amenable to gene transfer yet retained the essential characteristics of synaptic transmission in more intact tissue, including robust, PTX-sensitive presynaptic inhibition. Accordingly, hippocampal neurons are dissociated from newborn rats and are plated onto microislands of permissive substrate, isolated from one another by a nonpermissive background. This arrangement obliges neurons to form synapses (autapses) onto themselves. The culture protocol follows several published methods (Furshpan et al., 1986; Segal and Furshpan, 1990), with minor variations.

The central area ($\sim$25-mm-diameter spot) of 35-mm culture dishes is coated with a thin layer of 0.15% agarose (Sigma A-9918) solution using a sterile cotton applicator and is then air dried in a laminar flow hood for 10 min. A glass chromatography sprayer (Kontes) is used to spray onto agarose-coated dishes a fine mist of substrate solution containing high molecular weight poly-D-lysine (B-D Collaborative) at 0.4 mg ml^{-1} and rat tail collagen (B-D Collaborative) at 0.65 mg ml^{-1}. This procedure yields randomly distributed substrate microislands $\sim$50–200 μm in diameter. After appropriate euthanasia, a newborn Sprague–Dawley rat brain is dissected quickly into ice-cold dissection medium [Earle's balanced salt solution (EBSS) containing 10 mM HEPES and 1 mM pyruvate, pH 7.3]. The hippocampi are isolated carefully, cleaned of meninges and blood vessels, chopped into small (~1-mm) pieces, and then transferred to a flask containing 5 ml minimal essential medium (MEM) with papain (Worthington, PAP2) at 25 units ml^{-1}. The flask is placed on a platform rocker in an incubator (37°, 5% CO_2) and rocked for ~30–40 min. The papain solution is drawn off, the tissue is washed twice with 10% fetal bovine serum (FBS; Hyclone) in MEM, and is then triturated gently with two to four fire-polished Pasteur pipettes with decreasing bore.

After centrifugation, cells are resuspended in growth medium [MEM containing 2% B-27 (GIBCO), 0.1% mito serum extender (B-D Collaborative), 5% FBS, 0.6% glucose, 1 mM pyruvate, 1 mM L-glutamine, 50 IU ml^{-1} penicillin, and 50 μg ml^{-1} streptomycin. To inhibit proliferation of glial cells, fluorodeoxyuridine (2.5 μM) is added to the growth medium. Uridine (2.5 μM) is also included to prevent inhibition of RNA synthesis. Fifty thousand cells in 2 ml growth medium are plated onto each culture dish. In order to increase the chance that cells settle down on the substrate spots, dishes are placed on the rocker in the incubator and rocked ~10 times per minute for 2 h. Afterward, cell cultures are maintained stationary in the incubator. After every 7–10 days in culture, 0.5–1 ml of growth medium supplemented with 2.5 μM fluorodeoxyuridine, 2.5 μM uridine, and 100 μM aminophosphonovaleric acid (APV) is added to each dish. Using this procedure, after 1–2 weeks in culture, a few or more than 10 microislands occupied by a single neuron with elaborate neurites with or without a glial cell(s) are present in most of the dishes. Neurons can be maintained for at least 3 weeks using this procedure and are used for experiments after 2–3 weeks in culture.

Several features of this preparation relevant to the type of experiment described later are worth mentioning. First, because neurites are prevented from contacting adjacent neurons by the nonpermissive agarose background, neurons are forced to make a large number (hundreds) of synaptic contacts (autapses) onto themselves. The result of this overconnection is a large-amplitude synaptic response, which provides a favorable signal-to-noise ratio and low variability. However, such large synaptic responses can be problematic when recording in voltage-clamp mode, as the unavoidably inadequate space- and voltage-clamp can degenerate into unacceptable voltage escape. Synaptic strength tends to increase with time in these cultures, thus there is an optimum window for making high-quality recordings (~2–3 weeks). Despite this problem, autaptic responses are unquestionably monosynaptic, thus ensuring that the synaptic response recorded reflects neurotransmitter release from the same (infected) cell. In addition, recordings can be made with a relatively simple recording apparatus (i.e., a single patch-clamp amplifier and micromanipulator) compared to recordings between synaptically coupled pairs of neurons. This preparation offers the additional advantage that the entire cell surface area (constrained by the diameter of the permissive spot) can be encompassed easily by a local perfusion device, thus facilitating rapid solution exchange at a large number of synapses. A major disadvantage of this culture preparation is the delicate nature of neurons on microislands. Procedures as simple as exchanging growth media can kill these cells unless great care is taken to avoid excitotoxicity. For this reason, the glutamate receptor antagonist APV is present whenever the growth medium is changed or supplemented or when neurons are exposed to viral vectors (see later). Finally, even the best microisland preparations will yield at most a few dozen useful (isolated) neurons in each 35-mm dish. This low density requires the use of a highly efficient gene transfer technique.

Generation of Recombinant Adenoviruses and Neuronal Infection

Nonviral transfection of primary neurons is now performed routinely using a variety of techniques and reagents (see later). However, none of these techniques reliably transfects neurons with more than 10% efficiency. In contrast, several viral vectors have been used to transform primary neurons (including microisland cultures) with high efficiency (up to 100%). The advantages and disadvantages of various viral vectors for neuronal gene transfer have been reviewed in detail elsewhere (Washbourne and McAllister, 2002). Recent improvements in vector toxicity and construction methods have made these tools more attractive and widely accessible, and it is likely that this trend will continue. In order to express exogenous $G\alpha$ subunits in hippocampal neurons, we constructed recombinant, replication-defective adenoviruses. Adenoviruses will infect nondividing cells such as neurons with extremely high efficiency (essentially 100% at a sufficient titer). Several systems for the production of recombinant adenoviruses are available commercially. For these experiments, we used the AdEasy system, which was developed by He *et al.* (1998) and is currently available from ATCC (www.atcc.org). The key feature of this system is that viruses are produced by recombination in bacteria rather than in eukaryotic cells. This obviates the need for plaque purification and reduces the time required to produce viruses. Virus production is monitored by expression of the enhanced green fluorescent protein (EGFP), which is encoded by the viral shuttle vector.

Fragments containing the coding sequences for the G-protein α subunits $G\alpha_{i1}$, $G\alpha_{i2}$, $G\alpha_{i3}$, and $G\alpha_o$ are amplified from templates provided by R. R. Reed (Johns Hopkins University). The reverse primers for these reactions were designed to introduce a point mutation of the -4 cysteine residue to isoleucine (C352I for $G\alpha_{i2}$, C351I for the remainder), thus producing PTXi subunits. The forward primers were designed such that the start codon is under the optimal context for translation initiation. Fragments encoding double-mutant, RGSi/PTXi subunits are amplified using the same primers and RGSi templates ($G\alpha_o$ G184S and $G\alpha_{i1}$ G183S) provided by R. Neubig (University of Michigan). The polymerase chain reaction products are digested with the restriction enzymes *KpnI* and *XbaI* and are then ligated into a shuttle vector (pAdTrack-CMV). The shuttle vector (with insert) is linearized with *Eco*RI and cotransformed with a viral backbone vector (pAdEasy-1) into electrocompetent *Escherichia coli* BJ5183 cells. Recombinants are selected with kanamycin (He *et al.*, 1998) and are confirmed by restriction analysis and nucleotide sequencing. Recombinant adenoviral plasmids are linearized with *PacI*, ethanol precipitated, and then transfected into HEK 293 cells in a ~80% confluent flask (75 cm^2). After ~12–24 h, cells are subcultured into four 75-cm^2 flasks. After 8–10 days, recombinant adenoviruses are purified by cesium chloride banding and dialysis. The titer is calculated from the number of EGFP-positive HEK 293 cells in 35-mm culture dishes infected with 1 to 10 serially diluted adenoviral stock 30 h after infection.

Hippocampal microisland cultures are infected with recombinant adenoviruses after at least 14 days *in vitro*. Approximately 1 h prior to infection the glutamate receptor antagonist APV is added to the culture medium to a final concentration of 50–100 μM. We find that without APV pretreatment, infected neurons often appear granulated and fail to produce synaptic responses. The appropriate amount of adenovirus stock (~1–3 × 10^7 infectious units) is diluted in 50–100 μl of growth medium and is added to the culture dish. Because the pAdTrack-CMV shuttle vector contains an EGFP expression cassette, neurons infected by recombinant adenovirus express both EGFP and the gene of interest. Successfully infected neurons are identified for electrophysiological recording using standard epifluorescence microscopy; EGFP is visible in infected neurons as early as 16 h after infection.

Electrophysiology

Whole cell patch-clamp recordings are made from isolated (one neuron per microisland) neurons expressing EGFP 40–72 h after infection. For recording of excitatory postsynaptic (autaptic) currents (EPSCs), neurons are held at −60 mV and are depolarized above 0 mV with pairs (50-ms interval) of 2-ms^2 commands delivered every 2 s. These commands evoke unclamped action currents, which propagate as action potentials and evoke EPSCs. Series resistance (after compensation) is monitored during recording and cells are discarded if a significant increase occurs. Currents are digitized and recorded using a multifunction I/O board (National Instruments) and WinWCP software (provided by Dr. J. Dempster, Strathclyde University). Patch electrodes are prepared by pulling borosilicate glass capillaries (World Precision Instrument) using a Flaming/Brown micropipette puller (Sutter Instrument Co.). The resistance of the electrodes is between 2 and 4 MΩ after filling with a solution containing (in mM) 140 K-gluconate, 5 KCl, 0.2 EGTA, 10 HEPES, 3 MgATP, and 0.3 Na$_2$GTP (pH 7.2, $\sim$ 295 mOsm kg^{-1} H$_2$O). The external solution contains (in mM) 145 NaCl, 2.5 KCl, 10 HEPES, 10 glucose, 1.5 CaCl$_2$, and 2.5 MgCl$_2$ (pH 7.2, $\sim$310 mOsm kg^{-1} H$_2$O). All recordings are made at room temperature. Drugs are applied via an array of gravity-fed fused silica tubes (internal diameter 200 μm). Rapid switching between drug solutions is accomplished by moving the array laterally using a stepper motor (Warner Instrument Corp.), which is controlled manually or with a digital stimulator (NeuroData). Recordings of junction potential changes evoked by switching between normal and dilute extracellular solutions indicate that solution exchange at the recording site is complete within a few hundred milliseconds using this apparatus.

Results: Presynaptic G$\alpha_{i/o}$

Hippocampal neurons express several native GPCRs that couple to G$\alpha_{i/o}$ proteins and, when activated, mediate robust presynaptic inhibition of neurotransmitter release. Among these are adenosine A$_1$ receptors A1Rs and GABA$_B$ receptors (GBRs). When selective agonists (adenosine and baclofen, respectively) are applied, the amplitude of EPSCs evoked in microisland cultures is inhibited by $\sim$80% (Chen and Lambert, 2000). In uninfected neurons or neurons infected with viruses expressing only EGFP, this presynaptic inhibition is essentially complete within 2 s of agonist application and reverses (recovers) with a monoexponential time constant of $\sim$7 s. Overnight incubation with PTX (100 ng ml^{-1}; List Biological) completely abolished A1R- and GBR-mediated presynaptic inhibition, as expected for responses mediated by native G$\alpha_{i/o}$ proteins. Infection with adenoviruses expressing PTXi Gα_{i1-3} or PTXi Gα_o subunits partially rescued presynaptic inhibition mediated by A1Rs, although this rescue was much greater for Gα_i subunits (Chen and Lambert, 2000). Interestingly, GBR-mediated presynaptic inhibition was not rescued by these PTXi Gα subunits. It is possible that the C to I mutation used to render these subunits PTXi disrupted coupling to GBRs (Franek et $al.$, 1999). Presynaptic inhibition mediated by PTXi Gα subunits was similar to that mediated by native Gα subunits, albeit less complete and somewhat slower in onset. In contrast, when presynaptic inhibition was rescued with double-mutant

RGSi/PTXi Gα subunits, the recovery from inhibition was prolonged by 5- to 10-fold (Fig. 1).

This result is consistent with the idea that endogenous RGS proteins are present in CNS presynaptic terminals and are required for the rapid termination of GPCR signals in these structures (Chen and Lambert, 2000). The onset of presynaptic inhibition was also dramatically slower when mediated by RGSi/PTXi Gα_o than by PTXi Gα_o subunits. The onset effects were less pronounced for RGSi/PTXi Gα_i subunits. The mechanism whereby RGS proteins regulate the onset of GPCR signals is not known, but this result is consistent with previous studies of RGS regulation of inwardly rectifying potassium (GIRK) channel activation (Doupnik *et al.*, 1997; Jeong and Ikeda, 2001; see also Doupnik *et al.*, 2004). In summary, adenoviral expression of PTXi Gα subunits rescued presynaptic inhibition following PTX pretreatment. This allowed us to substitute RGSi Gα subunits for RGS-sensitive subunits and thus determine the role of endogenous RGS proteins at presynaptic terminals.

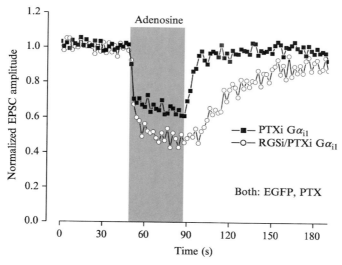

FIG. 1. Presynaptic inhibition of neurotransmitter release by pertussis toxin-insensitive (PTXi) and RGS-insensitive (RGSi)/PTXi Gα_{il} subunits. Normalized excitatory postsynaptic current (EPSC) amplitude recorded from hippocampal neurons grown on substrate microisland is plotted versus time. Native A$_1$ adenosine receptors (A1Rs) were activated by exogenously applied adenosine during the period indicated by the shaded rectangle. Responses from a cell infected with adenovirus expressing PTXi Gα_{il} and from a cell expressing RGSi/PTXi Gα_{il} (G183S) are shown superimposed. Both neurons also expressed the enhanced green fluorescent protein (EGFP) and were exposed to pertussis toxin for >12 h. The recovery from presynaptic inhibition mediated by RGSi/PTXi subunits is slower than that mediated by PTXi subunits.

Postsynaptic Excitation: Bypassing Endogenous $G\alpha_{q/11}$ Subunits with RGSi Chimeras

It is now well established that postsynaptic (e.g., GIRK-mediated) and presynaptic (e.g., presynaptic inhibition) signals mediated by PTX-sensitive $G\alpha_{i/o}$ subunits are regulated by endogenous RGS proteins. While it is well known that $G\alpha_{q/11}$ subunits mediate slow, GPCR-mediated synaptic excitation and that RGS proteins possess GAP activity at these subunits, it is not known whether endogenous RGS proteins regulate $G\alpha_{q/11}$-mediated synaptic responses. The importance of RGS proteins for $G\alpha_{q/11}$-mediated signaling is not as predictable as that for $G\alpha_{i/o}$-mediated signals. Whereas RGS proteins are the only known GAPs for $G\alpha_{i/o}$ subunits, both RGS proteins and the effector phospholipase C-β (PLCβ) can effectively accelerate the intrinsic GTPase activity of $G\alpha_{q/11}$ subunits (Ross and Wilkie, 2000). This raises an obvious question: is RGS GAP activity meaningful for signals mediated by $G\alpha_{q/11}$ and PLCβ? To approach this question, a strategy similar to that described for $G\alpha_{i/o}$ subunits can be used. Analogous point mutations to those used for $G\alpha_{i/o}$ subunits are introduced into $G\alpha_q$ subunits. Because endogenous $G\alpha_{q/11}$ subunits are not inactivated by PTX or other specific agents, an alternative means is needed to avoid signals mediated by these subunits. For this reason, RGSi point mutations are introduced into chimeric $G\alpha$ subunits consisting primarily of $G\alpha_q$ with the nine carboxyl-terminal amino acids exchanged for those found in $G\alpha_i$ (Conklin et al., 1993). These chimeric $G\alpha_{qi9}$ subunits stimulate PLCβ activity, but can be activated by $G\alpha_i$-coupled receptors (which normally fail to significantly activate this effector). The importance of endogenous RGS signals can thus be inferred by comparing signals mediated by "wild-type" (RGS-sensitive) $G\alpha_{qi9}$ subunits and those mediated by RGSi $G\alpha_{qi9}$ mutants.

Cerebellar Granule Neuron Cultures

In many neurons, $G\alpha_{q/11}$-mediated slow excitatory postsynaptic responses are generated by the closing of tonically active "leak" potassium channels. The best-known example of this type of response is the slow muscarinic excitatory postsynaptic potential in sympathetic ganglion neurons, but similar responses are generated in several CNS neurons. We have chosen cerebellar granule neurons (CGNs) as a model cell to study $G\alpha_{q/11}$-mediated regulation of postsynaptic ion channels. These cells are easy to prepare, maintain, and identify in a dissociated culture preparation. More importantly, they express endogenous m3 muscarinic receptors, activation of which inhibits a tonic potassium current ($I_{K,SO}$) mediated by two-pore domain potassium channels. Finally, these cells can be transfected reliably using standard reagents.

The protocol for preparing CGN cultures is a simplified version of the protocol described earlier for hippocampal microisland cultures. Apart from the brain region involved, the primary distinctions between methods are as follows. Cultures are prepared from 5- to 8-day-old animals and are plated onto 35-mm culture dishes that are first coated in their entirety with poly-L-lysine. The culture medium contains a final concentration of 25 mM potassium and is not changed or supplemented after the initial plating. Cultures are used for experiments after 7–12 days.

Generation of RGSi Gα_{qi9} and Polyethylenimine (PEI)-Mediated Transfection

Chimeric Gα_{qi9} is obtained from Dr. Bruce Conklin (University of California, San Francisco) and subcloned into the plasmid vector pcDNA3.1, which drives expression from the cytomegalovirus promoter. Point mutations that engender insensitivity to RGS protein action, analogous to those described earlier (e.g., G188S and S211D), are introduced using a commercially available site-directed mutagenesis kit (Quikchange, Stratagene) and are verified by nucleotide sequencing. CGNs are always contransfected with a plasmid expressing EGFP (pEGFP-N1, Clontech) to allow identification of transfected neurons.

Several nonviral transfection techniques are currently being used to transfect primary neurons, and it is likely that most or all of them would be applicable to CGNs. The protocol used for these experiments relies on polyethylenimine (PEI), a branched polycation that condenses DNA and may enhance plasmid DNA delivery to the nucleus (Boussif et al., 1995). A 0.1 M (of the monomer) stock solution of PEI is prepared by dissolving 45 mg of PEI (Aldrich; average molecular mass 25 kDa) in 10 ml of distilled water. After adjusting the pH to 7.0, the stock is sterile filtered (0.2 μm) and stored at 4° (for up to 1 year). We have found that, for CGNs, the optimal nitrogen/phosphate ratio is 10, meaning 1 μl of PEI stock is needed to complex 3.25 μg of plasmid DNA. To transfect ten 35-mm culture dishes, 2 μl of PEI stock is added to 250 μl of 150 mM NaCl in a microcentrifuge tube, and 6.25 μg of plasmid DNA is added to 250 μl of 150 mM NaCl in a separate tube. After vortexing, the PEI solution is added dropwise to the DNA solution, and 10 min are allowed for DNA complex formation. During this interval, the growth medium is removed from the culture dishes to be transfected and retained, and the neurons are washed once with MEM. The PEI/DNA complex solution is diluted 10-fold in MEM, and 1 ml is applied to each of the 10 dishes. Cultures are returned to the incubator for 1 h, after which time the transfection solution is removed and replaced with the original growth medium. The expression of exogenous proteins (e.g., EGFP) is evident 12 h after transfection.

Electrophysiology and Results: Postsynaptic $G\alpha_{q/11}$

Recordings are made from EGFP-expressing CGNs 12–24 h after transfection using methods similar to those described for hippocampal neurons. Recordings can be made using the standard whole cell recording configuration, but are more stable if made using the perforated patch technique. For perforated patch recordings, ~6 mg of the antibiotic amphotericin B (Sigma) is dissolved in 100 μl of dimethyl sulfoxide. This stock is diluted 1:200 into a pipette solution containing (in mM) 140 K-gluconate, 5 KCl, and 10 HEPES (pH 7.2, ~295 mOsm kg^{-1} H$_2$O) and is sonicated for ~1 min. This solution is usable for up to 1 h to completely fill patch pipettes. After formation of a high resistance seal, 5 min is allowed for the antibiotic to partition into the patch, and electrical access is monitored using square test voltage commands. Ion channels underlying $I_{K,SO}$ are blocked by low pH and are inhibited by divalent cations. Therefore, in order to maximize the amplitude of this current, the extracellular solution for recording is adjusted to pH 8.0, and the concentrations of Mg^{2+} and Ca^{2+} are lowered to 0.5 mM. Neurons are held in voltage-clamp mode at −20 mV, at which potential $I_{K,SO}$ is apparent as a tonic outward current.

As described originally by Watkins and Mathie (1996), application of the muscarinic acetylcholine receptor agonist carbamylcholine (carbachol) rapidly inhibits $I_{K,SO}$, presumably by acting at endogenous m3 receptors. Activation of other $G\alpha_{q/11}$-coupled receptors (e.g., P2Y, 5HT2) similarly inhibits $I_{K,SO}$, but neither endogenous nor heterologously expressed $G\alpha_{i/o}$-coupled receptors produce similar inhibition. However, if CGNs are first transfected with $G\alpha_{qi9}$, the activation of $G\alpha_{i/o}$-coupled receptors (e.g., A1Rs, α2 adrenoreceptors) inhibits $I_{K,SO}$. The kinetics of $I_{K,SO}$ inhibition mediated by $G\alpha_{qi9}$ subunits reasonably approximate inhibition mediated by native $G\alpha_{q/11}$ subunits; recovery of $I_{K,SO}$ after agonist washout proceeds over the course of 10–30 s, depending on the cotransfected receptor. If RGS protein-mediated GAP activity is critical for the termination of $G\alpha_{q/11}$ (or $G\alpha_{qi9}$)-mediated inhibition of $I_{K,SO}$, then such inhibition mediated by RGSi $G\alpha_{qi9}$ mutants should reverse very slowly compared to that mediated by wild-type $G\alpha_{qi9}$. However, as shown in Fig. 2, the reversal kinetics of $I_{K,SO}$ inhibition mediated by RGS-sensitive and RGSi $G\alpha_{qi9}$ subunits is comparable. This result is a stark contrast to analogous

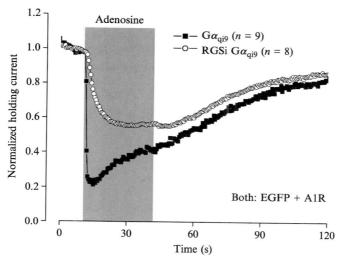

FIG. 2. Postsynaptic inhibition of a resting potassium current by RGS-sensitive and RGS-insensitive (RGSi) $G\alpha_{qi9}$ chimeras. Normalized holding current recorded from cerebellar granule neurons is plotted versus time. Heterologously expressed A1Rs were activated by exogenously applied adenosine during the period indicated by the shaded rectangle. Responses from a cell transfected with plasmid vectors expressing $G\alpha_{qi9}$ and from a cell expressing RGSi $G\alpha_{qi9}$ (G188S) are shown superimposed. Both neurons also expressed EGFP. The recovery of holding current inhibition mediated by RGSi $G\alpha_{qi9}$ subunits is comparable to that mediated by RGS-sensitive $G\alpha_{qi9}$ subunits.

experiments examining the deactivation of GIRK channels and recovery from presynaptic inhibition mediated by $G\alpha_{i/o}$ proteins. Interestingly, the onset of $I_{K,SO}$ inhibition mediated by RGSi subunits is slowed compared to that mediated by RGS-sensitive subunits. The mechanism of this difference is not known, but similar changes in response onset kinetics have been observed for $G\alpha_{i/o}$-mediated responses. In summary, the regulatory role of RGS proteins with respect to $G\alpha_{q/11}$-mediated synaptic responses may be very different than what would be predicted due to GAP activity alone. Indeed, the effector PLCβ may provide the primary GAP activity to regulate the kinetics of $G\alpha_{q/11}$-mediated synaptic responses.

Conclusions and Caveats

RGS-insensitive $G\alpha$ subunits have proven their worth as tools to elucidate the functional roles of endogenous RGS proteins. Nonetheless, the strategies outlined in this article have a few important limitations that should be pointed out. First, the mutations incorporated into $G\alpha$ subunits to render them RGSi could change other critical features of these proteins. Such changes could include the rate of receptor-catalyzed GDP release and/or heterotrimer dissociation, or activation of effectors by RGSi $G\alpha$ subunits. Biochemical studies of RGSi $G\alpha$ subunits have not revealed large changes in these parameters (Lan *et al.*, 1998), but such information is not available for every possible subunit. In the absence of biochemical data, several different RGSi mutations should be compared whenever possible. In any case, it should be kept in mind that mutations could change more than sensitivity to RGS proteins. Second, these strategies rely on reconstituting a substantial fraction of a signaling pathway in native cells. Because $G\alpha$ subunits must associate with $G\beta\gamma$ subunits in order to interact productively with GPCRs, it may be necessary to coexpress $G\beta\gamma$ subunits with RGSi $G\alpha$ subunits. However, the stoichiometry of G-protein subunit expression must be titrated carefully in order to prevent tonic $G\beta\gamma$ signaling (by excess free $G\beta\gamma$ subunits) or sequestration of $G\beta\gamma$ subunits (by excess $G\alpha$ subunits). This problem can be particularly vexing for $G\beta\gamma$-mediated signals. Fortunately in the case of presynaptic terminals, it was not necessary to coexpress $G\beta\gamma$ subunits. RGSi $G\alpha$ subunits presumably formed heterotrimers with native $G\beta\gamma$ subunits, perhaps even prior to transport to presynaptic terminals. In the case of $G\alpha$-mediated signals (e.g., activation of PLCβ by $G\alpha_{qi9}$), the stoichiometry appears to be less of an issue, probably due to the relatively inert nature of GDP-bound free $G\alpha$ subunits. Finally, both of the strategies outlined here rely on modified receptor G-protein coupling. In the case of $G\alpha_{i/o}$ signals, the mutation conferring PTX insensitivity can dramatically change coupling efficiency to various receptors (Bahia *et al.*, 1998). It is therefore important to match the receptor and mutant to approximate native signaling kinetics as closely as possible. Similarly, $G\alpha_{q/11}$-coupled receptors are bypassed entirely in the case of RGSi $G\alpha_{qi9}$. Here too we have found that the choice of receptor is critical for reconstituting signaling with kinetics comparable to the native situations.

In conclusion, RGSi $G\alpha$ subunits are useful tools for determining the general role of endogenous RGS proteins in CNS neurons. Combining site-directed mutagenesis, viral and nonviral gene transfer, and electrophysiology has shown that RGS proteins in CNS neurons are more than simple accelerators of GTPase activity.

Acknowledgments

We thank Drs. Rick Neubig, Randall Reed, Stephen Ikeda, and Bruce Conklin for providing cDNA clones used in these studies and John Dempster for providing data acquisition software. We thank Alex Pereira for expert technical assistance. This work was supported by grants from the NIH (NS36455) and a VA Merit Award.

References

Bahia, D. S., Wise, A., Fanelli, F., Lee, M., Rees, S., and Milligan, G. (1998). Hydrophobicity of residue351 of the G protein Gi1 alpha determines the extent of activation by the alpha 2A-adrenoceptor. *Biochemistry* **37,** 11555–11562.

Boussif, O., Lezoualc'h, F., Zanta, M. A., Mergny, M. D., Scherman, D., Demeneix, B., and Behr, J. P. (1995). A versatile vector for gene and oligonucleotide transfer into cells in culture and *in vivo:* Polyethylenimine. *Proc. Natl. Acad. Sci. USA* **92,** 7297–7301.

Chen, C. K., Burns, M. E., He, W., Wensel, T. G., Baylor, D. A., and Simon, M. I. (2000). Slowed recovery of rod photoresponse in mice lacking the GTPase accelerating protein RGS9-1. *Nature* **403,** 557–560.

Chen, H. M., and Lambert, N. A. (2000). Endogenous regulators of G protein signaling proteins regulate presynaptic inhibition at rat hippocampal synapses. *Proc. Natl. Acad. Sci. USA* **97,** 12810–12815.

Conklin, B. R., Farfel, Z., Lustig, K. D., Julius, D., and Bourne, H. R. (1993). Substitution of 3 amino-acids switches receptor specificity of G(Q)alpha to that of G(I)alpha. *Nature* **363,** 274–276.

DiBello, P. R., Garrison, T. R., Apanovitch, D. M., Hoffman, G., Shuey, D. J., Mason, K., Cockett, M. I., and Dohlman, H. G. (1998). Selective uncoupling of RGS action by a single point mutation in the G protein alpha-subunit. *J. Biol. Chem.* **273,** 5780–5784.

Dong, M. Q., Chase, D., Patikoglou, G. A., and Koelle, M. R. (2000). Multiple RGS proteins alter neural G protein signaling to allow C-elegans to rapidly change behavior when fed. *Genes Dev.* **14,** 2003–2014.

Doupnik, C. A., Davidson, N., Lester, H. A., and Kofuji, P. (1997). RGS proteins reconstitute the rapid gating kinetics of gbetagamma-activated inwardly rectifying K+ channels. *Proc. Natl. Acad. Sci. USA* **94,** 10461–10466.

Doupnik, C. A., Jaén, C. and Zhang, Q. (2004). Measuring the modulatory effects of regulator of G-protein signaling (RGS) proteins on G-protein-gated inwardly rectifying potassium channel gating. *Methods Enzymol.* **389,** 131–154.

Franek, M., Pagano, A., Kaupmann, K., Bettler, B., Pin, J. P., and Blahos, J. (1999). The heteromeric GABA-B receptor recognizes G-protein alpha subunit C-termini. *Neuro-pharmacology* **38,** 1657–1666.

Furshpan, E. J., Landis, S. C., Matsumoto, S. G., and Potter, D. D. (1986). Synaptic functions sympathetic neurons in microcultures. I. Secretion of norepinephrine and acetylcholine. *J. Neurosci.* **6,** 1061–1079.

He, T. C., Zhou, S. B., da Costa, L. T., Yu, J., Kinzler, K. W., and Vogelstein, B. (1998). A simplified system for generating recombinant adenoviruses. *Proc. Natl. Acad. Sci. USA* **95,** 2509–2514.

Ikeda, S. R., and Jeong, S. W. (2004). Use of regulators of G-protein signaling (RGS)-insensitive Gα subunits to study endogenous RGS protein action on G-protein modulation of N-type calcium channels in sympathetic neurons. *Methods Enzymol.* **389,** 170–189.

Jeong, S. W., and Ikeda, S. R. (2000). Endogenous regulator of G-protein signaling proteins modify N-type calcium channel modulation in rat sympathetic neurons. *J. Neurosci.* **20,** 4489–4496.

Jeong, S. W., and Ikeda, S. R. (2001). Differential regulation of G protein-gated inwardly rectifying K+ channel kinetics by distinct domains of RGS8. *J. Physiol. Lond.* **535,** 335–347.

Lan, K. L., Sarvazyan, N. A., Taussig, R., Mackenzie, R. G., DiBello, P. R., Dohlman, H. G., and Neubig, R. R. (1998). A point mutation in G alpha(o) and G alpha(il) blocks interaction with regulator of G protein signaling proteins. *J. Biol. Chem.* **273,** 12794–12797.

Natochin, M., and Artemyev, N. O. (1998). Substitution of transducin Ser(202) by Asp abolishes G-protein/RGS interaction. *J. Biol. Chem.* **273,** 4300–4303.

Neubig, R. R., and Siderovski, D. R. (2002). Regulators of G-protein signalling as new central nervous system drug targets. *Nature Rev. Drug Disc.* **1,** 187–197.

Ross, E. M., and Wilkie, T. M. (2000). GTPase-activating proteins for heterotrimeric G proteins: regulators of G protein signaling (RGS) and RGS-like proteins. *Annu. Rev. Biochem.* **69,** 795–827.

Segal, M. M., and Furshpan, E. J. (1990). Epileptiform activity in microcultures containing small numbers of hippocampal neurons. *J. Neurophysiol.* **64,** 1390–1399.

Tesmer, J. J. G., Berman, D. M., Gilman, A. G., and Sprang, S. R. (1997). Structure of RGS4 bound to AlF4⁻ activated G(i alpha 1): Stabilization of the transition state for GTP hydrolysis. *Cell* **89,** 251–261.

Washbourne, P., and McAllister, A. K. (2002). Techniques for gene transfer into neurons. *Curr. Opin. Neurobiol.* **12,** 566–573.

Watkins, C. S., and Mathie, A. (1996). A non-inactivating K+ current sensitive to muscarinic receptor activation in rat cultured cerebellar granule neurons. *J. Physiol.* **491,** 401–412.

Wu, L. G., and Saggau, P. (1997). Presynaptic inhibition of elicited neurotransmitter release. *Trends Neurosci.* **20,** 204–212.

Subsection D

Mouse Models of RGS Protein Action

[13]   *In Situ* Hybridization Analysis of RGS
mRNA Regulation and Behavioral Phenotyping
of RGS Mutant Mice

By STEPHEN J. GOLD and VENETIA ZACHARIOU

Abstract

To elucidate the functional role of regulators of G-protein signaling (RGS) *in vivo*, it will be critical to (i) determine how RGS activity is altered in response to a variety of manipulations and (ii) observe how the system is changed when RGS protein function is altered genetically. To facilitate studies of dynamic regulation of RGS protein activity, this article describes detailed methods for radioisotopic *in situ* hybridization for semi-quantitative analyses of RGS mRNA abundances. Toward characterizing the functional differences in mice with genetically altered RGS activities, this article describes a subset of behavioral tests suitable for assaying sensitivities to drugs of abuse. These protocols should provide valuable guidance for investigators to establish these methodologies independently in their own laboratories and, over time, increase our understanding of RGS function *in vivo*.

Introduction

Regulators of G-protein signaling (RGS) family of proteins negatively regulate G-protein signaling via their GTPase-accelerating activity (GAP) at Gα subunits of the heterotrimeric G-protein complex (Berman and Gilman, 1998; Ross and Wilkie, 2000; Siderovski *et al.*, 1999). At the very least, RGS proteins serve a role as negative feedback regulators on activated Gα subunits. They undoubtedly have other roles beyond their GAP activity via their diverse domain structures that include scaffolding (Ross and Wilkie, 2000; Siderovski *et al.*, 1999) and guanine nucleotide exchange activity (Kozasa *et al.*, 1998). To date, approximately 25 mammalian RGS genes have been identified and, via alternative splicing, there is likely to be several hundred different RGS protein isoforms expressed throughout the body. Although great strides have been made toward understanding the functional role of at least one RGS protein, RGS9 (Chen *et al.*, 2000; Rahman *et al.*, 2003), the field is only beginning to explore the roles of the great majority of RGS proteins. Many important questions remain, including the specificity of RGS modulation of signaling *in vivo*. For example, what is the specificity of RGS–Gα interactions *in vivo*? Also, are RGS proteins specific for Gα subunits activated by a particular G-protein-coupled receptor?

Undoubtedly, answers to these questions will take years of future research and will entail many experimental approaches. Two strategies that will be invaluable for determining RGS specificity and function at both the signaling complex and systems levels will be analyses of the dynamic regulation of RGS activity and the creation and characterization of RGS knockout (KO) mice. The first strategy, trying to understand the dynamic regulation of specific RGS proteins *in vivo*, is complicated by the fact that in any given tissue and cell type, there are likely to be several different RGS proteins expressed (Gold *et al.*, 1997; Grafstein-Dunn *et al.*, 2001). As a result, it will be nearly impossible to assay for changes in GAP activities of any one individual RGS subtype. Novel tools will be necessary to achieve these types of analyses.

In lieu of novel technologies for assaying activities of specific RGS proteins, one can use the following algorithm to partially resolve changes in RGS GAP activity. First, GAP assays on crude brain membranes provide a sensitive measure for detecting changes in GAP activity either at a given Gα subunit or at Gα subunits activated by specific receptors (Krumins and Gilman, 2002; Ross, 2002; Wang *et al.*, 1997). For example, via single-turnover assays, one can compare the relative amounts of GAP activity toward Gα$_i$ in brain membranes prepared from an animal treated with vehicle and an animal treated chronically with a drug of interest. Similarly, with steady-state GAP assays, one can assay the relative GAP activity toward receptor-activated Gα subunits in membranes prepared from animals treated with two or more different treatments. Second, one can obtain *indirect* measures of specific RGS subtype activity by assaying for changes in RGS protein concentration. If subtype-specific antibodies are available, changes in relative protein concentrations can be determined. If antibodies are not available, then one can obtain indirect measures of protein abundance by assaying for changes in mRNA abundance (Gall *et al.*, 1995; Seroogy and Herman, 1997). If using these approaches, both RGS protein abundance and net GAP activity are altered in the same direction, then one can make the argument that the GAP activity of that specific RGS protein has changed. The first half of this article focuses on the second aspect of these indirect measures of RGS activity, using the *in situ* hybridization technique to assay for changes in RGS mRNA expression. Specifically, isotopic *in situ* hybridization methods are detailed. The second half of this article describes a general algorithm for behavioral phenotyping of RGS KO mice. These experiments emphasize behavioral pharmacological assays that test for phenotypic changes relevant to psychostimulant and opiate addiction.

Background to *In Situ* Hybridization Studies

The mammalian central nervous system (CNS) is the most heterogeneous organ in mammals and includes hundreds of distinct cellular phenotypes that have unique patterns of gene expression. *In situ* hybridization histochemistry has become the standard technique for assaying mRNA expression with anatomical resolution. In recent years, however, several new techniques have become popular, including single cell reverse-transcribed polymerase chain reaction (RT-PCR), quantitative real-time PCR (QRT-PCR), and cDNA array analyses. All three of these techniques have advantages, including (i) the ability to correlate mRNA expression with electrophysiological and pharmacological characteristics of the cell (single cell RT-PCR); (ii) very precise measurements of relative mRNA concentrations (QRT-PCR), and (iii) the ability to survey the expression levels of thousands of mRNAs simultaneously, including novel uncharacterized mRNAs (cDNA array). Clearly, the most notable advantage of *in situ* hybridization histochemistry is the ability to analyze mRNA expression over a large number of brain structures while still enabling visualization of expression in single cells. A severe limit of the *in situ* hybridization histochemistry approach, however, is that the investigator is limited to surveying one to two mRNAs per section.

In situ hybridization protocols can be tailored to a range of tissue preparations, including fresh-frozen tissue, perfused-fixed tissue, and whole-mount preparations. Perfused, paraformaldehyde-fixed tissue provides the best resolution for morphological analyses in the same tissue sections. Nevertheless, perfusion introduces an additional procedural step that can contribute substantially to the interanimal variability in labeling densities. In addition, the background tends to be higher than that obtained with fresh-frozen tissue sections. The whole mount is particularly valuable for developmental studies where large portions of embryonic tissue need to be visualized. For quantification of relative mRNA levels, the approach preferred by our laboratories is the fresh-frozen technique. This provides for relatively quick tissue processing with low background signal and low experimenter-introduced interanimal variability (Fig. 1A and B).

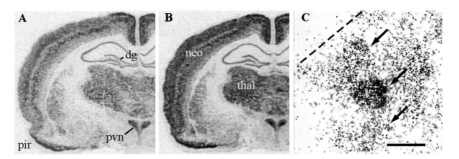

FIG. 1. Bright-field photomicrographs of film autoradiograms of rat coronal hemisections through the dorsal hippocampus hybridized to the ^{35}S-labeled RGS4 riboprobe (A, B). Note that in an animal killed 2 h after an electroconvulsive shock (B), RGS4 riboprobe labeling densities are increased in the neocortex (neo), dentate gyrus granule cells (dg), thalamus (thal), paraventricular nucleus of the hypothalamus (pvn), and piriform cortex (pir) relative to the control animal. (C) Bright-field photomicrograph of an emulsion autoradiogram of RGS2 riboprobe labeling illustrates the heterogeneity of labeling densities in three putative pyramidal cells (arrows) in layer 2 of the piriform cortex. Dashed line demarcates border between layers 1 and 2. Scale bar is 2.5 mm in A, B and 20 μm in C.

In Situ Hybridization Protocols

Tissue Preparation

Unanesthetized animals are killed by decapitation, the brain is removed from the skull, and the tissue block is cut flush for the desired plane of section and is then frozen rapidly on dry ice. When resting the brain on the bed of dry ice, we have found that we can substantially reduce the compression of the brain due to its own weight by sprinkling crushed dry ice over the top of the brain as it begins to freeze. This precaution is more important for rat brains, where the weight of the brain can produce substantial compression. As we have no hard data on rapid anesthesia-induced changes in RGS subtype mRNA abundance, we decapitate all animals unanesthetized per the approval of our institutional animal care and use committee. For storage, brains are wrapped in an inner layer of cellophane and an outer layer of aluminum foil and stored at $-80°$ until cutting.

Brains are cut in a cryostat (numerous quality manufacturers include Leica and Zeiss) at a thickness of 14 μm and thaw mounted onto Superfrost Plus glass microscope slides (Fisher, Pittsburgh, PA). When humidity is very low and static electricity is particularly high, we have found it advantageous to cut the sections without protective gloves. Bare hands allow one to stay electrically grounded and dramatically decreases the likelihood of sections flying away due to electrical fields created by the static electricity. There are less expensive alternatives to Superfrost Plus slides, including treating the slides with Vectabond (Vector Labs, Burlingham, CA). For adult brain, good sections can be obtained by maintaining the cryostat at $-16°$. Brains are mounted to the chuck in the appropriate orientation with mounting medium and are allowed to equilibrate to cryostat temperatures (equilibration time depends on the size of the sample). We prefer to maintain the slides in the cryostat at $-16°$ and position the cut section on a cold slide. Slides are labeled lightly with a #2 pencil, taking care to minimize graphite dust that can particularly affect analyses of emulsion autoradiography. Once the section is on the slide, positioned and flattened correctly, the section is thaw mounted by warming the back of the slide with a finger until the section becomes translucent. The slide is then placed back in the slide rack in the cryostat to chill.

For experiments where quantitative comparisons will be made between brains, it is imperative to cut the brains in the same orientation and plane of section. To gauge that this is the case, we make fine adjustments to the plane of section by visualizing the warmed, just-cut sections on a macroscope under dark-field illumination. Major gray and white matter landmarks can be observed easily in these unstained sections and fine adjustments made to the orientation of the brain via the x and y knobs controlling the chuck mount. With experience, this can readily be done by eye.

We use a sampling strategy whereby a series of spaced sections across a region of interest provide multiple representations whose labeling densities can be averaged to obtain a mean labeling density for that region. For example, to sample dorsal rat hippocampus, if 14-μm coronal sections are taken every 240 μm, this would provide four sections per brain for subsequent analyses. For adult rat sections, we routinely mount four sections per slide in a single row. For mouse sections, we usually mount eight sections per slide in two rows. If the slides will be processed for emulsion autoradiography, additional care should be taken to mount the sections a minimum of 3 mm from any edge of the slide, including the frosted writing surface. This will avoid complications that could arise from edge effects on emulsion thickness. Prior to pretreatments, cut sections are stored at $-80°$ in black, air-tight, 25 slide boxes (Becton-Dickinson, San Jose, CA).

Hybridization

The hybridization procedure can be broken down into four distinct steps: (i) tissue pretreatment, (ii) probe preparation, (iii) probe hybridization and, (iv) post-hybridization washes.

Tissue Section Pretreatment. Pretreatment of sections can be conducted immediately after removing the slides from the freezer. Alternatively, we have been successful letting sections thaw to room temperature for ~30 min with no apparent diminution of signal. Thawing is done with slides resting just off vertical with the frosted side down on a slide drying rack (high-density polyethylene, grooved slide holders from Fisher). For all of the following steps, we have two sets of dishes for processing referred to as day 1 and day 2 dishes. Day 1 dishes are never exposed to experimenter-introduced RNase activity. We prefer green Tissue Tek staining dishes (Sakura Finetek, Torrance, CA) that can each accommodate 24 slides. Unless noted, all steps described in green staining dishes are conducted at room temperature on an orbital shaker. All solutions described are made with MilliQ grade (Millipore, Billerica, MA) water. We have found that, with MilliQ grade water, it is unnecessary to treat water with diethyl-pyrocarbonate (DEPC) as no RNase-based degradation is apparent on ethidium bromide stained gels of *in vitro*-transcribed RNA. To better preserve the tissue sections through the processing steps, sections are first fixed in fresh, chilled, filtered, 4% paraformaldehyde in 0.1 M sodium phosphate buffer (pH 7.2) for 10 min in a fume hood without agitation, treated at room temperature on an orbital shaker with 0.1 M glycine in 0.1 M sodium phosphate buffer (pH 7.2; 2 × 3 min), washed in phosphate-buffered saline (pH 7.2; 2 × 5 min), and treated for 10 min with 0.25% acetic anhydride (v/v) in 0.1 M triethanolamine (pH 8.0) prepared imme-diately prior. The 0.1 M triethanolamine can be made from either concen-trated liquid or powder stocks. After it is in solution, adjust the pH, autoclave, and store in the dark for up to several weeks. The acetic anhydride step is followed immediately by dehydration and defatting with no agitation as follows: 50% EtOH (1 min), 70% EtOH (2 min), 95% EtOH (2 min), 100% EtOH (2 min), chloroform (spectrophotometric grade, 5 min in hood), 100% EtOH (1 min), and 95% EtOH (1 min) and then removed from the rack and dried with the frosted side down in a grooved rack. Slides can then be used that same day or stored at −20° for months with little decrease in signal. If stored in an amber bottle out of the light, chloroform can be reused until a film of fat can be seen floating on the surface. The alcohols can be reused three to five times.

Probe Preparation. We prefer to use *in vitro*-transcribed ^{35}S-labeled riboprobes for our *in situ* hybridization histochemistry. Oligonucleotide probes are far easier to obtain, but are less sensitive than riboprobes. Nonisotopic, hapten-based riboprobe technologies show great promise for dual-labeling fluorescent analyses (Rahman *et al.*, 2003). Perhaps within 2–3 years, this technology will be far enough along in development that it will be feasible for use on a regular basis.

The isotopic riboprobes are generated by *in vitro* transcription from plasmid constructs containing T7, T3, or Sp6 RNA polymerase recognition sequences adjacent to the multicloning sites. The most common vector we use for riboprobe templates is pBluescript SK (Stratagene, La Jolla, CA). Enough circularized template DNA is linearized to obtain approximately 0.5–1 $\mu g/\mu l$ of DNA template. We purify the template from the digestion with Quickspin columns (Qiagen, Valencia, CA). *In vitro* transcriptions are conducted in 0.65-ml microfuge tubes as described in Table I. An alternative, cloning-free approach for generating DNA templates for riboprobe synthesis is via PCR amplification of cDNA, where the T7 recognition sequence is appended to the 5' end of the reverse primer (Rhodes *et al.*, 1996).

Optimal lengths for riboprobes are 250–500 nucleotides. Nevertheless, longer probes are often adequate; in fact, if longer probes are fragmented by alkaline hydrolysis, then sensitivity can be increased (Cox *et al.*, 1984). After addition of all the reagents (Table I), the reaction proceeds for 1 h at 37° (42° for SP6), followed by a 10-min RNase-free DNase I treatment at 37° to degrade the template DNA. Yeast tRNA is added as a carrier (5 μl of a 10-$\mu g/\mu l$ frozen stock), the reaction volume is raised to 80 μl with STE buffer (Sambrook *et al.*, 1989) and the product is purified with an RNA Quickspin column (Roche, Indianapolis, IN). When purification is complete, the probe is stored temporarily on ice. Transcription and ^{35}S incorporation are confirmed by scintillation counting of 1 μl of purified product.

TABLE I

TRANSCRIPTION REACTION[a]

	Volume (μl)
Ingredients	
Riboprobe 5× transcription buffer	3.5
DTT (1 M frozen stock in water)	1
NTPs (2.5 mM each of CTP, ATP, and GTP and 10 μM UTP)	1.5
Linearized plasmid DNA	1–1.5
^{35}S-UTP[b] (~90 μCi)	7
RNase inhibitor (~35 U/μl)	1
Water (MilliQ)	If necessary
Polymerase (SP6, T7, or T3)	1
Total	17
One-hour transcription	
RNase-free DNase I (10 U/μl) for 10 min at 37°	1
Add yeast as carrier and purify with Quickspin column	

[a] All reagents are stored frozen at −20° except for the RNase inhibitor and RNA polymerases, which are stored at −20° in glycerol and radioactivity, which is stored frozen at −80°.

[b] ^{35}S-UTP (12.5 $\mu Ci/\mu l$) is purchased from either Amersham or NEN.

A yield of 80×10^6 cpm total counts must be achieved to be considered a successful transcription reaction. Moreover, our experience shows that, for a given riboprobe template, the higher the yield as measured by scintillation counting, the better signal-to-noise ratio for the *in situ* hybridization histochemistry. An alternative to ^{35}S-UTP as a tracer is ^{33}P-UTP. ^{33}P-UTP requires no additional shielding, but generally produces lower backgrounds than ^{35}S. A significant disadvantage of ^{33}P, which has biased us toward using strictly ^{35}S, is its very short, 25-day half-life.

Several parameters can have significant effects on the success of the transcription, including the amount of DNA and NTPs added and the duration of the reaction. For some templates, we have found that a 2-h incubation with additional RNA polymerase added at 1 h can lead to a significant increase in yield. Once it has been determined that sufficient transcription has occurred, the probe can either be aliquoted and stored frozen at $-80°$ or used immediately. For detection of mRNAs that are less abundant in brain, such as RGS5 or RGS6 (Gold *et al.*, 1997), we have found that using the probe immediately can result in a better signal-to-noise ratio.

Probe Hybridization. Once the probe is obtained, the hybridization cocktail can be made. The hybridization buffer is made in two steps, A and B (Table II). Hybridization buffer A can be made, aliquoted in 15-ml Falcon tubes, and stored for prolonged periods at $-20°$. Buffer B lasts for at least 2 weeks when stored at $-20°$. Note that 50% dextran sulfate is very viscous and should be added last. For measuring dextran sulfate, we have achieved very good results by carefully pipetting the

TABLE II
HYBRIDIZATION BUFFERS

	Volume (ml)
Hybridization buffer A	
Tris–HCl (1 M, pH 7.5)	1
EDTA, (0.5 M, pH 8.0)	0.1
Denhardt's solution (50× stock)	1
NaCl (5 M stock)	3.35
Water (MilliQ)	4.55
Total	10
Hybridization buffer B	
Hybridization buffer A	2
Deionized formamide	5
Dextran sulfate (50% stock solution)	2
Total	9

dextran sulfate into a 15-ml Falcon tube and using the graduation marks on the side of the tube for accurate measurement. Care should be taken not to touch the side of the tube with the pipette tip because a significant volume on the side of the tube will decrease the accuracy of the volumetric measurement greatly, usually increasing the net concentration of the dextran sulfate. To ensure that the dextran sulfate goes into solution, we heat the Falcon tube in a warmed beaker of water and then rotate it for 10 min, end over end on a rotomixer. To make up the hybridization cocktail from buffer B, to 9 parts buffer B add 0.3 parts of denatured salmon sperm DNA (10 mg/ml stock), 0.15 parts yeast tRNA (10 mg/ml stock), 0.4 parts dithiothreitol (DTT; 1 M stock), and radiolabeled probe to obtain 10×10^6 cpm/ml of hybridization buffer and water if needed to obtain 10 parts total.

The hybridization cocktail can be applied directly to the slide or indirectly to the coverslip (22 × 50-mm glass coverslips). We prefer the latter approach. Briefly, coverslips are arranged on a clean, preferably black benchtop, using a gloved finger. Each coverslip is held at the corner as 60 μl of cocktail is pipetted across the coverslip with the equivalent of a P200 or P100 micropipette so that solution extends along the length of the coverslip. The hybridization buffer is sufficiently viscous that any bubbles that result from the pipetting step can be removed easily with the micropipette. The microscope slide is then held in the frosted area, with a bent, coverslipping forceps so that the sections are horizontal and facing down. The slide is lowered slowly until the slide makes contact with the buffer on the coverslip. When the area of contact begins to increase, the slide is flipped over rapidly so that the coverslip is on top. The rapid flip creates centrifugal force that prevents the coverslip from sliding off. The coverslipped slide is then laid down in a square tissue culture dish (245 × 245 × 20 mm, Thomas Scientific, Swedesboro, NJ) that also contains two 50-ml Falcon tube caps filled with water. When the square dish is filled with slides (maximum 14 slides per dish), the dish is covered, sealed in a freezer grade, 1-gallon Zip-loc bag, and hybridized for 18 h in an oven at 60°. It is important that the shelf in the incubator/oven is level, as nonlevel shelves can lead to labeling asymmetries. After all the slides are in the oven, the lower shaft of the micropipette should be removed from the handle and rinsed well with water to remove radioactive contamination.

Washes. Posthybridization washes are a series of heated rinses in sodium citrate buffer (SSC) (Sambrook *et al.*, 1989) of decreasing concentrations resulting in increasing stringency. All SSC washes contain sodium thiosulfate (10 mM) as a reducing agent to maintain the integrity of the ^{35}S bound to the UTP. To initiate the washes, the culture plate is removed from the oven and its sealed plastic bag. The coverslips are taken off by immersing the slides in 4× SSC containing 10 mM sodium thiosulfate preheated to 60°. The slides are held with a coverslipping forceps with the frosted slide up and are immersed in the buffer in a 50-ml Falcon tube. Often the coverslip will fall right off the slide. However, if the coverslip tends to stick to the slide, then it can be nudged off gently with a gloved finger or a second forceps. Great care should be taken at this step to prevent the coverslip from damaging the tissue sections as it glides off the slide. The slide is then rinsed by dipping in a second Falcon tube of 4× SSC and is then placed in a Tissue Tek rack in a green staining dish containing preheated 4× SSC. Do not let the slide dry after removing the coverslip as this can produce nonspecific adherence of the ^{35}S probe to the section. Once all of the coverslips have been removed, sections are washed twice for 15 min in 4× SSC at 60° and then for 30 min in RNase A [20 μg/ml in 0.01 M Tris HCl, pH 8.0, 0.01 M EDTA, 3% NaCl (w/v)] at 45° followed by two washes in the following solutions: 2× SSC (15 min, room temperature), 0.5× SSC (15 min, 60°), and 0.1× SSC (15 min, 60°). The salt is then removed from the slides by a quick dip in MilliQ water, dipped rapidly in 95% EtOH, and dried on the grooved rack with writing side down.

Film Autoradiography

The bound riboprobe is visualized by emulsion autoradiography (detailed later) and/or film autoradiography. Should emulsion autoradiography be an end point, film autoradiography is done first to gauge the absolute magnitude of the signal and, more importantly, the signal-to-noise ratio. Slides are arranged and taped to a semi-rigid backing. We routinely use manila file folders trimmed to fit within a 23 × 28-cm autoradiography cassette. Avoid larger autoradiography cassettes as this leads to out-of-focus sections. Slides are taped as high up along the writing surface as possible to minimize the effects of the thickness of the tape on the film immediately adjacent to the sections. Specifically, if the tape is too close to the section, the adjacent section will be out of focus. If densitometric analyses are intended for the sections, then it is important to coexpose autoradiographic standards with each set of films to enable the creation of a standard curve relating relative optical density to known autoradiographic standards. Although there are published protocols for creating ^{14}C-labeled brain paste standards (Miller and Zahniser, 1987), we prefer to purchase plastic standards from either Amersham Pharmacia (Piscataway, NJ) or American Radiolabeled Chemicals (St. Louis, MO). Radioactivity values for the commercially available standards are usually expressed in radioactive units per unit mass of tissue (i.e., nCi/μg of tissue).

When the sections are ready to expose, the film is applied under appropriate safelight conditions in a darkroom. For maximal resolution and sensitivity, a grade of film equivalent to Biomax MR film (Kodak, Rochester, NY) should be used. Films of this quality tend to have autoradiographic emulsion coating on only one side of the film. Thus care should be taken to ensure that the film is exposed to the slides in the appropriate orientation: emulsion side toward the sections. The majority of RGS mRNAs of interest to our laboratories (RGS4, −7, and −9) are expressed at moderate to high densities in the brain. These expression levels allow shorter exposure times (1–3 days) so we use an automated film processor to develop the films of these mRNAs. Other less abundantly expressed RGS mRNAs, such as RGS5, −6, and −10, require longer exposure times (4–10 days). For these mRNAs, more time is lost if the developer malfunctions so we hand develop the films in a darkroom sink using GBX developer (Kodak) diluted 1 to 5 with water and fix with Rapid Fix (Kodak). At the time of writing, phosphor-imager screens do not approach the resolution of high-resolution autoradiographic films.

Emulsion Autoradiography

Cellular autoradiographic emulsion analysis, in combination with counterstaining, can provide valuable information as to the cellular phenotype of expression. In addition, if one wishes to resolve changes in mRNA abundance in a subpopulation of neurons expressed in a structure, then one will have to use slide emulsion autoradiography in conjunction with silver grain counting analyses to quantify cellular labeling densities. Figure 1C illustrates heterogeneity in cellular grain densities present between neurons. There are several good computer programs for the quantification of grain and grain cluster densities from autoradiographic emulsion-dipped slides. A detailed protocol for the quantification of grain densities on emulsion-dipped slides is beyond the scope of this article (see http://rsb.info.nih.gov/nih-image/download.html for information regarding downloading NIH Image freeware).

There are several manufacturers of autoradiographic emulsion. Our laboratories have had success with NTB2 emulsion from Kodak. The diameter of silver grains for this emulsion is small to medium, but the sensitivity is high. As the emulsion is extremely sensitive to radioactive emissions, great care should be taken to store the emulsion in appropriate refrigerators in the laboratory that are safe from radioactive sources.

After film autoradiography and prior to dipping into emulsion, the slides are removed from the support backing and placed into a slide rack. For even the most well-characterized *in situ* hybridization probes, it is important to incorporate autoradiographic test slides into the experiment. Test slides will be used to determine if the emulsion has been exposed for a sufficient period of time to provide for the desired grain density. For the test slides to have the most predictive value for the experimental slides, mount the sections on the test slides at the same density as for the experimental slides.

The slides are dipped in a small vial maintained at 42° via a low-rise water bath equivalent to the Tissue Float Bath model (Lab-line, Dubuque, IA). Note that thermometers with red liquid are not suitable for darkroom work and that any low-profile thermometer with a blue dye will be far easier to read under the yellow–red illumination. A desirable dipping vial that wastes the least volume of emulsion is barely larger than the dimensions of the slide. Dipping vials can be purchased from several suppliers, including Ted Pella Inc. (Redding, CA) and Electron Microscopy Sciences (Hatfield, PA). A low-budget vial that works quite well is a plastic slide mailer that opens along the short end. The NTB2 emulsion is shipped in a gelatinous form. On the day of dipping it must be diluted 1:1 with MilliQ grade water and then allowed to equilibrate to 42° for 1 h. Save time setting up the water bath by predetermining the 42° set point. Also, preheat the water in a microwave oven prior to filling the bath. For the purpose of diluting the emulsion, the water is measured at the laboratory bench into a 50-ml Falcon tube. Then in the darkroom and under safelight illumination, the emulsion is added with a clean plastic spoon to the Falcon tube until the meniscus of the water reads twice the volume of the water alone. The Falcon tube is then sealed and wrapped well in aluminum foil to protect it from the light. A preferred safelight for dipping is a sodium lamp equivalent to the Thomas Duplex safelight. Note that although NTB2 is not sensitive to sodium lamp emissions, other emulsions may be. A good alternative lamp is a standard Kodak safelight with a GBX-2 filter.

After 60 min, the diluted emulsion solution is rocked back and forth gently to mix it well. Avoid rapid shaking as this will introduce air bubbles into the emulsion solution, leading to irregularities in the emulsion coating. After mixing, the emulsion is poured into the dipping vial and dipping proceeds. Any time the dipping vial is replenished with fresh emulsion from the stock tube, residual bubbles should be removed by dipping a clean, blank slide into the emulsion vial. The general procedure for dipping is as follows: On the counter near the vial immersed in the water bath should be a dry paper towel, a wad of water-moistened paper towels, and a plastic drying tray lined with a level layer of paper towels (we use the same $245 \times 245 \times 20$-mm^2 tissue culture dishes for drying). Slides are dipped by dunking them once or twice into the emulsion. The slides are removed slowly from the vial so that the emulsion sheaths down the slide, resulting in a thin layer of emulsion. The bottom edge of the slide is then pressed four to five times in rapid succession along the moistened paper towels so that extra emulsion is wicked down off of the slide. The emulsion on the backside of the slide is then wiped with a dry paper towel. Additional excess emulsion is wicked away with the moistened paper towel and then the slide is placed section side up on the drying tray. The moistened paper towel should be replaced periodically with fresh paper towel, to ensure maximal wicking capacity.

When all of the slides are dipped, they will typically need 2–4 h to dry. Note that often the entire slide will be dry except for a small bead of liquefied emulsion at the end. This is dry enough for boxing and storage. The slides are stored in black 25-position slide boxes containing a small bag of desiccant. To make desiccant packets, roll a tablespoon of desiccant granules in a Kimwipe and tape closed. The desiccant packet should be separated from the emulsion-dipped slides via a blank glass slide. The boxes are sealed with black electrical or Time-med tape, wrapped in aluminum foil, and then stored at 4° in a refrigerator protected from any sources of radioactivity. Note that at this point it is convenient to box the test slides in a separate box. As exposure times can sometime extend to one or more months, it is important to label the boxes well immediately after wrapping them in foil.

Developing Slides

Before developing slides from the entire experiment, it is important to develop test slides to gauge whether the emulsion has been exposed for a sufficient period of time. Test slides are developed identically to regular slides, but a couple of precautions should be taken. Conditions such as the concentration of developer, temperatures of solutions, and number of slides per development should be standardized. The standard developing procedure is done in a "darkroom" set of Tissue Tek dishes and racks. Boxed slides should be removed from 4° at least 20 min prior to developing so that they can warm to room temperature. Slides are then developed with Developer D19 (Kodak) diluted 1:1 with water for 4 min at 17° with no agitation, followed by two brief washes in chilled water and then fixed for 10 min in chilled Kodak fixer with periodic agitation. Note that both the developer and the fixer are filtered prior to use. The developer is made and stored at full concentration at 4°. On the day of use, it can be diluted with room temperature water, which will expedite the time for the developer to reach temperature. Fixer is stored at room temperature. When setting up in the darkroom, it is usually sufficient to chill the rinse water and fixer for 10 min in their respective Tissue Tek dishes in a developer tray filled with ice water. Although the temperature of the developer is critical, we have not found it necessary to monitor its temperature during the 4-min develop time. As long as a stable 17° has been achieved prior to immersing the slides, grain densities will be very predictable between developments.

Another way to guarantee uniform developing temperatures between procedures is to develop a standard number of slides. We always develop a full rack of slides (24) and incorporate blank filler slides to fill up the rack. Superfrost Plus slides are very expensive, so for the purpose of blank filler slides, plain precleaned glass slides can be used. When the fix step is completed, slides are then light safe and can be transferred in the light to the main laboratory.

Slides are rinsed under a gentle stream of deionized water and then counterstained to aid in cellular identification. Most Nissl-type stains are sufficient, including thionin, toluidine blue, cresyl violet, and methylene green (all available through Sigma or Vector Laboratories). If low-power, dark-field images are the goal, then counterstaining should be very light. Slides are then dehydrated, cleared, and coverslipped with mounting medium.

Film exposure times are predictive of emulsion exposure times. A suitable *in situ* hybridization signal-to-noise ratio from a 1-day film exposure time would predict a 4- to 7-day exposure time on emulsion.

Densitometric Analysis of Film Autoradiograms

Densitometric analysis of *in situ* hybridization film autoradiograms permits quantitative comparisons of relative mRNA levels that can be used to resolve the effects of experimenter manipulations on mRNA abundance. There are both advantages and disadvantages to film analysis relative to grain analysis. The main advantage is the ease of obtaining autoradiographic films relative to emulsion autoradiograms. Another advantage is that once the autoradiograms have been generated, one can relatively quickly quantify relative expression levels for many brain regions. The major disadvantage is that film densitometry cannot resolve changes in mRNA abundance in subpopulations of cells in the structure of interest as one can do with grain counting.

There are several commercially available computerized densitometry analysis packages. There is a considerable hardware investment necessary to set this up that includes a digital camera, camera stand, lightbox, lenses, computer, monitor, and software. One can also assemble the components on one's own and use NIH Image freeware (http://rsb.info.nih.gov/nih-image/) for a considerable savings. All densitometry software packages use identical principles. So whether one decides on a commercial package or a homemade setup will be determined by the availability of funds. Although the actual commands will vary with the package that is chosen, the overall process should be identical and is described here in general terms.

An optimal exposure time is required for accurate quantification. Exposure times should result in relative optical densities (RODs) that are within the linear range of the film—there should be a range of shades of gray. In addition, RODs of the autoradiographic standards need to bracket the optical densities that are to be quantified. Because automated or hand developing of films, when done correctly, results in uniform backgrounds between films, one needs to expose a set of standards to only one of the films.

Optimal image capturing conditions are paramount for reproducible densitometry. While visualizing the live image on the computer screen, first adjust the height of the camera and focus the image with the camera lens. In doing so, confirm that the film is resting flat on the lightbox. If necessary, use weights to keep it flat. Set the camera height as low as possible to enable capturing the region of interest at the highest possible resolution. When the camera height is set and the image is focused, adjust the brightness and contrast by changing the lightbox intensity and f-stop settings on the lens. Most software packages have a "check for saturated pixels" function. Go through this procedure and then adjust light intensity so that there are no saturated pixels.

Autoradiographic standards coexposed to the film enable one to account for nonlinear response of the film. To do this, RODs of known radioactive standards are obtained with the software and a curve is generated relating the ROD to the standards. Either standard polynomial curve-fitting or custom-made algorithms are useful for this purpose. The NIH image software has the very useful RODBARD algorithm for generating this curve. In fitting the curve to the points, curves containing inflection points within the range of RODs of the region of interest should be avoided. Most densitometry programs will permit the removal of pixel values that are below a background threshold. The ROD of the background can be determined by measuring the average background over a part of the film where there are no brain section images. Once the curve and threshold are determined, the regions of interest can be captured and saved to disk. As it is difficult to reestablish identical lighting conditions, it is important to capture the images all in one sitting. We prefer to capture all of the calibrated images to disk and then analyze them afterward. The programs embed the calibration curve in the saved image files and permit measures of radioactive units per unit weight of tissue. This allows analysis of the images anytime thereafter.

When all of the images have been captured, sampling tools can be used to sample regions of interest. Sampling tools used most often in our laboratories are *box* and *circumscribe* tools. The *box* permits reproducible samples of identical area. The *circumscribe* tool permits outlining a region of interest with a freehand drawing. Because many brain regions are of an irregular shape, the *circumscribe* tool is the best way to survey reproducibly the entire structure. Densitometry packages vary in data summary and output qualities. For the NIH Image program, most users will need to transfer data manually using copy-and-paste commands from the densitometry program to a spreadsheet program.

Controls for Specificity of Hybridization

Several controls can be done to assay for the specificity of hybridization. The best way to ensure the specificity of hybridization is to generate two probes that are complementary to two distinct regions of the mRNA. If patterns of labeling generated with both probes are identical, this is strong evidence for the specificity of hybridization. Should the patterns be distinct, this does not necessarily mean that probe labeling is nonspecific. For example, it could result from regionally specific splicing of the gene product. Another very commonly used method for assaying specificity is via comparison of labeling with a sense riboprobe. The sense probe contains the same G/C content as the antisense probe, but should not hybridize to the mRNA. Thus, the labeling pattern for the sense riboprobe should be completely blank. In rare instances, a gene is encoded on the opposite strand of DNA whose mRNA is also expressed in the region of interest. If this is the case, one could very well see what appears to be distinct, specific labeling, but it would most likely be a very different pattern than that for the antisense probe. These two strategies, using probes complementary to two distinct regions and a sense probe, should provide adequate information to assay the specificity of hybridization (see Table III for troubleshooting).

Background to Behavioral Analyses of RGS Protein Mutant Mouse Strains

Studies of the regulation of RGS mRNA, as detailed earlier, should provide important leads toward the specific functional roles of individual RGS proteins in brain. Another important strategy for furthering our understanding of RGS protein function in the brain is the generation and analysis of RGS mutant mice. Genetic animal models have proven invaluable for elucidating RGS protein function. In *Saccharomyces cerevisiae* (Dohlman *et al.*, 1996), *Caenorhabditis elegans* (Koelle and Horvitz, 1996) and *Aspergillus nidulans* (Yu *et al.*, 1996), forward genetic screens identified RGS proteins critical for modulating mating behavior, egg laying, and sporulation, respectively. In addition, backward genetic approaches in mice have identified critical roles for RGS2 (Tang *et al.*, 2003) in vascular relaxation and for RGS9 in the kinetics of phototransduction (Chen *et al.*, 2000)

TABLE III
TROUBLESHOOTING

Problem	Solution
1. Probe labels but there is no hybridization signal above sense control	1a. Were reducing agents (dithiothreitol or sodium thiosulfate) added to the hybridization cocktail?
	1b. Try treating water used to make all solutions with diethylpyrocarbonate
2. Labeling pattern is like a Nissl stain—all cells are labeled	2a. Were the sections treated with RNase A in the posthybridization washes?
3. Probe will not label	3a. Use positive control template (cyclophilin, GAPDH) available from several manufacturers.
	3b. Run crude and purified transcription products on an agarose gel to verify cRNA synthesis.
	3c. Try fresh stock of RNA polymerase
4. Gradient of labeling from side to side of slides	4a. Is the hybridization chamber sealed?
	4b. Do the coverslips come off with minimal effort?
	4c. Is the hybridization chamber level?
5. Film image appears fuzzy and out of focus	5a. Are the films positioned immediately adjacent to the sections?

and in dopamine and opioid receptor modulation (Rahman *et al.*, 2003; Zachariou *et al.*, 2003). RGS mutant mice will undoubtedly continue to be critical tools for revealing the function of RGS proteins *in vivo*. The following section describes protocols for phenotyping RGS transgenic mice with an emphasis toward assaying psychiatric-related phenotypes such as sensitivity to psychostimulants and morphine, learning, and anxiety.

Behavioral Analysis Protocols

General Considerations

Most transgenic mice will appear grossly normal next to their wild-type littermates. Nevertheless, many phenotypic differences can be observed when they are challenged in appropriate behavioral assays. As the validity of many behavioral tests relies on intact sensory systems and motor abilities, it is critical to verify the integrity of these functions in mutant mice. Assays for sensory and motor abilities have been described in detail elsewhere (Crawley, 2003) and include the visual cliff, acoustic startle, locating hidden food, hot plate, tail flick, and rotarod tests. If the transgenic mice perform aberrantly in these sensory and motor screens relative to their wild-type littermates, then investigations into the mechanism(s) of how the genetic mutations impinge on sensory or motor system function are warranted.

Furthermore, for later behavioral screening, it will be critical to use assays that do not rely on the specific sensory or motor systems affected.

A wealth of behavioral genetic studies show overwhelmingly that genes do not work alone, but act via a complex set of genetic and epigenetic interactions. Thus, the background strain(s) will have a profound influence on data obtained in behavioral assays. This point is particularly relevant given that the C57B1/6 and 129/SvemJ strains, used most commonly for the generation of knockout and knockin mice, perform very differently on many of the behavioral assays described later. Therefore, if the mice being phenotyped in these assays are not backcrossed to an inbred strain of mice, the experimenter will need much greater numbers of animals per group to resolve the true effects of the gene of interest. In addition, one should choose an inbred strain of mice for backcrossing that performs well on the behavioral assays of interest (for a discussion of strain differences, see Crawley, 2000).

When possible, all behavioral apparatuses should be contained in rooms in close proximity to the colony rooms. Furthermore, the rooms should be small, quiet rooms isolated from outside sounds. White noise generators can be helpful to reduce the salience of outside sounds. Since the following behavioral assays are remarkably sensitive to many extraneous stimuli present within a typical vivarium, animals should be transferred to the behavior room at least 1 h before the beginning of the assay. For all assays, the experimenter should remain blind to the genotype of the mouse. Finally, for each round of behavioral testing, mice of all genotypes of interest should be represented.

Basal Locomotor Activity

Many central nervous system pathologies impact basal and drug-induced locomotor activity. Locomotor activity is monitored with an automated system in which the activity chambers consist of standard plastic mouse cages ($12 \times 18 \times 33$ cm) placed between a set of spaced infrared beams and sensors (seven pairs). When the animal interferes with the beam, the computer counts and registers the beam break. The experimenter determines the temporal window (bin) for each set of counts. Data from the locomotor apparatus can be exported to a spreadsheet program and then refined and summarized.

Locomotor Response to Psychostimulants

Animals are first habituated to the locomotor cages for 3 days. During the habituation days, mice are injected with saline and their locomotor activity is monitored for 1 h. On day 4, mice are again injected with saline and locomotor activity is monitored for 30 min, but immediately after, the mice are injected with cocaine or amphetamine and are placed back in the locomotor activity chambers for another 30 min. For psychostimulants, the doses used depend on the genetic background and range between 2–10 mg/kg (ip) and 1–5 mg/kg (ip) for cocaine and amphetamine, respectively (Downing et al., 2003; Ralph et al., 2001).

Locomotor Sensitization Assays

In order to study the role of an RGS protein in the development of locomotor sensitization to psychostimulants, mice are habituated for 3 days as described earlier and then, on days 4–10, are injected with the same dose of cocaine (7–20 mg/kg ip depending on the mouse strain) and locomotor activity is monitored for 30 min. In addition to the drug-treated groups, wild-type and KO mouse control groups must be included that receive saline throughout the study. Ten to 15 animals per group are required. Figure 2 illustrates a phenotypic difference in locomotor sensitization to cocaine between RGS9 KO mice and their wild-type littermates.

Place-Preference Conditioning to Morphine

Many drugs of abuse, including cocaine and morphine, have rewarding properties. The place conditioning test assesses the ability of the reward-ing properties of a drug to induce a place preference in mice. The apparatus consists of two large chambers each with discrete floor texture and wall color and pattern, separated by a smaller middle compartment. Vertical sliding doors allow the experimenter to confine the mouse to a subset of the compartments. Experiments are performed under dim illumination. The experimenter must confirm that mice show no bias for either of the large compartments by monitoring their activity for 20 min on day 1. A comput-er records the time spent in each compartment via infrared beams and sensors. Animals are excluded from the study if, on the baseline day, they stay in the central compartment for more than 8 min or more than 12 min in one of the large compartments. From days 2 to 7, mice alternate between days where they are injected with saline and confined to one of the large compartments for 20 min and days where they are injected with the desired dose of morphine and confined to the other large compartment. On the last session (day 8), mice are allowed to move freely between the different compartments for 20 min. Data are calculated as time spent in the drug paired side post- minus pre-conditioning. Several doses of morphine are used in order to compare the sensitivity of the rewarding properties to morphine between wild-type and mutant mice. The threshold morphine dose required to condition place preference may vary between different strains of mice. The lowest morphine dose that conditions place preference

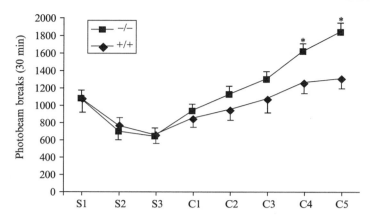

FIG. 2. RGS9 KO mice show an enhanced sensitization to the locomotor-inducing properties of cocaine. KO mice and wild-type littermates were given daily injections of saline for 3 days (S1–S3) followed by daily injections of cocaine (7.5 mg/kg) for 5 days (C1–C5). Locomotor activity was measured for 10 min after each injection. Data are expressed as mean ± SEM photocell disruptions over the 10-min test period. Wild-type ($n = 11$) and mutant ($n = 13$) mice showed small increases in locomotor activity after the first dose of cocaine, with no difference apparent between the two genotypes. Both groups of mice also showed significantly greater locomotor responses to the fifth dose of cocaine compared to the first dose (i.e., sensitization) ($p < 0.05$, ANOVA and t test), although KO mice showed significantly greater levels of sensitization ($p < 0.05$, ANOVA and t test). Data originally presented in Rahman et al. (2003).

in C57B1/6 mice is 5 mg/kg, whereas lower and higher doses are used in order to determine differences in morphine reward sensitivity between wild-type and mutant mice. A group of animals injected with saline in both compartments is always used as control. See Fig. 3 for an example of a phenotypic shift in a morphine dose-conditioning curve in RGS9 KO mice.

Virus-mediated gene transfer has also been used in order to evaluate the role of a particular protein in the nucleus accumbens or dorsal striatum in morphine reward. Bilateral microinjections of virus are delivered to the nucleus accumbens or dorsal striatum via stereotaxic surgery. As herpes simplex viruses (HSV) show maximal expression between days 2 and 5, behavioral assays have to be performed within this time window to examine the effects of the overexpressed protein. For animals infected with HSV, a modified place preference conditioning protocol is followed (Zachariou et al., 2003). In this case, mice are tested for preconditioning preference 1 day postsurgery and the following day they are conditioned to saline in the morning and to morphine in the afternoon. Place preference is tested after 2 days of conditioning, on the 4th day postsurgery.

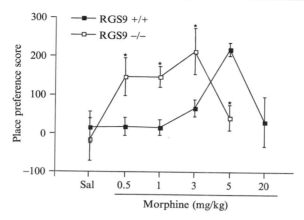

Fig. 3. RGS9 regulates morphine reward. RGS9 KO mice showed ~10-fold greater sensitivity to morphine over a full dose range in the place-preference conditioning assay. Scores are calculated as the difference between the time spent in the drug-paired side post-versus preconditioning. Morphine was injected sc at 0, 0.5, 1, 3, 5, and 20 mg/kg (n = 6–10). Data are expressed as mean ± SEM. *p < 0.05 between genotypes receiving the same treatment (ANOVA followed by probable least-squares difference posthoc test). Data originally presented in Zachariou *et al.* (2003).

Morphine Analgesia and Tolerance

To probe for alterations in pain sensitivity or in sensitivities to the analgesic properties of morphine, mice are tested in hot plate and tail-flick tests. For determining pain thresholds in the hot plate test, animals are placed on a 52°, or 56°, platform and the latency to paw lick or jump is monitored. A cutoff time of 40 s is set for each temperature in order to avoid tissue damage. For analgesia assays, following monitoring of baseline responses, mice are injected with different morphine doses (from 5 to 20 mg/kg sc) and 30 min later their responses to the hot plate are monitored again. Analgesia is expressed as percentage maximal possible effect (MPE), where MPE = (test − baseline latency)/(cutoff − baseline) × 100.

For morphine tolerance, repeated morphine injections (15 mg/kg, sc) are given daily for 5 days and analgesia is measured at baseline and 30 min after each drug administration. For continuous administration, morphine pellets [25 mg available through the National Institute on Drug Abuse (NIDA)] can be implanted subcutaneously under light isoflourane anesthesia, and the analgesic responses to morphine (15 mg/kg, sc) are monitored 6 h and 3 days after sc pellet implantation. Equalizing chronic administration doses in animals that have a weight phenotype is described later.

For the tail-flick test, mice are immobilized and focused radiant heat is applied to a certain part of the tail (usually 2 cm from the tip of the tail). A computer controls the heat intensity to produce a reaction time between 2.5 and 3.5 s (for computer-controlled tail-flick equipment, see IITC Lifescience, Woodland Hills, CA). A cutoff time of 10 s is set to minimize tissue damage. Morphine analgesia is assayed by comparing the baseline tail-flick latency to that observed 30 min following morphine sulfate injection (doses range from 0.5 to 5 mg/kg, sc). Analgesic response is calculated as MPE = (test − baseline latency)/(cutoff − baseline latency) × 100 as described previously.

Opiate Dependence

In order to determine if genetic manipulation influences morphine dependence, one can use a precipitated opiate withdrawal paradigm and monitor the severity of several signs of opiate withdrawal. For chronic morphine treatment, 25-mg morphine pellets are implanted (sc) with light isoflourane anesthesia and, 2.5 days later, a withdrawal syndrome is precipitated by injecting the opioid antagonist naloxone (1 mg/kg, sc). If there is a phenotypic weight difference between wild-type and KO mice, morphine injections are used instead of morphine pellets to ensure comparable dosing. In this case, increasing morphine doses (from 20 to 100 mg/kg, ip) are injected every 8 h for 2.5 days and naloxone is administered 2 h following the last morphine dose.

Withdrawal signs, including jumps, wet dog shakes, weight loss, diarrhea, ptosis, and tremor, are monitored for 30 min following naloxone administration. Either the number of times a behavior is observed (jumps, wet dog shakes, diarrhea, paw tremor) or the presence of a sign in each of the 5-min intervals of the recording time (tremor, ptosis) is monitored. Figure 4 illustrates enhanced physical dependence in RGS9 KO mice.

Morris Water Maze

The water maze is an excellent test of spatial learning and memory (Morris, 1984). This learning paradigm uses a round pool of colored water in a room containing visual cues on the wall. A movable platform is submerged beneath the surface. Mice are placed in the maze, and the time it takes for them to find the hidden platform is monitored with a camera suspended from above. The task is repeated for 8–10 days, and the mice learn the visual cues that allow them to find the platform. Learning is seen as a decrease over time in the latency to reach the platform. To ensure that visual acuity or motor performance deficits are not confounding the spatial learning component of the test, wild-type and mutant mice must exhibit the same latencies to swim to a *visible* platform. Differences in anxiety can also have profound effects on

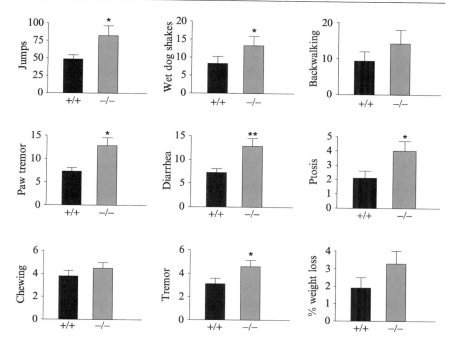

FIG. 4. RGS9 regulates morphine dependence. RGS9 KO mice show greater antagonist-precipitated morphine withdrawal compared with wild-type littermates ($n = 8$), an effect seen for all withdrawal behaviors examined. Data are expressed as mean ± SEM. *$p < 0.05$; **$p < 0.01$ between genotypes (ANOVA followed by PSTD test) (Zachariou *et al.*, 2003).

performance of this task. Therefore, potential differences in anxiety between wild-type and mutant mice should be assayed (Crawley, 2000).

Summary

Elucidating the functional roles of RGS proteins in mammalian brain will take decades of work. This article described two important strategies for achieving these goals. First, a detailed *in situ* hybridization method for assaying RGS mRNA levels in brain tissues was described. Studies of this sort should further our understanding of the stimuli that regulate RGS gene expression. Second, methods for behavioral phenotyping of RGS mutant mice were discussed with an emphasis on behaviors pertinent to drug addiction. These and other behavioral analyses will provide clues into the neural processes in which specific RGS proteins play a modulatory role. These two strategies will be important components in our path toward understanding RGS function in the brain.

References

Berman, D. M., and Gilman, A. G. (1998). Mammalian RGS proteins: Barbarians at the gate. *J. Biol. Chem.* **273**(3), 1269–1272.

Chen, C. K., Burns, M. E., He, W., Wensel, T. G., Baylor, D. A., and Simon, M. I. (2000). Slowed recovery of rod photoresponse in mice lacking the GTPase accelerating protein RGS9-1. *Nature* **403**, 557–560.

Cox, K. H., DeLeon, D. V., Angerer, L. M., and Angerer, R. C. (1984). Detection of mRNAs in sea urchin embryos by *in situ* hybridization using asymmetric RNA probes. *Dev. Biol.* **101**(2), 485–502.

Crawley, J. N. (2000). What's Wrong with My Mouse? Behavioral Phenotyping of Transgenic and Knockout Mice. Wiley, New York.

Crawley, J. N. (2003). Behavioral phenotyping of rodents. *Comp. Med.* **53**(2), 140–146.

Dohlman, H. G., Song, J., Ma, D., Courchesne, W. E., and Thorner, J. (1996). Sst2, a negative regulator of pheromone signaling in the yeast *Saccharomyces cerevisiae*: Expression, localization, and genetic interaction and physical association with Gpa1 (the G-protein alpha subunit). *Mol. Cell. Biol.* **16**, 5194–5209.

Downing, C., Rodd-Henricks, K., Marley, R. J., and Dudek, B. C. (2003). Genetic variation in the psychomotor stimulant properties of cocaine in *Mus musculus*. *Psychopharmacology (Berlin)* **167**, 159–166.

Gall, C. M., Lauterborn, J. C. and Guthrie, K. M. (1995). *In situ* hybridization: A sensitive measure of activity dependent changes in neuronal gene expression. In "*In Vitro/In Vivo* Autoradiography and Correlative Imaging" (W. Stumpf and H. Solomon, eds.), pp. 379–399. Academic Press, San Diego.

Gold, S. J., Ni, Y. G., Dohlman, H. G., and Nestler, E. J. (1997). Regulators of G-protein signaling (RGS) proteins: Region-specific expression of nine subtypes in rat brain. *J. Neurosci.* **17**, 8024–8037.

Grafstein-Dunn, E., Young, K. H., Cockett, M. I., and Khawaja, X. Z. (2001). Regional distribution of regulators of G-protein signaling (RGS) 1, 2, 13, 14, 16, and GAIP messenger ribonucleic acids by *in situ* hybridization in rat brain. *Brain Res. Mol. Brain Res.* **88**(1–2), 113–123.

Koelle, M. R., and Horvitz, H. R. (1996). EGL-10 regulates G protein signaling in the *C. elegans* nervous system and shares a conserved domain with many mammalian proteins. *Cell* **84**, 115–125.

Kozasa, T., Jiang, X., Hart, M. J., Sternweis, P. M., Singer, W. D., Gilman, A. G., Bollag, G., and Sternweis, P. C. (1998). p115 RhoGEF, a GTPase activating protein for Galpha12 and Galpha13. *Science* **280**(5372), 2109–2111.

Krumins, A. M., and Gilman, A. G. (2002). Assay of RGS protein activity *in vitro* using purified components. *Methods Enzymol.* **344**, 673–685.

Miller, J. A., and Zahniser, N. R. (1987). The use of 14C-labeled tissue paste standards for the calibration of 125I-labeled ligands in quantitative autoradiography. *Neurosci. Lett.* **81**(3), 345–350.

Morris, R. (1984). Developments of a water-maze procedure for studying spatial learning in the rat. *J. Neurosci. Methods* **11**(1), 47–60.

Rahman, Z., Schwarz, J., Gold, S. J., Zachariou, V., Wein, M. N., Choi, K. H., Kovoor, A., Chen, C. K., DiLeone, R. J., Schwarz, S. C., Selley, D. E., Sim-Selley, L. J., Barrot, M., Luedtke, R. R., Self, D., Neve, R. L., Lester, H. A., Simon, M. I., and Nestler, E. J. (2003). RGS9 modulates dopamine signaling in the basal ganglia. *Neuron* **38**(6), 941–952.

Ralph, R. J., Paulus, M. P., and Geyer, M. A. (2001). Strain-specific effects of amphetamine on prepulse inhibition and patterns of locomotor behavior in mice. *J. Pharmacol. Exp. Ther.* **298,** 148–155.

Rhodes, K. J., Monaghan, M. M., Barrezueta, N. X., Nawoschik, S., Bekele-Arcuri, Z., Matos, M. F., Nakahira, K., Schechter, L. E., and Trimmer, J. S. (1996). Voltage-gated K+ channel beta subunits: Expression and distribution of Kv beta 1 and Kv beta 2 in adult rat brain. *J. Neurosci.* **16**(16), 4846–4860.

Ross, E. M. (2002). Quantitative assays for GTPase-activating proteins. *Methods Enzymol.* **344,** 601–616.

Ross, E. M., and Wilkie, T. M. (2000). GTPase-activating proteins for heterotrimeric G proteins: Regulators of G protein signaling (RGS) and RGS-like proteins. *Annu. Rev. Biochem.* **69,** 795–827.

Sambrook, J., Maniatis, T., and Fritsch, E. F. (1989). "Molecular Cloning: A Laboratory Manual." Cold Spring Harbor Laboratory Press, Cold Spring Harbor, New York.

Seroogy, K. B., and Herman, J. P. (1997). *In situ* hybridization approaches to the study of the nervous system. *In* "Neurochemistry: A Practical Approach" (A. J. Turner and H. S. Bachelard, eds.), pp. 121–150. Oxford Univ. Press, Oxford.

Siderovski, D. P., Strockbine, B., and Behe, C. I. (1999). Whither goest the RGS proteins? *Crit. Rev. Biochem. Mol. Biol.* **34**(4), 215–251.

Tang, M., Wang, G., Lu, P., Karas, R. H., Aronovitz, M., Heximer, S. P., Kaltenbronn, K. M., Blumer, K. J., Siderovski, D. P., Zhu, Y., and Mendelsohn, M. E. (2003). Regulator of G-protein signaling-2 mediates vascular smooth muscle relaxation and blood pressure. *Nature Med.* **9**(12), 1506–1512.

Wang, J., Tu, Y., Woodson, J., Song, X., and Ross, E. M. (1997). A GTPase-activating protein for the G protein Galphaz: Identification, purification, and mechanism of action. *J. Biol. Chem.* **272,** 5732–5740.

Yu, J. H., Wieser, J., and Adams, T. H. (1996). The Aspergillus FlbA RGS domain protein antagonizes G protein signaling to block proliferation and allow development. *EMBO J.* **15**(19), 5184–5190.

Zachariou, V., Georgescu, D., Sanchez, N., Rahman, Z., DiLeone, R., Berton, O., Neve, R. L., Sim-Selley, L. J., Selley, D. E., Gold, S. J., and Nestler, E. J. (2003). Essential role for RGS9 in opiate action. *Proc. Natl. Acad. Sci. USA* **100**(23), 13656–13661.

[14] RGS-Insensitive G-Protein Mutations to Study the Role of Endogenous RGS Proteins

By Ying Fu, Huailing Zhong, Masakatsu Nanamori, Richard M. Mortensen, Xinyan Huang, Kengli Lan, and Richard R. Neubig

Abstract

Regulator of G-protein signaling (RGS) proteins are very active GTPase-accelerating proteins (GAPs) *in vitro* and are expected to reduce signaling by G-protein coupled receptors *in vivo*. A novel method is presented to assess the *in vivo* role of RGS proteins in the function of a

G protein in which Gα subunits do not bind to RGS proteins or respond with enhanced GTPase activity. A point mutation in the switch I region of Gα subunits (G184S Gα_o and G183S Gα_{i1}) blocks the interaction with RGS proteins but leaves intact the ability of Gα to couple to $\beta\gamma$ subunits, receptors, and downstream effectors. Expression of the RGS-insensitive mutant G184S Gα_o in C6 glioma cells with the μ-opioid receptor dramatically enhances adenylylcyclase inhibition and activation of extracellular regulated kinase. Introducing the same G184S Gα_o protein into embryonic stem (ES) cells by gene targeting allows us to assess the functional importance of the endogenous RGS proteins using *in vitro* differentiation models and in intact mice. Using ES cell-derived cardiocytes, spontaneous and isoproterenol-stimulated beating rates were not different between wild-type and G184S Gα_o mutant cells; however, the bradycardiac response to adenosine A$_1$ receptor agonists was enhanced significantly (seven-fold decrease EC$_{50}$) in Gα_oRGSi mutant cells compared to wild-type Gα_o, indicating a significant role of endogenous RGS proteins in cardiac automaticity regulation. The approach of using RGS-insensitive Gα subunit knockins will reveal the role of RGS protein-mediated GAP activity in signaling by a given G$_{i/o}$ protein. This will reveal the full extent of RGS regulation and will not be confounded by redundancy in the function of multiple RGS proteins.

Introduction

    Regulator of G-protein signaling (RGS) proteins modulate the strength and duration of G-protein-mediated signaling by acting as GTPase-accelerating proteins (GAP) though their conserved RGS domain. In addition to their role as negative regulators, RGS proteins also serve as versatile modulators and integrators of G-protein signaling via their diverse functional domains and structural motifs (Hollinger and Hepler, 2002). Considerable knowledge has been gained regarding the biochemical mechanisms of RGS protein stimulation of the GTPase activity of Gα (Ross and Wilkie, 2000). In addition, studies have begun to reveal the physiological roles of RGS proteins (reviewed by Hollinger and Hepler, 2002; Zhong and Neubig, 2001). Much previous work, however, uses overexpression of the individual RGS protein, which complicates conclusions regarding the functional importance of *endogenous* RGS proteins *in vivo*. A few RGS knockout mice (i.e., RGS2, RGS9, and RGS14) have been generated (Chen *et al.*, 2000; Oliveira-dos-Santos *et al.*, 2000; Siderovski, personal communication). RGS2 and RGS9 knockout phenotypes are consistent with a negative regulatory function on the vasoconstrictor (Heximer *et al.*, 2003; Tang *et al.*, 2003) and dopamine (Martemyanov *et al.*,

2003; Rahman *et al.*, 2003) signaling, respectively. However, RGS proteins may be functionally redundant, which may make single knockouts of limited value in assessing the overall role of RGS protein function *in vivo*. Thus, knowledge of the role of RGS proteins *in vivo* is still in its infancy.

Characterization of an RGS-Insensitive Mutation

The defining feature of RGS proteins is the ~120 amino acid-conserved RGS domain, which binds preferentially to the transition state of a Gα subunit and accelerates its intrinsic GTPase activity (Berman *et al.*, 1996; Hepler *et al.*, 1997; Hunt *et al.*, 1996). Yeast also has an RGS-regulated G-protein signaling system that is involved in mating pheromone responses. Disruption of the SST2 ("supersensitivity to pheromone-2") gene leads to enhanced pheromone signaling in terms of both ligand concentration and duration of response (Chan and Otte, 1982; Dohlman *et al.*, 1995). Using a genomic mutagenesis screen, DiBello *et al.* (1998) found that a single glycine-to-serine mutation in the yeast Gα subunit Gpa1 demonstrated the same phenotype as loss of the RGS protein (i.e., sst2 mutation). Enhanced function was also demonstrated for the G → S mutant of Gα_q. Lan *et al.* (1998) reported that the homologous G → S mutation in purified mammalian inhibitory G proteins, Gα_{i1} and Gα_o, also leads to loss of negative regulation by RGS proteins. The G184S mutant Gα_o showed no GTPase acceleration by RGS4 or RGS7 in single turnover GTP hydrolysis assays, while the intrinsic GTPase activity of Gα_o remained the same (Fig. 1) (Lan *et al.*, 1998). The mechanism of RGS insensitivity was

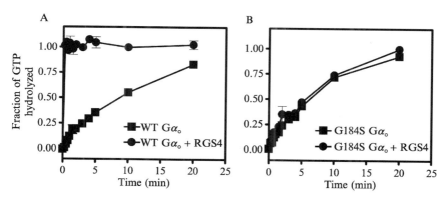

FIG. 1. Effect of RGS4 on single turnover GTP hydrolysis by mutant Gα_o. Purified wild-type Gα_o (WT, A) and RGSi Gα_o (G184S, B) were tested without (■) and with (●) 100 n*M* RGS4. Modified with permission from Lan *et al.* (1998)

established in flow cytometry protein–protein interaction studies, which showed a 30- to 100-fold reduction in affinity of RGS for the AlF_4^--induced transition state of the $G\alpha_o$ subunit (Lan *et al.*, 1998). This result is consistent with the crystal structure of the $G\alpha_{i1}$/RGS4 complex, in which the mutated Gly184 is located in the switch I region of the $G\alpha$ subunit and constitutes a critical interaction site with the conserved RGS domain (Tesmer *et al.*, 1997). Thus, the G $\rightarrow$ S mutation in switch I of the $G\alpha$ subunit shows a general ability to block $G\alpha$–RGS interactions (i.e., yeast Gpal, G_i family, and G_q family G proteins).

The G $\rightarrow$ S mutation also leaves other functions of the G protein intact. The ability of mutant $G\alpha_{i1}$ to couple to the receptor, as measured by the reconstitution of high-affinity agonist binding to the α_2-adrenergic receptor, was similar to that of wild-type $G\alpha$ (Fig. 2A). Moreover, binding of $G\alpha$ to $\beta\gamma$ subunits was not affected by the introduction of this point mutation (Fig. 2B). Also, its ability to regulate downstream effectors (i.e., adenylyl cyclase VI, Fig. 2C) was unaffected. In summary, the G $\rightarrow$ S mutation in the switch I region of $G\alpha_{i/o}$ specifically inhibits its ability to interact with RGS proteins without altering other aspects of $G\alpha$ function. Therefore, we will refer to this G $\rightarrow$ S mutation in $G\alpha$ subunits as the RGS-insensitive (RGSi) mutation.

As nearly 30 mammalian proteins with RGS domains have been identified to date and the majority of them demonstrate specificity toward G_i and G_q family G proteins (Hollinger and Hepler, 2002), it is clear that a certain degree of functional redundancy is likely to exist. Thus, the RGSi mutation that blocks RGS binding and GAP activity of all RGS proteins could be a valuable tool for examining the overall contribution of endogenous RGS proteins in physiological responses mediated by a specific G protein.

Use of RGSi $G\alpha$ to Unmask the Role of Endogenous RGS Proteins

Studying the Function of Endogenous RGS Proteins by Transfecting the RGSi $G\alpha_o$ Mutant into Cells

Assessing the role of transfected G proteins in the face of endogenous G proteins is difficult. This problem has been solved by the use of pertussis toxin-insensitive (PTXi) mutants of the $G_{i/o}$ family G proteins (Milligan, 1988). PTX treatment allows examination of responses mediated through the exogenous, expressed G protein without interference by endogenous $G_{i/o}$ proteins. Using cultured hippocampal neurons, Chen and Lambert (2000) employed doubly mutated RGSi/PTXi $G\alpha$ proteins expressed by adenoviral infection. They demonstrated that presynaptic inhibition induced by adenosine was enhanced markedly with RGSi/PTXi $G\alpha$

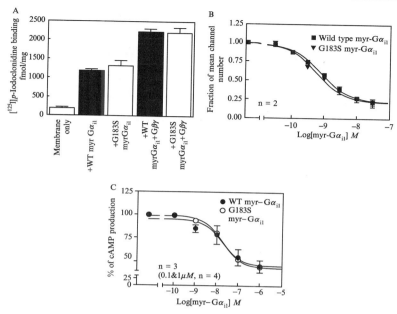

FIG. 2. Lack of effect of G183S mutation in $G\alpha_{i1}$ on non-RGS-related functions. (A) Binding of ^{125}I p-iodoclonidine to the $G\alpha_{i1}$-coupled α_{2A}-adrenergic receptor. Sf9 cell membranes (12 μg) with the α_{2A}-adrenergic receptor expressed were reconstituted with either wild-type (solid bar) or G183S (open bar) myr-$G\alpha_{i1}$ in the presence or absence of bovine brain $G\beta\gamma$. $[^{125}I]PIC$ binds to the high-affinity state of the α_2-adrenergic receptor, which is stabilized when the receptor is coupled to heterotrimeric $G\alpha\beta\gamma$. The addition of either wild-type $G\alpha_{i1}$ or G183S $G\alpha_{i1}$ produced similar increases in $[^{125}I]PIC$ binding compared to control (membrane only). Modified with permission from Lan (1999). (B) Affinity of $G\alpha_{i1}$ for biotinylated bovine brain $G\beta\gamma$. Competition by either wild-type or G183S myr-$G\alpha_{i1}$ with FITC-$G\alpha_o$ for biotinylated $G\beta\gamma$ was measured by flow cytometry (Lim et al. 2001; Sarvazyan et al., 1998). Data were fit to a one-site competition function. IC_{50} values for wild-type and G183 $G\alpha_{i1}$ to compete with FITC-labeled $G\alpha_o$ for binding to biotinylated $G\beta\gamma$ immobilized on avidin-coated beads are 0.8 ± 0.1 and 0.6 ± 0.1 nM, respectively. Thus, no difference was observed in the ability of wild-type $G\alpha_{i1}$ and G183 $G\alpha_{i1}$ to bind $G\beta\gamma$ subunits. Modified with permission from Lan (1999). (C) Inhibition of forskolin-stimulated type VI adenylyl cyclase activity by $G\alpha_{i1}$. The activity of forskolin-stimulated type VI adenylyl cyclase in Sf9 membranes was measured in the presence of various concentrations (0–1 μM) of myristoylated GTPγS-activated wild-type or G183S myr-$G\alpha_{i1}$. The production of ^{32}P cAMP was calculated as the percentage of cAMP produced in the absence of $G\alpha_{i1}$. Both GTPγS-activated wild-type and G183S $G\alpha_{i1}$ achieved about 50% of maximal inhibition of cAMP production. Furthermore, no difference was observed in the EC_{50} values of wild-type and mutant $G\alpha_{i1}$ to inhibit adenylyl cyclase (19 ± 8 vs 19 ± 7 nM). Modified with permission from Lan (1999).

expression compared to that of PTXi Gα infected cells. A similar strategy had been used previously by Jeong and Ikeda (2000) who demonstrated that α_2-adrenergic inhibition of N-type calcium currents is markedly inhibited by endogenous RGS proteins in rat sympathetic ganglia. Importantly, dose–response curves for norepinephrine showed a six- to eight-fold increase in potency in RGSi $G\alpha_o$ expressing cells compared to controls (Jeong and Ikeda, 2000). Similar results were shown for cells expressing a mutation in the switch II region of $G\alpha_o$ (S → D), which also confers insensitivity to RGS proteins (Natochin and Artemyev, 1998), confirming the effects of endogenous RGS proteins on the α_2-adrenergic regulation of $I_{Ca,N}$ in neurons.

It is known that the μ-opioid receptor, of interest for clinical analgesics and opioid drug abuse, can couple to all members of the $G\alpha_{i/o}$ family, with little selectivity for particular $G\alpha$ subunits (Laugwitz *et al.* 1993). In order to study the function of RGS proteins in μ-opioid receptor-mediated signaling, Clark *et al.* (2003) took advantage of the RGSi $G\alpha_o$ mutant and stably expressed it in C6 glioma cells (as $G\alpha_o$ RGSi/PTXi). As compared to wild-type (PTXi) $G\alpha_o$-mediated inhibition of adenylyl cyclase (AC), there was a dramatic enhancement of potency and efficacy in response to morphine and DAMGO with the RGSi mutant (Fig. 3), indicating a clear role of RGS proteins in regulating μ agonist sensitivity. In addition, ERK activation was also enhanced significantly in cells expressing

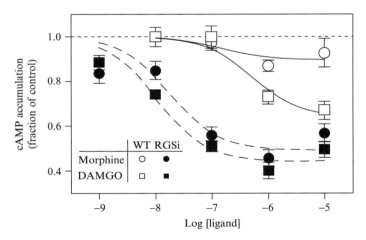

FIG. 3. Potentiation of μ-opioid inhibition of adenylyl cyclase (AC) by RGSi $G\alpha_o$. Stably expressed PTXi mutants of wild-type (WT) $G\alpha_o$ (open symbols) or RGSi $G\alpha_o$ (filled symbols) restored AC inhibition to PTX-treated C6 glioma cells. Responses to both the full agonist DAMGO (squares) and partial agonist morphine (circles) are enhanced. (Modified with permission from Clark *et al.*, 2003).

the RGSi Gα$_o$ mutant while an intracellular calcium response was affected less strongly (Clark *et al*, 2003). The strongly enhanced agonist response with RGSi Gα subunits supports the idea that RGS inhibition could be a useful pharmacologic approach (Neubig and Siderovski, 2002).

Although these studies provided convincing evidence for the role of endogenous RGS proteins, exogenously overexpressed G proteins may affect the fidelity of the observed responses or their amplitude. Therefore, the expression of an RGSi Gα subunit controlled by an endogenous promoter would be preferred.

Studying the Function of Endogenous RGS Proteins by Targeting RGSi Gα$_o$ into Embryonic Stem (ES) Cells

Gene Targeting Strategy. In addition to transfecting RGSi mutant G proteins into cells, this single point mutation can also be introduced into genomic DNA to allow expression of the mutant G protein under the control of its endogenous promoter. This should provide a better system to look at the function of endogenous RGS proteins *in vivo*.

A Gα$_o$ targeting construct was designed that contains the G → S point mutation in exon V, along with a diagnostic *Pvu*I restriction site. Between exons 5 and 6, a neomycin resistance gene flanked by loxP sites was inserted (Fig. 4A). The targeting construct also contains a negative selection marker (hsv-TK) to reduce the frequency of random genetic insertions and to enrich for homologous recombinant colonies (Mortensen and Seidman, 1994). The targeting construct was introduced into D3 mouse ES cells by electroporation, and 0.3 mg/ml G418 was used in the initial selection of clones that incorporated the targeting construct. Southern blot screening of genomic DNA was performed to identify homologously targeted ES cells, which demonstrate an increase in the size of a 10-kb *Xho*I fragment due to insertion of the 2-kb *neo* cassette (Fig. 4B).

To obtain homozygous knockin cell lines, heterozygous ES cells were screened for loss of heterozygocity in which both Gα$_o$ alleles contain *neo* genes, which make the cells more resistant to G418 (Mortensen and Seidman, 1994). To accomplish this, the heterozygous Gα$_o$ (G184S +/−) cells were plated at 10^5 per 100-mm dish containing 0.8, 1.0, and 1.5 mg/ml active G418 (note: G418 in solution is basic and must be neutralized to near pH 7.4 before adding to the cell culture medium). The cells are cultured for 7 to 9 days with daily medium changes until the colonies are detected easily. More than 40 colonies survived the selection at 0.8 mg/ml G418. Five colonies survived at 1.0 mg/ml G418. No colonies are detected in 1.5 mg/ml G418 cultures up to 10 days after plating. The 45 ES colonies were picked and cultured for further screening using Southern blot analysis

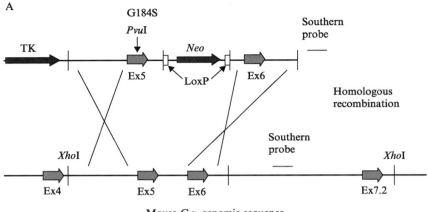

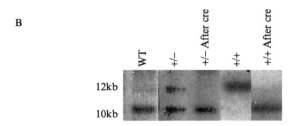

FIG. 4. (A) Targeting construct for genomic integration of the $G\alpha_o$ G184S mutation. (Top) The construct contains two fragments (4.5 and 2.2 kb) of mouse genomic $G\alpha_o$ DNA with a single mutation at exon 5, the negative selection marker (hsv-TK), and the *neo*-resistance gene flanked by loxP sites is inserted in intron 5, 435 bp from the point mutation. Homologous recombination with genomic DNA (bottom) would allow incorporation of the G → S mutation into the genomic DNA. A short distance between the point mutation and the *neo* cassette is preferred to obtain the desired homologous recombination in which the *neo* resistance gene is in *cis* with the point mutation, allowing selection of targeted colonies using neomycin. (B) Southern blot screen for the $G\alpha_o$ RGSi knockin mutant. *Xho*I digestion of targeted DNA shows a 12-kb fragment due to the inserted *neo* cassette while the wild-type allele is 10 kb. Cre recombinase expression in the targeted ES cell lines (+/− and +/+) allows for elimination of the *neo* cassette.

(Fig. 4B). Two of the 45 colonies were found to be homozygous for the *neo* allele, one each from the 0.8- and 1.0-mg/ml G418 selections.

In our original construct design, the *neo* marker between exons 5 and 6 was flanked with loxP sites to permit removal of the *neo* cassette if expression or splicing of the mutant $G\alpha_o$ RNA was affected by the presence of the 2-kb *neo* marker. loxP sites can be recognized by Cre recombinase, which carries out homologous recombination, resulting in the deletion of the *neo* resistance gene from genomic DNA, leaving only a single 34-bp loxP site plus 145 additional flanking residues. Preliminary studies (see later) indicated that both $G\alpha_o$ mRNA and protein were reduced in ES cells expressing mutant $G\alpha_o$ when the *neo* cassette was intact. Thus the pMC-Cre vector (Gu *et al.*, 1993), kindly provided by Dr. K. Rajewsky (The University of Cologne), is introduced into targeted heterozygous and homozygous G184S $G\alpha_o$ ES cells by electroporation (20 μg plasmid per 10^7 cells). Cells are initially plated at high density (5×10^6 per 10-cm plate). After 24 h, cells are trypsinized and plated at 3000 cells per 10-cm plate and allowed to grow in ES medium [high-glucose Dulbecco's modified Eagle's medium supplemented with 15% fetal calf serum (FCS), L-glutamine (2 mM), β-mercaptoethanol, nonessential amino acids, 100 U/ml of penicillin, 0.1 mg/ml of streptomycin, and 10 μg/ml of leukemia inhibitory factor (LIF)] without antibiotic selection. Single clones are picked after a week and subjected to Southern blotting to screen for the loss of the 2-kb *neo* cassette. Four out of 24 clones from G184S $G\alpha_o$ heterozygous knockin ES cells were identified as having eliminated the *neo* cassette, whereas only 3 out of 96 homozygous clones were found to contain only the 10-kb band. Incomplete elimination of the *neo* cassette (i.e., one band at 10 kb and one band at 12 kb) was found in more than 15 clones isolated from Cre-transfected homozygous ES cells.

Expression of RGSi $G\alpha_o$ Protein. In order to study the function of endogenous RGS proteins in $G\alpha_o$-mediated responses, we needed to examine the expression level of the mutant G protein in cells. It is known that $G\alpha_o$ has a very specific tissue distribution. It is absent from most tissues and cell types but is very abundant in neurons (Sternweis and Robishaw, 1984; Strittmatter *et al.*, 1990). Furthermore, a small amount of $G\alpha_o$ is also expressed in the heart and has been demonstrated to play a role in the regulation of cardiac $I_{Ca, L}$, which is crucial for the normal regulation of automaticity and contractility of the heart (Ye *et al.*, 1999).

Cardiocytes derived from ES cells by *in vitro* differentiation express receptors and ion channels comparable to functional cardiac cells *in vivo* (Boheler *et al.*, 2002). A hanging drop protocol (Metzger *et al.*, 1994) was used to generate spontaneously contracting cardiocytes from ES cell. Cells are trypsinized at about 70% confluence and resuspended in LIF-free medium. Twenty-microliter drops containing 2000 cells each are plated on 60-mm tissue culture dishes (Falcon), which are flipped over to allow the cells to aggregate without contact with the tissue culture plate surface. Hanging drops are cultured for 48 h, the plates are flipped back to the normal orientation, and LIF-free medium is added to suspend cell aggregates. Medium is not changed for about 5 days to avoid washing off

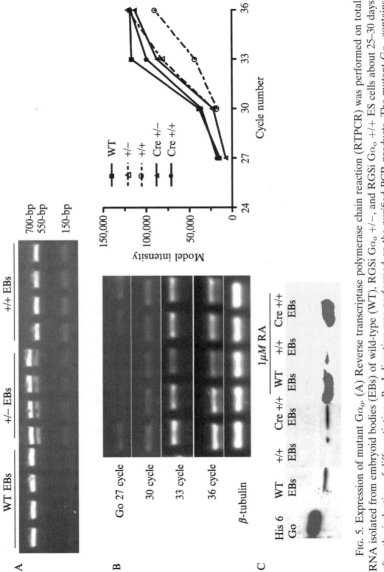

Fig. 5. Expression of mutant $G\alpha_o$. (A) Reverse transcriptase polymerase chain reaction (RTPCR) was performed on total RNA isolated from embryoid bodies (EBs) of wild-type (WT), RGSi $G\alpha_o$ +/−, and RGSi $G\alpha_o$ +/+ ES cells about 25–30 days after the induction of differentiation. PvuI digestion was performed on the purified PCR products. The mutant $G\alpha_o$ contains the recognition site for PvuI and therefore is cleaved into 550- and 150-bp fragments. In contrast, wild-type $G\alpha_o$ only shows the intact 700-bp PCR fragment after digestion. In heterozygous EBs, it is evident that the amount of G184S $G\alpha_o$ mRNA is less than that of the wild-type $G\alpha_o$ (about 40%). (B) PCR products measured at different cycle numbers. The 700-bp PCR product

cell aggregates. Once cells attach to the culture dish, they start to differentiate and eventually form three-dimensional structures called embryoid bodies (EB). Spontaneously contracting regions in embryoid bodies can be identified 9 days after LIF withdrawal, which serves as an indicator for differentiated cardiac cells. The spontaneous contractions continue for up to 30 days with LIF-free medium changes every 2 days. With this protocol, approximately 60–80% of the EB from D3 ES cells show at least one spontaneously contracting focus.

Total mRNA is extracted from 20–40 EBs, and reverse transcriptase polymerase chain reaction (PCR) is performed to examine the expression of Gα_o mRNA. Enzymatic digestion by $PvuI$ is used to confirm the expression of mutant G184S Gα_o mRNA whose amplified product is cleaved to 550- and 150-bp fragments, whereas that of wild-type Gα_o is left untouched at 700 bp. Figure 5A shows that, in four different heterozygous clones identified, there is less mRNA for the mutant Gα_o allele as compared to the wild-type allele, suggesting a potential defect in gene expression. Figure 5B shows that the amplified PCR product from heterozygous and homozygous mutant Gα_o expressing cells appeared at a later cycle number compared to wild-type ES cells. The abundance of mutant Gα_o mRNA after Cre-mediated *neo* elimination was restored to normal levels. Due to the difficulty of detecting Gα_o protein in these spontaneous cardiocyte differentiated cultures, we induced neuronal differentiation by treating EBs with 1 μM retinoic acid (RA) during the initial 48-h aggregation period. EBs are collected and Western blots performed using the whole cell lysate (Fig. 5C). RA treatment significantly increased Gα_o expression compared to spontaneous differentiation. With or without RA treatment, the G184S Gα_o homozygotes with the *neo* cassette intact expressed less Gα_o than cells with wild-type Gα_o. Homozygous cell lines after Cre-mediated elimination of the *neo* cassette have a comparable level of expression of mutant Gα_o protein to that seen in wild-type ES cells.

Endogenous RGS Proteins Play a Role in Gα_o-Mediated Regulation of Cardiac Automaticity. To characterize spontaneously beating ES cell-derived cardiocytes, we tested β-adrenergic stimulation with isoproterenol (Iso). Consistent with the observations by others (Hescheler *et al.*, 1997;

(without $PvuI$ digestion) from the targeted +/− and +/+ cell lines appeared at later cycle numbers than that from wild-type ES cells, indicating that less mRNA was present. Amplified products from cell lines after Cre-mediated elimination of the *neo* cassette showed similar amounts to that of WT. (C) Western blot for Gα_o in cell lysates of EBs with and without 1 μM retinoic acid. The expression of Gα_o in homozygous knockin cell lines is lower than that in WT. Cre recombinase expression restores the expression of mutant Gα_o to a level comparable to that of wild-type Gα_o.

Sachinidis *et al.*, 2003). ES cell-derived cardiocytes do not respond to Iso until later stages of differentiation, approximately 17 days after LIF withdrawal. In contrast, the basal beating rate can be suppressed by adenosine receptor agonists at a much early stage (i.e., 12 days after LIF withdrawal).

Taking advantage of RGSi $G\alpha_o$-targeted ES cells and the rhythmic contraction phenotype of differentiated cardiocytes, we examined the inhibition of beating on adenosine receptor activation. In order to have consistent Iso-stimulated responses, all studies were conducted from day 20 to day 24 after LIF withdrawal. All three cell lines, wild type, heterozygous $G\alpha_o$ G184S knockin, and homozygous $G\alpha_o$ G184S knockin, were studied in parallel on the same day. The beating rate measurement was conducted at room temperature in the high-glucose Dulbecco's modified Eagle's medium supplemented only with 100 U/ml of penicillin and 0.1 mg/ml of streptomycin. There is no difference in basal beating rate (beats per minute) among cell lines (61 ± 4.7 vs 62 ± 6.0 vs 57 ± 7.3, $n = 10$). Furthermore, they responded to Iso stimulation in a similar manner with no difference in maximal stimulation (3.4-fold vs 3.2-fold) or EC_{50} (20 nM vs 15 nM) for wild-type and homozygous G184S knockin cells. When a high concentration of Iso (i.e., 500 nM) was used, the beating rate increased dramatically and then slowed down within 5 min without washing off the agonist, indicating β-adrenergic receptor desensitization. However, 50 nM Iso produced a significant increase in beating rate and no desensitization was observed for up to 2 h. In order to study inhibition of the beating rate by adenosine receptor activation, 50 nM Iso was replenished each time with increasing concentrations of the adenosine receptor agonist N^6-(R-phenylisopropyl)adenosine (R-PIA) to generate a stable stimulation. The beating rate was counted for 30 s, 5 min after medium change to allow the response to reach the steady state. Interestingly, the inhibition of beating rate upon adenosine receptor activation showed significantly increased agonist potency for ES lines expressing the RGS-insensitive $G\alpha_o$. For cells before Cre recombinase transfection, EC_{50} values for R-PIA were 37 ± 12, 20 ± 3.7, and 11 ± 2.8 nM ($n = 9$, $p<0.05$, ANOVA) for wild-type, G184S $G\alpha_o$ (+/−), and G184S $G\alpha_o$ (+/+), respectively (Fig. 6A). The enhanced expression of the RGSi $G\alpha_o$ mutant after Cre transfection led to an even more pronounced functional response with EC_{50} values of 29 ± 7.8, 9.4 ± 1.9, and 4.2 ± 1.0 nM ($n = 7$, $p < 0.005$, ANOVA) for wild-type, G184S $G\alpha_o$ (+/−), and G184S $G\alpha_o$ (+/+), respectively (Fig. 6B). These data show that endogenous RGS proteins play a substantial role in modulating adenosine receptor/G_o–mediated bradycardiac responses and that inhibition of RGS proteins could significantly enhance G_o and possibly G_i signals in heart.

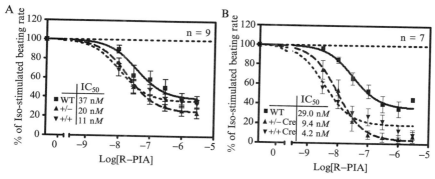

FIG. 6. A_1-adenosine receptor agonist potency is enhanced in $G\alpha_o$ RGSi knockin cells. (A) R-PIA-induced inhibition of the isoproterenol-stimulated beating rate occurred at a two- to three fold lower concentration in RGSi $G\alpha_o$ cells. (B) This enhancement in agonist sensitivity was increased further after the expression of mutant protein $G\alpha_o$ was improved by Cre recombinase expression, which shows a seven-fold reduction in IC_{50} for the homozygous knockin mutant.

In summary, the $G\alpha$ RGSi mutation is a useful tool in evaluating the combined effect of all endogenous RGS proteins on a particular $G\alpha$ subunit *in vivo*. Gene targeting of RGSi $G\alpha$ mutants into embryonic stem cells can be used to study the function of RGS proteins in a wide variety of cell types using *in vitro* differentiation models or to establish mouse germ-line transmission of the mutation. Present data show that elimination of the *neo* selection marker, in either setting, will be essential, as it appears to interfere with gene expression or RNA splicing, which could mask the enhanced functional responses. Obviously, the *in vivo* physiological phe-notype will be of substantial interest and those experiments are underway.

Acknowledgments

Supported by NIH Grants GM39561 (RRN), T32HL007853 (XH), MDRTC Grant (NIH P60 DK20572), and an AHA Predoctoral fellowship (YF). We thank Min Liu, Elizabeth Hughes, and Ginny Zawistowski for their technical support.

References

Berman, D. M., Kozasa, T., and Gilman, A. G. (1996). The GTPase-activating protein RGS4 stabilizes the transition state for nucleotide hydrolysis. *J. Biol. Chem.* **271**(44), 27209–27212.

Boheler, K. R., Czyz, J., Tweedie, D., Yang, H. T., Anisimov, S. V., and Wobus, A. M. (2002). Differentiation of pluripotent embryonic stem cells into cardiomyocytes. *Circ. Res.* **91**(3), 189–201.

Chan, R. K., and Otte, C. A. (1982). Isolation and genetic analysis of *Saccharomyces cerevisiae* mutants supersensitive to G1 arrest by a factor and alpha factor pheromones. *Mol. Cell Biol.* **2**(1), 11–20.

Chen, H., and Lambert, N. A. (2000). Endogenous regulators of G protein signaling proteins regulate presynaptic inhibition at rat hippocampal synapses. *Proc. Natl. Acad. Sci. USA* **97**(23), 12810–12815.

Clark, M. J., Harrison, C., Zhong, H., Neubig, R. R., and Traynor, J. R. (2003). Endogenous RGS protein action modulates mu-opioid signaling through $G\alpha$: Effects on adenylyl cyclase, extracellular signal-regulated kinases, and intracellular calcium pathways. *J. Biol. Chem.* **278**(11), 9418–9425.

DiBello, P. R., Garrison, T. R., Apanovitch, D. M., Hoffman, G., Shuey, D. J., Mason, K., Cockett, M. I., and Dohlman, H. G. (1998). Selective uncoupling of RGS action by a single point mutation in the G protein alpha-subunit. *J. Biol. Chem.* **273**(10), 5780–5784.

Dohlman, H. G., Apaniesk, D., Chen, Y., Song, J., and Nusskern, D. (1995). Inhibition of G-protein signaling by dominant gain-of-function mutations in Sst2p, a pheromone desensitization factor in *Saccharomyces cerevisiae*. *Mol. Cell Biol.* **15**(7), 3635–3643.

Gu, H., Zou, Y. R., and Rajewsky, K. (1993). Independent control of immunoglobulin switch recombination at individual switch regions evidenced through Cre-loxP-mediated gene targeting. *Cell* **73**, 1155–1164.

Hepler, J. R., Berman, D. M., Gilman, A. G., and Kozasa, T. (1997). RGS4 and GAIP are GTPase-activating proteins for Gq alpha and block activation of phospholipase C beta by gamma-thio-GTP-Gq alpha. *Proc. Natl. Acad. Sci. USA* **94**(2), 428–432.

Hescheler, J., Fleischmann, B. K., Lentini, S., Maltsev, V. A., Rohwedel, J., Wobus, A. M., and Addicks, K. (1997). Embryonic stem cells: A model to study structural and functional properties in cardiomyogenesis. *Cardiovasc. Res.* **36**(2), 149–162.

Heximer, S. P., Knutsen, R. H., Sun, X., Kaltenbronn, K. M., Rhee, M. H., Peng, N., Oliveira-dos-Santos, A., Penninger, J. M., Muslin, A. J., Steinberg, T. H., Wyss, J. M., Mecham, R. P., and Blumer, K. J. (2003). Hypertension and prolonged vasoconstrictor signaling in RGS2-deficient mice. *J. Clin. Invest.* **111**(4), 445–452.

Hollinger, S., and Hepler, J. R. (2002). Cellular regulation of RGS proteins: Modulators and integrators of G protein signaling. *Pharmacol Rev.* **54**(3), 527–559.

Hunt, T. W., Fields, T. A., Casey, P. J., and Peralta, E. G. (1996). RGS10 is a selective activator of G alpha i GTPase activity. *Nature* **383**(6596), 175–177.

Jeong, S. W., and Ikeda, S. R. (2000). Endogenous regulator of G-protein signaling proteins modify N-type calcium channel modulation in rat sympathetic neurons. *J. Neurosci.* **20**(12), 4489–4496.

Lan, K. L. (1999). "Mechanism and Specificity of G Protein Regulation by RGS Proteins." Dissertation, University of Michigan.

Lan, K. L., Sarvazyan, N. A., Taussig, R., MacKenzie, R. G., DiBello, P. R., Dohlman, H. G., and Neubig, R. R. (1998). A point mutation in $G\alpha_o$ and $G\alpha_i 1$ blocks interaction with regulators of G protein signaling. *J. Biol. Chem.* **273**, 12794–12797.

Laugwitz, K. L., Offermanns, S., Spicher, K., and Schultz, G. (1993). Mu and delta opioid receptors differentially couple to G protein subtypes in membranes of human neuroblastoma SH-SY5Y cells. *Neuron.* **10**(2), 233–242.

Lim, W. K., Myung, C. S., Garrison, J. C., and Neubig, R. R. (2001). Receptor-G protein γ specificity: $\gamma 11$ shows unique potency for A_1 adenosine and $5\text{-}HT_{1A}$ receptors. *Biochemistry* **40**, 10532–10541.

Martemyanov, K. A., Hopp, J. A., and Arshavsky, V. Y. (2003). Specificity of G protein-RGS protein recognition is regulated by affinity adapters. *Neuron.* **38**(6), 857–862.

Metzger, J. M., Lin, W. I., and Samuelson, L. C. (1994). Transition in cardiac contractile sensitivity to calcium during the *in vitro* differentiation of mouse embryonic stem cells. *J. Cell Biol.* **126**(3), 701–711.

Milligan, G. (1988). Techniques used in the identification and analysis of function of pertussis toxin-sensitive guanine nucleotide binding proteins. *Biochem. J.* **255**(1), 1–13.

Mortensen, R. M., and Seidman, J. G. (1994). Inactivation of G-protein genes: Double knockout in cell lines. *Methods Enzymol.* **237**, 356–366.

Natochin, M., and Artemyev, N. O. (1998). A single mutation Asp229-Ser confers upon Gsα the ability to interact with regulators of G protein signaling. *Biochemistry* **37**, 13776–13780.

Neubig, R. R., and Siderovski, D. P. (2002). Regulators of G-protein signaling as new central nervous system drug targets. *Nature Rev. Drug. Discov.* **1**(3), 187–197.

Oliveira-dos-Santos, A. J., Matsumoto, G., Snow, B. E., Bai, D., Houston, F. P., Whishaw, I. Q., Mariathasan, S., Sasaki, T., Wakeham, A., Ohashi, P. S., Roder, J. C., Barnes, C. A., Siderovski, D. P., and Penninger, J. M. (2000). Regulation of T cell activation, anxiety, and male aggression by RGS2. *Proc. Natl. Acad. Sci. USA* **97**(22), 12272–12277.

Rahman, Z., Schwarz, J., Gold, S. J., Zachariou, V., Wein, M. N., Choi, K. H., Kovoor, A., Chen, C. K., DiLeone, R. J., Schwarz, S. C., Selley, D. E., Sim-Selley, L. J., Barrot, M., Luedtke, R. R., Self, D., Neve, R. L., Lester, H. A., Simon, M. I., and Nestler, E. J. (2003). RGS9 modulates dopamine signaling in the basal ganglia. *Neuron.* **38**(6), 941–952.

Ross, E. M., and Wilkie, T. M. (2000). GTPase-activating proteins for heterotrimeric G proteins: Regulators of G protein signaling (RGS) and RGS-like proteins. *Annu. Rev. Biochem.* **69**, 795–827.

Sachinidis, A., Fleischmann, B. K., Kolossov, E., Wartenberg, M., Sauer, H., and Hescheler, J. (2003). Cardiac specific differentiation of mouse embryonic stem cells. *Cardiovasc. Res.* **58**(2), 278–291.

Sarvazyan, N. A., Remmers, A. E., and Neubig, R. R. (1998). Determinants of G$_{i1}$α and βγ binding. *J. Biol. Chem.* **73**(14), 7934–7940.

Sternweis, P. C., and Robishaw, J. D. (1984). Isolation of two proteins with high affinity for guanine nucleotides from membranes of bovine brain. *J. Biol. Chem.* **259**(22), 13806–13813.

Strittmatter, S. M., Valenzuela, D., Kennedy, T. E., Neer, E. J., and Fishman, M. C. (1990). Go is a major growth cone protein subject to regulation by GAP-43. *Nature* **344**(6269), 836–841.

Tang, M., Wang, G., Lu, P., Karas, R. H., Aronovitz, M., Heximer, S. P., Kaltenbronn, K. M., Blumer, K. J., Siderovski, D. P., Zhu, Y., and Mendelsohn, M. E. (2003). Regulator of G-protein signaling-2 mediates vascular smooth muscle relaxation and blood pressure. *Nature Med.* **9**(12), 1506–1512.

Tesmer, J. J., Berman, D. M., Gilman, A. G., and Sprang, S. R. (1997). Structure of RGS4 bound to AlF4-activated G(i alpha1): Stabilization of the transition state for GTP hydrolysis. *Cell* **89**(2), 251–261.

Ye, C., Sowell, M. O., Vassilev, P. M., Milstone, D. S., and Mortensen, R. M. (1999). Galpha(i2), alpha(i3) and Galpha(o) are all required for normal muscarinic inhibition of the cardiac calcium channels in nodal/atrial-like cultured cardiocytes. *J. Mol. Cell Cardiol.* **31**(9), 1771–1781.

Zhong, H., and Neubig, R. R. (2001). Regulator of G protein signaling proteins: Novel multifunctional drug targets. *J. Pharmacol. Exp. Ther.* **297**(3), 837–845.

Subsection E
Methods of RGS Protein Inhibition

[15] Ribozyme- and siRNA-Mediated Suppression of RGS-Containing RhoGEF Proteins

By QIN WANG, MIN LIU, TOHRU KOZASA, JEFFREY D. ROTHESTEIN, PAUL C. STERNWEIS, and RICHARD R. NEUBIG

Abstract

Given recent efforts to determine the sequence information on thousands of genes in the human genome, the current challenge is to identify the functions of these genes, including those encoding the regulator of G-protein signaling protein gene superfamily, and to establish their roles in particular signaling pathways in a native system. Increasingly, reverse genetic approaches are being used to address these questions. This article compares two powerful approaches [ribozyme and "short interfering" RNA (siRNA) techniques] under identical conditions for the first report on the suppression of endogenous RGS domain-containing RhoGEFs. The siRNA technique was found to be much more potent than ribozyme targeting at the same mRNA site of RGS-RhoGEFs. Also, the three siRNAs targeting LARG, PDZ-RhoGEF, and p115-RhoGEF are able to discriminate the closely related sequences within this RGS-RhoGEF gene family.

Introduction

With reverse genetic strategies, gene sequence information can now be used to disrupt gene function. Increasingly, anti-mRNA agents are valuable tools for inhibiting the expression of a target gene in a sequence-specific manner. They may be used in functional genomics, target validation, and even for therapeutic purposes. There are three basic approaches available to date: antisense oligonucleotides, ribozymes, and RNA interference (RNAi). Among these three technologies, RNAi has rapidly become the new method of choice, replacing conventional antisense and ribozyme technologies, to determine gene function for a wide range of biomedical applications. Even though antisense oligonucleotide techniques have been improved dramatically by modifying base structure, the poor specificity due to the secondary action of RNase H degradation of nontarget mRNA sharing as little as a

5-bp match limits its application (Branch, 1998; Stein, 2000). This article focuses on comparing RNA-based ribozyme and "short interfering" RNA (siRNA) inhibition of regulator of G-protein signaling (RGS) domain containing RhoGEFs with an emphasis on siRNA technology.

Ribozymes

Ribozymes are unique RNA enzymes that can recognize complementary mRNA targets and cleave them at "GUC" sites in a sequence-specific fashion (Cech et al., 1981; Guerrier-Takada and Altman, 1984). This makes ribozymes powerful tools for studying the functional consequences of suppressed gene expression. They have been widely used to inhibit gene expression in HIV infection (Taylor et al., 1992), cancer (William and Aymen, 1995), and human type I collagen gene mutations (Grassi et al., 1997) and with G-protein γ subunits to define the functional role of the $\gamma7$ subunit in β-adrenergic receptor signaling (Wang et al., 1997, 1999). Among the different types of ribozymes, hammerhead ribozymes, which we have been using successfully for years, are the most extensively characterized with enhanced catalytic turnover and stability (Taylor et al., 1992).

RNA Interference

RNA interference is a newly discovered, powerful approach for silencing endogenous gene expression in mammalian cell systems (Elbashir et al., 2001a,b; Paddison et al., 2002). In this phenomenon, a double-stranded "short interfering" RNA acts as the triggering agent to initiate the degradation of targeted mRNAs in a gene-specific fashion, thereby silencing their expression (Fire, 1999; Fire et al., 1998). RNAi is a naturally occurring, catalytic gene regulation system thought to have evolved primarily as a defense mechanism against molecular pathogens. Used as a research tool and potential therapeutic approach, RNAi offers unprecedented advantages: (1) high potency, specificity, and scalability; (2) wide cross-species applicability; and (3) excellent experimental reproducibility.

siRNA are 21–23 nucleotide double-stranded RNA molecules that trigger the formation of a multicomponent nuclease complex (siRNP) that carries out sequence-specific target recognition and degradation of the targeted mRNA (Zamore, 2001). Most recently, the RNAi effect has been demonstrated in mammalian cells with gene expression silenced more than 95% (Elbashir et al., 2001a,b; Paddison et al., 2002). Of further note, siRNAs are effective in quantities much lower than antisense or ribozyme-based strategies.

Before the RNAi technique emerged, we demonstrated that ribozyme approaches could be used to suppress selectively members of gene families. Because ribozymes and RNAi act through different mechanisms, we were interested in comparing these two approaches in terms of efficacy and specificity. In this article, we have chosen the RGS domain containing RhoGEFs as targets for designing both ribozymes and siRNA for our study.

RGS Proteins as Targets of Ribozyme and siRNA Knockdown Methods

RGS Proteins

The regulators of G-protein signaling proteins make up a diverse protein family best known for their ability to bind directly to Gα subunits in their active GTP-bound state and stimulate GTP hydrolysis. Thus they act as GTPase-activating proteins (GAPs), turning off G-protein signaling (Berman et al., 1996; De Vries et al., 2000; Ross and Wilkie, 2000; Siderovski et al., 1999). The RGS protein family has been grouped into nine groups depending on their domain architecture (Neubig and Siderovski, 2002). RGS proteins have specificity for discrete Gα subfamilies and selectivity for the modulation of particular receptor actions as well. To elucidate functional roles for specific endogenous RGS proteins, we developed a ribozyme approach to inactivate individual RGS proteins and demonstrated specific roles of RGS3 and RGS5 in inhibiting ERK activation by muscarinic and angiotensin receptors, respectively (Wang et al., 2002). This receptor-selective effect of RGS protein function provided new evidence for a unique targeting of RGS action to specific cellular responses.

RGS Domain Containing RhoGEFs

The newly recognized RGS subfamily (family F, GEF) includes three RGS domain-containing RhoGEFs: p115-RhoGEF, PDZ-RhoGEF, and leukemia-associated RhoGEFs (LARG), which contain both DH/PH and RGS domains (Fukuhara et al., 1999, 2000; Hart et al., 1996, 1998; Kozasa et al., 1998). This group of RGS protein binds to and accelerates the intrinsic GTPase activity of G12 family Gα subunits (and possibly Gαq as well) (Booden et al., 2002; Vogt et al., 2003). Guanine nucleotide exchange factors (GEFs) control GTP exchange on the small G proteins (Seasholtz et al., 1999). A family of RhoGEFs, including lbc, lsc, and lfc, was first identified as oncogenes (Glaven et al., 1996; Toksoz and Williams, 1994). They contain a Dbl homology (DH) domain responsible for Rho exchange activity and a Pleckstrin homology (PH) domain thought to be involved in subcellular localization. On binding an activated, GTP-bound $G\alpha$ subunit, the GEF activity of these RhoGEFs is increased, leading to RhoA activation. The Rho family of small G proteins is well recognized as a mediator of cell growth, actin cytoskeletal rearrangement, and gene transcription in mammalian cells (Aspenstrom, 1999). The discovery that extracellular stimuli activate Rho-mediated effects indicates a role for Rho in GPCR-mediated signal transduction (Sah *et al.*, 2000; Seasholtz *et al.*, 1999). In contrast to the general role of RGS proteins as negative regulators of G-protein signaling, RGS domain-containing RhoGEFs play a positive role in GPCR regulation via GEF activation through binding of activated $G\alpha 13$ (Hart *et al.*, 1998; Wells *et al.*, 2002). This results in downstream effects such as cytoskeletal rearrangement, transcriptional upregulation, cell growth, cell migration, and smooth muscle contraction. Evidence suggests that LARG responds to $G\alpha_q$ as well as to $G\alpha 12/13$ subunits (Booden *et al.*, 2002). This provides a mechanism by which GPCRs that couple to $G\alpha 13$ or $G\alpha_q$ could activate RhoA and its downstream responses through a RhoGEF. Thus, RGS domain-containing RhoGEFs are a direct molecular bridge between hetero-trimeric $G\alpha$ subunits and small G-protein RhoA activation.

Selection of Target Sites for RGS-RhoGEF siRNA and Ribozymes

In order to compare the two RNA-based reverse genetic technologies, ribozymes and siRNA, in suppressing the expression of RGS-RhoGEF proteins, we decided to select the same region of mRNA for designing both chemically synthesized siRNA and ribozymes. Selection of target sites and designing ribozymes require reliable identification of a cleavage site within the targeted RNA molecule that is amenable to the formation of hybrid complexes between the ribozyme and the mRNA template. This process is still largely empirical, even though available computer programs for determining the secondary structure of mRNA may be helpful in chosing the target site (Amarzguioui and Prydz, 1998; Zuker, 1994). Selection of a target site for a hammerhead ribozyme requires the identification of a cleavage site "GUC" trinucleotide, which is the most efficient cleavage substrate (Robishaw *et al.*, 2002, 2003). We then designed an siRNA around the same sequence. The siRNA contained a 19 nucleotide recognition sequence plus a 3′–TT overhang based on the criteria for designing siRNA as shown in Fig. 1. Initial empirical rules have been established by the Tuschl lab for the design of siRNAs and are available at http://www.rockefeller.edu/labheads/tuschl/sirna.html (Elbashir *et al.*, 2001b).

The general rules for design of siRNAs include the following. (1) Find a sequence starting with "AA-", pick the following 19 nucleotides, and add

siRNA/RZ	Sequences	Inhibition of luc activity
1. PDZ-A:	5′ GTAGCCTGTCTTCTCTGGG 3′	–
2. PDZ-B:	5′ ACTGAAGTCTCGGCCAGCT 3′	++++
3. PDZ-Binv:	5′ TCGACCGGCTCTGAAGTCA 3′	–
4. PDZ-B/RZ:	5′ GGAGAAACTGAAGTCTCGGCCAG 3′	
5. LARG-A:	5′ CACACAGTCTACTATCACC 3′	–
6. LARG-B:	5′ GAAACTCGTCGCATCTTCC 3′	++++
7. LARG-Binv:	5′ CCTTCTACGCTGCTCAAAC 3′	–
8. LARG-B/RZ:	5′ GAAACTCGTCGCATCTTCC 3′	
9. p115-A:	5′ CGAGCTGGAGACAAACTCA 3′	+++
10. p115-B:	5′ CATACCATCTCTACCGACG 3′	++++
11. p115-Binv:	5′ GCAGCCATCTCTACCATAC 3′	–
12. p115-C:	5′ GGTTCCCGTCAGCATCATC 3′	–
13. p115-D:	5′ GAAAAGAGTGCTGCCGTGG 3′	–
14. p115-E:	5′ GGCCTTTTGCAAGAAGGAG 3′	+++

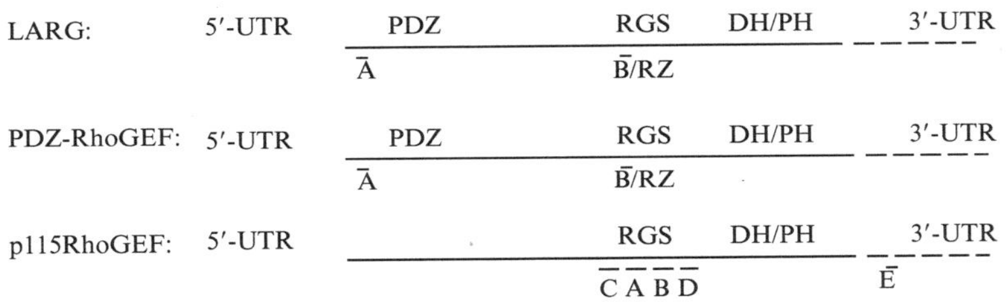

FIG. 1. Targeting sites for human RGS-RhoGEF siRNA and ribozymes. Different targeting sites for each siRNA (lettered A through E) and ribozymes (RZ) are shown. "inv" indicates mutant siRNA obtained by inverting the sequence of siRNA. Inhibition of SRE. L-promoter luciferase (Luc) activity was performed as described in the text for screening siRNA function at different targeting sites. The bottom schematic map indicates the relative location of each siRNA/ribozyme on each RGS-RhoGEF gene structure.

2 nucleotides of dTdT at the 3′ end to generate an AA(N19)TT motif. (2) Select a target sequence to have around 50% GC content. Many siRNA with 60% GC content are effective, but siRNAs with $\geq$70% GC are not as effective. (3) Choose an estimated T_m of 55°. (4) Avoid more than three Gs or three Cs in a row, which can form agglomerates, thereby interfering with the siRNA-silencing mechanism. All candidate target sites are then checked for gene specificity by a homology search of the nucleic acid databases with the NCBI Blast tool to avoid other possible mRNAs that may cross-react within the closely related members in the gene family as well as unrelated gene products. In most situations, there is partial homology with other unrelated genes. If there is homology within the gene family for which the function is being evaluated, it is especially important to select a different region of the mRNA.

Emerging experimental evidence suggests that secondary structure and accessibility of the target RNA sequence are important factors in determining the potency of an siRNA. Because synthetic siRNA duplexes are expensive, methods to improve siRNA design are clearly needed. There are more programs available for the design of 21 nucleotide siRNA and more companies are providing service for the synthesis of siRNA. For example, Qiagen, Ambion, Dharmacon, and Invitrogen Life Technology, Inc. provide such services. The Zuker lab at Rensselaer Polytechnic Institute in New York has an RNA folding program for determining RNA secondary structure at www.bioinfo.rpi.edu/applications/mfold/old/rna (Zuker, 1994, 2003). The Sfold software combines secondary structure and accessibility predictions with the established empirical design rules to improve siRNA design and is available through Web servers at http://sfold.wadsworth.org/ and http://www.bioinfo.rpi.edu/applications/sfold.

In our study, the RGS domains of human LARG and PDZ-RhoGEF were chosen for designing both siRNA and ribozymes as shown in Fig. 1. We initially selected two target sites for LARG and PDZ-RhoGEFs and, in each case, one of them turned out to have good activity as an siRNA (PDZ-B and LARG-B) as determined by the luciferase assay (see later). However, for human p115-RhoGEF, out of five target sites, only one very good siRNA (p115-B) was obtained while two others (p115-A and p115-E) had slightly lower efficiency. Therefore, we selected one good siRNA for each RGS-RhoGEF (PDZ-B siRNA for targeting PDZ-RhoGEF; LARG-B siRNA for targeting LARG; and p115-B for targeting p115-RhoGEF) for later experiments. An example of the structures of these siRNA and ribozymes is given for the PDZ-RhoGEF target (PDZ-B) in Fig. 2.

Synthesis of Synthetic RGS-RhoGEF siRNA and Ribozymes

For generating RGS-RhoGEF ribozymes, chimeric DNA–RNA ribozymes targeted to human LARG and PDZ-RhoGEF are synthesized chemically and modified by introducing two phosphorothioate linkages at the 3′ end to improve their resistance to intracellular endonucleases (Shimayama et al., 1993). Synthesis is done using an ABI 394 DNA/RNA synthesizer (Applied Biosystems, CA) and then ribozymes are further prepared by desilylation (deprotection), desalting, and HPLC purification, as described previously (Wang et al., 1997). The inactive control ribozymes are prepared in the same way, but by scrambling the two flanking regions and introducing two point mutations in the catalytic core region, which has been shown previously to prevent ribozyme cleavage activity *in vitro* (Ruffner et al., 1990). This study used an inactive ribozyme targeted to the human G-protein γ7 subunit (γ7RZml) as a negative control (Wang et al., 2001).

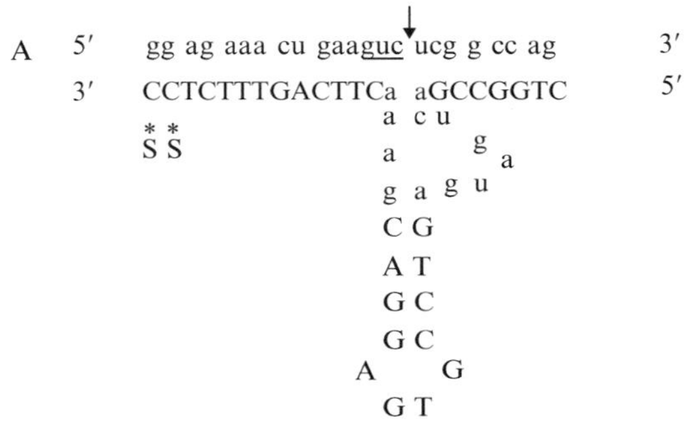

A   5'   gg ag aaa cu gaaguc ucg g cc ag   3'
    3'   CCTCTTTGACTTCa aGCCGGTC   5'
          * *   a c u
          S S   a   g
                                      a
                              g a g u
                              C G
                              A T
                              G C
                              G C
                              A   G
                              G T

B   PDZ-siRNA: 5'   acugaagucucggccagcuTT   3'
               3'   TTugacuucagagccggucga   5'

C   PDZinv: 5'   ucgaccggcucugaagucaTT   3'
               3'   TTagcuggccgagacuucagu   5'

FIG. 2. Structure of human PDZ-RhoGEF hammerhead ribozyme and siRNA. (A) Hammerhead structure of PDZ-RhoGEF ribozyme. The targeted PDZ-RhoGEF mRNA transcript is shown in the 5' to 3' direction (*upper sequence*), and the constructed DNA–RNA chimeric ribozyme is shown in the 3' to 5' direction (*lower sequence*). The targeted GUC trinucleotide in the mRNA is underlined, and the cleavage site is indicated by the arrow. Ribonucleotides required for ribozyme activity in the core region are shown in lowercase letters, and complementary flanking regions of deoxyribonucleotides are shown in capital letters. The two phosphorothioate linkages at the 3' end of ribozyme to enhance the stability are indicated by asterisks. (B) Structure of the PDZ-RhoGEF double-stranded siRNA "PDZ-B." A 19 nucleotide sequence from PDZ-RhoGEF cDNA is selected and synthesized chemically with 3' "TT" overhangs and annealed as double-stranded siRNA. (C) Mutant PDZ-RhoGEF siRNA (PDZinv). The mutant siRNA is designed by inverting the siRNA sequence from the 3' to 5' direction and then annealing to form the double-stranded RNA duplex as a negative control.

All siRNAs are made commercially through Qiagen siRNA synthesis service (Xenogene Inc., MD) as standard siRNA duplex and IE-HPLC purified. To anneal sense and antisense strands to form double-stranded RNA, the two synthesized siRNAs are resuspended at 0.3 mg/ml (total) in 1.0 ml buffer containing 100 mM potassium acetate, 30 mM HEPES–KOH, and 2 mM magnesium acetate, pH 7.4, and are heated to 90° for 1 min and then incubated at 37° for 1 h. This treatment disrupts possible aggregates formed during the original lyophilization and freezing and reforms duplex RNA. Aliquots are then placed into small vials to make siRNA stock solutions at 20 μM (about 300 ng/μl) and are frozen at $-20°$. Because both siRNA and ribozyme are RNAs, precautions for handling these RNA reagents are needed for decreasing the rapid degradation of RNA such as always using sterile tips and tubes and DEPC-treated water for buffers and for resuspension of ribozymes.

Delivery of Synthetic RGS-RhoGEF siRNA and Ribozymes into Cells

HEK293T cells were chosen for testing the synthetic RGS-RhoGEF siRNA and ribozymes due to their high efficiency of transfection and the fact that they express all three RGS-RhoGEFs endogenously. We have found over 90% transfection efficiency with Lipofectamine 2000 (Invitrogene Life Technology Inc.) by use of a green fluorescent protein construct (pEGFP from Clontech). One day before transfection, HEK293T cells are plated such that they are about 80–90% confluent at the time of transfection. The mixture of Lipofectamime 2000 (about 3 μl/μg RNA) with RhoGEF siRNA or ribozyme in OPTI-MEM (Invitrogen Corp.) is incubated for 20 min at room temperature and is then added directly into the dishes with complete medium with 10% fetal bovine serum but without antibiotics for at least 5 h at 37°. Then fresh complete medium is added until harvesting at 48–72 h posttransfection.

Comparison of RGS-RhoGEF siRNA vs Ribozyme Activity in Cells

The mechanisms of siRNA and ribozyme action in cells are different. We were interested in comparing the efficacy of LARG and PDZ-RhoGEF siRNA and ribozymes generated from the same targeting site. The RGS-RhoGEF siRNA and ribozymes were introduced into HEK293T cells under identical conditions, including same amount of siRNA or ribozyme, Lipofectamine 2000, and transfection time and then the inhibition of both protein expression and RhoGEF-mediated luciferase activity was studied.

Effect on Protein Suppression by Western Blotting Analysis

To determine the effect of ribozyme and siRNA transfection on RhoGEF protein suppression, 2 μg of ribozyme (LARG-B/RZ or PDZ-B/RZ) or siRNA (LARG-B or PDZ-B) and their mutant controls [inverted siRNA as mutant control for siRNA (LARG-Binv or PDZ-Binv) and γ7RZml as mutant control for ribozyme] is mixed with 8 μl of Lipofectamine 2000 and introduced into HEK293T cells in a 6-well plate. Protein lysates are prepared 72 h posttransfection. Cells are washed with phosphate-buffered saline (PBS), suspended in 0.3 ml of RIPA lysis buffer containing 1× PBS, 1% Igepal CA-630 (Sigma), 0.5% sodium deoxycholate, 0.1% sodium dodecyl sulfate (SDS) with a fresh addition of 0.15–0.3 TIU/ml aprotinin (Sigma), 1 mM sodium orthovanadate, 1 mM

benzamidine, 1 μg/ml pepstatin, and 2 μg/ml leupeptin and passed through a 21-gauge needle 10 times on ice. After adding 10 μg/ml phenylmethyl sulfonyl fluoride, the mixture is incubated for 1 h on ice and is finally centrifuged at 10,000g for 10 min. The supernatant is saved as total cell lysate. Protein samples (80 μg of total cell lysate) are subjected to SDS–polyacrylamide gel electrophoresis (PAGE) on an 8% minigel and transferred to an Immobilon-P transfer membrane. Prestained SDS–PAGE protein standards (Kaleidoscope from Bio-Rad) are used to determine the size of detected proteins. The membrane is blocked with 1.0% bovine serum albumin (BSA) in TBS (10 mM Tris–HCl, pH 8.0, 150 mM NaCl) without Tween 20 for 1 h and is incubated with RGS-RhoGEF-specific antibodies in 1.0% BSA for another 1 h. The anti-LARG antibody, purchased from Santa Cruz (N-14), was used at 1:400 dilution. The affinity-purified polyclonal anti-PDZ-RhoGEF antibody was a kind gift of Dr. J. Rothstein (Johns Hopkins University, Baltimore, Maryland) and was generated from a synthetic peptide corresponding to an epitope of rat PDZ-RhoGEF (GTRAP48, KTPERTSPSHHRQPSD) (Jackson et al., 2001). It is used at 1:150 dilution.

After three consecutive washes with TBS plus 0.05% Tween 20 (10 min for each), the membrane is incubated with the secondary antibody at 1:10,000 dilution for 1 h at room temperature (Santa Cruz, bovine anti-goat IgG-HRP for LARG and goat anti-rabbit IgG-HRP for PDZ-Rho-GEF). Proteins are visualized by chemiluminescence with SuperSignal West Dura (Pierce). After stripping, the blots are reprobed with an anti-G-protein β subunit antibody (Santa Cruz, sc-378, 1:2000 dilution) as an internal control for the determination of protein loading. The intensity of bands is quantified by densitometry analysis of X-ray films in the linear range of exposure using an HP ScanJet scanner.

From our preliminary data, both anti-LARG and anti-PDZ-RhoGEF antibodies detected a band around the size of ∼170 kDa, which has been confirmed by detection of the appropriate RhoGEF in HEK293T cells transfected with a plasmid DNA. Figure 3 shows that siRNAs targeted toward either LARG or PDZ-RhoGEF mRNAs are much more effective than ribozymes to the same target site. The marginal suppression by the ribozymes might result from several causes, which are explored later.

Effect on SRE Transcription in Luciferase Assay

RGS-RhoGEFs are thought to be the bridge between Gα12/13/q and RhoA activation and RhoA-stimulated enhanced gene transcription. A system has been established using a modified serum response element (SRE.L)-driven luciferase reporter construct to detect RhoA activation

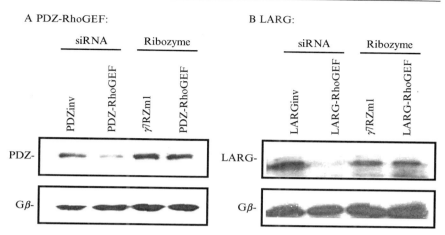

FIG. 3. Comparison of RhoGEF siRNA and ribozyme activity on RhoGEF protein suppression in cells. (A) Effect of PDZ-RhoGEF ribozyme vs siRNA on PDZ-RhoGEF protein suppression. PDZ-RhoGEF siRNA (PDZ-B), mutant siRNA (PDZinv), PDZ-RhoGEF ribozyme, or control ribozyme (γ7RZml) (2 μg/well) was transfected separately into HEK293T cells by Lipofectamine 2000 (8 μl) in a 6-well plate. Protein lysates were prepared 72 h posttransfection and subjected to Western blot analysis with the anti-PDZ-RhoGEF-specific antibody. (B) Effect of LARG ribozyme vs siRNA on LARG protein suppression. LARG siRNA (LARG-B), mutant siRNA (LARG-Binv), LARG ribozyme (LARG-B/RZ), or control ribozyme (γ7RZml) (2 μg/well) was transfected, separately and protein lysates were prepared as described in A and then subjected to Western blot analysis with the anti-LARG-specific antibody. The same membrane is stripped and reblotted with the anti-G protein β antibody as the loading control.

(Hill *et al.*, 1995). This promoter is a derivative of the c-fos SRE, which is activated by the serum response factor (SRF) and its coactivator MAL but not SRF with the ternary complex factor (TCF). A signaling pathway controlled by the activation of RhoA induces actin polymerization through two effector pathways: the ROCK-LIM kinase-cofilin pathway that stabilizes F-actin (Geneste *et al.*, 2002), while the mDial pathway promotes its assembly (Copeland and Treisman, 2002). MAL protein is a G-actin-associated signal-regulated SRF coactivator, which is redistributed from the cytoplasm to the nucleus in response to RhoA-induced actin polymerization and mediates SRF activation of SRE.L transcription (Miralles *et al.*, 2003).

To determine the effect of ribozymes and siRNA to suppress RhoGEF-mediated RhoA activation, 30 ng of either ribozyme or siRNA targeted to PDZ-RhoGEF-B or LARG-B, their mutant controls (described earlier), and RhoGEF plasmid DNA (supplied by Dr. T. Kozasa) (Suzuki *et al.*,

2003) (5 ng/well), together with luciferase reporters [SRE.L (firefly luciferase reporter) and pRL-TK (Renilla luciferase control) at a 30:3 ratio], are introduced into HEK293T cells in a 96-well plate at the density of 35,000 cells/well in 100 μl of complete medium with Lipofectamine 2000. Twenty-four hours after transfection, cells are changed to medium containing only 0.5% fetal bovine serum. Lysates are prepared 48 h posttransfection by suspending them in 100 μl of 1× Passive lysis buffer supplied with the Promega Dual Luciferase assay kit. Luciferase activity is measured in a Wallac Victor 1420 multilabel counter (Perkin Elmer Life Sciences) by injecting 30 μl of each firefly and then the Renilla substrate into wells containing 30 μl of cell lysates. Counts are read from 1 to 10 s after injection. The ratio of firefly to Renilla luciferase activity is calculated and used to compare to the control with mutant siRNA or ribozyme.

There were 11-, 15-, and 13-fold increases in the firefly/Renilla ratio upon transfection of LARG, PDZ-RhoGEF, and p115-RhoGEF expression vector plasmid DNA, respectively, over the basal signal from untransfected cells. The suppression of RhoGEF-mediated SRF activity by its matching siRNA (LARGsiRNA and PDZ-RhoGEFsiRNA) is more dramatic than suppression by ribozyme, directed at the same target region, which shows almost no suppression (Fig. 4).

Ribozymes have proven to be an effective tool for knocking down gene expression before the RNAi technique emerged. So far, there has been no side-by-side comparison of these two approaches. In preliminary studies, we found that the siRNA act at much lower concentrations than ribozymes to suppress RhoGEF protein levels and RhoGEF-mediated transcription under identical conditions (data not shown). Such marginal suppression by ribozymes might result from several causes in this current study:

Transfection Efficiency. In previous experiments (Wang *et al.*, 1997) for ribozyme transfection, serum-free medium was used in order to decrease ribozyme degradation and repeated ribozyme additions to cells were used. In our current experiment, we only added ribozyme once and it was in complete medium with 10% serum, which may cause quicker degradation.

Differing Structures. siRNA is a 21 nucleotide double-stranded RNA, which is shorter and linear. However, the ribozyme has a hammerhead structure with substantial secondary structure. Given this structural difference, ribozymes may be harder to get into cells than shorter siRNA.

Nonoptimized Conditions for Ribozyme Treatment. This study optimized the conditions for siRNA treatment first, which may not be optimal for ribozyme treatment.

Purity. The ribozymes and siRNA were synthesized by different companies using different equipment and methods. Even though both were HPLC purified, it is possible that the purity of the siRNA and ribozyme

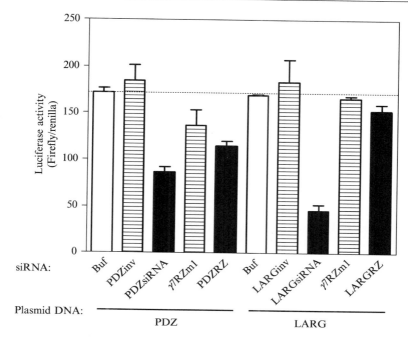

FIG. 4. Comparison of RhoGEF siRNA and ribozyme actions on suppression of RhoGEF-mediated SRE.L luciferase activity. PDZ-RhoGEF siRNA (PDZ-B), LARG siRNA (LARG-B), mutant siRNA (PDZ-Binv and LARG-Binv), ribozymes (PDZ-B/RZ and LARG-B/RZ), or control ribozyme (γ7RZml) (30 ng/well), together with either PDZ-RhoGEF or LARG plasmid DNA (5 ng/well) and dual luciferase reporters (SRE.L firefly and pRL-TK Renilla at 30 and 3 ng/well, respectively), was cotransfected into HEK293T cells by Lipofectamine 2000 (0.2 μl/well) in a 96-well plate. At 30 h posttransfection, cells were serum starved (0.5 % serum). At 48 h, cells were lysed, and lysates were analyzed with the Promega dual luciferase assay kit in a Wallac Victor 1420 multilabel counter. The ratio of firefly to Renilla luciferase activity was calculated and expressed as the percentage of the control value (i.e., transfection with buffer only). Data shown are means ± SEM of three independent experiments conducted in triplicate.

preparations is different. Based on the good results achieved with siRNA, we decided to focus on the use of RhoGEF siRNA for subsequent studies.

Efficacy and Gene Specificity of RGS-RhoGEF siRNAs

Well-designed siRNAs should be specific for their target mRNA and encoded protein. Typically, the effects of siRNA knockdown are judged by the levels of their target mRNA and protein decreasing much more than those of control mRNAs and proteins. High concentrations of cationic lipid transfection reagents can be cytotoxic. Therefore, proper controls are important to ensure that any suppression effect observed is not due to general nonspecific cytotoxicity. In addition, an RGS-RhoGEF siRNA should only suppress its target RGS-RhoGEF mRNA and protein and not the other two RGS-RhoGEFs. The failure of the inactive siRNA (inverted siRNA) as designed in Fig. 2 to cause these changes is another important control. For this purpose, methods for measuring the levels of RGS-RhoGEF mRNA, protein, and functional readouts are needed, as described later.

RT-PCR Analysis

To determine the effect of siRNA treatment on cellular RGS-RhoGEF mRNA levels, HEK293T cells in 6-well plates (2×10^6/well) are transfected with or without a panel of RGS-RhoGEF-specific siRNAs (2 μg/well) and Lipofectamine 2000 (8 μl/well). At 48 h posttransfection, total RNA is prepared using the RNeasy minikit (Qiagen) following the manufacturer's protocol. The prepared RNA is subjected to reverse transcription polymerase chain reaction (RT-PCR) with the SuperScript One-Step RT-PCR system according to the supplier's manual (Invitrogen Life Technologies, Inc.) using RhoGEF-specific primers. Primer pairs generated from the flanking sequence of the RGS domain of RGS-RhoGEF cDNA are used to amplify each RGS-RhoGEF (GenBank accession numbers for human LARG, PDZ-RhoGEF, and p115-RhoGEF mRNA sequences used are NM_015313, AB002378, and BC011726, respectively). The primer sequences used in RT-PCR are as follows:

LARG sense strand: 5' ATGCTTGTCGGAAGTCCCTCA 3'
LARG antisense strand: 5' GGAGACACCTTAGATGGCACACC
    TTAG 3'
PDZ-RhoGEF sense strand: 5' ATGGGCTCGGATGCAGCAGTC 3'
PDZ-RhoGEF antisense strand: 5' GAGGATGATCTGGGCCAG
    TAG 3'
p115-RhoGEF sense strand: 5' ATGGTTCCCGTCAGCATCATC 3'
p115-RhoGEF antisense strand: 5' CTTGAACCAGAAGAGCCTCC
    CGGCTAA 3'

(Each pair of primers is first checked by amplifying RGS-RhoGEF plasmid DNA to make sure that the correct size of PCR product is achieved.) Briefly, 200 ng total RNA is used with 0.3 μM of each primer and 1.2 mM MgSO$_4$ in a 25-μl volume. Reverse transcription is performed at 45° for 30 min, denatured at 95° for 5 min, and followed by PCR with 35 cycles at 95° for 30 s, 62° for 45 s, and 72° for 15 s. The addition of 4%

dimethyl sulfoxide and 2.4 mM MgSO$_4$ is required for amplifying p115Rho-GEF. PCR products are separated on a 1.0% agarose gel, stained with ethidium bromide, and a gel image is taken with KODAK ImageStation and quantified by densitometry analysis using Kodak One-dimensional 3.5 software. The expected sizes of PCR product are PDZ-RhoGEF, 1257 bp; p115RhoGEF, 1119 bp; and LARG, 1110 bp. All three RGS-RhoGEFs are detected in HEK293T cells, although p115-RhoGEF is the least abundant. The elongation factor EF1α is amplified under identical conditions in a separate tube and run together onto the agarose gel as an internal control.

The RT-PCR analysis for the activity of RGS-RhoGEF-specific siRNAs is shown in Fig. 5. When introduced into HEK293T cells, each RGS-RhoGEF siRNA suppressed the levels of their corresponding RGS-RhoGEF mRNA without affecting the other two RGS-RhoGEF mRNAs,

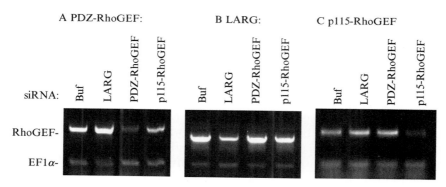

FIG. 5. Suppression of endogenous RhoGEF mRNA levels in RhoGEF siRNA-transfected cells. A panel of different RhoGEF siRNAs (LARG-B, PDZ-B, and p115-B) or buffer (Buf) control (2 μg/well) was introduced into HEK293T cells by transfection with Lipofectamine 2000 (8 μl) in a 6-well plate. Total RNA samples were prepared at 48 h posttransfection and subjected to RT-PCR analysis using RGS-RhoGEF-specific primers as described in the text. (A) Effect of different RhoGEF siRNAs on PDZ-RhoGEF mRNA suppression. The level of PDZ-RhoGEF mRNA was decreased to 40% of the control (Buf) level by PDZ-RhoGEF siRNA but not by other RhoGEF siRNAs. (B) Effect of different RhoGEF siRNAs on LARG mRNA suppression. The level of LARG mRNA was decreased to 42% of the control (Buf) value by LARG siRNA but not by other RhoGEF siRNAs. (C) Effect of different RhoGEF siRNAs on p115-RhoGEF mRNA suppression. The level of p115-RhoGEF mRNA was decreased to 51% of the control (Buf) level by p115-RhoGEF siRNA but not by other RhoGEF siRNAs. The elongation factor (EF1α) is amplified under the same conditions in a separate tube and run together onto the agarose gel as the loading control. These data indicate the efficiency and gene specificity of RhoGEF siRNA on suppression of their own mRNA. Percentage inhibitions indicated are means of three independent experiments.

except for PDZ-RhoGEF siRNA, which may have a slightly nonspecific effect on p115-RhoGEF. PDZ-RhoGEF siRNA reduced PDZ-RhoGEF mRNA to 40% of the basal level, LARG siRNA suppressed LARG mRNA to 42% of the basal level, and p115-RhoGEF siRNA decreased p115-RhoGEF mRNA to 51% of the basal level, respectively. This RT-PCR analysis indicates that the designed siRNA can be used effectively to reduce the levels of RGS-RhoGEF mRNA in a gene-specific manner.

Western Blotting Analysis

It is important to confirm that the siRNA-induced suppression of the target RGS-RhoGEF mRNA is paralleled by the reduction of protein levels. For this purpose, protein lysates are prepared from HEK 293T cells in 6-well plates at 72 h posttransfection with controls without siRNA, different RhoGEF siRNAs, and mutant siRNA with an inverted sequence. All oligonucleotides are added at 2 μg/well with Lipofectamine 2000 (8 μl/well). Protein expression is then evaluated by Western blot analysis with RGS-RhoGEF-specific antibodies. The p115-RhoGEF antibody is from Santa Cruz (C-19, 1:400 dilution), and the secondary antibody used is bovine anti-goat IgG-HRP (1:8000 dilution). PDZ-RhoGEF and LARG are analyzed as described earlier.

To determine the efficacy and gene specificity of RhoGEF siRNAs, a panel of different RhoGEF siRNAs (LARG-B siRNA, PDZ-RhoGEF-B siRNA, p115-RhoGEF-B siRNA) and mutant siRNAs (LARG-Binv, PDZ-RhoGEF-Binv, p115-RhoGEF-Binv) (2 μg/well), including buffer control, is evaluated by Western blotting with RGS-RhoGEF subtype-specific antibodies. As shown in Fig. 6, the siRNA efficiently blocked the corresponding protein expression without affecting other RGS-RhoGEF proteins. The mutant siRNAs failed to suppress protein expression, indicating the specificity of siRNA effect. After stripping, the same membrane was blotted with the anti-G protein β subunit as the protein-loading control. G-protein β subunit expression was not affected by the transfection of different siRNAs. These data clearly indicate that each RGS-RhoGEF siRNA can efficiently suppress its own corresponding endogenous protein in a gene-specific manner.

We have studied the time course of the effect of LARG siRNA treatment on LARG protein expression (data not shown). Cell lysates were prepared at both 48 and 72 h posttransfection with either LARG siRNA or buffer control and then subjected to Western blot analysis with the anti-LARG antibody. The LARG siRNA suppressed LARG protein expression at very similar levels to about 30% of control level at both 48 and 72 h posttransfection. It is also notable that protein suppression is greater than

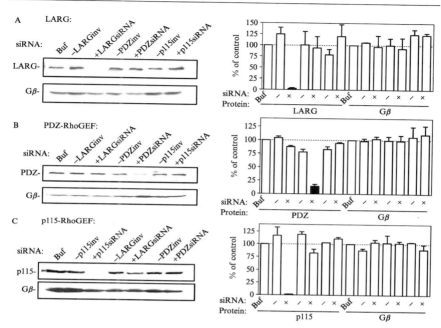

FIG. 6. Suppression of endogenous RhoGEF protein levels by transfection of RhoGEF siRNA into HEK293T cells. A panel of different RhoGEF siRNAs [LARG-B, PDZ-B, p115-B, buffer (Buf) control] and mutant siRNAs (LARG-Binv, PDZ-Binv, p115Binv) (2 μg/well) were individually introduced into HEK293T cells by Lipofectamine 2000 (8 μl) in a 6-well plate, and protein lysates were prepared at 72 h posttransfection. Western blot analysis was done with RGS-RhoGEF-specific antibodies. (A) Effect of different RhoGEF siRNAs on LARG protein suppression. The expression of LARG protein is almost completely suppressed (2.8% of control level) by LARG siRNA but is not affected by other RhoGEF siRNAs or by the mutant LARG siRNA (LARG-Binv). (B) Effect of different RhoGEF siRNAs on PDZ-RhoGEF protein suppression. The expression of PDZ-RhoGEF is suppressed to about 12% of the buffer control level by PDZ siRNA but is not affected by other RhoGEF siRNAs or the mutant PDZ siRNA (PDZ). (C) Effect of different RhoGEF siRNAs on p115-RhoGEF protein suppression. The expression of p115-RhoGEF protein is completely eliminated (1.4% of the buffer control level) by p115-RhoGEF siRNA (p115-B) but is not affected by other RhoGEF siRNAs or by the mutant p115-RhoGEF siRNA (p115-Binv). Then, the same membrane was stripped and reblotted with the anti-G protein β subunit antibody as loading control, which shows relatively even loading. All bar graphs on the right represent the mean $\pm$ SEM of two to three independent experiments. Data indicate that each RGS-RhoGEF siRNA can suppress its cognate endogenous protein efficiently and in a gene-specific manner.

A LARG siRNA:

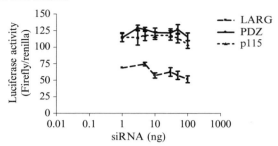

B PDZ-RhoGEF siRNA:

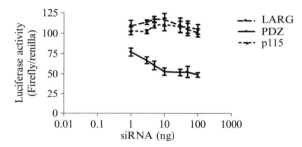

C p115-RhoGEF siRNA:

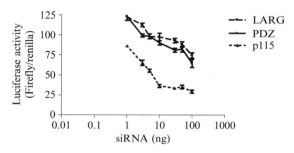

FIG. 7. Dose response and gene specificity of RhoGEF siRNA actions on RhoGEF-mediated SRE.L luciferase activity. One of the RGS-RhoGEF siRNAs (LARG-B, PDZ-B, p115-B) and mutant siRNAs (LARG-Binv, PDZ-Binv, p115-Binv) in amounts of 0, 1, 3, 5, 10, 30, 50, and 100 ng/well, together with one of three RhoGEF expression vectors [p115-RhoGEF (2 ng), PDZ-RhoGEF (5 ng), and LARG (5 ng)], plus the dual luciferase reporters (SRE.L and pRL-TK at 30 and 3 ng/well, respectively), was introduced into HEK293T cells in a 96-well plate with Lipofectamine 2000 (0.2 μl/well). At 24 h posttransfection, cells were serum starved (0.5% serum) for another 24 h. At 48 h, cells were lysed with 1× Passive lysis buffer (100 μl/well) and analyzed with the Promega dual luciferase assay kit in a Wallac Victor 1420 multilabel counter (Perkin Elmer Life Sciences) with 30 μl of lysate and 30 μl of

the suppression of mRNA by siRNA. Protein suppression may even last longer than 72 h, as there is no sign of protein level recovery at 72 h but we have not yet examined longer times.

Luciferase Assay

The next step of our study was to determine how the siRNA-mediated suppression of the RGS-RhoGEF affects the RhoA activation pathway. The specific role of an RGS-RhoGEF is demonstrated if the siRNA-mediated loss of RGS-RhoGEF results in significant attenuation of the RGS-Rho-GEF-activated luciferase activity. In this regard, a method to measure the efficiency of RGS-RhoGEF siRNA has been developed, as shown in Fig. 7.

The suppression of RGS-RhoGEF-mediated luciferase activity was observed with as little as 1 ng of siRNA per well and reached maximal suppression by 10 ng/well (Fig. 7). Importantly, each siRNA suppressed only responses induced by its own RhoGEF plasmid DNA with almost no effect on responses mediated by the other two RhoGEF plasmids. For p115RhoGEF siRNA, there is a slight nonspecific effect at higher amounts of siRNA, but at 10 ng/well, the p115-RhoGEF siRNA also shows specific suppression. These data show excellent efficiency and gene specificity of all three RhoGEF siRNAs to suppress RhoGEF-mediated SRF activity, indicating that RGS-RhoGEF siRNA can be used as a powerful tool for defining the functional role of endogenous RGS-RhoGEF protein.

Potential Problems

siRNA-Specific Rather than Target-Specific Signature

Although siRNA has been increasingly employed successfully, it is difficult to select an siRNA sequence that will be absolutely specific for the target of interest. It has been thought that RNAi requires near identity between the siRNA and its cognate mRNA. To determine if cross-hybridization happens with transcripts containing partial identity to the siRNA sequence, which may elicit phenotypes reflecting silencing of unintended transcripts in addition to the target gene, Jackson *et al.* (2003) showed that 15 nucleotides, perhaps as few as 11 contiguous nucleotides, of sequence identity are sufficient to direct the silencing of nontargeted transcripts. Using gene

---

each substrate counting for 10 s. The ratio of firefly to renilla luciferase counts was calculated and used to compare to the controls (without siRNA transfected) and was expressed as percentage of the control value. Data shown are means ± SEM of three independent experiments conducted in triplicate.

expression profiling in cultured human cells (HeLa cells), the result suggests that it may be difficult to select an siRNA sequence that will be absolutely specific for the target of interest.

Accessibility of Targeting Site

Ribozymes and siRNA are thought to provide a significantly higher potency to the inhibition of transcript levels as compared to the traditional antisense approaches, even though in our case the ribozyme seems to be less efficient than siRNA. The accessibility of target sites to ribozyme and siRNA intervention is equally hard to predict. For this purpose, either ribozymes or siRNA designed at different sites along the mRNA of a particular gene can be used together to target multiple sites in the target mRNA, thereby increasing the accessibility as well as specificity. In case of functional redundancy of genes (e.g., the gene subfamily of RGS-RhoGEFs), one could combine three siRNAs, one targeted to each RGS-RhoGEF, in order to knock down the expression of all members of this gene family if functional redundancy between them is a cause of partial suppression of function.

Instability of Ribozymes and siRNA in Cell Culture

Because both ribozymes and siRNA are RNAs, they are easily accessible to RNA degradation in the cell culture. For ribozymes, we have synthesized modified forms by adding 3' end phosphorothioate linkages with the RNA–DNA chimera. Currently, for the synthesis of siRNA, modified nucleotides are introduced to the ends of both strands of the siRNA (Amarzguioui *et al.*, 2003). It has been shown that an siRNA with two 2'-*O*-methyl RNA nucleotides at the 5' end and four methylated monomers at the 3' end is as active as its unmodified counterpart and leads to a prolonged silencing effect in cell culture.

Short-Term Transient Transfection vs Long-Term Stable Introduction of siRNA or Ribozymes into Cells

Although gene silencing by siRNA has a longer duration than ribozyme effects (normally a couple of days to a week), longer effects may be required for studying long-lived proteins or certain complex functions. The current approach of transiently transfecting siRNA or ribozymes used might not be practical. The silencing by synthetic siRNAs is transient. This limitation can be overcome by stably introducing a gene encoding the ribozyme or siRNA into cells to produce a sustained expression of the siRNA or ribozyme. Paddison *et al.* (2002) and Brummelkamp *et al.* (2002) have stably expressed short hairpin RNAs (shRNAs), which are processed intracellularly by Dicer into siRNAs. Currently, retroviral vectors including lentiviral systems are used most commonly for stable expression in both cell culture and animals (Rubinson *et al.*, 2003). There are several advantages to the lentiviral system, including a wide host range, stable integration into both dividing and nondividing cell types, which are hard to transfect, and the use of replication-deficient viral strains, which reduce the immune response.

Acknowledgment

This work was supported by NIH Grant GM39561 awarded to R.R.N.

References

Amarzguioui, M., Holen, T., Babaie, E., and Prydz, H. (2003). Tolerance of mutations and chemical modifications in siRNA. *Nuclec Acids Res.* **31**, 589–595.

Amarzguioui, M., and Prydz, H. (1998). Hammerhead ribozyme design and application. *Cell. Mol. Life Sci.* **54**, 1998.

Aspenstrom, P. (1999). The Rho GTPases have multiple effects on the actin cytoskeleton. *Exp. Cell Res.* **246**, 20–25.

Berman, D. M., Wilkie, T. M., and Gilman, A. G. (1996). GAIP and RGS4 are GTPase-activating proteins for the Gi subfamily of G protein alpha subunits. *Cell* **86**, 445–452.

Booden, M. A., Siderovski, D. P., and Der, C. J. (2002). Leukemia-associated Rho guanine nucleotide exchange factor promotes Gαq-coupled activation of RhoA. *Mol. Cell. Biol.* **22**, 4053–4061.

Branch, A. D. (1998). A good antisense molecule is hard to find. *Trends Biochem. Sci.* **23**, 45–50.

Cech, T. R., Zaug, A. J., and Grabowski, P. J. (1981). *In vitro* splicing of the ribosomal RNA precursor of Tetrahymena: Involvement of a guanosine nucleotide in the excision of the intervening sequence. *Cell* **27**, 487–496.

Copeland, J. W., and Treisman, R. (2002). Activation of SRF by the diaphanous related forming mDial is mediated by its effects on actin polymerization. *Mol. Biol. Cell* **13**, 2002.

De Vries, L., Zheng, B., Fisher, T., Elenko, E., and Farquhar, M. G. (2000). The regulator of G protein signaling family. *Annu. Rev. Pharmacol. Toxicol.* **40**, 235–271.

Elbashir, S. M., Harborth, J., Lendeckel, W., Yalcin, A., Weber, K., and Tuschl, T. (2001a). Duplexes of 21-nucleotide RNAs mediated RNA interference in cultured mammalian cells. *Nature* **411**, 494–498.

Elbashir, S. M., Lendeckel, W., and Tuschl, T. (2001b). RNA interference is mediated by 21 and 22 nt RNAs. *Genes Dev.* **15**, 188–200.

Fire, A. (1999). RNA-triggered gene silencing. *Trends Genet.* **15**, 358–363.

Fire, A., Xu, S., Montgomery, M. K., Kostas, S. A., Driver, S. E., and Mello, C. C. (1998). Potent and specific genetic interference by double-stranded RNA in *Caenorhabditis elegans*. *Nature* **391**, 806–811.

Fukuhara, S., Chikumi, H., and Gutkind, J. S. (2000). Leukemia-associated Rho guanine nucleotide exchange factor (LARG) links heterotrimeric G proteins of G12 family to Rho. *FEBS Lett.* **485**, 183–188.

Fukuhara, S., Murga, C., Zohar, M., Igishi, T., and Gutkind, J. S. (1999). A novel PDZ domain containing guanine nucleotide exchange factor links heterotrimeric G proteins to Rho. *J. Biol. Chem.* **274,** 5868–5879.

Geneste, O., Copeland, J. W., and Treisman, R. (2002). LIM kinase and diaphanous cooperate to regulate serum response factor and actin dynamics. *J. Cell Biol.* **157,** 831–838.

Glaven, J. A., Whitehead, I. P., Nomanbhoy, T., Kay, R., and Cerione, R. A. (1996). Lfc and Lsc oncoproteins represent two new guanine nucleotide exchange factors for the Rho GTP-binding protein. *J. Biol. Chem.* **271,** 27374–27381.

Grassi, G., Forlino, A., and Marini, J. C. (1997). Cleavage of collagen RNA transcription by hammerhead ribozyme in vitro is mutation-specific and shows competitive binding effects. *Nucleic Acids Res.* **25,** 3451–3458.

Guerrier-Takada, C., and Altman, S. (1984). Catalytic activity of an RNA molecule prepared by transcription in vitro. *Science* **223**(4633), 285–286.

Hart, M., Sharma, S., elMasry, N., Qiu, R. G., McCabe, P., Polaki, P., and Bollag, G. (1996). Idetification of a novel guanine nucleotide exchange factor for the Rho GTPase. *J. Biol. Chem.* **271,** 25452–25458.

Hart, M. J., Jiang, X., Kozasa, T., Roscoe, W., Singer, W. D., Gilman, A. G., Sternweis, P. C., and Bollag, G. (1998). Direct stimulation of the guanine nucleotide exchange activity of p115 RhoGEF by Gα13. *Science* **80,** 2112–2114.

Hill, C. S., Wynne, J., and Treisman, R. (1995). The family GTPases RhoA, Rac1, and CDC42Hs regulate transcriptional activity by SRF. *Cell* **81,** 1159–1170.

Jackson, A. L., Bartz, S. R., Schelter, J., Kobayashi, S. V., Burchard, J., Mao, M., Li, B., Cavet, G., and Linsley, P. S. (2003). Expression profiling reveals off-target gene regulation by RNAi. *Nature Biotechnol.* **21,** 635–637.

Jackson, M., Song, W., Liu, M. Y., Jin, L., Dykes-Hoberg, M., Lin, C. I., Bowers, W. J., Federoff, H. J., Sternweis, P. C., and Rothstein, J. D. (2001). Modulation of the neuronal glutamate transporter EAAT4 by two interacting proteins. *Nature* **410,** 89–93.

Kozasa, T., Jiang, X., Hart, M. J., Sternweis, P. M., Singer, W. D., Gilman, A. G., Bollag, G., and Sternweis, P. C. (1998). p115 RhoGEF, a GTPase activating proten for Gα12 and Gα13. *Science* **280,** 2109–2112.

Miralles, F., Posern, G., Zaromytidou, A. I., and Treisman, R. (2003). Actin dynamics control SRF activity by regulation of its coactivator MAL. *Cell* **113,** 329–342.

Neubig, R. R., and Siderovski, D. P. (2002). Regulators of G protein signaling as new central nervous system drug targets. *Nature Rev. Drug Disc.* **1,** 187–197.

Paddison, P. J., Caudy, A. A., and Hannon, G. J. (2002). Stable suppression of gene expression by RNAi in mammalian cells. *Proc. Natl. Acad. Sci. USA* **99,** 2002.

Robishaw, J. D., Guo, Z.-P., and Wang, Q. (2003). Ribozyme as tools for suppression of G proteinγ subunits. *Methods Mol. Biol.* **237,** 169–179.

Robishaw, J. D., Wang, Q., and Schwindinger, W. F. (2002). Ribozyme-mediated suppression of G proteinγ subunits. *Methods Enzymol.* **344,** 435–451.

Ross, E. M., and Wilkie, T. M. (2000). GTPase-activating proteins for heterotrimeric G proteins: Regulators of G protein signaling (RGS) and RGS-like proteins. *Annu. Rev. Biochem.* **69,** 795–827.

Rubinson, D. A., Dillon, C. P., Kwiatkowski, A. V., Sievers, C., Yang, L., Kopinja, J., Rooney, D. L., Ihrig, M. M., McManus, M. T., Gertler, F. B., Scott, M. L., and Van Parijs, L. (2003). A lentivirus-based system to functionally silence genes in primary mammalian cells, stem cells and transgenic mice by RNA interference. *Nature Genet.* **33,** 401–406.

Ruffner, D. E., Stormo, G. D., and Uhlenbeck, C. (1990). Sequence requirement of the hammerhead RNA self-cleavage reaction. *Biochemistry* **29,** 10695.

Sah, V. P., Seasholtz, T. M., Sagi, S. A., and Brown, J. H. (2000). The role of Rho in G protein-coupled receptor signaling transduction. *Annu. Rev. Pharmacol. Toxicol.* **40,** 459–489.

Seasholtz, T. M., Majumdar, M., and Brown, J. H. (1999). Rho as a mediator of G protein-coupled receptor signaling. *Mol. Pharm.* **55,** 949–956.

Shimayama, T., Nishikawa, F., Nishikawa, S., and Taira, K. (1993). Nuclease-resistant chimeric ribozymes containing deoxyribonucleotides and phosphorothioate linkages. *Nucleic Acid Res.* **21,** 2605–2611.

Siderovski, D. P., Strockbine, B., and Behe, C. I. (1999). Whither goest the RGS proteins? *Crit. Rev. Biochem. Mol. Biol.* **34,** 215–251.

Stein, C. A. (2000). Is irrelevant cleavage the price of antisense efficacy? *Pharmacol. Ther.* **85,** 231–236.

Suzuki, N., Nakamura, S., Mano, H., and Kozasa, T. (2003). Galpha 12 activates Rho GTPase through tyrosine-phosphorylated leukemia-associated RhoGEF. *Proc. Natl. Acad. Sci. USA* **100,** 733–738.

Taylor, N. R., Kaplan, B. E., Swiderski, P., Li, H., and Rossi, J. (1992). Chimeric DNA-RNA hammerhead ribozymes have enhanced *in vitro* catalytic efficiency and increased stability *in vivo*. *Nucleic Acid Res.* **20,** 4559–4565.

Toksoz, D., and Williams, D. A. (1994). Novel human oncogene lbc detected by transfection with distinct homology regions to signal transduction products. *Oncogene* **9,** 621–628.

Vogt, S., Grosse, R., Schultz, G., and Offermanns, S. (2003). Receptor-dependent RhoA activation in G12/G13-deficient cells: Genetic evidence for an involvement of Gq/G11. *J. Biol. Chem.* **278,** 28743–28749.

Wang, Q., Jolly, J. P., Surmeier, J. D., Mullah, B., Lidow, M. S., Bergson, C. M., and Robishaw, J. D. (2001). Differential dependence of the D1 and D5 dopamine receptors on the G protein $\gamma 7$ subunit for activation of adenylylcyclase. *J. Biol. Chem.* **276,** 39386–39393.

Wang, Q., Lin, M., Mullah, B., Siderovski, D. P., and Neubig, R. R. (2002). Receptor-selective effects of endogenous RGS3 and RGS5 to regulate MAP kinase activation in rat vascular smooth muscle cells. *J. Biol. Chem.* **277,** 24949–24958.

Wang, Q., Mullah, B., Hansen, C., Asundi, J., and Robishaw, J. D. (1997). Ribozyme-mediated suppression of the G protein $\gamma 7$ subunit suggests a role in hormon regulation of adenylylcyclase activity. *J. Biol. Chem.* **272,** 26040–26048.

Wang, Q., Mullah, B., and Robishaw, J. D. (1999). Ribozyme approach identifies a functional association between the G protein $\beta 1 \gamma 7$ subunits in the β-adrenergic receptor signaling pathway. *J. Biol. Chem.* **274,** 17365–17371.

Wells, C. D., Liu, M. Y., Jackson, M., Gutowski, S., Sternweis, P. M., Rothstein, J. D., Kozasa, T., and Sternweis, P. C. (2002). Mechanism for reversible regulation between G13 and Rho exchange factors. *J. Biol. Chem.* **277,** 1174–1181.

William, J., and Aymen, A. S. (1995). RNA enzymes as tools for gene ablation. *Curr. Opin. Biotech.* **6,** 44–49.

Zamore, P. D. (2001). RNA interference: Listening to the sound of silence. *Nature Struct. Biol.* **8,** 746–750.

Zuker, M. (1994). Prediction of RNA secondary structure by energy minimization. *Methods Mol. Biol.* **25,** 1994.

Zuker, M. (2003). Mfold web server for nucleic acid folding and hybridization prediction. *Nucleic Acids Res.* **31,** 3406–3415.

[16] Structure-Based Design, Synthesis, and Activity of Peptide Inhibitors of RGS4 GAP Activity

By Yafei Jin, Huailing Zhong, John R. Omnaas, Richard R. Neubig, and Henry I. Mosberg

Abstract

One of the principal roles of the multifunctional regulator of G-protein signaling (RGS) proteins is to terminate G-protein-coupled receptor (GPCR) signaling by binding to the G-protein $G\alpha$ subunit, thus acting as GTPase-activating proteins (GAPs). In principle, then, selective inhibitors of this GAP function would have potential as therapeutic agents, as they could be used to augment the effects of endogenous or exogenous GPCR agonists. Using the published RGS4-$G_{i\alpha1}$ X-ray structure, we have designed and synthesized a series of cyclic peptides, modeled on the $G_{i\alpha}$ switch I region, that inhibit RGS4 GAP activity, presumably by blocking the interaction between RGS4 and $G_{i\alpha1}$. These compounds should prove useful for elucidating RGS-mediated activity and serve as a starting point for the development of a novel class of therapeutic agent.

Introduction

Recognition of regulators of G-protein signaling (RGS) proteins as a distinct family of highly diverse, multifunctional signaling proteins stemmed from the observation of a homologous 120 residue domain (defined as the RGS domain) simultaneously in several proteins (De Vries *et al.*, 1995; Druey *et al.*, 1996; Koelle and Horvitz, 1996; Siderovski *et al.*, 1996) and the subsequent demonstration that this domain is responsible for binding of the RGS protein to the G-protein α subunit. Upon binding to $G\alpha$, RGS proteins act as GTPase-activating proteins (GAPs), reducing the lifetime of bound GTP and thereby attenuating G-protein signaling by the rapid turnoff of the G-protein-coupled receptor (GPCR) signaling pathway (De Vries *et al.*, 2000; Hepler, 1999; Siderovski *et al.*, 1999). Thus, RGS proteins represent a novel drug target, as inhibitors of RGS acceleration of GTP hydrolysis by G_{α} could, in principle, potentiate the effects of agonist GPCR ligands (Neubig and Siderovski, 2002; Zhong and Neubig, 2001).

In 1997, Tesmer and colleagues opened the door to structure-based design of RGS inhibitors with their report of a 2.8-Å resolution crystal structure of the RGS protein, RGS4, complexed with $G_{i\alpha 1}$–Mg^{2+}–GDP-AlF_4^-. In this complex the GDP-AlF_4^- mimics the transition state of GTP during its hydrolysis to GDP, inducing the conformation of G_α thought to be stabilized by RGS binding. The $G\alpha$-binding, RGS domain of RGS4 forms a nine α-helix bundle that interacts with three distinct "switch" regions of $G_{i\alpha 1}$, so named because they undergo large conformational changes during transition from the inactive GDP-bound form of the G_α subunit to the active GTP-bound form (Lambright et al., 1994). These conformational changes allow the switch I and II regions to contact the γ phosphate of GTP and play an essential role in GTP hydrolysis by G_α. The crystal structure of the complex reveals that RGS4 does not contribute catalytic residues that interact directly with either GDP or AlF_4^-. Instead, RGS4 appears to promote rapid hydrolysis of GTP primarily by stabilizing the switch regions of $G_{i\alpha 1}$ in the transition state conformation.

The RGS4–$G_{i\alpha 1}$ crystal structure provides an excellent starting point for the structure-based design of inhibitors of the RGS interaction with $G\alpha$ subunits, which would be expected to block RGS GAP activity and thus increase the signal transduction efficiency of the associated $G\alpha$ subunit. Such RGS inhibitors could then be used, alone, to potentiate the functions of endogenous agonists, similar to the action of benzodiazepines at the ionotropic GABA-A receptor (Macdonald and Olsen, 1994), or could be used to augment the effects of exogenous GPCR agonist drugs. Jin et al. (2004) have described initial results of structure-based inhibitors of RGS4 GAP activity, which provide a "proof of concept." These initial results are summarized later, along with additional studies that further demonstrate the feasibility of this approach.

Results and Discussion

Examination of the crystal structure of the RGS4–$G_{i\alpha 1}$–Mg^{2+}–GDP-AlF_4^- complex (PDB file 1agr) indicates that the functional binding site for RGS4 on the surface of $G_{i\alpha 1}$ is formed by residues in the three switch regions of $G_{i\alpha 1}$: residues 179–185 in switch I, residues 204–213 in switch II, and residues 235–237 in switch III. Of these regions, switch I of $G_{i\alpha 1}$ interacts with three-fourths of the RGS4-binding pocket (Tesmer et al., 1997). The importance of switch I is further suggested by the observation that two surface residues of this region (Thr-182 and Gly-183) appear to be essential for high-affinity $G\alpha$–RGS interaction (Lan et al., 1998; Tesmer et al., 1997). Consequently, the switch I region appears to be the most promising starting point for the development of small peptide ligands designed to hinder the interaction between RGS4 and $G_{i\alpha 1}$ and thus inhibit the GAP activity of RGS4.

As shown in Fig. 1, the RGS4-binding region of switch I, residues 179–185, has the amino acid sequence Val-Lys-Thr-Thr-Gly-Ile-Val. Additionally, Glu-186, which is conserved in most G_α subunits, is positioned such that it may interact with nearby positively charged residues of RGS4 (Arg-172 and/or Arg-167). Therefore, we chose the octapeptide, Val[1]-Lys-Thr-Thr-Gly-Ile-Val-Glu[8], as the starting point for the design of analogs of the $G_{i\alpha1}$ switch I region. Unconstrained, linear octapeptides, like this example, can be expected to be very flexible. Consequently, this lead switch 1 peptide would be unlikely to populate highly the desired native conformation observed in the RGS4–$G_{i\alpha1}$ X-ray structure. Indeed, the linear switch 1 peptide has been prepared and was found to be devoid of inhibitory activity (Lan and Neubig, unpublished result). Hence, we focused our attention on cyclic peptide analogs, as cyclization is expected to reduce conformational freedom and enhance binding affinity by mimicking

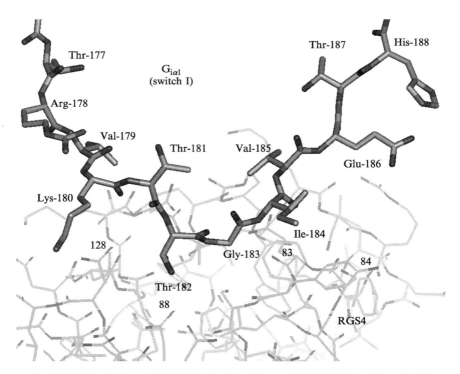

FIG. 1. Interface between RGS4 (thin lines) and $G_{i\alpha1}$ switch I (thick lines) from the X-ray structure of Tesmer et al. (1997).

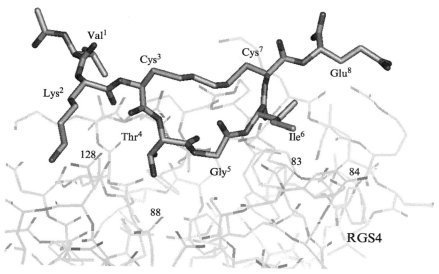

FIG. 2. Interface between RGS4 (thin lines) and designed ethylene-bridged dithio-ether peptide, **1** (thick lines). Cysteine residues are indicated by their sequence positions, 3 and 7.

or inducing bound structure motifs. Specific cyclic peptide targets were determined by inspection of the RGS4–$G_{i\alpha 1}$ crystal structure (Fig. 1), which reveals that the side chains of T181 and V185 of $G_{i\alpha 1}$ are pointing toward each other and have no direct interaction with RGS4. These residues are thus ideal candidates for substitution by amino acids that allow side chain–side chain cyclization. Accordingly, we designed an initial peptide in which the native Thr^3 and Val^7 residues in the model octapeptide are replaced by cysteine.

Cyclization could then be effected by linking these two Cys residues through a disulfide or dithioether bridge. Figure 2 depicts a model of the ethylene dithioether-containing cyclic analog, **1**, in the RGS4-binding site. Although cyclization via an ethylene dithioether provides the best spatial fit to the observed distance between the α carbons T181 and V185 in the X-ray structure (8.665 Å in X-ray vs 8.486 Å in the modeled ethylene dithioether), the optimal geometry for a small peptide ligand may differ from that of the corresponding region of the much larger $G\alpha$ subunit. Consequently, a structurally diverse set of peptide scaffolds, shown in Scheme 1, was sampled as possible mimics of the $G_{i\alpha 1}$ switch 1 region:

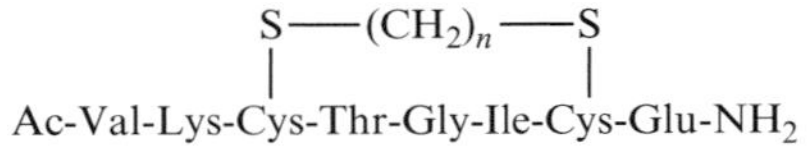

$$S\text{---}(CH_2)_n\text{---}S$$

$$\text{Ac-Val-Lys-Cys-Thr-Gly-Ile-Cys-Glu-NH}_2$$

SCHEME 1.

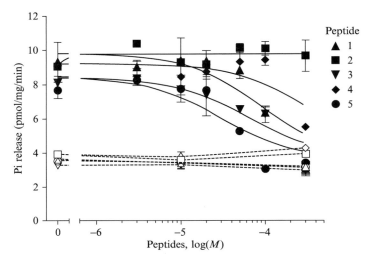

FIG. 3. Inhibition by peptides **1–5** of RGS stimulation of steady-state GTPase activity. P_i release as a function of peptide concentration in the presence (closed symbols) or absence (open symbols) of GST–RGS4 is shown.

where $S\text{-}(CH_2)_n\text{-}S$ denotes intramolecular cyclization through the cysteine side chain sulfurs via disulfide ($n = 0$) or dithioether ($n = 1–3$) linkages and where amino-terminal acetylation (Ac) and carboxy-terminal amidation (-NH$_2$) were chosen to best mimic the corresponding region of $G_{i\alpha1}$.

Figure 3 presents the initial observations, specifically the concentration dependence of the inhibition of RGS4 GAP activity exhibited by the set of cyclic peptides described in Scheme 1, along with a control linear peptide in which the cysteines in positions 3 and 7 were both converted to methylthioethers are depicted. The inhibitory potencies of these peptides are summarized in Table I.

As seen in Table I, compound **1**, the lead ethylene dithioether-containing analog, does indeed inhibit RGS acceleration of $G_{i\alpha1}$ GTPase activity, with an IC$_{50}$ of 88 μM. Compound **1** did not inhibit the steady-state GTPase activity of the membrane-bound receptor/G-protein complex, indicating that the effect is on RGS4 and not on the receptor or the G protein. Consistent with the earlier observation that the linear octapeptide

TABLE I

IC$_{50}$ VALUES OF PEPTIDE INHIBITORS OF RGS4 GAP ACTIVITY

#	Structure	IC$_{50}$ (μM)
1	Ac-Val-Lys-Cys-Thr-Gly-Ile-Cys-Glu-NH$_2$ (Et)[a]	88 ± 12
2	Ac-Val-Lys-Cys (Me)-Thr-Gly-Ile-Cys (Me)-Glu-NH$_2$	$\gg 300$
3	Ac-Val-Lys-Cys-Thr-Gly-Ile-Cys-Glu-NH$_2$ (Me)	79 ± 6
4	Ac-Val-Lys-Cys-Thr-Gly-Ile-Cys-Glu-NH$_2$ (Pr)	61% at 300 μM
5	Ac-Val-Lys-Cys-Thr-Gly-Ile-Cys-Glu-NH$_2$ (SS)	26 ± 2
6	Ac-Val-Lys-Cys (SH)-Thr-Gly-Ile-Cys(SH)-Glu-NH$_2$	$\gg 300$
7	H-Val-Lys-Cys-Thr-Gly-Ile-Cys-Glu-NH$_2$ (Et)	$\gg 300$
8	Ac-Val-Lys-Cys-Thr-Gly-Ile-Cys-Glu-OH (Et)	$\gg 300$

[a] Cyclization type between the two Cys residues denoted in parentheses: (SS) = -SS-, (Me) = -SCH$_2$S-, (Et) = -SCH$_2$CH$_2$S-, and (Pr) = -SCH$_2$CH$_2$CH$_2$S-.

corresponding to the switch 1 region is inactive, compound 2, the linear analog of 1 (with each Cys sulfur converted to a methyl thioether), displays no inhibition of GTPase activity at the highest concentration tested, 300 μM. Compounds 3–5 further explore the effect of ring size in the cyclic RGS inhibitor series. As seen from Table I, 3, the methylene dithioether (dithioacetal), in which the ring is one carbon smaller than in 1, displays comparable inhibitory potency as 1. In contrast, 4, the propylene dithioether, with ring size one carbon larger than 1, displays considerably lower inhibitory potency (61% inhibition at 300 μM). Interestingly, 5, the disulfide-containing analog, whose ring size is two carbon atoms smaller than the lead compound 1, displays approximately threefold higher potency (IC$_{50}$ = 26 μM).

To examine the possibility that the improved potency of 5 is due to the reduced, linear sulfhydryl-containing species, 6, the free sulfhydryl-containing precursor of dithioethers 1, 3, and 4 and of disulfide 5, was evaluated. As shown in Table I, the linear 6 displayed no RGS inhibitory activity at 300 μM, the highest concentration tested, indicating that contribution of the free sulfhydryl form of the peptide to the observed potency of 5 is unlikely. Finally, the design assumption that N-acetylation and C-terminal amidation, which most closely resemble the bounded fragment in switch 1, represent the optimal starting point for analogs in this series was evaluated by preparing and testing the free amino containing analog, 7, and the free carboxylate-containing analog, 8. As predicted, neither of these analogs inhibited RGS GAP activity at 300 μM.

As a first attempt to improve inhibitory potency in the switch 1 peptides, we further analyzed the switch 1 region of the RGS4–G$_{i\alpha 1}$ X-ray structure, seeking possible interactions that could be facilitated by substitutions of the switch 1 peptide residues 4–6, that is, within the loop formed by cyclization of the cysteine residues in positions 3 and 7. Examination of the lagr PDB file suggested that residues 5 and 6 were the most promising candidates for variation. As illustrated in Fig. 4, substitution of Gly^5 by a D-amino acid orients the side chain of this D-amino acid toward RGS4 residues 83–85. Of these proximal residues, Glu^{83} and Tyr^{84} represent potential hydrogen bonding or other electrostatic partners for appropriate switch 1 peptide, Gly^5 substitutions. Accordingly, we prepared analogs 9–11 using the dithioether containing lead compound 1 as a template, in which Gly^5 was replaced by D-1,4-diaminobutyric acid (D-A_2bu, analog 9), D-1,3-diamino-propionic acid (D-A_2pr, analog 10), or D-Ser (analog 11). As shown in Table II, compounds 9 and 10 were poor inhibitors of RGS activity, perhaps due to adverse steric interactions between the D-A_2bu and the D-A_2pr side chains and RGS4. In contrast, analog 11, with the smaller (and uncharged) D-Ser side chain, retains equivalent inhibitory potency as 1. This suggests that the choice of D-amino acids for substitution at this position is appropriate.

Further examination of the RGS4–$G_{i\alpha1}$ X-ray structure revealed the proximity between Tyr-84 of RGS4 and the Ile^6 side chain of lead compound 1. This proximity, illustrated in Fig. 5, suggested that replacement of Ile^6 by residues with aromatic side chains might enable aromatic–aromatic interactions with Tyr-84, resulting in improved binding affinity and hence improved inhibitory potency. Toward this end, the aliphatic side chain of Ile^6 was replaced by the aromatic side chains of Phe^6 (analog 12) and Hfe^6 (homophenylalanine, analog 13). The extra (compared to Phe) side chain methylene that separates the aromatic side chain of Hfe from the backbone allows evaluation of the preferred spatial positioning of the aromatic. As shown in Table II, both 12 and 13 show modest improvement in RGS4

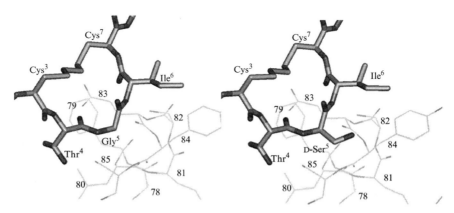

FIG. 4. Interface between RGS4 (thin lines) and the Gly^5 region of the parent peptide inhibitor, 1 (*left*), and its D-Ser^5 analog (*right*).

TABLE II
INHIBITORY POTENCIES OF RESIDUE 5 AND 6 ANALOGS OF LEAD COMPOUND **1**

#	Structure	Relative potency[a]
9	Ac-Val-Lys-Cys-Thr-D-A$_2$bu-Ile-Cys-Glu-NH$_2$ (Et)	<0.1
10	Ac-Val-Lys-Cys-Thr-D-A$_2$pr-Ile-Cys-Glu-NH$_2$ (Et)	<0.3
11	Ac-Val-Lys-Cys-Thr-D-Ser-Ile-Cys-Glu-NH$_2$ (Et)	1.0
12	Ac-Val-Lys-Cys-Thr-Gly-Phe-Cys-Glu-NH$_2$ (Et)	1.4
13	Ac-Val-Lys-Cys-Thr-Gly-Hfe-Cys-Glu-NH$_2$ (Et)	2.1

[a] Potency relative to compound **1** = 1.0.

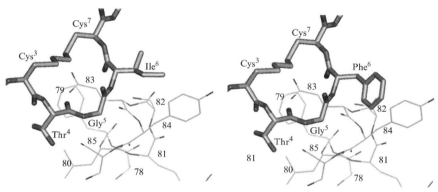

FIG. 5. Interface between RGS4 (thin lines) and the Ile[6] region of the parent peptide inhibitor, **1** (*left*), and its Phe[6] analog (*right*).

GAP inhibitory potency compared to the lead compound, **1**, consistent with the hypothesis that aromatic–aromatic interactions between Tyr-84 and peptide inhibitors can improve potency.

The aforementioned results lead to three related conclusions. First, cyclic (but not flexible, linear) peptides based on the switch 1 region of G$_{i\alpha}$ inhibit the interaction between RGS4 and G$_{i\alpha}$ and, in so doing, diminish RGS GAP activity. Second, D-amino acid substitutions for Gly-5 in the lead inhibitory cyclic peptide 1, which are predicted from analysis of the RGS4–G$_{i\alpha1}$ X-ray structure to be well tolerated, do indeed maintain inhibitory potency. Finally, substitution of Ile-6 of the lead inhibitory cyclic peptide 1 by amino acids with aromatic side chains can improve inhibitory potency, consistent with the prediction that such substitutions may promote an aromatic–aromatic interaction with the RGS4 Tyr-84 side chain. These results thus lend strong support to the hypothesis that the structure-based design of inhibitors of RGS protein GAP activity modeled on the RGS binding conformation of $G_{i\alpha1}$ is a promising, viable approach. Further, the cyclic peptides described here represent the first examples of any rationally designed RGS inhibitors. Efforts to improve potency within this series are in progress.

Materials and Methods

Solid-Phase Peptide Synthesis

All peptides are synthesized by solid-phase methods (Wellings and Atherton, 1997) on an ABI Model 431A solid-phase peptide synthesizer (Applied Biosystems, Foster City, CA) using Fmoc-protected amino acids obtained from Advanced ChemTech (Louisville, KY). Rink resin (Advanced ChemTech, Louisville, KY) is used as the solid support for C-terminal carboxamide peptides, and preloaded PEG-PS resin (Applied Biosystems, Foster City, CA) is used for the C-terminal carboxylic acid peptide. Peptide elongation on the peptide–resin involved treating resin with piperidine (Aldrich, Milwaukee, WI) to cleave the Fmoc-protecting group, followed by coupling of the next amino acid with O-benzotriazol-1-yl-N,N,N',N'-tetramethyl uronium hexafluorophosphate (HBTU) and 1-hydroxybenzotriazole (HOBt) (Applied Biosystems). Trifluoroacetic acid/H_2O/dithioethane (9:0.5:0.5,v/v/v) is used to cleave the linear peptide from the resin and simultaneously remove the side chain-protecting groups (Guy and Fields, 1997). The peptide solution is filtered from the resin and is then subjected to preparative RP-HPLC to afford the linear disulfhydryl-containing peptide.

General Method for Disulfide Cyclization of Peptides

To obtain the disulfide-cyclized peptide, the linear disulfhydryl-containing peptide is dissolved in a 1% (v/v) HOAc in H_2O solution (saturated with N_2) at 5° (1 mg linear peptide/ml of aqueous HOAc solution). The pH of the peptide solution is raised to 8.5 using NH_4OH, followed by the addition of 4 mol equivalents of $K_3Fe(CN)_6$. The reaction mixture is stirred for 1 min and is then quenched by adjusting the pH to 3.5 with HOAc. The mixture is then subjected to preparative RP-HPLC to afford the disulfide-cyclized peptide. Yields of desired, monomeric disulfide-containing peptides range from 45 to 80%.

General Method for Dithioether Cyclization of Peptides

To form dithioether-containing cyclic peptides, linear disulfhydryl peptide is added to dimethylformamide and maintained at 5° under a N_2 atmosphere (0.1 mg linear peptide/ml dimethylformamide). Five mol equivalents of potassium *tert*-butoxide are added to the peptide solution, followed by the addition of 2.5 mol equivalents of Br-$(CH_2)_n$-Br ($n =$ 1, 2, or 3). The reaction is quenched with 2 mL HOAc after 2 h, and the solvent is removed *in vacuo*. The residue is dissolved in water, filtered, and then subjected to preparative RP-HPLC to afford the alkyl dithioether-cyclized peptide. Yields of desired, monomeric dithioether-containing peptides range from 25 to 60%.

All final product peptides are >95% pure as assessed by analytical RP-HPLC in two solvent systems and all display the appropriate molecular weights as determined by mass spectrometry.

Biological Materials

[γ-^{32}P]GTP (30 Ci/mmol) is from New England Nuclear (NEN) Life Science Products (Boston, MA). A Chinese hamster ovary (CHO) cell line with stable expression of an HA-epitope tagged porcine α_{2a}AR adrenoceptor (α_{2a}AR-CHO, 10–20 pmol/mg) is cultured, and cell membranes are prepared as described (Wade *et al.*, 1999). The GST fusion protein containing rat RGS4 is prepared as described by Lan *et al.* (2000).

[^{32}P]GTPase Assay

The ability of RGS proteins to stimulate steady-state [^{32}P]GTPase activity of receptor-stimulated G proteins in CHO cell membranes expressing high (>5 pmol/mg) concentrations of α_{2a}AR has been described by Zhong *et al.* (2003) and has also been reported with α_{2a}AR–Gα_o fusion proteins (Cavalli *et al.*, 2000). To assess peptide inhibition of RGS-stimulated [^{32}P]GTPase, measurements are done in a reaction mixture (100 μl) containing 0.2 mM ATP, 1 μM GDP, 50 units/ml creatine phosphokinase, 50 mM phosphocreatine, 20 mM NaCl, 2 mM MgCl$_2$, 0.2 mM EDTA, 10 mM Tris–HCl, 1 mM dithiothreitol (DTT), and 0.1 μM [^{32}P]GTP (pH 7.6). All components are preincubated for ~10 min on ice along with 4 μg of α_{2a}AR-CHO membrane protein, the α_{2a}AR adrenoceptor agonist, UK 14,304, (10 μM), 1–300 μM synthesized peptide, with or without 300 $n M$ GST-RGS4 protein. The reaction is started by the addition of [γ-^{32}P]GTP to the preincubation mixture, and GTP hydrolysis is allowed to proceed for 10 min at 30°. The reaction is then terminated by adding 1 ml of 50% (w/v) ice-cold activated charcoal slurry in 20 mM phosphoric acid, followed by incubation on ice for 30 min. Reaction tubes are then centrifuged at 4000g for 20 min at 4°, and 200 μl of supernatant fluid containing the free [^{32}P$_i$] is withdrawn and counted by liquid scintillation counting. Each peptide is tested in three separate experiments. Data are fit by nonlinear least-squares analysis in Graph Pad Prism 3.0 (San Diego, CA) to the equation

$$Y = Y_{max} + (Y_{min} - Y_{max})/(1 + 10^{\log IC_{50} - X});$$

where X is the logarithm of peptide concentration, Y is GTPase activity, Y_{max} is the maximum, and Y_{min} is minimum GTPase activity. Y_{min} was constrained to equal the GTPase activity in the absence of RGS4.

Acknowledgments

We are grateful to Dr. Irina Pogozheva for many helpful discussions and Leighton Janes for assistance with GTPase assays. This study was supported by NIH Grants DA03910 (HIM) and GM39561 (RRN).

References

Cavalli, A., Druey, K. M., and Milligan, G. (2000). The regulator of G protein signaling RGS4 selectively enhances alpha 2A-adrenoreceptor stimulation of the GTPase activity of Go1alpha and Gi2alpha. *J. Biol. Chem.* **275,** 23693–23699.

De Vries, L., Mousli, M., Wurmser, A., and Farquhar, M. G. (1995). GAIP, a protein that specifically interacts with the trimeric G protein G alpha i3, is a member of a protein family with a highly conserved domain. *Proc. Natl. Acad. Sci. USA* **92,** 11916–11920.

De Vries, L., Zheng, B., Fischer, T., Elenko, E., and Farquhar, M. G. (2000). The regulator of G protein signaling family. *Annu. Rev. Pharmacol. Toxicol.* **40,** 235–271.

Druey, K. M., Blumer, K. J., Kang, V. H., and Kehrl, J. H. (1996). Inhibition of G-protein mediated MAP kinase by a new mammalian gene family. *Nature* **379,** 742–746.

Guy, C. A., and Fields, G. B. (1997). Trifluoroacetic acid cleavage and deprotection of resin-bound peptides following synthesis by Fmoc chemistry. *Methods Enzymol.* **289,** 67–83.

Hepler, J. R. (1999). Emerging roles for RGS proteins in cell signalling. *Trends Pharmacol. Sci.* **20,** 376–382.

Jin, Y., Zhong, H., Omnaas, J. O., Neubig, R. R., and Mosberg, H. I. (2004). Structure-based design, synthesis, and pharmacological evaluation of peptide RGS4 inhibitors. *J. Peptide Res.* **63,** 141–146.

Koelle, M. R., and Horvitz, H. R. (1996). EGL-10 regulates G protein signaling in the *C. elegans* nervous system and shares a conserved domain with many mammalian proteins. *Cell* **84,** 115–125.

Lambright, D. G., Noel, J. P., Hamm, H. E., and Sigler, P. B. (1994). Structural determinants for activation of the alpha-subunit of a heterotrimeric G-protein. *Nature* **369,** 621–628.

Lan, K.-L., Sarvazyan, N. A., Taussig, R., Mackenzie, R. G., DiBello, P. R., Dohlman, H. G., and Neubig, R. R. (1998). A point mutation in Go and Gi1 blocks interaction with regulator of G protein signaling proteins. *J. Biol. Chem.* **273,** 12794–12797.

Lan, K. L., Zhong, H. L., Nanamori, M., and Neubig, R. R. (2000). Rapid kinetics of regulator of G-protein signaling (RGS)-mediated Gi and Go deactivation: G specificity of RGS7 and RGS4. *J. Biol. Chem.* **275,** 33497–33503.

Macdonald, R. L., and Olsen, R. W. (1994). GABA (A) receptor channels. *Annu. Rev. Neurosci.* **17,** 569–602.

Neubig, R. R., and Siderovski, D. P. (2002). Regulators of G-protein signaling as new central nervous system drug targets. *Nature Rev. Drug Disc.* **1,** 189–199.

Siderovski, D. P., Hessel, A., Chung, S., Mak, T. W., and Tyers, M. (1996). A new family of regulators of G protein-coupled receptors? *Curr. Biol.* **6,** 211–212.

Siderovski, D. P., Strockbine, B., and Behe, C. I. (1999). Whither goest the RGS proteins? *Crit. Rev. Biochem. Mol. Biol.* **34,** 215–251.

Tesmer, J. J. G., Berman, D. M., Gilman, A. F., and Sprang, S. R. (1997). Structure of RGS4 bound to AlF4-activated G(i alpha 1): Stabilization of the transition state for GTP hydrolysis. *Cell* **89,** 251–261.

Wade, S. M., Lim, W., Lan, K. L., Chung, D. A., Nanamori, M., and Neubig, R. R. (1999). Gi activator region of 2A-adrenergic receptors: Distinct basic residues mediate Gi versus Gs activation. *Mol. Pharmacol.* **56,** 1005–1013.

Wellings, D. A., and Atherton, E. (1997). Standard Fmoc protocols. *Methods Enzymol.* **289,** 44–67.

Zhong, H. L., and Neubig, R. R. (2001). RGS proteins: Novel multifunctional drug targets. *J. Pharmacol. Exp. Ther.* **297,** 837–845.

Zhong, H., Wade, S. M., Woolf, P. J., Linderman, J. J., Clark, M. J., Traynor, J. R., and Neubig, R. R. (2003). A spatial focusing model for G protein signals: Regulator of G protein signaling (RGS) protein-mediated kinetic scaffolding. *J. Biol. Chem.* **278,** 7278–7284.

[17]   Yeast-Based Screening for Inhibitors of RGS Proteins

By Kathleen H. Young, Yuren Wang, Corey Bender, Seena Ajit, Fernando Ramirez, Adam Gilbert, and Bart W. Nieuwenhuijsen

Abstract

This article provides information on two screening platforms for the identification of regulators of G-protein signaling (RGS) protein modulators. Utilization of the yeast pheromone response pathway enabled the creation of a functional screen for RGS4 modulators. The RGSZ1-focused screen employs advances in yeast two-hybrid screening technologies and targets the protein–protein interface of the RGS domain/Gα interaction. Moreover, the RGSZ1 screen provides the opportunity to multiplex the screening of two targets of interest, given the development of two different luciferase reporter genes that enabled sequential determination and intra-assay controls. The screen formats were validated, implemented, and conducted as automated 384-well, liquid-based, high-throughput small molecule screens. Primary "hits" were confirmed using benchtop 96-well formats of these assays and advanced to *in vitro* functional evaluation assays. The yeast-based assay platforms provide robust cellular assays that result in the identification of small molecule modulators for both RGS targets. These molecules can serve both as tools with which to probe biological implications of RGS proteins and as potential starting points toward the development of novel modulators of G-protein signaling pathways. Such modulators may show potential for controlling and treating diseases resulting from inappropriate activity of G-protein signaling pathways.

Introduction

The G-protein signaling pathway is one of the most important signaling cascades for relaying extracellular signals such as neurotransmitters, hormones, odorants, and light. Such pathways have been identified in diverse organisms, including yeast and mammals. Classically, the system is composed of three major components: G-protein-coupled receptors (GPCRs), heterotrimeric G proteins having α, β, and γ subunits, and intracellular effectors (Gilman, 1987). The G-protein signaling pathway commences upon activation of a GPCR, which is characterized by seven transmembrane domains. When stimulated, the GPCR intracellular loops and C-terminal tail interact with an associated G protein (Wieland and Chen, 1999). The $G\alpha$ subunit of the G protein then releases guanosine diphosphate (GDP) and binds guanosine triphosphate (GTP) in its place. Binding of the GTP alters the conformation of the $G\alpha$ subunit, resulting in dissociation of the heterotrimer into a GTP-bound $G\alpha$ subunit and a $G\beta\gamma$ dimer. The released subunits induce downstream signaling events, ultimately mediating biochemical responses, changes in cellular physiology, or other specific cellular responses. Signaling is terminated when the $G\alpha$ subunits hydrolyze GTP, returning to the GDP-bound state, and reassemble with $G\beta\gamma$ subunits to form the inactive heterotrimer (Kehrl, 1998).

To elicit an appropriate cellular response, the strength of the intracellular signals must be regulated tightly. While there are a number of types of regulation of the system, such as phosphorylation of GPCRs, receptor-binding proteins, and $G\beta\gamma$-trapping proteins, many investigators have focused on GTPase-activating proteins (GAPs) (Wieland and Chen, 1999). GAPs accelerate the rate of $G\alpha$-mediated GTPase hydrolysis, thereby reducing the signal generated in the pathway and "desensitizing" the system (De Vries and Gist Farquhar, 1999).

Regulators of G-protein signaling (RGS) proteins represent a relatively new class of GAPs. The first member of the family was obtained from the yeast *Saccharomyces cerevisiae* (Dohlman *et al.*, 1996; Wieland and Chen, 1999). Haploid mutants were identified that were hypersensitive to pheromone-induced cell cycle arrest, a response mediated by a GPCR pathway. The gene product mutated in the hypersensitive yeast, "supersensitivity to pheromone-2" (Sst2), normally interacts with the yeast $G\alpha$ subunit (Gpa1) as a GAP and thus serves as a negative regulator of the system. Subsequently, many more members of the RGS family have been characterized in a number of species, including over 20 different members in mammals (Zheng *et al.*, 1999). It is hypothesized that the repertoire of RGS proteins is increased greatly by alternative splicing (Sierra *et al.*, 2002). Although all proteins contain the 120 amino acid "RGS domain," the protein family has been defined further into subfamilies, based primarily on structural features (Sierra *et al.*, 2000; Zheng *et al.*, 1999). RGS proteins provide a mechanism by which cells can fine-tune both the duration and the magnitude of a signal generated through the G-protein pathways (Kehrl, 1998).

Because of the size of the RGS family and the crucial role of RGS proteins in regulating G-protein signaling pathways, research has focused on mechanisms by which individual RGS proteins achieve their specificity. One level of specificity results from the expression pattern of different RGS proteins; some RGS proteins are expressed in particular tissues, whereas others are expressed ubiquitously (Grafstein-Dunn *et al.*, 2001; Zheng *et al.*, 1999). Additionally, different RGS proteins may have differing specificity for individual $G\alpha$ subunits (Kehrl, 1998). This would enable certain RGS proteins to preferentially modulate certain G-protein signal pathways over others (Zeng *et al.*, 1998) or bias a dual $G\alpha$ response from a single GPCR (Neubig and Siderovski, 2002).

Many cell-based research platforms link the desired effect of a gene or drug of interest to a change in cell phenotype through the use of reporter genes. In yeast systems, for example, reporter genes have commonly focused on auxotrophy genes for cell growth on selective media or the *lacZ* gene for colorometric end point using assays that detect β-galactosidase activity. To extend reporter gene utility in yeast, we developed luciferase genes from *Renilla reniformis* and *Photinus pyralis* as reporter genes in yeast. Luciferase reporters increase assay sensitivity, speed, ease, signal-to-noise ratios, and provide high-quality quantitative data to yeast-based assays for a myriad of target identification and drug discovery applications. Use of the luciferase reporter gene in yeast provides substantial improvements to yeast-based assays and, as described here, is ideal for robust assays for modulators of RGS proteins.

RGS Proteins as Small Molecule Targets

The interaction between the RGS domain and the $G\alpha$ protein has been characterized functionally, biochemically, and structurally (Dohlman *et al.*, 1996; Hepler, 1999; Koelle, 1997; Koelle and Horvitz, 1996; Neubig and Siderovski, 2002; Ross and Wilkie, 2000; Tesmer *et al.*, 1997). The particular amino acids of the RGS domain that engage the switch regions on $G\alpha$

have been well characterized (Tesmer *et al.*, 1997). Moreover, mutational studies have further narrowed the functionality of specific amino acids within the RGS core domain that control specificity of an RGS protein to function effectively as a GAP for different Gα subtypes (Heximer *et al.*, 1999). An additional region on the RGS domain may be amenable to small molecule modulation because phosphatidylinositol 3,4,5-trisphosphate (PIP$_3$) and calmodulin are believed to regulate, in an allosteric fashion, the GAP activity of RGS4 (Popov *et al.*, 2000). These sites have been summarized by Neubig (2002). In addition, both RGS3 and RGS12 have regions predicted to assume a coiled coil structure. Such domains often mediate interactions with proteins of the cellular cytoskeleton (Kehrl, 1998) and may enable RGS proteins to fluctuate between membrane-associated and cytosolic pools, thus altering the availability of the RGS proteins at a given time and/or influencing the type of Gα subunit the RGS protein modulates (De Vries and Gist Farquhar 1999). Receptor impact on selective RGS proteins by particular GPCRs has also been suggested (Bernstein *et al.*, 2004).

These studies and others (Moy *et al.*, 2001) suggest that small molecules could be identified that perturb the function of RGS proteins via multiple avenues: prevention of RGS protein association at the G-protein/receptor complex, prevention or disruption of RGS domain interaction with the switch regions of GTP-bound Gα, or via allosteric or kinetic modulation. Kinetic modulation follows the model of Ross and Wilkie (2000) and suggests that inhibition of the RGS protein function would maintain the Gα protein GTPase kinetics in the "outer" or slower kinetic parameter cycle.

This article describes the application of yeast-based screening technology in two platforms, both using full-length RGS proteins, for the identification of compounds capable of regulating RGS protein function. Both platforms use modified yeast cells and methods incorporating these cells in both automated high-throughput screening (HTS) and manual benchtop assays to provide an efficient and specific screening system to detect compounds capable of regulating RGS activity. The RGS4 screen uses the endogenous pheromone-responsive GPCR signal transduction pathway in yeast and enables the identification of modulators with various mechanisms of action. The RGSZ1 screen employs advances on YTH technology and focuses on the RGS domain/Gα protein–protein interaction, thus having the potential to identify disruptors or allosteric modulators of RGS protein GAP activity. Such compounds may have therapeutic value, as they could potentially modulate one or a variety of G-protein pathways that mediate a vast array of biological processes and underlie several diseases. In the near term, RGS protein-modulating compounds can provide tools to further our understanding of RGS proteins and the biological repercussions of their modulation.

RGS4 Yeast Pheromone Response Assay

All yeast media are obtained from Qbiogene (Palo Alto, CA) and prepared according to the manufacturer's recommendation or are obtained as prepoured agar plates from Teknova (Half Moon Bay, CA).

Generation of Reagents

Yeast Strains—SST2 Knockout. To generate a yeast strain in which the endogenous RGS protein, SST2, is deleted, the plasmid pEK139/138 (Price *et al.*, 1995) is digested with *Xho*I and *Sac*I and the resulting 4.0-kb *sst2*-NEO-*sst2* cassette is gel isolated and purified. To enhance integration frequency, this cDNA fragment is cotransformed with the uracil-auxotropy marker *URA3* of plasmid pRS416 (Stratagene, La Jolla, CA) into two separate yeast strains: CY770 (MATa leu2-3,112 ura3-52 trp1-901 his-200 ade2-101 gal4 gal80 lys2::GALuas-HIS3 cyh^R (Ozenberger and Young, 1995) and yeast strain YPH499 (American Type Culture Collection, Manassas, VA). Following transformation, yeast cells are grown in YPD medium and plated on SC–Ura ("synthetic complete lacking uracil") medium. Resultant URA$^+$ yeast colonies are sequentially replica-plated onto YPD medium containing G418 (Geneticin, Invitrogen, CA) at 50 μg/ml and then at 100 μg/ml. To verify *sst2*-NEO-*sst2* construct integration, G418-resistant CY770 yeast colonies are confirmed by polymerase chain reaction (PCR) analysis using primers 5'-TATCGAGTCAATGGGGCAGGC-3', and Kx25, 5'-CGAAACGTGGATTGGTGAAAG-3'. A 4.2- or 4.9-kb PCR product is generated for the confirmed knockout strain or wild-type yeast strain, respectively. Primers Kx26 (5'-ATTCGGCTATGACTGGGCA-CAAC-3') and Kx27 (5'GTAAAGCACGAGGAAGCGGTCAG-3') are also used for PCR verification and generated an expected 2.6-kb PCR product from a confirmed knockout stain, while no PCR product is generated (as expected) from the wild-type strain. Verified Δsst2 yeast colonies are then plated on SC-Ura plates to confirm the loss of the pRS416 plasmid used to facilitate integration. The CY770 Δsst2 strain is designated yKY103 (MATa leu2-3,112 ura3-52 trp1-901 his-200 ade2-101 gal4 gal80 Lys2::GALuas-HIS3 cyh^R, sst2, G418R), and the YPH499 Δsst2 strain is designated yKY113 (MATa ura3-52 lys2-801^{amber} ade2-101^{ochre} trp1-Δ63 his3-Δ200 leu2-Δ1, sst2). Both strains are functionally confirmed by demonstrating a lack of recovery from cell cycle arrest, in comparison to wild-type parental strains, following α factor stimulation using pheromone-response halo assays (Dohlman *et al.*, 1996).

FUS1 Promoter-Linked Reporter Genes. We first generated a FUS1-lacZ reporter gene. The FUS1-promoter region (GenBank M16717) is generated by PCR as a 1.095-kb *Eco*RI–*Sal*I fragment; the β-galactosidase gene (GenBank CVU89671) is generated as a 3.0-kb *SalI–EagI* fragment, and the pRS424 plasmid (Stratagene) is digested with *XhoI* and *EagI*. A three-way, directional ligation is conducted between the prepared pRS424 vector, the FUS1 and the β-galactosidase DNA fragments, using standard methods, to generate plasmid Kp27 (FUS1-lacZ, 2 μM origin of replication, URA3 selection marker). Recombinant DNA is transformed into *Escherichia coli*, and DNA from selected transformants is prepared using standard methods. The Kp27 construct is confirmed by restriction endonuclease digestion and sequence analyses. Kp27 is transformed into yKY103 and yKY113 to generate yKY120 and yKY114, respectively, and tested for dose-dependent increases in β-galactosidase activity in response to α-factor pheromone stimulation (data not shown). This confirmed that the plasmid backbone and pheromone-responsive FUS1 promoter would link pheromone signaling to reporter gene expression to facilitate the study of RGS protein activity on the pheromone signaling pathway.

We then generated the luciferase reporter gene linked downstream to the FUS1 promoter by replacing the *lacZ* gene. This luciferase system provides certain advantages over the *lacZ* reporter gene, most notably increased speed and ease of monitoring gene expression. The FUS1-luciferase reporter cassette is constructed by a *NcoI–XbaI* digestion of plasmid pGL (Promega, Madison WI) to isolate a 1.7-kb fragment containing the firefly (*Photinis photolus*) luciferase gene and is blunt ended and purified using standard techniques. The Kp27 vector serves as the base vector and is digested with *BamHI–NotI* (to remove the *lacZ* gene), dephosphorylated, blunt ended, and purified. The prepared vector and luciferase fragment are ligated to generate plasmid Kp120. This cloning scheme results in an expected four extra amino acid residues (methionine, alanine, glycine, serine) from the original Kp27 vector that are fused in frame to the N terminus of the luciferase open reading frame. The *BamHI* and *NcoI* restriction sites are retained. The plasmid Kp120 construct is confirmed by DNA sequencing using primers Kx45 (5′-ATATAAGCCATCAAGTTTCTG-3′) and Kx46 (5′-CTCACTAAAGG-GAACAAAAG-3′). The FUS1-luciferase reporter cassette is then excised from Kp120 using *KpnI* and *SacI* restriction enzymes. The 2.8-kb fragment containing the FUS1 promoter and luciferase reporter cassette is gel purified, blunt ended, and ligated into pRS416 (URA3 marker, CEN-based vector; Stratagene) to generate Kp131.

RGS4 Expression Vector. The cDNA encoding full-length rat RGS4 (Genbank AF117211) is obtained by PCR using Kx13(forward): 5′-GA*CGTCT*CCCATGTGCAAAGGACTCG-3′, which contains an embedded *BsmBI* restriction site (in *italics*) and Kx41(reverse): 5′-CG*GGAT-CC*TTATTAGGCACACTGAGGGACTAGGGAAG-3′, which contains an extra stop codon and the embedded *Bam*HI restriction site (in *italics*). The resultant ~650-bp PCR product is digested with *Bsm*BI and *Bam*HI and is ligated into a similarly prepared Kp46 vector to generate phosphoglucose kinase (PGK) promoter-driven expression of an untagged RGS4 protein. We also generated a hemagglutinin-tagged ("HA") RGS4 open reading frame using a different 3′primer Kx42(reverse): 5′-GA*GGAT-CC*GGCACACTGAGGGACTAGGGAAG-3′ which lacks the stop codon, but contains an embedded *Bam*HI restriction site (in *italics*). The resultant ~650-bp PCR product (using Kx13 and Kx42) is digested with *Bsm*BI and *Bam*HI and is ligated into a similarly prepared Kp57 vector, which contains a 3′ in-frame sequence encoding the HA tag. The resultant plasmids Kp118 (RGS4-Kp46) and Kp119 (RGS4-HA-Kp57) are transformed into bacterial cells, and recombinant DNA is prepared and confirmed by sequence analysis.

FUS1-Luciferase Reporter Yeast-Based Assay for RGS4 Activity

Generation of Yeast Strains. To test the ability of the exogenously expressed RGS4 protein to complement the loss of the Sst2 protein and reduce pheromone signaling (see schematic in Fig. 1), pheromone-responsive, RGS4-expressing test and control yeast strains are generated using the lithium-acetate transfection method (Rose *et al.*, 1990) and subsequent plating on appropriate plasmid retention ("synthetic dropout") medium. Cotransformation of plasmids Kp118 (RGS4 expression plasmid) and Kp131 (FUS1-luc reporter plasmid) into yeast strain yKY103 generates yeast strain yKY115. Control yeast strains are prepared by cotransforming the empty expression plasmid Kp46, which lacks an RGS protein-encoding sequence, and the reporter plasmid Kp131 into strain yKY103 to generate yeast strain yKY116. Plasmids Kp118 and Kp131 are cotransformed into yKY113 to generate RGS test yeast strain yKY117. Plasmids Kp46 (empty expression plasmid) and Kp131 (FUS1-luc reporter plasmid) are cotransformed into yKY113 to generate control strain yKY118.

Yeast Strain Testing. In comparison to control strains, the expression of RGS4 protein decreased pheromone-induced transcription of the luciferase protein when strains were treated with α factor (e.g., Fig. 2), suggesting that the exogenously expressed RGS4 protein is able to negatively regulate the yeast Gα subunit (Gpa1p). This is in agreement with previous studies (Dohlman *et al.*, 1996; Druey *et al.*, 1996). Pheromone-responsive luciferase activity was observed in yeast strains generated from both base yeast strains, yKY103 and yKY113. These findings correlated with our previous findings on β-galactosidase activity; dose–response curves for the luciferase

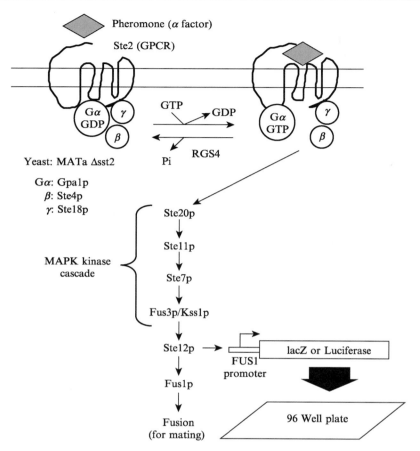

Fig. 1. Schematic of yeast pheromone response screen for RGS4 modulatory compounds. Activation of the GPCR-induced pheromone response pathway in yeast is linked operationally to a FUS1-luciferase reporter gene.

system revealed that luciferase reporter gene worked as well as lacZ as the reporter gene. We did observe greater sensitivity of the luciferase reporter gene, in comparison to the β-galactosidase reporter gene, as noted by a higher-fold increase in response to similar α-factor stimulation (data not shown). Moreover, the response of the yKY113-based yeast strain was slightly stronger (data not shown) and therefore yKY113-based strains were further developed into screening strains.

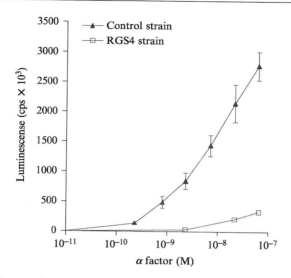

FIG. 2. RGS4 expression attenuates α factor pheromone-induced luciferase activity. Yeast strains expressing RGS4 are prepared as described in the text. Treatment of test or control yeast strains with the α factor results in activation of the pheromone response pathway and the FUS1 promoter-linked luciferase reporter gene. In the control strain [yKY118], the absence of SST2 (the yeast RGS protein that normally acts as a GAP for Gpa1p) results in a lack of negative regulation of the G-protein signal pathway, leading to increased luminescence with increasing concentrations of α factor. Expression of exogenous RGS4 in the yeast strain [yK117] functionally complements the Δsst2 phenotype, regulates G-protein signaling negatively, attenuates the pheromone-induced response of the FUS1-luciferase reporter gene, and thus is observed to decrease luminescence activity. Data shown here are means ± SE of triplicate measurements.

Yeast Pheromone Response Screening Methods for RGS4 Inhibitors

Preparation of Frozen Yeast Cells

1. Yeast strains yKY117 (RGS4 and luciferase gene) and control strain yKY118 (empty vector and luciferase gene) are used for screening.
2. Inoculate a single colony of each strain into 5 ml of SC-leu-trp medium and grow overnight at 30° with shaking.
3. Inoculate 200 ml of SC-leu-trp medium with an overnight culture of each yeast strain and grow for 8 h at 30° with shaking.
4. Transfer the culture to 1 liter of SC-leu-trp medium and shake at 30° overnight
5. Overnight growth results in yeast cells with an optical density ($OD_{600 \, nm}$) of about 1.5.

6. Spin down yeast cells (2000g, 15 min, at 25°) and resuspend in SC-leu-trp medium with 25% glycerol to an OD_{600} of 5.

7. Freeze and store aliquoted cells at −80°.

Automated (384 Well) HTS Small Molecule Screening Conditions

1. Obtain a frozen cell aliquot of RGS4-expressing strain yKY117 and control strain yKY118 and resuspend to 0.1 OD_{600}/ml in SC-leu-trp medium.

2. Mix well and dispense 15 μl per well into nonsterile 384-well plates (Corning, white) containing 5 μl of test compound at 50 μg/ml (~30 μM final concentration) with a dimethyl sulfoxide (DMSO) concentration of 3%.

3. Incubate for 45 min.

4. Add 15 μl of α factor (final concentration 25 nM) to each well.

5. Incubate overnight for approximately 16–24 h.

6. Mix LucScreen reagent (Tropix) bottles A and B together and store in a dark container to retard light exposure.

7. Add 30 μl of mixed LucScreen reagent to each well and incubate in the dark at room temperature for 90 min.

8. Seal plates with Topseal A (Perkin Elmer Life Sciences, Boston, MA) and measure luciferase activity on a Topcount (Perkin Elmer Life Sciences) at a setting of 10 s/well, + one counting cycle, and a count delay of 1 min for dark adaptation. Compound activities (raw data) are recorded as counts per second (CPS).

9. Calculate fold induction for each well by dividing the activity of each sample by the average activity of the entire plate. Fold induction = sample activity (CPS)/average of the plate (CPS).

Luciferase Reporter, Single-Dose Compound Assay—Benchtop Protocol

1. Thaw a frozen aliquot of yeast strains yKY117 and yKY118 cells and dilute to 0.1 OD_{600} in SC-leu-trp medium.

2. Add 100 μl of mixed cells to each well of 96-well plates (Perkin-Elmer Life Science).

3. Thaw compounds (96-well plates) to room temperature and mix gently.

4. Using a 96-pin microplate replicator (Boekel Scientific, Feaster-ville, PA), add ~1 μl of compound to 100 μl yeast cells in well.

5. Incubate cells with the compound for 3 h at room temperature with shaking.

6. Add 10 μl of α factor (Sigma) to each well (final concentration, 100 nM) and mix gently by vortexing the plates at low speed.

(Initial experiments included a "no α-factor" control and "medium-only" control.

7. Shake plates at room temperature for 1 h.
8. Prepare LucLight substrate according to manufacturer's instructions (Perkin-Elmer Life Science) and add 100 μl to each well.
9. Seal plates tightly with TopSeal A (Perkin-Elmer Life Science) and shake at room temperature in the dark for 50 min.
10. Determine luminescence (dark delay 2 min and count 2 s) using a Topcount.

Luciferase Reporter, Dose–Response Compound Assay—Benchtop Protocol

1. Prepare yeast strains yKY117 and yKY118 cells as described earlier for the single-dose assay.
2. Add compound to each well of three wells within the first row (row A) of a 96-well plate to a final concentration of 300 μM.
3. Add 150 μl of yeast cells to row A and 100 μl cells to rest of the wells (rows B through H) of the 96-well plate.
4. After adding the cells, vortex the plates gently to mix the compound and cells in row A and take out 50 μl from each well using a multichannel pipettor.
5. Transfer 50 μl from each well of row A to the corresponding well in row B.
6. Transfer 50 μl from row B to the corresponding well in row C, and repeat the process until row G.
7. Discard the 50 μl taken out from row G.
8. Leave the wells in the last row H without any compound. This will result in a series of compound concentration ranging from 300, 100, 33, 11, 3.7, 1.2, 0.4, and 0 μM from rows A to H, respectively.
9. Follow the luciferase protocol for the single-dose assay mentioned earlier from step 5 onward.

RGS4 HTS Results

The pheromone-responsive luciferase reporter assay for RGS4 inhibitors was conducted as a single-dose assay at a final compound concentration of approximately ~30 μM following the protocol as outlined previously. Over 600,000 random and chemically diverse small molecule compounds were screened. Compounds that resulted in a >2.0-fold increase in luciferase activity in the RGS4 test strain were identified as putative positives; 825 compounds advanced as putative positives. This list was further triaged for

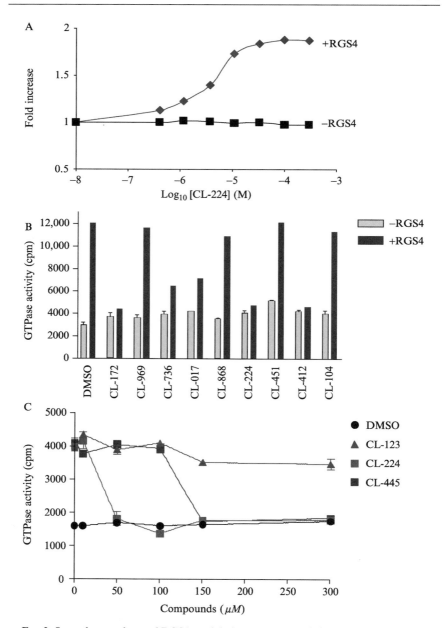

FIG. 3. Secondary analyses of RGS4 modulating compounds. (A) Cells from both RGS4-expressing (yKY117) and control (yKY118) yeast strains were incubated with compound (3 h). Cells were then stimulated with 100 nM α factor (2 h), and FUS1 promoter-dependent

drug-like properties (e.g., Lipinski *et al.*, 1997; reactive properties etc.), and a subset of compounds was advanced for further testing in secondary screening. Some initial prioritization was conducted via dose–response assays in similar yeast cells; however, all available putative positives were evaluated using an RGS protein/$G\alpha_i$ biochemical single turnover GTPase assay (Ross, 2002; Wang *et al.*, 1998, 2002). Approximately 300 compounds from the initial screening were evaluated in these *in vitro* GTPase assays, with approximately 10% of the compounds showing interesting characteristics in the modulation of RGS4 function, and thus retained as initial lead compounds of interest.

Representative results of secondary assay progression are depicted in Fig. 3. The primary screen was formatted as a "rescue" screen, wherein a putative positive compound would be identified by an increase in luminescence. As shown in Fig. 3A, compound CL-224 demonstrated a dose-dependent increase in luminescence in the RGS4-containing yeast strain, but not in the control strain. In secondary testing, compounds were evaluated at 300 μM in a single-dose, ^{32}P-based single turnover GTPase assay (Ross, 2002; Wang *et al.*, 2002) using the recombinant RGS4 core domain and $G\alpha_{i1}$ proteins (Fig. 3B). Several compounds demonstrated ~100% inhibition of RGS4 activity (e.g., CL-172, CL-224, CL-412), whereas others demonstrated partial (e.g., CL-736, CL-017) or no activity (e.g., CL-969, CL-451). Dose–response results from the single turnover assay (Fig. 3C) further defined the ability of the compounds to affect RGS4 function in accelerating the intrinsic GTPase activity of $G\alpha_{i1}$. Representative data (Fig. 3C) show CL-224 having good activity, with an estimated EC$_{50}$ of ~25 μM. Testing of additional small molecule screening hits, and available analogs of confirmed RGS protein inhibitor compounds, has enabled the identification of additional compounds with estimated EC$_{50}$ values in the low micromolar range (5–12 μM), with some compounds showing

---

luciferase activity (CPS) was measured using a TopCount after incubation with substrate (1 h). Data were plotted by calculating the fold changes of CPS on the compound incubation over 1% DMSO control. (B) Characterization of RGS4 modulatory compounds in a single turnover GTPase assay. The GTPase activity of $G\alpha_{i1}$ (100 nM) in the presence of each compound (300 μM), plus and minus 50 nM recombinant RGS4 protein, was conducted in 30-s periods at 0°. RGS4 was preincubated (30 min) with or without compounds (300 μM). All compounds were dissolved in DMSO. Data shown are means ± SE of triplicate measurements in two independent experiments. (C) Dose-dependent inhibition of RGS4 GAP activity on $G\alpha_{i1}$ by selected compounds. The RGS4 protein (50 nM) was pre-incubated with or without compounds in the indicated concentration for 30 min. Thereafter, GTP hydrolysis from 100 nM $G\alpha_{i1}$ protein was measured in 30 s. Data shown here are means ± SE of triplicate measurements in two independent experiments. CL-224 appears to be more potent than CL-445. CL-123 had no appreciable effect.

further specificity for RGS4 when tested in single turnover GTPase assays using other RGS protein subtypes.

RGSZ1 Yeast Two-Hybrid Screen

Generation of RGSZ1 Reagents

Yeast Strains for RGSZ1 Inhibitor Screening. The yeast two-hybrid (YTH) strains for the $G\alpha_z$/RGSZ1 and Kv4.3/KChIP1 (An *et al.*, 2000) interactions were generated as described elsewhere (Nieuwenhuijsen *et al.*, 2003). The 2-μm-based firefly and *Renilla* luciferase reporter gene plasmids were constructed as described previously (Nieuwenhuijsen *et al.*, 2003; Wang *et al.*, 2002). Initially, the entire HTS screen and the first half of the single-dose compound screening were performed using yeast strains generated using the parent yeast strain CY770 (Ozenberger and Young, 1995). The "established" assay used yeast strains yBN70L (RGSZ1 strain) and yBN77 (KChIP1 strain), with the $G\alpha_z$/RGSZ1 interaction activating the *Renilla* luciferase reporter gene and the Kv4.3/KChIP1 interaction activating the firefly reporter gene. The "Switched" assay (luciferase reporter genes switched between the two YTH interactions) used the yeast strains yBN70 (RGSZ1 strain) and yBN77L (KChIP1 strain) (Nieuwenhuijsen *et al.*, 2003).

Integration of YTH Luciferase Reporter Genes by Homologous Recombination. Yeast strains were developed that contained integrated luciferase reporter genes. The parent yeast strain for the integrated luciferase reporter genes was Y187 (Harper *et al.*, 1993). The firefly luciferase reporter gene (FF-Luc) or the *Renilla* luciferase reporter gene (RenLuc) was integrated into the genome of Y187 at the LYS1 gene locus using homologous recombination and blasticidin (Bsd) selection. The homologous recombination cassettes are constructed as follows: A 125-bp fragment from the 3′-untranslated region of the *S. cerevisiae* LYS1 gene locus is PCR amplified using these primers: 3′-LYS1 forward primer 5′-AAC TGC AGT TAA AGT ATA TAC ATT ATC GTT ACG–3′ and 3′-LYS1 reverse primer 5′-CCC AAG CTT TAT AAT GAA AAG ACA GTA CAC TTA–3′. After purification of the PCR product, the fragment is digested with *Pst*I and *Hind*III (restriction sites underlined) and cloned into yeast shuttle vector pTEF/Bsd (Invitrogen, Carlsbad, CA). A fragment, approximately 200 bp in length from the 5′-untranslated region of the *S. cerevisiae* LYS1 gene locus, is PCR amplified using the 5′-LYS1 forward primer 5′-CCG CTC GAG TTT GCG GTT GTG TGA AAA ATA AAG C-3′ and 5′-LYS1 reverse primer 5′-GAA GAT CTC GAG AAA CCA CGA TGA AAT ATA–3′. After purification, the PCR fragment is digested sequentially with *Bgl*II and then with *Xho*I (restriction sites underlined). The pTEF/

Bsd-3'-LYS1 construct is also digested with $BglII$ and $XhoI$ and dephosphorylated. Following the ligation of the 5'-LYS1 fragment, STBL2-competent cells (Invitrogen) are transformed with the pTEF/Bsd-[5'-LYS1-3'-LYS1] construct. The resulting construct is verified by sequence analysis. Gal4 promoter-linked firefly luciferase and Gal4 promoter-linked $Renilla$ luciferase reporter cassettes are released from pEK1 plasmids (Ozenberger and Young, 1995) by $EcoRI$ and $XbaI$ digestion and are gel purified. The pTEF/Bsd-[5'-LYS1-3'-LYS1] construct is also digested with $EcoRI$ and $XbaI$, dephosphorylated, and gel purified. Following ligation of the Gal4-firefly luciferase or the Gal4-$Renilla$ luciferase cassette, STBL2 cells are transformed. Final constructs are verified by restriction endonuclease digest analysis.

The 5'-Lys1–Gal4-luciferase–3'-LYS1–Bsd cassettes are released from the pTEF/Bsd constructs by digestion with $SfiI$ and $AatII$. Following gel purification of the cassettes, Y187 yeast cells are transformed using the Li-acetate method (Rose $et~al.$, 1990). Transformed Y187 yeast cells are plated on YPD agar plates containing 80 μg/ml blasticidin and incubated for 3 days at 30°. The Y187 yeast strains with integrated luciferase reporter genes have the following genotype: Y187-RenLuc = MATα, ura3-52, his3-200, ade2-101, trp1-901, leu2-3, 112, gal4Δ, gal80Δ, met$^-$, lys1Δ, URA3::GAL1$_{UAS}$-GAL1$_{TATA}$-LacZ, Bsd::Gal4$_{UAS}$-RenLuc. Y187-FF-Luc = MATα, ura3-52, his3-200, ade2-101, trp1-901, leu2-3, 112, gal4Δ, gal80Δ, met$^-$, lys1Δ, URA3::GAL1$_{UAS}$-GAL1$_{TATA}$-LacZ, Bsd::Gal4$_{UAS}$-FF-Luc. Y187-RenLuc and Y187-FF-Luc yeast cells are transformed with Gα_z(Q205L)-pGBKT7 (i.e., constitutively active, GTPase-deficient Gα_z) and RGSZ1-pACT2 or with Kv4.3-pGBT9 and KChIP1-pACT2 plasmids using the Li-acetate method, and the resulting colonies are tested for luciferase activity. Validation experiments with the integrated luciferase reporter genes demonstrated that integrating the luciferase reporter genes into the yeast genome gave less variability in luminescence output and resulted in a lower error rate. However, it was not possible to use the integrated firefly reporter gene for the Gα_z(Q205L)/ RGSZ1 interaction, as the luminescence signal was not robust enough, therefore the yeast strain expressing the two-hybrid plasmids from the 2μ vectors was employed (yBN70) for the "established" assay.

RGSZ1 YTH Luciferase-Screening Methods

Preparation of Frozen Yeast Cells

1. Pick a small colony of RGSZ1 YTH strain and inoculate 3 ml SC-ura-leu-trp medium. Note that one should set up a minimum of five independent colonies; smaller colonies tend to have a higher plasmid copy number and therefore higher reporter gene expression levels.

2. Grow cultures overnight at $30°$ with shaking.

3. The next morning, determine the optical density ($OD_{600 nm}$) of the overnight cultures. Take an aliquot of each culture and dilute to an OD_{600} of 0.15.

4. Transfer 100 μl of diluted aliquot to a 96-well plate (in duplicate).

5. Let the remainder of each culture grow if the OD_{600} is less than 1 or put on ice when OD_{600} is approximately 1.

6. Meanwhile, measure luciferase activity of the diluted aliquots: incubate a 96-well plate for 3 h at $30°$ and then add 100 μl luciferase reagent 1 (dual luciferase kit, Promega). Incubate at room temperature for 1 h while shaking. Measure the resultant firefly luciferase signal using a Topcount (Perkin-Elmer Life Sci.). Optional: add 100 μl of Stop&Glow reagent to strains containing *Renilla* luciferase and measure *Renilla* luciferase activity.

7. Determine which culture has the highest luciferase response: this culture will be used to start a large overnight culture.

8. Inoculate several 50 ml SC-ura-leu-trp medium cultures (in vented T75 culture flasks) with the 3-ml starter cultures. Remeasure the OD_{600}. The final OD_{600} of the 50-ml culture should be close to 0.0125. [*Note.* If the culture is initiated midafternoon, the OD_{600} will be approximately 0.8 following overnight incubation.]

9. Incubate the 50-ml cultures overnight in a shaking incubator at $30°$.

10. The next morning, measure the OD_{600}, which should not be greater than 1. If the cells have grown too fast (i.e., $OD_{600} > 1.3$), discard and do not use.

11. If OD_{600} is correct ($\sim$1.0), pool cultures into a 250-ml spin bottle and centrifuge at $\sim$4000g for 10 min at room temperature.

12. Discard supernatant and add the following to the pellet: $0.1 \times$ original culture volume of 50% glycerol and $0.1 \times$ original culture volume of fresh SC-ura-leu-trp medium. This will concentrate the cells approximately fivefold to an OD_{600} of $\sim$5.

13. Resuspend the cells and transfer to microfuge tubes in 1-ml aliquots (or less if desired).

14. Freeze cells in an ethanol-dry ice bath and store at $-80°$.

15. For assays, remove tubes from $-80°$ and let thaw at room temperature. Dilute with fresh medium to the desired OD_{600}.

RGSZ1 YTH Automated (384 Well) High-Throughput Small Molecule Screening Conditions

1. Mix test and control yeast strains in SC-leu-trp-ura medium and dilute to OD_{600} of 0.2.

2. Add 25 μl of diluted mixed yeast strains to compound-containing 384-well plates (10 μl compound (dissolved in 100% DMSO) per well, 14 $\mu g/ml$ final concentration).

3. Incubate yeast and compound for 3 h at room temperature, add 25 μl of firefly luciferase reagent (Promega, Madison, WI) to each well, and incubate for 1 h.

4. Measure luminescence (firefly luciferase reporter) on a Viewlux plate reader (Perkin Elmer Life Sci.).

5. After measuring firefly luciferase activity, add 25 μl *Renilla* luciferase reagent (Stop & Glow, Promega) to each well individually and determine luminescence immediately (within seconds, due to the short half-life of the *Renilla* luminescence) using a Victor2 plate reader (Perkin Elmer, Wellesley, MA) equipped with an injection system and interfaced in a Thermo-CRS stacker-based robotic system (Thermo-CRS Ltd., Burlington, Ontario, Canada).

Note. A glow-type *Renilla* luciferase substrate has become available (Dual Glow Luciferase Assay System, Promega, Madison, WI), allowing more flexibility in HTS logistics. For example, immediate plate reads are not necessary with the glow-type substrate.

RGSZ1 Single-Dose Dual Luciferase Assay—BenchTop Protocol

1. Thaw aliquots of frozen yeast strains RGSZ-FF [yBN70] and RGSZ-RL [yBN70L or yBN177], KChIP-FF [yBN77 or yBN178], and KChIP-RL [yBN77L or yBN179]. Dilute each yeast strain in SC-ura-leu-trp [for yBN70(L) and yBN77(L)] or in SC-leu-trp (for yBN177, yBN178 and yBN179) to a final OD_{600} of 0.01.

2. "Established assay": mix equal amounts of KChIP-FF [yBN77 or yBN178] and RGSZ-RL [yBN70L or yBN177] cells at OD_{600} of 0.01 and dispense into 96-well plates (100 μl per well). For the "switched reporter assay," mix equal amounts of RGSZ-FF [yBN70] and KChIP-RL [yBN77L or yBN179] cells at OD_{600} of 0.01 and dispense into 96-well plates.

3. Transfer 1 μl of each compound (dissolved in DMSO, 3 mM stock) to the yeast cell plates using a 96-pin replicator (Boekel Scientific, Feasterville, PA) to a final concentration of 30 μM.

4. Incubate compounds with cells for 24–48 h at room temperature (holes need to be poked in the plate sealers for aeration).

5. After incubation with compounds, add 100 μl dual luciferase assay substrate (Promega) to each well and incubate the plates for 1 h at room temperature.

6. Read firefly luciferase counts on a Topcount plate reader (Perkin Elmer, MA).

7. Finally, add 100 μl Stop&Glow (*Renilla* luciferase) substrate to each well and read luminescence.

RGSZ1 Dose–Response Luciferase Assay—Benchtop Protocol

1. Thaw aliquots of frozen yeast strains RGSZ-FF [yBN70] and RGSZ-RL [yBN70L or yBN177], KChIP-FF [yBN77 or yBN178], KChIP-RL [yBN77L or yBN179]. Dilute each yeast strain to a final OD_{600} of 0.01.

2. "Established assay": mix equal amounts of KChIP-FF [yBN77 or yBN178] and RGSZ-RL [yBN70L or yBN177] cells at OD_{600} of 0.01. For the "switched reporter assay," mix equal amounts of RGSZ-FF [yBN70] and KChIP-RL [yBN77L or yBN179] cells at OD_{600} of 0.01.

3. Add DMSO to mixed yeast strains (1% final DMSO) and add 100 μl mixed yeast strains to each well in rows B–H of a 96-well plate (each compound will be tested in both "established" and "switched" luciferase dose–response assays on separate 96-well plates).

4. For each compound to be tested, add 15 μl compound (dissolved in DMSO at a concentration of 30 mM) to 500 μl mixed yeast strains ("established" and "switched" reporter assay) and vortex.

5. Dispense 150 μl of the mixed yeast strains plus compound (in triplicate) to row A of the 96-well plate. A serial 1:3 dilution is made across row A through H by transferring 50 μl of the previous row to the next row and mixing.

6. Seal plate with plate sealer, and poke holes over each well (for aeration) using a sterile 26-gauge needle.

7. Incubate cells with compound at 30° for 24–48 h.

8. Following incubation of cells with compounds, add 100 μl dual luciferase assay substrate (Promega) to each well and incubate the plates for 1 h at room temperature.

9. Read firefly luciferase counts on a Topcount plate reader (Perkin Elmer, MA).

10. Finally, add 100 μl Stop&Glow (*Renilla* luciferase) substrate to each well and read luminescence.

RGSZ1 Screen Results

HTS Screening Results and HTS Data Analysis

The multiplexed dual luciferase YTH assay was used to screen approximately 360,000 chemically diverse and random compounds to screen for modulators of two independent protein–protein interactions. In our HTS

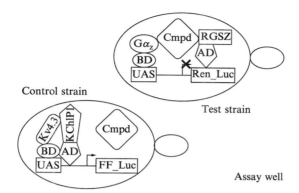

Control strain

Test strain

Assay well

FIG. 4. RGSZ1 yeast screen schematic for the interaction between constitutively active Gα$_z$(Q205L) and RGSZ1. BD and AD indicate the Gal4 DNA-binding domain and activation domain fusion proteins, respectively. Ren-Luc and FF-Luc represent yeast two-hybrid (YTH) *Renilla* luciferase and firefly luciferase reporter genes, respectively. "Cmpd" indicates compound being screened for its ability to disrupt binding between Gα$_z$ and RGSZ1. From Nieuwenhuijsen *et al.* (2003), copyright © 2003. Reprinted by permission of Sage Publications.

assay, the Gα$_z$(Q205L)-pGBKT7 and RGSZ1-pACT2 YTH interaction was linked functionally to increased *Renilla* luciferase luminescence ("test strain"), and the Kv4.3-pGBT9 and KChIP1-pACT2 YTH interaction was linked to increased firefly luciferase luminescence ("control strain") (Fig. 4). A SAS-based Excel program was developed in-house to analyze the two sets of luminescence data and has been described previously (Nieuwenhuijsen *et al.*, 2003). This program performs regression analysis on residual values of the *Renilla* luciferase [Gα$_z$(Q205L)-pGBKT7 and RGSZ1-pACT2 YTH interaction] and the firefly luciferase (Kv4.3-pGBT9 and KChIP1-pACT2 YTH interaction) data sets, respectively. Regression analyses enable detection of automation-induced mechanical bias, which may be caused by photomultiplier tube configurations or plate fill patterns, for example. Representative examples of regression analysis plots are shown in Fig. 5A–C. The residual value for each compound is then calculated by subtracting the plate mean from the bias-corrected compound value. Residual values for test and control strains are plotted (Fig. 5D), and the plot is then used to select compounds for further analysis. We selected approximately 3000 compounds for further testing based on residual values that fell within user-defined ranges of firefly luciferase residual and *Renilla* luciferase residual values. This set of compounds was reduced to 850

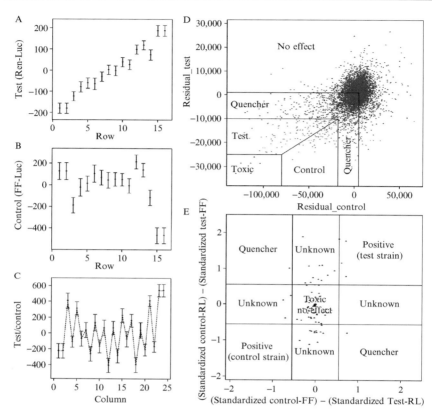

FIG. 5. Regression analysis of *Renilla* luciferase and firefly luciferase raw data from test and control strains identified automation-induced mechanical bias. Means and standard errors were calculated for test strain and control strain data and were plotted by plate row and column. Samples in the outer wells of the plate exhibited a substantially different pattern from other wells in the plate (A, B, and C). Analysis of test data by row (*Renilla* reporter) identified a drift in baseline (A). A typical up-and-down movement for both test and control strain was detectable in data analyzed by column (both test and control strain, C). A plot of residual values for both test and control strain (D) was used to select compounds for additional testing. Single-dose assay data were analyzed and plotted (E) by a custom SAS-based Excel program. This program grouped compounds into six categories, *i.e.*, positive (test strain), positive (control strain), no effect, toxic/nonspecific, quencher, or unknown based on preset cutoff values. Compound data from single-dose experiments were normalized and plotted. Compounds that were able to affect the test protein–protein interaction will have a substantially greater luminescence difference between test and control strain and were considered positive interacting compounds. From Nieuwenhuijsen *et al.* (2003), copyright © 2003. Reprinted by permission of Sage Publications.

compounds by applying Lipinski rules for drug-like characteristics (Lipinski et al., 1997) and other filter criteria.

RGSZ1 Single-Dose Small Molecule Assay Results

The set of 850 compounds was tested in a single-dose assay (30 μM compound) in two formats. In addition to the assay described earlier (test strain with *Renilla* luciferase and control strain with firefly luciferase reporter genes), the compounds were also tested in a YTH assay with switched reporter genes (test strain with firefly and control strain with *Renilla* luciferase reporter gene, respectively). This allowed us to identify compounds that have a direct (inhibiting) effect on luminescence (false positives/quenchers). An in-house SAS-Excel program was used to normalize compound data by the plate average and to group the compounds into six categories based on pre-set cutoff values: positive (test strain), positive (control strain), no effect, toxic/nonspecific, quencher, or unknown (unclear data); see plot in Fig. 5E. This single-dose compound assay verified 75 compounds for the $G\alpha_z$/RGSZ1 interaction and 65 compounds for the control strain interaction, corresponding to a combined confirmation rate of 16.5% (140/850). The verified compounds were prioritized further in parallel dose–response assays using the established and switched reporter gene formats described earlier. Dose–response data analysis identified 7 confirmed compounds (from 75) that were able to inhibit the interaction between $G\alpha_z$(Q205L) and RGSZ1.

Functional Confirmation of RGSZ1 Hits from YTH Luciferase Screen

The seven confirmed modulators of the RGSZ1/$G\alpha_z$(Q205L) interaction were then tested in a single turnover GTPase assay to test the ability of the compounds to inhibit the GAP activity of RGSZ1. Two compounds demonstrated functional inhibition of RGSZ1 GAP activity, which, in turn, impacted the GTPase activity of $G\alpha_z$ in a dose-dependent manner. One compound in particular (WAY-243) showed activity consistent with the inhibitory effects first shown in yeast. At increasing concentrations, WAY-243 was able to decrease GTP hydrolysis only in the presence of RGSZ1 (Fig. 6A). The IC_{50} of WAY-243 is estimated to be approximately 50 μM. In addition, we demonstrated that WAY-243 selectively inhibits RGSZ1 GAP activity on $G\alpha_z$ by testing the ability of this compound to inhibit RGS4-mediated GAP activity on $G\alpha_{i1}$ (Fig. 6B). Results using the single turnover GTPase assay demonstrate that WAY-243 is highly selective for RGSZ1 and has only a very small effect on RGS4 GAP function, with a predicted IC_{50} of >300 μM.

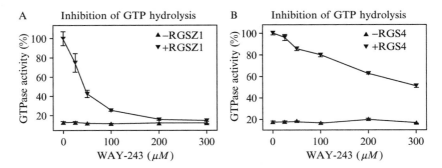

FIG. 6. Characterization of RGSZ1 compounds via single turnover GTPase analysis. WAY-243 inhibits RGSZ1-mediated acceleration of GTP hydrolysis by $G\alpha_z$ in a single turnover GTPase assay. An increasing amount of WAY-243 was incubated with either (A) 40 nM RGSZ1 and 15 nM $G\alpha_z$ at 4° and reactions were stopped after 10 min or (B) 50 nM RGS4 and 50 nM $G\alpha_{i1}$ at 0° and reactions were stopped after 30 s. Dose-dependent inhibition of RGS protein-accelerated GTP hydrolysis in the presence of WAY-243 is depicted as percentage GTPase activity [ratio of GTPase activity (cpm) with and without RGS protein]. Data shown are means ± SE of triplicate measurements. From Nieuwenhuijsen *et al.* (2003), copyright © 2003. Reprinted by permission of Sage Publications.

Summary

G-protein-coupled receptors regulate many biological responses and, therefore, are of primary interest as targets in the development of new therapeutics across disease indications. With increased sophistication in technology and chemistry, enhanced understanding of protein–protein interfaces, and innovative screening approaches, one should be able to manipulate G-protein signaling pathways at sites independent of the extracellular ligand-binding site. The combination of these disciplines can facilitate the development of novel therapeutics directed at intracellular targets, previously considered to be nondrugable, thus opening a new vantage point to alleviate disease states. This article described the use of two different formats using robust, yeast-based, automated high-throughput screening to identify modulators of RGS protein action. Although the screens were both directed to the RGS domain/$G\alpha$ interaction, these two screens do provide important examples of both functional and protein–protein interaction-based screen designs. Using these screening paradigms, we identified small molecule inhibitors for both RGS4 and RGSZ1, with estimated IC_{50} values in the low micromolar range. Moreover, these small molecules demonstrate selectivity among other RGS protein isoforms. These results add support to the idea of modulating RGS protein function with small molecule inhibitors, toward the modulation of G-protein signaling pathways *in vivo*, and have also provided tools toward the further investigation of RGS protein biology, as well as a basis for potential novel mechanisms in future therapeutic design.

Acknowledgments

The authors thank J. Cao for the generation of RGS4 initial screen design elements and characterization; D. Chen for DNA sequence support; J. Zhang for initial screen validation; J. Morin for helpful discussion on screen implementation; D. Lombardo, G. Kalgoankar, and L. Heydt for HTS analysis; Y. Huang for statistical analysis, W. Edris for protein purification; and S. O'Connor for helpful and thoughtful discussions.

References

An, F., Bowlby, M., Betty, M., Cao, J., Ling, H. P., Mendoza, G., Hinson, J. W., Mattsson, K. I., Strassle, B. W., Trimmer, J. S., and Rhodes, K. J. (2000). Modulation of A-type potassium channels by a family of calcium sensors. *Nature* **403,** 533–556.

Bernstein, L. S., Ramineni, S., Hague, C., Cladman, W., Chidiac, P., Levey, A. I., and Hepler, J. R. (2004). RGS2 binds directly and selectively to the M1 muscarinic acetylcholine receptor third intracellular look to modulate Gq/11α signaling. *J. Biol. Chem.* [Feb19 epub].

De Vries, L., and Gist Farquhar, M. (1999). RGS proteins: More than just GAPs for heterotrimeric G proteins. *Trends Cell Biol.* **9,** 138–144.

Dohlman, H. G., Song, J., Ma, D., Courchesne, W. E., and Thorner, J. (1996). Sst2, a negative regulator of pheromone signaling in the yeast *Saccharomyces cerevisiae*: Expression, localization, and genetic interaction and physical association with Gpa1 (the G-protein alpha subunit). *Mol. Cell. Biol.* **16,** 5194–5209.

Druey, K. M., Blumer, K. J., Kang, V. H., and Kehrl, J. H. (1996). Inhibition of G-protein mediated MAP kinase activation by a new mammalian gene family. *Nature* **379,** 742–746.

Gilman, A. G. (1987). G-proteins: Transducers of receptor-generated signals. *Annu. Rev. Biochem.* **56,** 615–649.

Grafstein-Dunn, E., Young, K. H., Cockett, M., and Khawaja, X. (2001). Regional distribution of regulators y G-protein signaling (RGS) 1, 2, 13, 14, 16 and Gaie messenger ribonucleic acids by *in situ* hybridization in rat brain. *Mol. Brain Res.* **88**(1–2), 113–123.

Harper, J. W., Adami, G. R., Wei, N., Keyomarko, K., and Elledge, S. (1993). The p21 Cdk-interacting protein Cip1 is a potent inhibitor of G1 cyclin-dependent kinases. *Cell* **75,** 805–816.

Hepler, J. R. (1999). Emerging roles for RGS proteins in cell signalling. *Trends Pharmacol. Sci.* **20,** 376–382.

Heximer, S. P., Srinivasa, S. P., Bernstein, L. S., Bernard, J. L., Linder, M. E., Hepler, J. R., and Blumer, K. J. (1999). G protein selectivity is a determinant of RGS2 function. *J. Biol. Chem.* **274,** 34253–34259.

Kehrl, J. H. (1998). Heterotrimeric G protein signaling: Roles in immune function and fine-tuning by RGS proteins. *Immunity* **8,** 1–10.

Koelle, M. R. (1997). A new family of G-protein regulators—The RGS proteins. *Curr. Opin. Cell Biol.* **9,** 143–147.

Koelle, M. R., and Horvitz, H. R. (1996). EGL-10 regulates G protein signaling in the *C. elegans* nervous system and shares a conserved domain with many mammalian proteins. *Cell* **84,** 115–125.

Lipinski, C. A., Lombardo, F., and Dominy, B. W. (1997). Experimental and computational approaches to estimate solubility and permeability in drug discovery. *Adv. Drug Deliv. Rev.* **23**, 3–25.

Moy, F. J., Haraki, K., Mobilio, D., Walker, G., Powers, R., Tabei, K., Tong, H., and Siegel, M. M. (2001). MS/NMR: A structure-based approach for discovering protein ligands and for drug design by coupling size exclusion chromatography, mass spectrometry, and nuclear magnetic resonance spectroscopy. *Anal. Chem.* **73**, 571–581.

Neubig, R. R. (2002). Regulators of G protein signaling (RGS proteins): Novel central nervous system drug targets. *J. Peptide Res.* **60**, 312–316.

Neubig, R. R., and Siderovski, D. P. (2002). Regulators of G-protein signalling as new central nervous system drug targets. *Nature Rev. Drug Disc.* **1**, 187–197.

Nieuwenhuijsen, B. W., Huang, Y., Wang, Y., Ramirez, F., Kalgaonkar, G., and Young, K. H. (2003). A dual luciferase multiplexed high-throughput screening platform for protein-protein interactions. *J. Biomol. Screen.* **8**, 676–684.

Ozenberger, B. A., and Young, K. H. (1995). Functional interaction of ligands and receptors of the hematopoietic superfamily in yeast. *Mol. Endocrinol.* **9**, 1321–1329.

Popov, S. G., Krishna, U. M., Falck, J. R., and Wilkie, T. M. (2000). Ca^{2+}/calmodulin reverses phosphatidylinositol 3,4,5-trisphosphate-dependent inhibition of regulators of G protein-signaling GTPase-activating protein activity. *J. Biol. Chem.* **275**, 18962–18968.

Price, L. A., Kajkowski, E. M., Hadcock, J. R., Ozenberger, B. A., and Pausch, M. H. (1995). Functional coupling of a mammalian somatostatin receptor to the yeast pheromone response pathway. *Mol. Cell. Biochem.* **11**, 6188–6195.

Rose, M. D., Winston, R., and Heiter, P. (1990). "Methods in Yeast Genetics." Cold Spring Harbor Laboratory Press, Cold Spring Harbor, NY.

Ross, E. M. (2002). Quantitative assays for GTPase-activating proteins. *Methods Enzymol.* **344**, 601–617.

Ross, E. M., and Wilkie, T. M. (2000). GTPase-activating proteins for heterotrimeric G proteins: Regulators of G protein signaling (RGS) and RGS-like proteins. *Annu. Rev. Biochem.* **69**, 795–827.

Sierra, D. A., Gilbert, D. J., Householder, D., Grishin, N. V., Yu, K., Ukidwe, P., Barker, S. A., He, W., Wensel, T. G., Otero, G., Brown, G., Copeland, N. G., Jenkins, N. A., and Wilkie, T. M. (2002). Evolution of the regulators of G-protein signaling multigene family in mouse and human. *Genomics* **79**, 177–185.

Sierra, D. A., Popov, S., and Wilkie, T. M. (2000). Regulators of G-protein signaling in receptor complexes. *Trends Cardiovasc. Med.* **10**, 263–268.

Tesmer, J. J., Sunahara, R. K., Gilman, A. G., and Sprang, S. R. (1997). Crystal structure of the catalytic domains of adenylyl cyclase in a complex with Gsalpha. GTPgammaS. *Science* **278**, 1907–1916.

Wang, J., Ducret, A., Tu, Y., Kozasa, T., Aebersold, R., and Ross, E. M. (1998). RGSZ1, a Gz-selective RGS protein in brain: Structure, membrane association, regulation by Galphaz phosphorylation, and relationship to a Gz GTPase-activating protein subfamily. *J. Biol. Chem.* **273**, 26014–26025.

Wang, Y., Ho, G., Zhang, J. J., Nieuwenhuijsen, B., Edris, W., Chanda, P. K., and Young, K. H. (2002). Regulator of G protein signaling Z1 (RGSZ1) interacts with Galpha i subunits and regulates Galpha i-mediated cell signaling. *J. Biol. Chem.* **277**, 48325–48332.

Wieland, T., and Chen, C. K. (1999). Regulators of G-protein signaling: A novel protein family involved in timely deactivation and desensitization of signaling via heterotrimeric G-proteins. *Naunyn. Schmiedebergs Arch. Pharmacol.* **360**, 14–26.

Zheng, B., De Vries, L., and Gist Farquhar, M. (1999). Divergence of RGS proteins: Evidence for the existence of six mammalian RGS subfamilies. *Trends Biochem. Sci.* **24,** 411–414.

Zeng, W., Xu, X., Popov, S., Mukhopadhyay, S., Chidiac, P., Swistok, J., Danho, W., Yagaloff, K. A., Fisher, S. L., Ross, E. M., Muallem, S., and Wilkie, T. M. (1998). The N-terminal domain of RGS4 confers receptor-selective inhibition of G protein signaling. *J. Biol. Chem.* **273,** 34687–34690.

Section II

G-Protein Regulators of Model Organisms

[18]   Genetic Analysis of RGS Protein Function in *Caenorhabditis elegans*

By DANIEL L. CHASE and MICHAEL R. KOELLE

Abstract

Caenorhabditis elegans has close homologs or orthologs of most mammalian (RGS) and G proteins, and mutants for all the RGS and G-protein genes of *C. elegans* have been generated. *C. elegans* RGS proteins can be matched to the specific $G\alpha$ proteins they regulate *in vivo* by comparing the defects in animals lacking or transgenically overexpressing an RGS protein with defects in a specific $G\alpha$ mutant. Transgenic expression of mutated RGS proteins or subdomains in *C. elegans* has also been used to carry out structure/function studies of RGS proteins. We propose that similar strategies can be used to understand the function of RGS proteins from other organisms by expressing them in *C. elegans*. This article describes general considerations regarding such experiments and provides detailed protocols for quantitatively measuring G-protein signaling phenotypes in *C. elegans*.

Introduction

The physiological functions of most mammalian RGS proteins remain poorly understood. A current dilemma concerns reconciling the fact that many RGS proteins act as GTPase-activating proteins (GAPs) for the same $G\alpha$ proteins *in vitro* with the expectation that different RGS proteins should have different functions *in vivo*. For example, biochemical analyses indicate that most RGS proteins function as GAPs for $G\alpha_{i/o}$ and $G\alpha_q$ class $G\alpha$ proteins, but not for $G\alpha_s$ or $G\alpha_{12/13}$. The results of such experiments, however, can be difficult to interpret, as the *in vitro* $G\alpha$ target specificity of certain RGS proteins can vary depending on the assay method used, or the inclusion of additional proteins (reviewed by Ross and Wilkie, 2000). In addition, certain RGS proteins are efficient *in vitro* GTPase activators for $G\alpha$ proteins with which they are apparently never coexpressed in nature. Such results raise doubts about the ability of *in vitro* experiments alone to assign physiologically relevant $G\alpha$ targets and functions to RGS proteins. In fact, with the exception of the RGS9–1 protein that regulates the visual $G\alpha_t$ protein in the retina (He *et al.*, 1998), the identities of the specific $G\alpha$ proteins targeted *in vivo* by most mammalian RGS proteins remain in

METHODS IN ENZYMOLOGY, VOL. 389

doubt. A further puzzle is that while all RGS proteins contain the ~120 amino acid RGS domain that functions as a Gα GTPase activation domain, many RGS proteins contain additional domains that could serve to regulate RGS activity and/or Gα selectivity, but the biological functions of such domains remain largely undefined (Burchett, 2000).

Genetic experiments have the potential to identify definitively the biological functions of RGS proteins, identify their physiologically relevant Gα targets, and clarify the functions of the various parts of multidomain RGS proteins. Such genetic experiments in *Caenorhabditis elegans* have identified the *in vivo* Gα targets and biological functions for several RGS proteins (Dong *et al.*, 2000; Hajdu-Cronin *et al.*, 1999; Koelle and Horvitz, 1996). Transgenic experiments in *C. elegans* have also allowed structure/function analysis of RGS proteins by expressing RGS protein subdomains and assessing their functions *in vivo* (Patikoglou and Koelle, 2002). Because mammalian proteins can easily be transgenically expressed in *C. elegans*, such methods could be applied to use *C. elegans* as a genetic assay system to examine the Gα target preferences of mammalian RGS proteins and to investigate the functions of their subdomains. This article describes the general considerations in using *C. elegans* to assess experimentally RGS function *in vivo*. Such experiments involve quantitatively measuring changes in animal behavior due to perturbations in G-protein signaling, and so we also provide detailed protocols for assessing G-protein signaling defects.

RGS and G Proteins Present in *Caenorhabditis elegans*

Caenorhabditis elegans has 12 RGS proteins, including members of four of the five RGS subfamilies identified by sequence analysis of the mammalian proteins (Sierra *et al.*, 2002). An example of the conservation between *C. elegans* and human RGS proteins is EGL-10 and its mammalian ortholog RGS7. EGL-10 and human RGS7 share the same multidomain structure, are up to 69% identical within their conserved domains, and both proteins are expressed specifically in the nervous systems of their respective organisms (Koelle and Horvitz, 1996). *C. elegans* also has 20 Gα proteins, including at least one ortholog of each of the four major classes of mammalian Gα proteins ($G\alpha_{i/o}$, $G\alpha_q$, $G\alpha_{12/13}$, and $G\alpha_s$) (Jansen *et al.*, 1999). As an example, mammalian $G\alpha_o$ is 80% identical to its *C. elegans* ortholog, GOA-1, and both proteins are also expressed specifically in the nervous systems of their respective organisms (Mendel *et al.*, 1995; Ségalat *et al.*, 1995).

This extraordinary conservation suggests that studies of *C. elegans* RGS and G proteins could provide insights into the functions of the corresponding mammalian proteins. For example, genetic studies showing that EGL-10 functions as a specific regulator of GOA-1 in *C. elegans* (Koelle and Horvitz, 1996) suggest that mammalian RGS7 protein could specifically target $G\alpha_o$. This conservation also suggests that mammalian RGS or G proteins could be transgenically expressed in *C. elegans* to complement mutations in the corresponding *C. elegans* orthologs, thus providing a genetic system that can be used to gain insight into $G\alpha$ target selectivity, as well as for performing structure/function studies of the mammalian RGS proteins.

Function of RGS and G Proteins in *C. elegans*

The genes encoding each $G\alpha$ protein in *C. elegans* have been disrupted by mutation and the resulting mutants have been examined for developmental and behavioral defects. Mutations in the *C. elegans* orthologs of mammalian $G\alpha_o$ and $G\alpha_q$ (GOA-1 and EGL-30, respectively) cause dramatic behavioral defects. Mutations in GOA-1 cause animals to move faster and to engage in egg-laying behavior more frequently than wild-type animals (Mendel *et al.*, 1995; Ségalat *et al.*, 1995). Mutations in EGL-30 cause the precise opposite behavioral defects as *goa-1* mutants: *egl-30* partial loss-of-function mutants move slower and engage in egg-laying behavior less frequently than wild-type animals (Brundage *et al.*, 1996). These results indicate that $G\alpha_o$ and $G\alpha_q$ function antagonistically in *C. elegans*. Null mutations in EGL-30 are lethal, as are mutations in the $G\alpha_s$ ortholog GSA-1 (Korswagen *et al.*, 1997). No defects have yet been identified for mutations in the *C. elegans* ortholog of the mammalian $G\alpha_{12}$ protein, GPA-12 (Sierra *et al.*, 2002).

Of the 12 *C. elegans* RGS proteins, mutants for four, EGL-10, RGS-1, RGS-2, and EAT-16, have been analyzed so far. In each case, the defects caused by mutations in these genes were used to identify their specific $G\alpha$ targets. Three of these RGS proteins (EGL-10, RGS-1, and RGS-2) function to inhibit signaling by the $G\alpha_o$ ortholog GOA-1. Loss of EGL-10 causes too much GOA-1 signaling, resulting in animals that move slowly and rarely engage in egg-laying behavior (Koelle and Horvitz,1996). This suggests that EGL-10 functions to constitutively inhibit GOA-1 signaling. RGS-1 and RGS-2 are redundant with each other, and a defect in signaling is seen only when both are mutated. This defect is relatively subtle: double-mutant animals are unable to increase properly their frequency of egg laying when fed after starvation (Dong *et al.*, 2000). Under this specific condition, it appears that RGS-1/RGS-2 activity is normally induced to inhibit GOA-1 signaling and alter egg-laying behavior. The fourth *C. elegans* RGS protein that has been analyzed, EAT-16, functions to inhibit the $G\alpha_q$ ortholog EGL-30 (Hajdu-Cronin *et al.*, 1999). Loss of EAT-16 causes too much EGL-30 signaling, resulting in animals that move more rapidly and engage in egg-laying behavior more frequently than wild-type animals. Interestingly, when EGL-10 and EAT-16 are both expressed in the same neurons, they cause opposite effects on behavior by specifically targeting two different $G\alpha$ proteins, providing perhaps the best demonstration to date of distinct *in vivo* $G\alpha$ target specificity by two different RGS proteins. Beyond the four RGS genes just described, the remaining eight *C. elegans* RGS genes have also been knocked out (Hess and Koelle, unpublished result). We expect that analysis of these mutants will reveal additional *in vivo* functions for RGS proteins.

The EGL-10 and EAT-16 proteins belong to the R7 subfamily of RGS proteins, whereas RGS-1 and RGS-2 belong to the RZ subfamily (Sierra *et al.*, 2002). Because the behavioral defects associated with mutation of these RGS proteins are already well defined and their $G\alpha$ protein targets are established (and highly conserved in mammals), they represent excellent candidates for transgenic complementation with R7 and RZ class RGS proteins of other organisms.

Transgenic Expression of RGS Proteins in *C. elegans* as an Experimental System for Studying RGS Function

Caenorhabditis elegans provides unique advantages for the study of RGS protein function. These include the ability to use genetic screens to identify novel components of G-protein signaling pathways, an approach that was used to originally help identify the family of RGS proteins and their role as G-protein inhibitors (Koelle and Horvitz, 1996). A second advantage is the ease with which reverse genetic experiments can be carried out in *C. elegans*. RNA interference (RNAi), originally discovered in *C. elegans*, provides a very rapid and convenient method to knock down gene function and has been used to analyze certain G-protein functions in *C. elegans* (Gotta and Ahringer, 2001). A limitation to this approach is that most G proteins in *C. elegans* are expressed in neurons, which are less sensitive to RNAi than other cell types (Timmons *et al.*, 2001). Another reverse genetic approach that overcomes this limitation involves random chemical mutagenesis followed by PCR screening to identify mutants in which a gene of interest has suffered a deletion mutation. This gene knockout technology, along with traditional genetic screens, has been used to generate mutants for every G protein and RGS protein gene of *C. elegans* (Dong *et al.*, 2000; Hess and Koelle, unpublished result; Jansen *et al.*, 1999).

While the approaches described previously (genetic screens, RNAi, and gene knockouts) are limited to analyzing function of endogenous *C. elegans*

RGS proteins, and are not addressed further here, one can also use genetic analysis in *C. elegans* to examine the function of RGS proteins from other organisms by expressing these RGS proteins in *C. elegans* using transgenes. Two types of experiment can be performed. First, one can look for phenotypes induced by overexpressing an RGS protein in wild-type *C. elegans*. This strategy can identify the Gα protein targeted by the RGS protein *in vivo*. For each of the four *C. elegans* RGS proteins that have been analyzed in detail, overexpression strongly and specifically inhibits the Gα protein normally regulated by that RGS protein, as overexpression of the RGS protein induces a phenotype similar to a null mutation for that Gα protein (Dong *et al.*, 2000; Hajdu-Cronin *et al.*, 1999; Koelle and Horvitz, 1996). Based on the high degree of conservation of Gα proteins in *C. elegans* and mammals, mammalian RGS proteins expressed in *C. elegans* might be expected to recognize the *C. elegans* ortholog of their physiological Gα target and thus cause behavioral defects similar to those of the corresponding *C. elegans* Gα protein mutants. Direct comparison of the behavioral defects caused by the overexpression of a particular RGS protein with those caused by Gα mutants will indicate which Gα protein is targeted by the RGS protein.

In a second transgenic approach, one can express an RGS transgene in a *C. elegans* mutant lacking an endogenous *C. elegans* RGS protein and look for functional complementation of the mutant phenotype. Mutations in a number of *C. elegans* genes have been complemented by transgenic expression of their mammalian orthologs (Hengartner and Horvitz, 1994; Levitan *et al.*, 1996), and we have shown that the mammalian RGS7 protein can functionally substitute for its *C. elegans* ortholog, EGL-10 (Patikoglou and Koelle, unpublished result). Once functional complementation has been demonstrated, deletions and mutations can be made in the transgene to identify regions of the RGS protein important for its function *in vivo*. We have used this strategy to analyze *in vivo* functions of various subdomains of EGL-10 and EAT-16, as well as chimeras between these two RGS proteins (Patikoglou and Koelle, 2002).

Both types of experiment outlined earlier require the expression of an RGS transgene in *C. elegans*. Because most *C. elegans* G proteins are found in neurons, one strategy has been to express RGS cDNAs using a promoter expressed in all *C. elegans* neurons. A general vector for expressing cDNAs in all neurons has been developed and used successfully to rescue the defects of the *egl-10* and *eat-16* RGS mutants with cDNAs from the corresponding genes (Patikoglou and Koelle, 2002). Generating transgenic animals is a straightforward and almost trivial exercise for a *C. elegans* researcher (Mello and Fire, 1995). However, it does require special expertise and equipment, including a microinjection apparatus, and we therefore suggest that researchers with no *C. elegans* experience generate transgenic strains with the help of an experienced worm lab. The phenotypic assays described later that can be used to assess RGS function in transgenic animals, however, require no special equipment other than a dissecting microscope.

The most common type of transgene used in *C. elegans* is produced by microinjecting plasmid DNAs into the gonads of adult animals. The injected DNA forms heritable, high-copy extrachromosomal arrays. These transgenes are lost at some frequency during cell division, so a marker plasmid (e.g., that expresses green fluorescent protein or that induces a visible phenotype) is coinjected with the plasmid of interest (Mello and Fire, 1995). The transgene arrays thus formed carry both the plasmids, and animals that retain the transgene array can be selected by picking animals that retain the marker phenotype. Such animals, however, are genetic mosaics that can randomly lose the transgene in certain cell lineages, leading to some animal-to-animal variability in the effects of the transgene. Thus the effects of a transgene are always assessed by assaying at least 10–30 animals per transgenic line and averaging the results. An additional variable is that each extrachromosomal array produced by a microinjection will carry a slightly different number of plasmids and thus produce different expression levels of the transgenes carried on the plasmids. To control for this effect, we typically generate at least five independent transgenic lines from each microinjection and average the results from all lines. Thus by analyzing at least 10 animals from each of at least five independent transgenic lines, the effects of variations due to mosaicism and copy number inherent in *C. elegans* extrachromosomal transgenes can be minimized. In some cases, it is desirable to highly overexpress a transgene. This can be done by increasing the concentration of the injected plasmid DNA, thus increasing the number of plasmids per array and the expression level of the transgene. To highly overexpress a transgene, one simply injects a plasmid at a high (50–100 ng/μl) concentration, while one might use a lower concentration ($\sim$10 ng/μl) to achieve more modest expression levels.

The following assays have been used to analyze the effects of RGS proteins on $G\alpha_o$ (GOA-1) and $G\alpha_q$ (EGL-30) signaling in *C. elegans*. While knockout mutations of $G\alpha_s$ (GSA-1) also have dramatic effects in *C. elegans*, it is not clear that $G\alpha_s$ is the target of RGS regulation, and we thus do not present assays of $G\alpha_s$ signaling defects. We also do not present assays for effects of RGS proteins on $G\alpha_{12/13}$ signaling, because at this point no defects have been observed in knockouts of the *C. elegans* $G\alpha_{12/13}$ homolog GPA-12. *C. elegans* has an additional family of $G\alpha$ proteins that are expressed in chemosensory neurons and that are nematode specific (i.e., they do not have close mammalian orthologs). Although these have not yet been shown to be regulated by RGS proteins, one could express

RGS proteins in *C. elegans* chemosensory neurons and assay for chemotaxis defects to investigate this possibility. For more information on *C. elegans* chemotaxis, we refer readers to Mori (1999).

Caenorhabditis elegans $G\alpha_o$ and $G\alpha_q$ are expressed in most or all neurons (Brundage *et al.*, 1996; Mendel *et al.*, 1995; Ségalat *et al.*, 1995). $G\alpha_o$ and $G\alpha_q$ mutants appear to have many behavioral defects, but the most obvious of these are defects in the rates of locomotion and egg-laying behaviors. Quantitative assays of these behaviors have been developed and used successfully in a number of laboratories to examine signaling by $G\alpha_o$ and $G\alpha_q$. The following sections present details of one locomotion assay and two egg-laying assays.

Quantitation of Locomotion Rate by Counting Body Bends per Minute

Overview

Caenorhabditis elegans moves forward by initiating body bends near the head that propagate in a sinusoidal fashion toward the tail. Animals occasionally reverse these sinusoidal waves to move backward. In the laboratory, *C. elegans* is typically cultured on agar plates seeded with a bacterial lawn that the animals use as food. Movement on agar plates is strongly affected by the presence and quality of a bacterial lawn, by temperature, and by mechanical stimulation (e.g., due to a tap to the plate). Thus all of these factors must be strictly controlled to obtain reproducible measurements of locomotion.

Animals with defects in G-protein signaling can show either of two opposite defects in locomotion. The GOA-1 null mutant has a hyperactive locomotion phenotype, which includes more frequent and deeper body bends and more frequent spontaneous reversals of direction. EGL-30 loss-of-function mutants have a sluggish locomotion phenotype, which includes infrequent, shallow body bends, and few reversals. RGS protein mutations or transgenic expression of RGS proteins can cause either the hyperactive or sluggish locomotion phenotype depending on which G protein is targeted by the RGS protein under study.

G-protein signaling effects on locomotion are typically quantitated by measuring the rate at which body bends are initiated near the head of the animal. This is preferred over, for example, measuring the actual distance animals travel per unit time. The latter is a complex function of the properties of many subbehaviors, including the rate that body bends are initiated, the depth, speed, and coordination with which the bends are propagated, and the frequency of reversals.

Locomotion assays are typically carried out on young adult animals. G-protein signaling affects locomotion in larvae as well as in adults; however, larvae go through a period of lethargic locomotion at each molt and using adult animals avoids this potential complication. In adults, G-protein signaling affects egg laying as well as locomotion. Perturbation of G-protein signaling can thus give rise to adults whose bodies are packed with unlaid eggs, which could indirectly affect locomotion behavior. To avoid this problem, assays are carried out on precisely staged adults that are 24 h past the late fourth larval (L4) stage. Such young adults have only recently begun producing fertilized eggs, and thus these animals can only carry a small number of unlaid eggs at the time of the assay regardless of any defects they may have in egg-laying behavior.

The locomotion assay measures the average number of body bends per minute. In general, quantitation of animal behaviors is challenging because most behaviors show considerable variability on short time scales. This is particularly true for C. elegans locomotion. Only by taking the average of many measurements over a long time scale can an accurate assessment of locomotion behavior for a particular strain be made. We typically measure body bends over a 3-min interval and repeat this measurement on 30 animals to obtain an average locomotion rate for a given strain. The standard deviation, a measure of the variation among the three minute measurements, remains high even as the number of measurements increases, reflecting the short time scale variability that exists in data. However, as the number of 3-min measurements increases, the average number obtained converges to the true mean rate for the strain. This can be seen by calculating the 95% confidence interval of the mean, which decreases as the number of 3-min measurements increases.

Typical results obtained from combining 30 three-minute measurements are shown in Table I for three strains: (1) the wild type; (2) a null mutant lacking the egl-10 RGS gene that functions to inhibit GOA-1; and (3) a strain carrying the vsIs1 transgene, which overexpresses the RGS protein RGS-1, which also acts as an inhibitor of GOA-1 signaling.

TABLE I
TYPICAL RESULTS FOR C. elegans LOCOMOTION ASSAYS

Genotype	Mean body bends/min	Standard deviation	95% confidence interval
Wild type	16.6	10.6	3.9
egl-10(md176)	6.3	4.2	1.6
vsIs1(rgs-1 overexpressor)	32.9	19.8	7.4

As can be seen in the example, the effects of RGS proteins on locomotion are twofold or greater and can be easily demonstrated to be highly significant using a *t* test.

Locomotion Assay Protocol

1. Culture animals on standard NGM agar plates seeded with OP50 bacteria at 20°. See Sulston and Hodgkin (1988) for a description of standard *C. elegans* culture media and methods.

2. One day prior to the assay, seed NGM plates with OP50 solution. Place the freshly seeded plates in a covered box at 37° overnight to allow a reproducibly thin lawn of OP50 to grow. Allow the plates to cool to room temperature before placing worms on them. To control conditions as carefully as possible, all assays that are compared to each other should be carried out on the same day using seeded plates prepared in a single batch.

3. For each strain to be assayed, pick 30 late L4 animals from an unstarved plate to the newly seeded plates, one worm per plate. At the late L4 stage, animals have relatively large, dark bodies in which the developing vulva appears in the dissecting microscope as a white crescent on the ventral surface of the animal (Fig. 1). Animals appear this way for only a few hours, and selecting such animals thus allows precise staging. Place the selected animals in a 20° incubator to age for 24 ± 1 h, at which time they will be young adults.

4. Assays are typically performed at room temperature (22°), and a room with steady temperature is required.

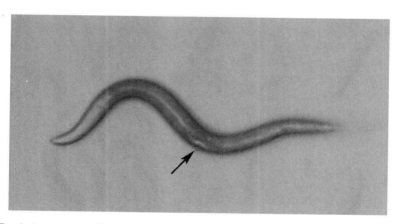

Fig. 1. Appearance of late-stage L4 larvae. Photograph of a late L4 larval stage animal as viewed through a Wild M420 dissecting microscope. Arrow indicates position of the developing vulva.

5. Gently place a plate containing a single staged animal on a dissecting microscope (so as not to stimulate the worm mechanically). Start a 3-min timer, and begin counting body bends. Every time the part of the worm just behind the pharynx reaches a maximum bend in the opposite direction from the bend last counted, advance the count by one. For example, if the worm moves forward, then reverses spontaneously, and the region just behind the pharynx initially bends during the reversal in the same direction it had just bent during the previous forward movement, do not advance the count.

Movement changes when animals reach the edge of the bacterial lawn. Do not use measurements if an animal remains at the edge of the lawn for more than 15 s during a 3-min measurement. In such a case, put the plate aside for a few minutes and come back to it for measurement when the worm is away from the lawn edge.

6. Repeat for 30 animals per strain.

7. Calculate average body bends per minute for each worm. Calculate an average, standard deviation, and 95% confidence interval for a strain using the values obtained for the 30 individual measurements.

Quantitative Assessment of Egg-Laying Behavior by Counting Unlaid Eggs

Overview

Caenorhabditis elegans produces fertilized eggs internally and retains them temporarily in the uterus, during which time they initiate development. Eggs are eventually laid through a behavior in which the animal contracts muscles to squeeze eggs out of the uterus through the vulva. Wild-type adult animals engage in an episode of egg-laying behavior about every 20 min, during which several eggs are laid (Waggoner *et al.*, 1998). The number of unlaid eggs present in the uterus at any given time is thus a function of the rates of egg production and of egg laying. Perturbations in G-protein signaling typically have strong effects on the frequency of egg-laying behavior and typically small effects on egg production, so that the steady-state number of unlaid eggs in the uterus can go up or down by manyfold. Thus measuring the number of unlaid eggs in the uterus provides an indirect measure of the frequency of egg-laying behavior. To be certain that defects in egg production are not responsible for changes in the accumulation of unlaid eggs, the brood size of any strain tested using this assay should be determined to ascertain that a near-normal number of fertilized eggs are produced.

Animals with defects in G-protein signaling can show either of two opposite defects in egg laying. The GOA-1 null mutant has a hyperactive egg-laying phenotype, in which animals engage in egg-laying behavior too frequently and thus accumulate very few unlaid eggs. Whereas wild-type adults have 11–15 unlaid eggs present in their uterus at any one time, strong hyperactive egg layers can have only 1 to 3 unlaid eggs in their uterus. EGL-30 loss-of-function mutants have the egg-laying defective (Egl) phenotype in which animals rarely engage in egg-laying behavior and thus accumulate many unlaid eggs. Strongly Egl animals may have >50 unlaid eggs. Because such a large number of unlaid eggs can accumulate, this assay is a particularly sensitive measure of the Egl phenotype. The other egg-laying assay described in this article, in which the stage of freshly laid eggs is quantitated, provides a more sensitive measure of the hyperactive egg-laying phenotype.

Unlaid eggs are typically counted by dissolving adult animals in a drop of bleach solution. Fertilized eggs are protected by a bleach-resistant egg shell, and bleach treatment thus leaves a pile of eggs that are counted easily using a dissecting microscope (Fig. 2). Carefully staged adult animals must be used in the assay, as the absolute number of unlaid eggs that accumulate in the uterus changes over time. Staged adult animals are obtained by

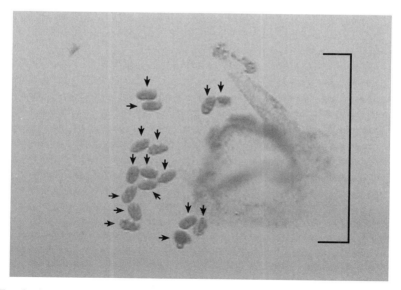

FIG. 2. Appearance of animal after a 10-min treatment with bleach solution. Bracket indicates position of partially dissolved carcass. Arrows indicate bleach-resistant eggs.

selecting animals at the late L4 stage and then aging them for typically 30 h prior to the assay.

The accumulation of unlaid eggs represents a time-averaged record of an animal's egg-laying behavior over a period of many hours. It thus smooths out any short time scale fluctuations in egg-laying rates and provides a readout of egg-laying behavior that is remarkably consistent between individual animals of the same genotype. Most other assays of behavior, such as the locomotion assay described earlier, show highly variable results reflecting the fact that most animal behaviors are highly variable over short time scales. This variability typically makes it necessary that many individual measurements be averaged to obtain means with acceptable levels of error. *C. elegans* egg-laying behavior is thus unusual relative to most animal behaviors for the ease and accuracy with which its frequency can be quantitated. This unique feature makes egg-laying behavior an especially convenient quantitative readout for effects on G-protein signaling. Egg-laying assays give results that are reproducible enough that data sets obtained on different days or by different researchers can be compared directly.

Unlaid Egg Assay Protocol

1. Culture animals on standard NGM agar plates seeded with OP50 bacteria at 20° (Sulston and Hodgkin, 1988).

2. One day prior to the assay, transfer 30 late L4 animals to a new seeded plate. Late L4 animals are recognized as described in the locomotion assay protocol. Place the animals at 20°.

3. Assays are typically carried out 30 h after late L4. Assays can also be carried out at 36 or 40 h after late L4, but all assays that will be compared with each other must be done using animals of the same age. For experiments involving strains that are strongly Egl, 30 h is preferred. At later times, some unlaid eggs in these strains can hatch inside of the uterus, and the hatched larvae can damage the mother's gonad, disrupting egg production and thus decreasing the ability of the mother to accumulate eggs.

4. Prepare a solution of 20% commercial bleach in water. Commercial bleach is a solution of 5 to 6.15% sodium hypochlorite.

5. Assays are carried out on the lid of a 96-well tissue culture dish. This is a flat piece of plastic with 96 imprinted circles. In each of 30 of the circles, place a 7-μl drop of the diluted bleach. Using a platinum worm pick (Sulston and Hodgkin, 1988) and working under a dissecting microscope, transfer one staged adult into each drop.

6. After about 10 min in the bleach solution, the adult animals will have dissolved, leaving piles of bleach-resistant fertilized eggs that can be

counted easily under a dissecting microscope (Fig. 2). There may be one or two "ghosts"—partly dissolved eggs that were fertilized immediately before bleach treatment and that had not yet become fully bleach resistant. Because these newly fertilized eggs dissolve more quickly in bleach than later-staged eggs, their presence is not always detected in this assay and thus for consistency we do not include "ghosts" in the egg count. Because all of the eggs will eventually begin to dissolve after about 25 min in bleach, complete the egg counts before this time.

7. Calculate an average, standard deviation, and 95% confidence interval for the sample of 30 measurements.

Quantitative Assessment of Egg-Laying Behavior by Measuring the Stage of Freshly Laid Eggs

Overview

A second assay of egg-laying behavior exists in which the developmental stages of freshly laid eggs are examined. This assay overcomes two weaknesses of the unlaid egg assay. First, while the unlaid egg assay detects the Egl phenotype very sensitively, it cannot easily detect weak hyperactive egg-laying phenotypes. Second, the unlaid egg assay cannot be used to measure egg-laying behavior in strains that have severe defects in egg production: such strains cannot accumulate eggs simply because they produce few eggs. Measuring the stage of freshly laid eggs can demonstrate the hyperactive egg-laying phenotype even in strains that produce few eggs and is thus mostly used to measure the hyperactive egg-laying phenotype.

The freshly laid egg assay measures the delay between the time a newly produced egg is fertilized and the time it is laid. During this delay, fertilized eggs undergo the first cell divisions of development. In wild-type animals, the delay is about 2.5 h, and almost all the eggs have reached about the 100-cell stage of development at the time they are laid. Animals with a strong hyperactive egg-laying phenotype engage in egg-laying behavior so frequently that their eggs are laid very soon after fertilization, so that almost all eggs are laid at the 1-cell stage (within ~35 min of fertilization), 2-cell stage (~35–50 min), or the 4- to 8-cell stages (~50–65 min). By counting the percentage of freshly laid eggs that are at the early (1- to 8-cell) stages, a quantitative measure of the hyperactive egg-laying phenotype is obtained.

Unlike the other assays described in this article, results obtained in this assay are relatively insensitive to the age of the adult animals used. We nevertheless routinely use precisely staged animals.

Freshly Laid Egg Assay Protocol

1. Culture animals on standard NGM agar plates seeded with OP50 bacteria at 20° (Sulston and Hodgkin, 1988).

2. One day prior to the assay, transfer about 30 late L4 animals to a new seeded plate. Late L4 animals are recognized as described in the locomotion assay protocol. Place the animals at 20°.

3. Assays are typically carried out 36 h after late L4, a time at which egg production is at its peak. Assays can also be carried out at 30 h, but all assays that will be compared with each other should be done using animals of the same age.

4. Using a platinum worm pick, transfer the staged animals to a new seeded plate, being careful not to transfer any laid eggs with the animals. A convenient method is to place all 30 animals on the same spot, and after they move away use the pick to remove any laid eggs that were transferred with the adults.

5. Set the plate at 20° for exactly 30 min to allow eggs to be laid. At the end of this period there should be 25–100 freshly laid eggs on the plate.

6. Stage the eggs under a high-magnification dissecting microscope. We use a Wild M420 microscope, which has a zoom objective and allows a total of 1100× magnification. It is important that the microscope have high-quality optics so that the individual cells of early-stage embryos can be discerned easily.

To stage the eggs, place the petri dish in an inverted dish lid on which parallel black lines have been drawn about every 2.5 mm. With the microscope set at a low magnification, use the pattern of lines to systematically scan across the plate from top to bottom to identify each laid egg exactly once. Examine each egg identified using higher magnification and categorize it as having 1, 2, 3–4, 5–8, or >8 cells. Identifying and categorizing all the eggs on a plate should take <5 min.

7. Repeat the assay with fresh worms until a sample of at least 100 eggs is counted.

8. An important control is to show that all strains being analyzed undergo the first few cell divisions of embryogenesis with approximately the same timing. This guarantees that differences between the strains reflect differences in egg-laying rates, not differences in embryogenesis. Do this by simply watching embryos undergo development from fertilization to the eight cell stage under a microscope. Place an adult animal in a drop of M9 buffer (Sulston and Hodgkin, 1988) on a glass coverslip. Use a 25-gauge hypodermic needle to slice the animal in half and place the coverslip (animal side down) on top of a flat agarose pad on a microscope slide. The animal will typically contain one or two freshly fertilized eggs, identifiable

as still containing two pronuclei, visible in a microscope equipped with Nomarski optics. Observe the cell divisions in a microscope, measure the times at which the first divisions occur, and compare to the wild type.

References

Brundage, L., Avery, L., Katz, A., Kim, U. J., Mendel, J. E., Sternberg, P. W., and Simon, M. I. (1996). Mutations in a *C. elegans* Gqα gene disrupt movement, egg laying, and viability. *Neuron* **16**, 999–1009.

Burchett, S. A. (2000). Regulators of G protein signaling: A bestiary of modular protein binding domains. *J. Neurochem.* **75**, 1335–1351.

Dong, M. Q., Chase, D., Patikoglou, G. A., and Koelle, M. R. (2000). Multiple RGS proteins alter neural G protein signaling to allow *C. elegans* to rapidly change behavior when fed. *Genes Dev.* **14**, 2003–2014.

Gotta, M., and Ahringer, J. (2001). Distinct roles for Gα and Gβγ in regulating spindle position and orientation in *Caenorhabditis elegans* embryos. *Nature Cell Biol.* **3**, 297–300.

Hajdu-Cronin, Y. M., Chen, W. J., Patikoglou, G. A., Koelle, M. R., and Sternberg, P. W. (1999). Antagonism between Goα and Gqα in *Caenorhabditis elegans*: The RGS protein EAT-16 is necessary for Goα signaling and regulates Gqα activity. *Genes Dev.* **13**, 1780–1793.

He, W., Cowan, C. W., and Wensel, T. G. (1998). RGS9, a GTPase accelerator for phototransduction. *Neuron* **20**, 95–102.

Hengartner, M. O., and Horvitz, H. R. (1994). *C. elegans* cell survival gene *ced-9* encodes a functional homolog of the mammalian proto-oncogene *bcl-2*. *Cell* **76**, 665–676.

Jansen, G., Thijssen, K. L., Werner, P., van der Horst, M., Hazendonk, E., and Plasterk, R. H. (1999). The complete family of genes encoding G proteins of *Caenorhabditis elegans*. *Nature Genet.* **21**, 414–419.

Koelle, M. R., and Horvitz, H. R. (1996). EGL-10 regulates G protein signaling in the *C. elegans* nervous system and shares a conserved domain with many mammalian proteins. *Cell* **84**, 115–125.

Korswagen, H. C., Park, J. H., Ohshima, Y., and Plasterk, R. H. (1997). An activating mutation in a *Caenorhabditis elegans* Gs protein induces neural degeneration. *Genes Dev.* **11**, 1493–1503.

Levitan, D., Doyle, T. G., Brousseau, D., Lee, M. K., Thinakaran, G., Slunt, H. H., Sisodia, S. S., and Greenwald, I. (1996). Assessment of normal and mutant human presenilin function in *Caenorhabditis elegans*. *Proc. Natl. Acad. Sci. USA* **93**, 14940–14944.

Mello, C., and Fire, A. (1995). DNA transformation. *In* "Methods in Cell Biology" (D. C., Shakes and H. F., Epstein, eds.), Vol. 48, pp. 451–482. Academic Press, San Diego.

Mendel, J. E., Korswagen, H. C., Liu, K. S., Hajdu-Cronin, Y. M., Simon, M. I., Plasterk, R. H., and Sternberg, P. W. (1995). Participation of the protein Go in multiple aspects of behavior in *C. elegans*. *Science* **267**, 1652–1655.

Mori, I. (1999). Genetics of chemotaxis and thermotaxis in the nematode *Caenorhabditis elegans*. *Annu. Rev. Genet.* **33**, 399–422.

Patikoglou, G. A., and Koelle, M. R. (2002). An N-terminal region of *Caenorhabditis elegans* RGS proteins EGL-10 and EAT-16 directs inhibition of Gαo versus Gαq signaling. *J. Biol. Chem.* **277**, 47004–47013.

Ross, E. M., and Wilkie, T. M. (2000). GTPase-activating proteins for heterotrimeric G proteins: Regulators of G protein signaling (RGS) and RGS-like proteins. *Annu. Rev. Biochem.* **69**, 795–827.

Ségalat, L., Elkes, D. A., and Kaplan, J. M. (1995). Modulation of serotonin-controlled behaviors by Go in *Caenorhabditis elegans*. *Science* **267**, 1648–1651.

Sierra, D. A., Gilbert, D. J., Householder, D., Grishin, N. V., Yu, K., Ukidwe, P., Barker, S. A., He, W., Wensel, T. G., Otero, G., Brown, G., Copeland, N. G., Jenkins, N. A., and Wilkie, T. M. (2002). Evolution of the regulators of G-protein signaling multigene family in mouse and human. *Genomics* **79**, 177–185.

Sulston, J., and Hodgkin, J., (1988). Methods *In* "The Nematode *Caenorhabditis elegans*" (Wood and the Community of *C. elegans* Researchers eds.), pp. 587–606. Cold Spring Harbor Laboratory, Cold Spring Harbor, NY.

Timmons, L., Court, D. L., and Fire, A. (2001). Ingestion of bacterially expressed dsRNAs can produce specific and potent genetic interference in *Caenorhabditis elegans*. *Gene* **263**, 103–112.

Waggoner, L. E., Zhou, G. T., Schafer, R. W., and Schafer, W. R. (1998). Control of alternative behavioral states by serotonin in *Caenorhabditis elegans*. *Neuron* **21**, 203–214.

[19] Purification and *In Vitro* Functional Analysis of the *Arabidopsis thaliana* Regulator of G-Protein Signaling-1

By Francis S. Willard and David P. Siderovski

Abstract

The model organism *Arabidopsis thaliana* contains a restricted set of heterotrimeric G-protein subunits, with only one canonical Gα subunit (AtGPA1), one Gβ subunit (AtAGB1), and two Gγ subunits (AtAGG1 and AtGG2) identified. We have identified a novel additional component of heterotrimeric G-protein signaling in the *A. thaliana* genome, regulator of G-protein signaling-1 (AtRGS1). This protein has the predicted topology and structure of a G-protein-coupled receptor in that it contains seven transmembrane domains, but AtRGS1 also contains a unique C-terminal extension, namely a regulator of G-protein signaling domain (RGS box). This article describes methods for the purification and *in vitro* functional analysis of the RGS box of AtRGS1.

Introduction

Yeast, plants, and metazoan organisms all use heterotrimeric G-protein signaling for signal processing and homeostasis. However, the model organism *Arabidopsis thaliana* (thale cress) contains an unusually restricted complement of G-protein signaling pathway components. *Arabidopsis* contains only one canonical Gα subunit (AtGPA1), one Gβ subunit (AtAGB1), and two Gγ subunits (AtAGG1 and AtGG2) (Jones, 2002). Enigmatically, neither a G-protein-coupled receptor (GPCR) nor a direct Gα or Gβγ effector has been described for *Arabidopsis* (reviewed in Jones, 2002). We have completed *in vitro* and *in vivo* functional analyses of the *Arabidopsis* regulator of G-protein signaling-1 [AtRGS1 (Chen *et al.*, 2003)]. AtRGS1 has predicted topological and structural similarity to GPCRs in that it contains seven transmembrane domains, but uniquely it contains a C-terminal regulator of G-protein signaling domain (or "RGS box"). This article describes methods for the identification, purification, and *in vitro* functional analysis of the RGS box of AtRGS1. AtGPA1 and the RGS box of AtRGS1 can be purified following overexpression in *Escherichia coli*. Complex formation between AtGPA1 and the AtRGS1 RGS box can be quantified using coprecipitation and surface plasmon resonance approaches. AtRGS1 RGS box-catalyzed GTPase-accelerating protein (GAP) activity on AtGPA1 can be measured in single turnover GAP assays using a fluorescent biosensor for inorganic phosphate. This article also describes fluorescence-based methods for using $2'$-(or-$3'$)-O-(N-methylanthraniloyl)guanosine $5'$-triphosphate (MANT-GTP) to measure AtRGS1-catalyzed GAP activity on AtGPA1.

Materials

Unless otherwise specified, all chemicals are of the highest purity obtainable from Sigma (St. Louis, MO) or Fisher Scientific (Pittsburgh, PA).

Bioinformatic Identification and Analysis of AtRGS1

A search for *A. thaliana* open reading frames (ORF) with similarity to an RGS box-hidden Markov model was performed using the Simple Modular Architecture Research Tool (SMART) web server (http://smart.embl-heidelberg.de) (Letunic *et al.*, 2002). This search revealed a ORF of 459 amino acids, hereafter called *AtRGS1*, that encodes an extended N-terminal region with seven predicted transmembrane domains and a C-terminal RGS box (Fig. 1A). A database search of *Arabidopsis* ORFs using the N- and C-terminal regions of AtRGS1 failed to yield homologs; thus, it appears that AtRGS1 represents the single member of this type of protein in the *Arabidopsis* genome.

The first 250 amino acids of AtRGS1 are predicted to span the membrane with a topology reminiscent of GPCRs, that is, an extracellular N terminus, seven transmembrane domains, and an intracellular C terminus (Fig. 1B). The N-terminal 250 residues of AtRGS1 were submitted to the SMART server for searches of homology to known protein classes and

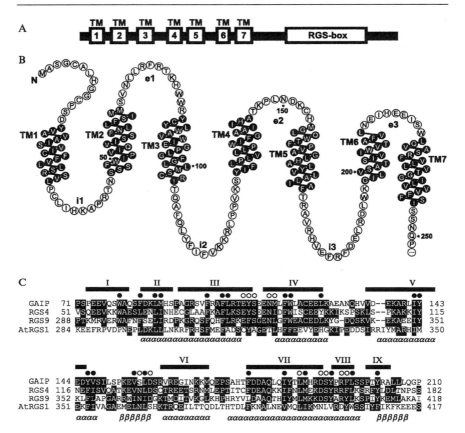

Fig. 1. Domain architecture of the AtRGS1 protein. (A) Schematic of AtRGS1 domain architecture. Positions of transmembrane (TM) and RGS box regions are annotated. (B) Diagram of the predicted overall topology of the AtRGS1 N-terminal region (amino acids 1 to 251) as generated by the residue-based diagram editor (RbDe) web server (Konvicka et al., 2000). (C) Multiple sequence alignment of the RGS box regions of human GAIP (RGS19), rat RGS4, bovine RGS9, and *Arabidopsis* AtRGS1 proteins. Regions of predicted α-helical and β-strand secondary structure within the AtRGS1 RGS box based on the PSI-Pred algorithm (McGuffin et al., 2000) are denoted underneath the AtRGS1 sequence. Conserved amino acids identified by ClustalW (Thompson et al., 1997) are boxed. Closed circles denote conserved residues forming the RGS box hydrophobic core, and open circles highlight conserved residues making direct contacts with Gα in the RGS4/Gα_{i1} crystal structure (Tesmer et al., 1997). The nine α helices observed within the NMR solution structure of human GAIP (de Alba et al., 1999) are numbered in roman numerals and overlined. Primary sequences in the alignment are human GAIP (SwissProt accession number P49795), rat RGS4 (P49799), bovine RGS9 (O46469), and AtRGS1 (GenPept NP_189238).

returned similarity to a Pfam model of family C GPCRs with an expect value (E) of 2.8×10^{-1} (Bateman *et al.*, 2002). Thus, the seven transmembrane (7TM) region of AtRGS1 has weak overall similarity to the metabotropic glutamate GPCR subfamily ("family C") that also includes odorant, pheromone, calcium-sensing, and γ-aminobutyric acid type B (GABA$_B$) receptors. However, the large, ligand-binding ectodomain found at the N termini of most members of this GPCR subfamily is clearly not present in the AtRGS1 ORF (Fig. 1B). The predicted topology of AtRGS1 (Fig. 1B) places cysteine residues at the entry to transmembrane domain 3 (TM3; Cys-84) and the second extracellular loop (e2; Cys-153), respectively, in similar positions to a common disulfide bridge found frequently in GPCRs (Gether, 2000).

The presence of an RGS box in the C terminus of AtRGS1 (GenPept accession number NP_189238) was verified by applying the 3D-PSSM protein-fold recognition algorithm (http://www.sbg.bio.ic.ac.uk/~3dpssm/) (Kelley *et al.*, 2000) to the final 211 amino acids of AtRGS1 (amino acids 249–459). The top three predictions for protein-fold matches were all RGS box structures: human GAIP/RGS19 (PDB code 1 cmz; E value 8.7×10^{-5}), rat RGS4 (1agr; E value 2.7×10^{-3}), and bovine RGS9 (1fqj; E value 2.7×10^{-3}). Overlap of predicted α-helical secondary structure regions, and conservation of critical core residues, between the AtRGS1 RGS box and these three mammalian RGS box proteins is detailed in Fig. 1C.

Purification of AtGPA1 and AtRGS1

AtGPA1 and AtRGS1 are purified following overexpression in *E. coli* as hexahistidine (His$_6$) and glutathione-*S*-transferase (GST) fusion proteins, respectively. Using conventional recombinant DNA methods, the ORF of AtGPA1 is force directionally cloned into the *EcoRI* and *XhoI* sites of pPROEXHTb (Invitrogen, Carlsbad, CA). pPROEXHT vectors provide fusion proteins with a N-terminal tobacco etch virus (TEV)-protease cleavable His$_6$ tag. Amino acids 249–459 encompassing the RGS box of AtRGS1 are cloned into the Gateway vector pDEST15 (Invitrogen) to provide a N-terminal GST fusion. The resultant vectors are transformed into a protease-deficient *E. coli* strain containing the λDE3 lysogen [BL21(DE3); Stratagene, La Jolla, CA] and plated onto LB agar containing 50 μg/ml carbenicillin to select for stable transformants.

A single colony is isolated and grown overnight (225 rpm, 37°) in 100 ml of LB medium supplemented with ampicillin (50 μg/ml). This overnight culture is used to inoculate four 1-liter volumes of LB medium (50 μg/ml ampicillin) in baffled 6-liter flasks, which are incubated at 37°

with constant shaking (225 rpm) until reaching a density of $OD_{600\ nm}$ = 0.6–1.0. Fusion protein expression is induced by the addition of 1 mM isopropyl-β-D-thiogalactopyranoside (IPTG, Promega, Madison, WI). In the case of His_6-AtGPA1, the cultures are then incubated at 27° for 6 h, whereas for GST-AtRGS1(249–459), the cultures are incubated at 20° for 9 h. Cells are pelleted by centrifugation at 3000g (4°, 30 min) and stored frozen at −20°.

Frozen cell pellets are thawed slowly on ice-water, brought up to a volume of 40 ml (with 50 mM Tris–HCl, pH 7.5; 100 mM NaCl), and 50 μg/ml phenylmethanesulfonyl fluoride (Sigma) is added. Cell suspensions are lysed by three passes through an Emulsiflex-C5 homogenizer (Avestin, Ottawa, Canada) at a pressure of 75–150 MPa. The bacterial cell lysate is clarified by centrifugation at 100,000g (4°, 30 min). Both His_6-AtGPA1 and GST-AtRGS1(249–459) proteins are purified from the 100,000g supernatant using a fast protein liquid chromatography system (Akta FPLC, Amersham Bioscience, Uppsala, Sweden). Automated chromatography systems provide significant advantages in terms of experimental convenience and data analysis, but the methods described herein would still be amenable to nonautomated chromatographic methods.

His_6-AtGPA1 is purified using metal affinity chromatography with 5 ml HiTrap chelating HP (Amersham) columns. Chelating columns are stored, charged, and maintained according to manufacturer's directions. A Ni^{2+}-loaded HiTrap chelating column is equilibrated with 5 column volumes (CV) of N1 buffer (50 mM Tris–HCl, pH 7.5; 100 mM NaCl; 5% (v/v) glycerol; 10 mM imidazole; 50 μM GDP; 5 mM $MgCl_2$; 30 μM $AlCl_3$; 20 mM NaF). The bacterial supernatant is then injected onto the column. Unbound protein is removed by washing with 5 CV of N1. Elution is performed with sequential 5 CV steps of 3, 30, 70, and 100% (v/v) N2 [50 mM Tris–HCl, pH 7.5; 100 mM NaCl; 5% (v/v) glycerol; 1 M imidazole; 50 μM GDP; 5 mM $MgCl_2$; 30 μM $AlCl_3$; 20 mM NaF]. Protein elution can be observed by UV-light detection (280 nm absorbance) and subsequent resolution of fractions by SDS–polyacrylamide gel electrophoresis (SDS–PAGE). The His_6-AtGPA1 protein elutes with 30% (v/v) N2. Note that the His_6-AtGPA1 protein is maintained in the AlF_4^--induced transition state through the first chromatographic step, as both N1 and N2 buffers contain 5 mM $MgCl_2$, 30 μM $AlCl_3$, and 20 mM NaF.

Fractions containing His_6-AtGPA1 are pooled and subjected to gel-exclusion chromatography on a HiPrep 26/60 Sephacryl S200 HR column (Amersham). The S200 column is equilibrated with 1.3 CV of S200 buffer [50 mM Tris–HCl, pH 7.5; 150 mM NaCl; 1 mM dithiothreital (DTT); 5% (v/v) glycerol; 50 μM GDP], and then pooled NiNTA fractions are injected onto the S200 column followed by isocratic elution with 1.7 CV of S200 buffer. Chromatogram peak fractions (as detected by 280 nm absorption) are then analyzed by SDS–PAGE. The S200 column is calibrated with molecular weight standards (Sigma; MW-GF-1000), allowing estimation of apparent elution molecular weights. Fractions containing monodisperse His_6-AtGPA1 are pooled, concentrated using a Centriprep YM-30 (Millipore, Bedford, MA), aliquoted, snap frozen on dry ice, and stored at $-80°$. The final purified protein is greater than 90% pure as determined by SDS–PAGE (e.g., Fig. 2A). Purification to homogeneity has also been achieved using sequential Ni^{2+}-NTA affinity chromatography, TEV protease cleavage, dialysis, anion-exchange chromatography, and gel-exclusion chromatography;

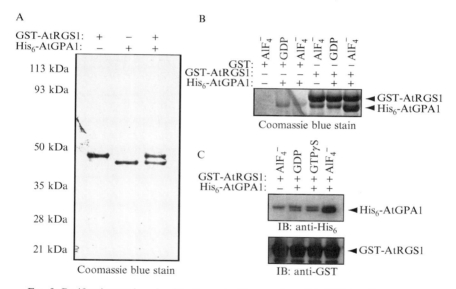

FIG. 2. Purification and nucleotide-dependent interaction of AtGPA1 and AtRGS1. (A) Purified His_6-AtGPA1 and GST-AtRGS1(249–459) protein samples were resolved by 12% (w/v) SDS–PAGE and Coomassie Blue staining. Poor electrophoretic separation between His_6-AtGPA1 and GST-AtRGS1(249–459) proteins is illustrated. (B) His_6-AtGPA1 and GST or GST-AtRGS1(249–459) proteins were incubated in GDP or GDP · AlF_4^- containing buffers and were then affinity precipitated with glutathione agarose. A nucleotide-dependent interaction between His_6-AtGPA1 and GST-AtRGS1(249–459) was observed upon separation of bound proteins by 12% (w/v) SDS–PAGE and Coomassie Blue staining. Positions of His_6-AtGPA1 and GST-AtRGS1 are indicated at right. (C) His_6-AtGPA1 and GST-AtRGS1(249–459) proteins were incubated in GDP, GTPγS, or GDP · AlF_4^- containing buffers and were then affinity precipitated with glutathione agarose. A nucleotide-dependent interaction between His_6-AtGPA1 and GST-AtRGS1(249–459) was observed upon separation of bound proteins by 12% (w/v) SDS–PAGE and immunoblot analysis (IB) with fusion specific antibodies as indicated.

however, this more laborious scheme of purification is unnecessary for most biochemical applications.

GST-AtRGS1(249–459) is purified by affinity chromatography using 5 ml HiTrap glutathione Sepharose 4 Fast Flow (GSTrap) columns (Amersham), which are stored and maintained according to the manufacturer's directions. A GSTrap column is equilibrated with 5 CV of buffer G1 [50 mM Tris–HCl, pH 7.5; 100 mM NaCl; 1 mM DTT; 5% (v/v) glycerol]. The bacterial supernatant is then injected onto the column. Unbound protein is removed by washing with 5 CV of G1. Protein is eluted isocratically with 6 CV of 100% (v/v) G2 buffer [50 mM Tris–HCl, pH 7.5; 100 mM NaCl; 1 mM DTT; 5% (v/v) glycerol; 10 mM glutathione]. The remainder of the purification scheme is identical to that detailed earlier for His$_6$-AtGPA1. Note that only a small fraction of GST-AtRGS1(249–459) expressed in $E.$ $coli$ is soluble. Accordingly, final yields of protein are low and, for this reason, a simple two-column approach is used. Therefore, preparations generally contain GST breakdown products. Purification to homogeneity using ion-exchange chromatography has not yet been attempted.

Proteins are quantitated by denaturation with 8 M guanidine HCl (Sigma) and absorbance measurements at 280 nm. The calculated molecular masses and extinction coefficients of His$_6$-AtGPA1 and GST-AtRGS1 (249–459) are 48.2 kDa, 47,440 $M^{-1}cm^{-1}$ and 52.2 kDa, 62,660 $M^{-1}cm^{-1}$, respectively. Figure 2A is an example of SDS–PAGE analysis of purified His$_6$-AtGPA1 and GST-AtRGS1(249–459).

Protein Interaction Assays

Coprecipitation Assay

The nucleotide-dependent interaction between GST-AtRGS1(249–459) and His$_6$-AtGPA1 can be determined using coprecipitation with glutathione agarose (known idiomatically as the "GST-pulldown" assay). RGS proteins have a characteristic high affinity toward $G\alpha \cdot GDP \cdot AlF_4^-$ but low affinity for either the GDP- or the GTPγS-bound forms of $G\alpha$. RGS proteins act catalytically by stabilizing the transition state for nucleotide hydrolysis (Berman et $al.$, 1996). The $G\alpha \cdot GDP \cdot AlF_4^- \cdot Mg^{2+}$ complex mimics the transition state for nucleotide hydrolysis and thus is bound most avidly by RGS proteins in $vitro$ (Druey and Kehrl, 1997; Popov et $al.$, 1997; Wieland et $al.$, 1997).

Coprecipitations are performed in the nucleotide-specific buffers P1 (50 mM Tris–HCl, pH 7.5; 50 mM NaCl; 2 mM DTT; 50 μM GDP), P2 (50 mM Tris–HCl, pH 7.5; 50 mM NaCl; 2 mM DTT; 50 μM GTPγS;

5 mM MgCl$_2$), and P3 (50 mM Tris–HCl, pH 7.5; 50 mM NaCl; 2 mM DTT; 50 μM GDP; 10 mM NaF; 30 μM AlCl$_3$; 5 mM MgCl$_2$). Forty micrograms of GST or GST-AtRGS1(249–459) is mixed with 80 μg His$_6$-AtGPA1 in the requisite buffer P1, P2, or P3 to a total volume of 950 μl. An excess of Gα subunit to RGS protein is used to facilitate the purification of a stoichiometric complex. Glutathione agarose beads (Sigma) are swollen in 50 mM Tris–HCl, pH 7.5, and washed three times in the same buffer. Fifty microliters of a 1:1 slurry is then added to coprecipitation samples, which are then incubated for 1 h at room temperature with gentle rocking.

Following incubation, samples are loaded onto empty micro Bio-Spin chromatography columns (Bio-Rad, Hercules, CA), which are placed in an empty P1000 pipette tip box. The immobilized beads are then washed by gravity flow with 5 × 1 ml of the appropriate buffer (P1, P2, or P3). Any residual buffer is removed by centrifugation at 14,000g (1 min at room temperature). Samples are then eluted from the beads with 50 μl Laemmli buffer and brief centrifugation (14,000g, 1 min at room temperature). Glutathione agarose beads can also be washed by centrifugation and aspiration of the supernatant (Zhang *et al.*, 1995), but we have found the use of microspin columns less time-consuming and consequently having a higher throughput. Twenty microliters of samples is resolved by 12% SDS–PAGE and complex formation is detected by Coomassie brilliant blue staining (e.g., Fig. 2B) or Western blot analysis (e.g., Fig. 2C).

Figure 2 illustrates the AlF$_4^-$ dependence of the AtGPA1/AtRGS1 interaction, indicative of *in vitro* RGS box activity. Note that GST-AtRGS1(249–459) and His$_6$-AtGPA1 proteins migrate in close proximity to each other on SDS–PAGE (Fig. 2A and B). Moreover, in some preparations, GST-AtRGS1 breakdown products run at the same mobility as His$_6$-AtGPA1 (Fig. 2B). For this reason, immunoblot analysis [using anti-GST (Sigma) and anti-His$_6$ (Covance, Berkeley, CA) antibodies] should be used to provide an unequivocal result (Fig. 2C). We use the His$_6$-Gα_{i1}/GST-RGS4 interaction as a positive control for these studies, as purification of both proteins is facile (Kimple *et al.*, 2001), and their AlF$_4^-$ selective interaction is well described (Berman *et al.*, 1996). The use of GST alone serves as an appropriate negative control to detect nonspecific binding (Fig. 2B).

Surface Plasmon Resonance

Surface plasmon resonance (SPR) experiments to test for nucleotide-dependent interactions between AtGPA1 and AtRGS1 are conducted with an immobilized "ligand" (in this case, GST fusion protein bound to an anti-GST antibody sensor surface) and binding is measured in resonance units (RU) as an interacting "analyte" (in this case, purified G-protein α subunit) is injected over the biosensor surface. To obtain reliable kinetic data, low levels of ligand (<200 RU) must be bound to the sensor chip to prevent rebinding events during the dissociation phase. However, to maximize the signal-to-noise ratio in more qualitative binding experiments, upward of 1000 RU of GST fusion protein is loaded onto the sensor chip.

Biosensor surfaces (flow cells 1–4 or "Fc1–Fc4") are created and analyzed for AtGPA1-binding affinity using the BIAcore 3000 (Biacore Inc., Piscataway, NJ). Carboxymethylated-dextran (CM5, Biacore) sensor chips with covalently bound goat anti-GST IgG antibody surfaces are created by amine coupling with N-hydroxysuccinimide and N-ethyl-N-(di-methylaminopropyl)carbodiimide according to the manufacturer's instructions (GST kit for fusion capture, Biacore). Antibody coupling is followed by injection of 1 M ethanolamine hydrochloride to inactivate any remaining N-hydroxysuccinimide groups.

Binding assays are conducted using filtered (0.22 μm) and degassed "Wittinghofer" running buffer [buffer W: 10 mM HEPES, pH 7.4; 150 mM NaCl; 5 mM MgCl$_2$; 0.005% (v/v) NP-40]. As described in the literature (Lenzen et $al.$, 1998), buffer W maximizes the longevity of covalently linked anti-GST antibodies. Buffer W is added to the intake lines and the pumps are flushed using the PRIME command. A new sensorgram is initiated using the RUN command, specifying an initial flow rate of 5 μl/min, multichannel detection (Fc1,2,3,4), and multichannel flow path (Fc1-Fc2-Fc3-Fc4). GST alone is loaded onto Fc1 and Fc2, up to a level of 1000 RU, using the FLOWPATH (Fc1-Fc2) command and the INJECT command. Similarly, GST-AtRGS1(249–459) is loaded onto Fc3 and Fc4 up to a level of 1000 RU. In practice we have found that 10-μl injections of 5 μM GST-fusion protein are sufficient to saturate binding sites on anti-GST CM5 chips. To obtain specific nucleotide conformations of AtGPA1, His$_6$-AtG-PA1 is locked into the (A) GDP, (B) GTPγS, or (C) GDP $\cdot$ AlF$_4^-$ bound states by incubation for 20 min at 25° in buffer W containing (A) 50 μM GDP, (B) 50 μM GTPγS, or (C) 50 μM GDP, 10 mM NaF, and 30 μM AlCl$_3$, respectively. Note that AtGPA1 has a fast spontaneous rate of nucleotide binding (unpublished observations) and thus 20 min at room temperature is sufficient for quantitative exchange; other Gα subunits with slower nucleotide-binding rates will require more extensive exchange times at higher temperatures. In addition, it is important that these dilutions be carried out in the background of buffer W (i.e., the same buffer components as the "running buffer" for the instrument, except for differing nucleotide components) in order to minimize "buffer shifts" (immediate swings in observed resonance units) caused by refractive index differences between protein solvent and running buffer that are seen upon analyte injections during interaction trials.

To analyze guanine nucleotide-dependent binding between AtGPA1 and AtRGS1, an anti-GST antibody-coated CM5 chip is loaded with 1000 RU GST (Fc1-Fc2) and 1000 RU GST-AtRGS1(249–459) (Fc3-Fc4). Thirty microliters of 5 μM His$_6$-AtGPA1 · GDP is injected over the biosensor surface using the KINJECT command at a sensor chip temperature of 25°. A 300-s dissociation phase of the reaction is specified in the KINJECT command. Regeneration of the surface is then conducted by sequential injection of 5 μl of 10 mM glycine, pH 2.2 (Biacore), and 5 μl of 0.05% (w/v) SDS at a flow rate of 5 μl/min. The regeneration process removes both captured GST fusion proteins and nonspecifically adsorbed protein. GST fusion protein surfaces are then reloaded and analysis is performed with GTPγS and GDP · AlF$_4^-$-ligated His$_6$-AtGPA1. Sensorgrams generated by the GST control surfaces are subtracted from GST-AtRGS1-derived sensorgrams using BIAevaluation software version 3.2 (Biacore) to remove any components of nonspecific binding or bulk buffer shifts (Kimple *et al.*, 2001; Snow *et al.*, 2002). Figure 3A illustrates AlF$_4^-$-dependent binding of AtGPA1 to AtRGS1 in that a clear saturable association phase and dissociation phase can be resolved, whereas any GDP-dependent interaction, if present, is below the SPR detection limit (consistent with coprecipitation data shown in Fig. 2).

It is good practice to repeat these SPR experiments by switching the ligand-flow cell pairing to account for any differences in anti-GST capture and nonspecific binding between flow cells. In practice, we have observed only minimal differences; nevertheless, this is still an important consideration in the design of surface plasmon resonance experiments. Quantitation of maximal GDP- and AlF$_4^-$-dependent specific binding of GST-AtRGS1(249–459) to 5 μM His$_6$-AtGPA1 on six independent flow cells (Fc1, Fc2, Fc3 ×2, Fc4 ×2), following subtraction of nonspecific binding to matching GST control flow cells (Fc3, Fc4, Fc1 ×2, Fc2 ×2), is illustrated in Fig. 3B. AlF$_4^-$-dependent binding is significantly greater in magnitude than GDP-dependent binding ($P < 1 \times 10^{-5}$, Student's *t* test). Apparent dissociation constants for these interactions can be determined using concentration and kinetic analyses as described elsewhere (Morton and Myszka, 1998; O'Shannessy *et al.*, 1994). As a positive control for measuring Gα/RGS-box interactions using SPR, we have previously employed the Gα_{i1}/GST-RGS4 interaction in which a reproducible, high-affinity interaction is observed consistently with Gα_{i1}-GDP · AlF$_4^-$ but neither Gα_{i1}-GDP nor Gα_{i1}-GTPγS subunits (Popov *et al.*, 1997). To control for Gα nucleotide selectivity, GST-GoLoco motif fusion proteins can be used; as described previously, GST-GoLoco motif fusion proteins exhibit high-affinity

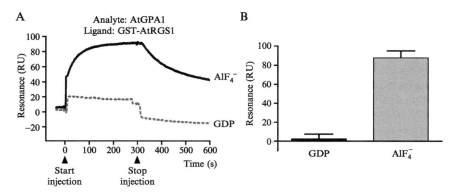

FIG. 3. Nucleotide-dependent interaction of AtGPA1 and AtRGS1 measured by surface plasmon resonance. (A) Independent flow cells of a CM5 sensor chip, derivatized with covalently coupled anti-GST antibody, were loaded separately with GST-AtRGS1(249–459) and GST. AtGPA1 · GDP (5 μM) or AtGPA1 · GDP · AlF$_4^-$ (5 μM) was injected over the surface for 300 s at a flow rate of 5 μl/min, followed by a 300-s dissociation phase in running buffer. Sensorgrams from negative control GST sensor surfaces were subtracted from GST-AtRGS1 sensorgrams to remove any nonspecific binding or bulk buffer shift components. (B) Quantitative analysis of maximal GDP and GDP · AlF$_4^-$ specific binding of GST-AtRGS1 to 5 μM AtGPA1 on six independent flow cells following subtraction of nonspecific binding to paired GST flow cells. AlF$_4^-$-dependent binding is significantly greater in magnitude than GDP-dependent binding ($P < 1 \times 10^{-5}$, Student's t test).

interactions solely with the GDP-liganded form of G$\alpha_{i/o}$ subunits in SPR assays (Colombo *et al.*, 2003; Kimple *et al.*, 2001).

GAP Assays

Single Turnover GAP Assays using E. coli *Phosphate-Binding Protein*

RGS boxes accelerate the GTPase activity of their cognate Gα partners. Standard assays of RGS protein-mediated GAP activity rely on the release of ^{32}P-labeled inorganic phosphate (P$_i$) from [γ-^{32}P]GTP-bound G-protein α subunits (Krumins and Gilman, 2002; Ross, 2002). Generally, this assay of GAP activity is conducted in a single turnover format, as nucleotide release from Gα, rather than GTP hydrolysis, is the rate-limiting step in the G-protein cycle in the absence of guanine nucleotide exchange factors (such as GPCRs) (Ferguson *et al.*, 1986; Higashijima *et al.*, 1987). Activated charcoal is used to separate ^{32}P$_i$ from [^{32}P]GTP, and subsequent quantitation of ^{32}P$_i$ is performed by liquid scintillation spectrometry.

We have instead developed a nonradioactive means of measuring RGS box-mediated GAP activity. The ability to measure inorganic phosphate production in real time has been described previously using the *E. coli* phosphate-binding protein (PBP) (Brune *et al.*, 1994). Recombinant PBP with a mutagenically generated cysteine (A197C) allows selective *in vitro* labeling of PBP with the fluorophore *N*-2[2-(1-maleimidyl)ethyl]-7-(diethylamino)coumarin-3-carboxamide (MDCC). Upon binding P_i, MDCC-PBP undergoes an 8-fold increase in fluorescence quantum yield. MDCC-PBP is an efficacious inorganic phosphate biosensor, as P_i binding is fast, of high affinity, and of subnanomolar sensitivity (Brune *et al.*, 1998). A 13-fold increase in fluorescence emission upon P_i binding to MDCC-PBP (Brune *et al.*, 1998) facilitates a highly sensitive, real-time assay of RGS box-catalyzed GTP hydrolysis (Kimple *et al.*, 2003). Expression, purification, and MDCC labeling of the *E. coli* PBP protein are part of an elaborate, multistep procedure described in detail by Brune *et al.* (1994, 1998).

Prior to conducting PBP-based GAP assays, GTP-loaded His_6-AtGPA1 is prepared by incubating His_6-AtGPA1 in 10 mM Tris–HCl, pH 8.0, 5 mM GTP, 2 mM EDTA (4°, 30 min). His_6-AtGPA1 is then purified from unbound GTP by rapid gel filtration using Sephadex G-25 columns (5 ml HiTrap desalting; Amersham) and is then quantitated using absorbance measurements at 280 nm (as described earlier). Measurements of MDCC-PBP fluorescence are made in a LS55 luminescence spectrometer (Perkin Elmer, Boston, MA) equipped with a four cell changer. This allows measurements to be made on four cuvettes, essentially simultaneously. The LS55 cuvette holder is water jacketed and thus can be connected to a circulating water bath to provide controlled temperature between 4 and 40°. MDCC fluorescence is excited at 425 nm and emission is recorded at 465 nm, with excitation and emission slit widths of 5 nm.

Two-milliliter quartz cuvettes (Fisher) with stirring bars (Fisher) are loaded with 1 ml buffer (10 mM Tris–HCl, pH 8.0; 1 mM EDTA) containing 10 μM MDCC-PBP. The magnetic stirrer is set to high for the duration of the experiment. Cuvettes are then equilibrated in the spectrometer at 30° for 5 min. Continuous fluorescence measurements are then initiated, and GTP-loaded His_6-AtGPA1 (~500 nM final concentration), GST-AtRGS1(249–459) (variable concentrations), and control diluent samples are added to cuvettes. Readings are allowed to stabilize for 100 s and then GTP hydrolysis reaction is initiated by the addition of 5 mM $MgCl_2$ (final concentration) to all cuvettes. The magnesium ion is a crucial cofactor for nucleotide hydrolysis by heterotrimeric G proteins (Sunyer *et al.*, 1984), hence His_6-AtGPA1 is loaded with GTP in the absence of Mg^{2+} (see earlier discussion).

The use of MDCC-PBP to measure, in real time, the GTP hydrolysis activity of AtGPA1 in the presence and absence of RGS box acceleration is illustrated in Fig. 4A. Substoichiometric amounts of GST-AtRGS1

(249–459) accelerate the intrinsic GTPase activity of AtGPA1 significantly. It is important to note that even at 30°, AtGPA1 is a comparatively slow GTPase (unpublished observations) in comparison with its mammalian orthologs. Following background subtraction, it can be observed that a molar ratio of 1:3.2 of GST-AtRGS1(249–459) to His$_6$-AtGPA1 accelerates the apparent initial rate of phosphate production by His$_6$-AtGPA1 more than 10-fold [$k_{obs(AtGPA1)} = 1.22 \times 10^{-4}$ s^{-1} versus $k_{obs(AtGPA1 + AtRGS1)} = 1.25 \times 10^{-3}$ s^{-1}]. The data analysis procedures used to generate rate constants are described later.

As MDCC-PBP is sensitive to subnanomolar levels of P$_i$, it is essential to have all solutions and cuvettes free of contaminating P$_i$, as this will, by definition, decrease the sensitivity of the assay. We have found that proteins purified via multiple chromatographic steps, using phosphate-free buffers, appear to function well with the MDCC-PBP biosensor. It is essential to have the appropriate controls to account for background phosphate levels unique to different protein preparations. As illustrated in Fig. 4A, a cuvette containing only buffer and MgCl$_2$ exhibits increasing fluorescence over time. This may be due in part to contaminating

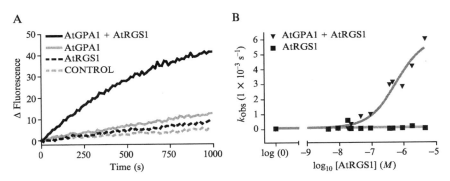

FIG. 4. Measurement of GTPase-accelerating protein (GAP) activity of AtRGS1 on AtGPA1 using fluorescent phosphate-binding protein (MDCC-PBP). (A) Single turnover GTPase assays using GTP-loaded His$_6$-AtGPA1 (575 nM) in the absence or presence of 182 nM GST-AtRGS1(249–459) were initiated with the addition of 5 mM MgCl$_2$ at 0 s. Inorganic phosphate release was measured in real time using coumarin-derivatized fluorescent phosphate-binding protein MDCC-PBP and is expressed as change in fluorescence units over time. Basal level of protein-independent inorganic phosphate production is also denoted (CONTROL). (B) Dose-dependent AtRGS1 GAP activity. Single turnover GTPase assays were performed as in A using 500 nM His$_6$-AtGPA1 and various concentrations of GST-AtRGS1(249–459) (0 to 4.6 μM). The apparent initial rate of phosphate production (k_{obs} in s^{-1}), calculated from changes in fluorescence, is plotted against the GST-AtRGS1(249–459) protein concentration.

phosphates, but is more likely due to the result of nonspecific fluorescence changes over time. This is potentially due to temperature-related effects on the stability of the MDCC-PBP sensor, as these GAP assays are performed at 30 or 37° to accommodate the slow GTPase kinetics of AtGPA1. Accordingly, experiments are conducted simultaneously with four buffer-filled cuvettes containing (a) no additional components ("CONTROL"), (b) AtGPA1, (c) AtRGS1, and (d) AtGPA1 + AtRGS1. Initial rates of fluorescence increases from the four cuvettes, as measured after the addition of $MgCl_2$, are fit to a single exponential function using Graph Pad Prism version 4.0 (Graph Pad Software, San Diego, CA). Background subtraction is calculated using the apparent rate constants (k) and the following equations to yield observed rate constants (k_{obs}):

Initial rate of P_i production by AtGPA1 alone:
(1) $k_{obs(AtGPA1)} = k_{AtGPA1} - k_{CONTROL}$
Initial rate of P_i production by AtRGS1 alone:
(2) $k_{obs(AtRGS1)} = k_{AtRGS1} - k_{CONTROL}$
Initial rate of P_i production by AtGPA1 in the presence of AtRGS1:
(3) $k_{obs(AtGPA1+AtRGS1)} = k_{AtGPA1+AtRGS1} - k_{AtRGS1}$

Thus, a dose–response relationship for AtRGS1-mediated GAP activity can be obtained by plotting $k_{obs(AtRGS1)}$ and $k_{obs(AtGPA1+AtRGS1)}$ versus input concentration of GST-AtRGS1(249–459) protein (0–4.6 μM; Fig. 4B).

GAP Assays Using Fluorescent Guanine Nucleotides

AtRGS1 acceleration of AtGPA1 GTPase activity can also be determined using 2'-(or-3')-*O*-(*N*-methylanthraniloyl)guanosine 5'-triphosphate (MANT-GTP; Molecular Probes, Eugene, OR). The MANT fluorophore is conjugated to guanine nucleotides via the 2' or 3' hydroxyl of the ribose moiety (Hiratsuka, 1983). MANT-conjugated guanine nucleotide analogs are available commercially from Molecular Probes or can be synthesized according to published protocols (Hiratsuka, 1983; Remmers *et al.*, 1994). The use of MANT-guanine nucleotides to study heterotrimeric G-protein signaling *per se* has been described elsewhere (McEwen *et al.*, 2002). When excited directly ($\lambda_{ex} = 360$ nm), MANT-GTP analogs show a weak increase in fluorescence upon binding Gα subunits ($\lambda_{em} = 440$ nm). However, if tryptophan residues in the Gα protein are excited ($\lambda_{ex} = 280$ nm), bound MANT-GTP undergoes a substantial fluorescence increase given fluorescence resonance energy transfer (FRET) from excited tryptophan residues to the MANT recipient ($\lambda_{em} = 440$ nm) (Lan *et al.*, 1998). The fluorescence enhancement provided by FRET is primarily due to (a) MANT-nucleotide excitation can only occur when a Gα–MANT-nucleotide complex is formed and (b) the separation of partially overlapping excitation and emission spectra providing lower background fluorescence. The advantageous property of MANT-GTP is that it has a higher quantum yield when bound to Gα than Gα-MANT-GDP, or MANT-GTP in solution (Fig. 5A) (Remmers *et al.*, 1994), thus it is a useful probe of the G-protein activation state and, consequently, GTPase activity.

Analysis of MANT–Nucleotide Interaction with AtGPA1. The relative fluorescence profiles of MANT-GDP and MANT-GTP binding to AtGPA1 can be determined by both FRET and direct excitation. Measurements of MANT fluorescence are conducted in an LS55 luminescence spectrometer as generally described in the previous section. MANT-nucleotide fluorescence is excited at 280 nm for FRET, or directly at 360 nm, and emission is recorded at 440 nm (with excitation and emission slit widths of 2.5 nm). Cuvettes are loaded with 1 ml buffer M (10 mM Tris–HCl, pH 8.0; 1 mM EDTA; 5 mM MgCl$_2$) In some instances, NaCl (100–300 mM) may be required for protein stability over extended time periods. Cuvettes containing buffer M are equilibrated at 37° for 5 min. Continuous fluorescence measurements are initiated on the spectrometer and, after 200 s of baseline recording, His$_6$-AtGPA1 (200 nM) or a vehicle control is added, as denoted in Fig. 5A and B. At 400 s, MANT-GDP or MANT-GTP (100 nM) is added to the cuvettes, as denoted with arrows in Fig. 5A and B. At approximately 3500 to 4000 s, 1 μM GTPγS is added to displace competitively MANT-nucleotide binding to AtGPA1. Using direct excitation (Fig. 5A), substantial fluorescence increases are observed upon the addition of MANT-nucleotides to cuvette buffer alone. Noticeably, in the presence of AtGPA1, the addition of MANT-GTP leads to a biphasic fluorescence profile, consistent with rapid nucleotide binding by AtGPA1 and subsequent slow hydrolysis of bound nucleotide to MANT-GDP (Fig. 5A). This is seen even more dramatically using FRET (Fig. 5B; compare ordinate scale with that of Fig. 5A), as only the combination of AtGPA1 and MANT-GTP provides a significant fluorescence increase above baseline, and this exhibits a biphasic response consistent with nucleotide binding and hydrolysis. These results clearly illustrate the utility of MANT-GTP as a real-time sensor for AtGPA1 GTPase activity. Furthermore, the superior signal-to-noise ratio of FRET over direct MANT excitation is clearly evident.

Use of MANT-GTP to Measure AtRGS1 GAP Activity. For convenience, these GTPase assays are performed using a molar excess of AtGPA1 to input MANT-GTP. This facilitates rapid loading of AtGPA1 and a pseudo-single turnover assay of GTPase activity. This assay is also

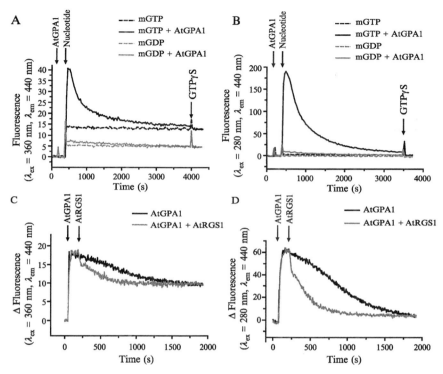

FIG. 5. Use of MANT-GTP to measure intrinsic (A, B), and AtRGS1-accelerated (C,D) GTPase activity of AtGPA1. (A) Direct excitation (λ_{ex} = 360 nm, λ_{em} = 440 nm) measurements of MANT-GDP and MANT-GTP binding to AtGPA1. The MANT-nucleotide (100 nM) was added to cuvettes in the presence or absence of 200 nM His$_6$-AtGPA1. At 4000 s, GTPγS (1 μM) was added to displace bound MANT-nucleotides. (B) Fluorescence resonance energy transfer (λ_{ex} = 280 nm, λ_{em} = 440 nm) measurements of MANT-GDP and MANT-GTP binding to His$_6$-AtGPA1. At 3500 s, GTPγS (1 μM) was added to displace bound MANT-nucleotides. (C) Direct excitation (λ_{ex} = 360 nm, λ_{em} = 440 nm) measurements of AtRGS1-catalyzed GAP activity on AtGPA1. At 0 s, 200 nM His$_6$-AtGPA1 was added to 100 nM MANT-GTP. At the peak of fluorescence (approximately 200 s), 200 nM GST-AtRGS1(249–459) (denoted "AtGPA1 + AtRGS1") or GST (denoted "AtGPA1") was added. (D) Fluorescence resonance energy transfer (λ_{ex} = 280 nm, λ_{em} = 440 nm) measurements of AtRGS1-catalyzed GAP activity on AtGPA1. The experiment was conducted in an identical fashion to that denoted for C. Note the greater separation in signal between intrinsic and AtRGS1-accelerated GTPase activity (diminution of fluorescence) using FRET between the AtGPA1 tryptophan residue(s) and the nucleotide MANT group (D) versus direct MANT excitation (C).

amenable to the single turnover protocol as described previously for PBP assays. To measure AtRGS1-mediated acceleration of AtGPA1 GTPase activity, cuvettes are first loaded with 1 ml buffer M and 100 nM MANT-GTP. His$_6$-AtGPA1 (200 nM) is added to the cuvette and, at the peak of fluorescence (approximately 230 s), 200 nM GST-AtRGS1(249–459) or a 200 nM GST control is added to cuvettes. Figure 5C and D illustrate the significant acceleration of AtGPA1 GTP hydrolysis kinetics provided by equimolar AtRGS1 (approximate $t_{1/2}$ values: 670 s [AtGPA1], 170 s [AtGPA1 + AtRGS1]). Furthermore, the FRET-based assay (Fig. 5D) is clearly more sensitive than direct nucleotide excitation (Fig. 5D). Fluorescent nucleotide analogs have been developed as probes of G-protein α subunit activation, given their convenience and their applicability to real-time measurement (Kimple et al., 2003). The presented method here illustrates this capability uniquely. To our knowledge, this is the first report applying fluorescent nucleotides to the measurement of RGS protein GAP activity. We have investigated the generality of this method for RGS box/ Gα pairs by measuring RGS4-catalyzed acceleration of Gα_o GTPase activity. We observed only a minor acceleration of Gα_o-MANT-GTP by RGS4 (unpublished observations), suggesting that protein complex-specific steric factors may negate the universal application of this assay of RGS protein function. Notwithstanding this, the presented MANT-GTP-based assay is the most expedient method for measuring AtRGS1 GAP activity that we have encountered.

Acknowledgments

We thank Dr. Martin R. Webb (National Institute of Medical Research, Mill Hill, UK) for provision of the PBP expression construct and Drs. Adam Shutes (UNC) and Krister Wennerburg (UNC) for assistance with the production of MDCC-PBP. Thanks also to Dr. Alan Jones (UNC) for the construct pDEST15-AtRGS1(249–459) and the UNC Department of Pharmacology Protein Core for provision of instrumentation. FSW is an American Heart Association Postdoctoral Fellow. This work was funded by NIH Grants R01 GM062338 and P01 GM065533. DPS is a recipient of the Burroughs Wellcome Fund New Investigator Award in the Pharmacological Sciences.

References

Bateman, A., Birney, E., Cerruti, L., Durbin, R., Etwiller, L., Eddy, S. R., Griffiths-Jones, S., Howe, K. L., Marshall, M., and Sonnhammer, E. L. (2002). The Pfam protein families database. Nucleic Acids Res. 30, 276–280.
Berman, D. M., Kozasa, T., and Gilman, A. G. (1996). The GTPase-activating protein RGS4 stabilizes the transition state for nucleotide hydrolysis. J. Biol. Chem. 271, 27209–27212.

Brune, M., Hunter, J. L., Corrie, J. E., and Webb, M. R. (1994). Direct, real-time measurement of rapid inorganic phosphate release using a novel fluorescent probe and its application to actomyosin subfragment 1 ATPase. *Biochemistry* **33,** 8262–8271.

Brune, M., Hunter, J. L., Howell, S. A., Martin, S. R., Hazlett, T. L., Corrie, J. E., and Webb, M. R. (1998). Mechanism of inorganic phosphate interaction with phosphate binding protein from *Escherichia coli. Biochemistry* **37,** 10370–10380.

Chen, J. G., Willard, F. S., Huang, J., Liang, J., Chasse, S. A., Jones, A. M., and Siderovski, D. P. (2003). A seven-transmembrane RGS protein that modulates plant cell proliferation. *Science* **301,** 1728–1731.

Colombo, K., Grill, S. W., Kimple, R. J., Willard, F. S., Siderovski, D. P., and Gonczy, P. (2003). Translation of polarity cues into asymmetric spindle positioning in *Caenorhabditis elegans* embryos. *Science* **300,** 1957–1961.

de Alba, E., De Vries, L., Farquhar, M. G., and Tjandra, N. (1999). Solution structure of human GAIP (Galpha interacting protein): A regulator of G protein signaling. *J. Mol. Biol.* **291,** 927–939.

Druey, K. M., and Kehrl, J. H. (1997). Inhibition of regulator of G protein signaling function by two mutant RGS4 proteins. *Proc. Natl. Acad. Sci. USA* **94,** 12851–12856.

Ferguson, K. M., Higashijima, T., Smigel, M. D., and Gilman, A. G. (1986). The influence of bound GDP on the kinetics of guanine nucleotide binding to G proteins. *J. Biol. Chem.* **261,** 7393–7399.

Gether, U. (2000). Uncovering molecular mechanisms involved in activation of G protein-coupled receptors. *Endocr. Rev.* **21,** 90–113.

Higashijima, T., Ferguson, K. M., Smigel, M. D., and Gilman, A. G. (1987). The effect of GTP and Mg^{2+} on the GTPase activity and the fluorescent properties of Go. *J. Biol. Chem.* **262,** 757–761.

Hiratsuka, T. (1983). New ribose-modified fluorescent analogs of adenine and guanine nucleotides available as substrates for various enzymes. *Biochim. Biophys. Acta* **742,** 496–508.

Jones, A. M. (2002). G-protein-coupled signaling in *Arabidopsis. Curr. Opin. Plant Biol.* **5,** 402–407.

Kelley, L. A., MacCallum, R. M., and Sternberg, M. J. (2000). Enhanced genome annotation using structural profiles in the program 3D-PSSM. *J. Mol. Biol.* **299,** 499–520.

Kimple, R. J., Jones, M. B., Shutes, A., Yerxa, B. R., Siderovski, D. P., and Willard, F. S. (2003). Established and emerging fluorescence-based assays for G-protein function: Heterotrimeric G-protein alpha subunits and regulator of G-protein signaling (RGS) proteins. *Comb. Chem. High Throughput Screen.* **6,** 399–407.

Kimple, R. J., De Vries, L., Tronchere, H., Behe, C. I., Morris, R. A., Gist Farquhar, M., and Siderovski, D. P. (2001). RGS12 and RGS14 GoLoco motifs are G alpha(i) interaction sites with guanine nucleotide dissociation inhibitor activity. *J. Biol. Chem.* **276,** 29275–29281.

Konvicka, K., Campagne, F., and Weinstein, H. (2000). Interactive construction of residue-based diagrams of proteins: The RbDe web service. *Protein Eng.* **13,** 395–396.

Krumins, A. M., and Gilman, A. G. (2002). Assay of RGS protein activity *in vitro* using purified components. *Methods Enzymol.* **344,** 673–685.

Lan, K. L., Remmers, A. E., and Neubig, R. R. (1998). Roles of G(o)alpha tryptophans in GTP hydrolysis, GDP release, and fluorescence signals. *Biochemistry* **37,** 837–843.

Lenzen, C., Cool, R. H., Prinz, H., Kuhlmann, J., and Wittinghofer, A. (1998). Kinetic analysis by fluorescence of the interaction between Ras and the catalytic domain of the guanine nucleotide exchange factor Cdc25Mm. *Biochemistry* **37,** 7420–7430.

Letunic, I., Goodstadt, L., Dickens, N. J., Doerks, T., Schultz, J., Mott, R., Ciccarelli, F., Copley, R. R., Ponting, C. P., and Bork, P. (2002). Recent improvements to the SMART domain-based sequence annotation resource. *Nucleic Acids Res.* **30,** 242–244.

McEwen, D. P., Gee, K. R., Kang, H. C., and Neubig, R. R. (2002). Fluorescence approaches to study G protein mechanisms. *Methods Enzymol.* **344,** 403–420.

McGuffin, L. J., Bryson, K., and Jones, D. T. (2000). The PSIPRED protein structure prediction server. *Bioinformatics* **16,** 404–405.

Morton, T. A., and Myszka, D. G. (1998). Kinetic analysis of macromolecular interactions using surface plasmon resonance biosensors. *Methods Enzymol.* **295,** 268–294.

O'Shannessy, D. J., Brigham-Burke, M., Soneson, K. K., Hensley, P., and Brooks, I. (1994). Determination of rate and equilibrium binding constants for macromolecular interactions by surface plasmon resonance. *Methods Enzymol.* **240,** 323–349.

Popov, S., Yu, K., Kozasa, T., and Wilkie, T. M. (1997). The regulators of G protein signaling (RGS) domains of RGS4, RGS10, and GAIP retain GTPase activating protein activity *in vitro*. *Proc. Natl. Acad. Sci. USA* **94,** 7216–7220.

Remmers, A. E., Posner, R., and Neubig, R. R. (1994). Fluorescent guanine nucleotide analogs and G protein activation. *J. Biol. Chem.* **269,** 13771–13778.

Ross, E. M. (2002). Quantitative assays for GTPase-activating proteins. *Methods Enzymol.* **344,** 601–617.

Snow, B. E., Brothers, G. M., and Siderovski, D. P. (2002). Molecular cloning of regulators of G-protein signaling family members and characterization of binding specificity of RGS12 PDZ domain. *Methods Enzymol.* **344,** 740–761.

Sunyer, T., Codina, J., and Birnbaumer, L. (1984). GTP hydrolysis by pure Ni, the inhibitory regulatory component of adenylyl cyclases. *J. Biol. Chem.* **259,** 15447–15451.

Tesmer, J. J., Berman, D. M., Gilman, A. G., and Sprang, S. R. (1997). Structure of RGS4 bound to AlF$_4^-$ activated G(i alpha 1): Stabilization of the transition state for GTP hydrolysis. *Cell* **89,** 251–261.

Thompson, J. D., Gibson, T. J., Plewniak, F., Jeanmougin, F., and Higgins, D. G. (1997). The CLUSTAL_X windows interface: Flexible strategies for multiple sequence alignment aided by quality analysis tools. *Nucleic Acids Res.* **25,** 4876–4882.

Wieland, T., Chen, C. K., and Simon, M. I. (1997). The retinal specific protein RGS-r competes with the gamma subunit of cGMP phosphodiesterase for the alpha subunit of transducin and facilitates signal termination. *J. Biol. Chem.* **272,** 8853–8856.

Zhang, X. F., Marshall, M. S., and Avruch, J. (1995). Ras-Raf complexes *in vitro*. *Methods Enzymol.* **255,** 323–331.

[20] AtRGS1 Function in *Arabidopsis thaliana*

By Jin-Gui Chen and Alan M. Jones

Abstract

Arabidopsis thaliana RGS1 is a novel "regulator of G-protein signaling" (AtRGS1) protein that consists of an N-terminal seven transmembrane domain characteristic of G-protein-coupled receptors and a C-terminal RGS box. AtRGS1 modulates plant cell proliferation. *Atrgs1* mutants are insensitive to glucose and less sensitive to fructose and sucrose, suggesting that sugar signaling in Arabidopsis involves AtRGS1. In addition, sugar metabolism and phosphorylation by hexokinase (HXK) are not required for AtRGS1-mediated sugar signaling, suggesting that AtRGS1 functions in a HXK-independent glucose signaling pathway.

Introduction

Glucose is not only a nutrient for eukaryotic cells, but also serves as a potent signal controlling growth and development (Rolland et al., 2001, 2002). In the yeast Saccharomyces cerevisiae, glucose-induced filamentous growth preceded by an increase in the intracellular cAMP level requires Gpr1p, a G-protein-coupled receptor (GPCR) and the Gα subunit Gpa2p (Colombo et al., 1998; Forsberg and Ljungdahl, 2001; Kraakman et al., 1999; Rolland et al., 2000, 2001; Versele et al., 2001).

Plants, like other eukaryotes, use heterotrimeric G-protein signaling in the regulation of growth and development (Assmann, 2002; Jones, 2002; Ullah et al., 2001, 2002, 2003); however, compared to metazoans, plants utilize a far simpler and atypical system. The Arabidopsis genome contains only one gene encoding a canonical G-protein α subunit (GPA1), one gene encoding a G-protein β subunit (AGB1), and two genes encoding G-protein γ subunits (AGG1 and AGG2) (Assmann, 2002; Jones, 2002). In contrast, there are an estimated 20 Gα, 6 Gβ, and 12 Gγ genes in mammals (Vanderbeld and Kelly, 2000). To date, a plant GPCR with its cognate ligand has not been identified.

Regulators of G-protein signaling (RGS) proteins accelerate the deactivation of G-protein α subunits to reduce GPCR signaling (Neubig and Siderovski, 2002). A novel RGS protein (AtRGS1) has been identified in Arabidopsis (Chen et al., 2003). The Arabidopsis AtRGS1 protein has an unusual modular construction: a predicted N-terminal seven transmembrane (7TM) domain characteristic of GPCRs and a C-terminal RGS box; therefore, AtRGS1 appears to be a structural hybrid of a GPCR and an RGS protein. The RGS box of AtRGS1 binds to the Arabidopsis GPA1 protein in a nucleotide-dependent manner, accelerates its intrinsic GTPase activity, and complements the pheromone supersensitivity phenotype of the yeast RGS mutant, sst2Δ, indicating that the RGS box of AtRGS1 is functional. AtRGS1 is a critical modulator of plant cell proliferation, and some evidence suggests that it may mediate sugar signaling (Chen et al., 2003). Atrgs1 mutants have altered sensitivity to high concentrations of glucose, raising the possibility that AtRGS1 is a cell surface sugar receptor coupled by GPA1.

In addition to the pathway involving the GPCR Gpr1, S. cerevisiae contains three other glucose-sensing pathways: one involving hexokinase

(HXK), one utilizing glucose carrier-like proteins and one using glycolytic intermediates as possible metabolic messengers (Rolland *et al.*, 2001). The GPCR pathway is desensitized by one of the four yeasts RGS proteins, RGS2 (Versele *et al.*, 1999). While less defined as in yeast, plants may also have at least three distinct glucose signal transduction pathways: a hexokinase (HXK)-dependent pathway (Jang *et al.*, 1997; Moore *et al.*, 2003; Xiao *et al.*, 2000), a HXK-independent pathway (Ciereszko *et al.*, 2001; Martin *et al.*, 1997; Mita *et al.*, 1997), and a glycolysis-dependent pathway that depend on the catalytic activity of HXK (Xiao *et al.*, 2000). These pathways can be dissected using sugar analogs that are differently transported and metabolized.

Materials and Methods

Plant Materials

The genotypes of T-DNA insertion mutants *Atrgs1-1* and *Atrgs1-2* are described in Chen *et al.* (2003). All mutants used here are in the ecotype Columbia (Col-0) *Atrgs1-1* is from the Torrey Mesa Research Institute Arabidopsis T-DNA "SAIL" (formerly GARLIC) Collection (www.tmri.org). SAIL is an insertion collection that has been generated from approximately 100,000 individual T-DNA mutagenized Arabidopsis plants (Columbia ecotype). It should be noted that effective as of January 1, 2004, Syngenta Biotechnology, Inc. (SBI) will no longer be distributing seeds from the SAIL collection to academic researchers (http://www.tmri.org/pages/collaborations/ garlic files/GarlicDescription.html). However, of the approximately 100,000 lines in the SAIL collection, about 90% are being donated to the Arabidopsis Biological Resource Center (ABRC) in Columbus, Ohio, so researchers will be able to access these publicly accessible lines directly through the ABRC (http://www.arabidopsis.org/abrc/). Sequence information for the entire SAIL collection will be deposited into GenBank and is also being provided to Joe Ecker at the Salk Institute to enter into his public site for finding Arabidopsis mutants (http://signal.salk.edu/tabout.html). *Atrgs1-2* was obtained from another publicly accessible strain repository, the Salk Institute sequence-indexed T-DNA insertion mutant collection (http:// signal.salk.edu/cgi-bin/tdnaexpress), which contains 141,486 T-DNA sequences, and the seeds were ordered from ABRC (http://www.arabidopsis. org/abrc/). Plants homozygous for *Atrgs1-1* and *Atrgs1-2* are isolated, and the insertion is confirmed by sequencing at UNC-Chapel Hill, using T-DNA left border primers (for *Atrgs1-1*: 5'-TAGCATCTGAATTTCA-TAACCAATCTCGATACAC-3', and for *Atrgs1-2*: 5'-GGCAAT-CAGCTGTTGCCCGTCTCACTGGTG-3'). Loss of detectable *AtRGS1* transcripts in *Atrgs1-1* and *Atrgs1-2* mutants is verified by reverse transcriptase polymerase chain reaction (PCR) (Chen *et al.*, 2003)

AtRGS1 Cloning and Expression

The entire open reading frame of *AtRGS1* (At3g26090) is amplified by PCR from a cDNA library made from seedlings grown in light for 10 days and cloned into the pENTR/D-TOPO vector (Invitrogen, Carlsbad, CA) and then subcloned into Gateway plant transformation destination vectors pB2GW7 and pK7FWG2, respectively (Karimi *et al.*, 2002) by LR recombination reaction. In both constructs, expressions of *AtRGS1* (in vector pB2GW7) or *AtRGS1-GFP* (in vector pK7FWG2) are driven by the 35S promoter of cauliflower mosaic virus. Both constructs are transformed into *A. thaliana* (Columbia–0 ecotype) by *Agrobacterium*-mediated transformation (Bechtold and Pelletier, 1998). At least three independent transgenic lines are analyzed for each construct.

Chemical Reagents

All sugars and sugar analogs are purchased from Sigma-Aldrich (St. Louis, MO). Murashige & Skoog (M&S) basal medium with Gamborg's vitamins is purchased from ICN Biomedicals Inc. (Aurora, OH). Phytoagar is purchased from Research Products International Corp. (Mt. Prospect, Ill).

Sugar Sensitivity Assay

Wild-type (Col-0) and *Atrgs1-1* and *Atrgs1-2* mutant seeds are sown, chilled, and light treated, and the plants are grown under identical condition until maturation. When the plants are dried and seeds turn brown, seeds are collected from each genotype separately by breaking the siliques and are cleaned by removing debris. The seeds are put into a small envelope (2.5 × 4.0 in.) and stored at 23° with low humidity (about 20%). For sugar sensitivity assays, seeds are sterilized with 80% ethanol for 2 min, followed by 30% bleach with 0.1% Tween 20 for 10 min, and then washed with sterile deionized water six times under sterile condition. Sterilized seeds are stratified at 4° by placing them in the dark for 48 h, sown on plates containing 1/2 M&S salts with Gamborg's vitamins (pH adjusted to 5.7 with 1 *N* KOH), 0.5% phytoagar, and different concentrations or types of sugars or sugar analogs, and germinated and grown horizontally at 23° with constant fluorescent light (75 μmol/m^2/s). Ten days later, the growth of seedlings is photographed by a digital camera linked with a dissection

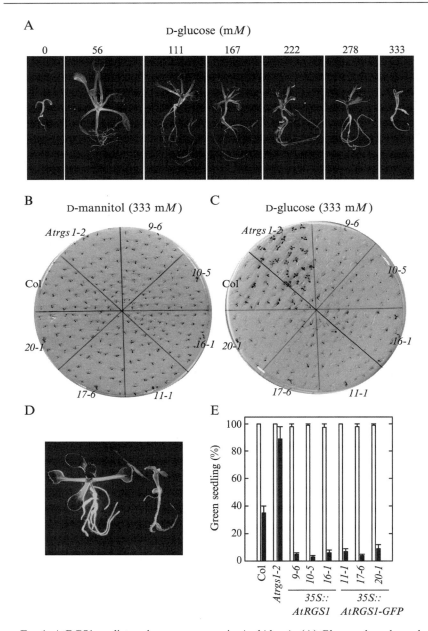

FIG. 1. AtRGS1 mediates glucose responses in *Arabidopsis*. (A) Glucose dose-dependent developmental responses of wild-type (Col-0) *Arabidopsis* seedlings. Seedlings were photographed after 14 days grown at 23°, constant light. (B) *Atrgs1* mutants have a wild-type

microscope, and the lengths of hypocotyls and primary root are also measured under a dissection microscope.

The timing for maximal differences between the genotypes may vary depending on the degree of seed dormancy. This variable is a function of the time from harvest and the temperature and relative humidity during seed storage. Germination is a complex trait. Full germination capacity is not realized until after several weeks of an after-ripening process for which temperature and humidity are important influences. Light and chilling (stratification) are typically required for germination and, as such, must also be controlled in comparative studies on germination. Finally, germination potential is lost over time and is influenced by storage conditions. Therefore, it is important to use seeds lots that are produced and harvested identically for germination or sugar sensitivity assay because sugar also affects seed germination. The percentage of green seedling is defined here as the number of green seedlings divided by the total number of seeds. Each experiment should be repeated at least twice. About 50 seeds are scored for each treatment of each genotype.

Experimental Results

Sugars affect both seed germination and early seedling development. At low concentrations, sucrose and D-glucose promote seedling development (Fig. 1A), but are not prerequisites for seed germination. At high concentrations, sugar inhibits both seed germination (Price *et al.*, 2003; Ullah *et al.*, 2002) and early seedling development. Arabidopsis seedling growth in the presence of 333 mM D-glucose (6%) is arrested, as evidenced by the absence of postembryonic leaves, inhibition of root elongation, accumulation of anthocyanin ("reddish" coloring), and lack of chlorophyll (Fig. 1A and D). This developmental arrest has been used for the genetic

---

response to D-mannitol. T-DNA insertion mutant *Atrgs1-2* and transgenic lines overexpressing *AtRGS1* (lines *9-6*, *10-5*, and *16-1*) or *AtRGS1-GFP* (lines *11-1*, *17-6*, and *20-1*) were sown on 1/2 MS plates containing 333 mM (6%) D-mannitol. In transgenic lines, expression of *AtRGS1* or *AtRGS1-GFP* was driven by the 35S promoter of cauliflower mosaic virus. (C) *Atrgs1* mutants are insensitive to high concentration of D-glucose. Seeds in B were sown on 1/2 MS plates containing 333 mM (6%) D-glucose. Seedlings in B and C were photographed after 10 days grown at 23°, constant light. (D) Green seedling *(left)* and development arrested seedling *(right)* on 333 mM (6%) D-glucose. Seedlings were photographed after 10 days grown at 23°, constant light. (E) Quantitative analyses of sugar sensitivities in *Atrgs1-2* mutants and transgenic lines overexpressing *AtRGS1* or *AtRGS1-GFP*. Open bar: 333 mM D-mannitol. Closed bar: 333 mM D-glucose. The percentages of green seedlings were scored after 10 days grown at 23°, constant light. Shown are means ± SE of three replicates. (See color insert.)

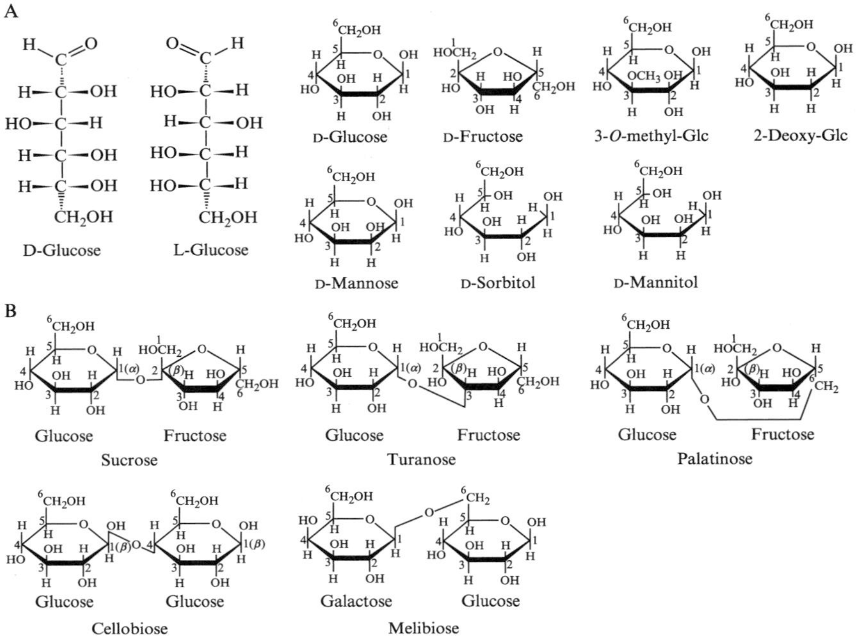

A

B

C

Compound	Structure	Transportable	Metabolizable	Phosphorylatable by HXK
Monosaccharide				
D-glucose		Yes	Yes	Yes
L-glucose		No	No	No
D-fructose		Yes	Yes	Yes
3-O-methyl-glucose		Yes	No	No
2-Deoxy-glucose		Yes	No	Yes
D-mannose		Yes	No	Yes
D-sorbitol		Yes	No	Yes
D-mannitol		Yes	No	Yes
Disaccharide				
Sucrose	Glc[1–2]Fru	Yes	Yes	Yes
Turanose	Glc[1–3]Fru	No	No	Yes
Palatinose	Glc[1–6]Fru	No	No	Yes
Cellobiose	Glc[1–4]Glc	Yes	Yes	Yes
Melibiose	Gal[1–6]Glc	Yes	No	Yes

FIG. 2. Sugar and sugar analogs used in *Atrgs1* sugar sensitivity assay. (A) Monosaccharides. (B) Disaccharides. (C) Some biological characteristics of these sugars and sugar analogs.

selection of mutants with an altered response to sugars (Rolland *et al.*, 2002; Smeekens, 2000).

Seedling development of *Atrgs1* mutants is less sensitive to high concentrations of sugar (Chen *et al.*, 2003), suggesting that sugar signaling in plants may involve the heterotrimeric G-protein complex. Consistent with this, overexpression of *AtRGS1* conferred hypersensitivity to glucose (Fig. 1C and E). The insensitivity of *Atrgs1* mutants to glucose is not due to osmotic stress because *Atrgs1* mutants and *AtRGS1* overexpressors have wild-type responses to 333 mM (6%) D-mannitol (Fig. 1B and E).

Sensitivities of *Atrgs1* mutants to different monosaccharide and disaccharide types (Fig. 2) were tested to dissect the three potential signaling pathways in plants described earlier. In all these experiments, the particular sugar/disaccharide or analogs were the sole sugar source. Each of these sugar analogs has altered characteristics in transport, metabolism, or phosphorylation compared to glucose and sucrose (Fig. 2C). L-Glucose is an unnatural enantiomer of D-glucose and is transported poorly (Gogarten and Bentrup, 1989; Oliveira *et al.*, 2002), metabolized poorly (Chevalier *et al.*, 1996; Koch *et al.*, 2000), and phosphorylated poorly (Tiessen *et al.*, 2003). L-Glucose at 333 mM completely inhibited seed germination both in wild-type and in *Atrgs1* mutants (Fig. 3). D-Fructose and D-glucose are constitutional isomers. *Atrgs1* mutants are less sensitive to 333 mM D-Fructose than wild type, evident by longer roots and hypocotyls, and the formation of true leaves, compared with stunted wild-type seedlings having only the cotyledons (the embryonic leaf pair). However, the resistance of *Atrgs1* mutants to D-fructose was less compared to the same concentration of D-glucose (Fig. 3). 3-*O*-Methyl-D-glucose is a nonmetabolizable D-glucose analog and is not a substrate for HXK. At 50 mM, 3-*O*-methyl-D-glucose inhibits seedling growth but not seed germination both in wild-type and in *Atrgs1* mutants. The phenotypes of seedlings grown in 3-*O*-methyl-D-glucose mimic the phenotypes of seedlings grown in the absence of sugar. This implies that sugar metabolism and phosphorylation by HXK are required for early seedling growth and development.

2-Deoxy-D-glucose and D-mannose are also nonmetabolizable D-glucose analogs, but they are substrates for HXK. Both 2-deoxy-D-glucose and D-mannose cause developmental arrest at relatively low concentrations. At 5 mM, 2-deoxy-D-glucose and D-mannose completely block seed germination, whereas it requires approximately 333 mM D-glucose to block seed germination. At 1 mM 2-deoxy-D-glucose or D-mannose, both wild-type and *Atrgs1* mutants can germinate, although at a low rate, and form green seedlings (Fig. 3). However, root production was blocked completely. This inhibition can be restored by metabolizable sugars (Pego *et al.*, 1999), suggesting that sugar metabolism is required for seedling development.

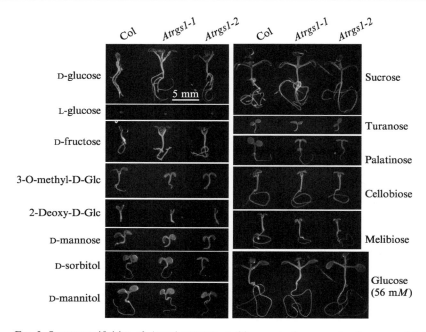

FIG. 3. Sugar specificities of *Atrgs1* mutants. D-Glucose, L-glucose, D-fructose, D-sorbitol, and D-mannitol were used at 333 mM. 3-O-Methyl-D-glucose (3-O-methyl-D-Glc) was used at 50 mM. 2-Deoxy-D-glucose (2-Deoxy-D-Glc) and D-mannose were used at 1 mM. All disaccharides (sucrose, turanose, palatinose, cellobiose, and melibiose) were used at 200 mM. Wild-type (Col) and *Atrgs1* mutants grown at 56 mM D-glucose were used as controls. In all these experiments, the particular sugar/disaccharide or analog was the sole sugar source. Seedlings were photographed after 10 days grown at 23°, constant light. Seeds on 333 mM L-glucose did not germinate. (See color insert.)

Because no significant difference in growth phenotype was observed between wild-type and *Atrgs1* mutants in either 3-O-methyl-D-glucose (transportable, nonmetabolizable, not a substrate for HXK) or 2-deoxy-D-glucose or D-mannose (both transportable, nonmetabolizable, substrates for HXK), sugar metabolism and phosphorylation by HXK are not responsible for the glucose insensitivity observed in *Atrgs1* mutants. AtRGS1 probably functions in an HXK-independent glucose-signaling pathway.

Both D-sorbitol and D-mannitol are reduced forms of D-glucose and are used as comparable osmotic solutes here. No difference between wild-type and *Atrgs1* mutants was observed when they were grown at 333 mM (6%) D-sorbitol or D-mannitol (Fig. 3).

Plants sense a wide variety of sugars, but among soluble carbohydrates, disaccharide sucrose is the predominant form. Sucrose may also act as a signaling molecule in plants, although no sucrose sensor has been identified so far. Because glucose and sucrose are interconverted in the plant cell metabolically, the effect of sucrose on seedling development could be attributed to the constituent hexoses (glucose and fructose), sucrose, or both. At 200 mM (6.85%) sucrose, the accumulation of chlorophyll (dark-green leaves) was evident in wild-type Arabidopsis seedlings. At this concentration, the phenotypic differences between wild-type and *Atrgs1* mutants were subtle. However, the sensitivity of wild-type and *Atrgs1* mutants on 300 mM (10.27%) sucrose was comparable to that on 333 mM glucose (data not shown).

Turanose and palatinose are structural isomers of sucrose composed of glucose and fructose with different glycosidic linkages (Fig. 2). They are not synthesized in higher plants and cannot be cleaved or transported by plant enzymes. Neither turanose nor palatinose competes for sucrose transport (Sinha *et al.*, 2002). However, in tomato suspension culture cells, both turanose and palatinose can specifically activate MAPK activity, whereas glucose and sucrose elicit only weak MAPK activation probably due to an osmotic effect because the same concentration of mannitol has a similar effect (Sinha *et al.*, 2002). These findings indicate that nonmetabolizable sucrose isomers such as turanose and palatinose can activate distinctly different signal transduction pathways from metabolizable sugars. Here we show that both turanose and palatinose at 200 mM inhibit seedling development dramatically (Fig. 3) and that turanose blocks root formation completely. No significant difference was observed between wild-type and *Atrgs1* mutants. These results further support the idea that sugar metabolism is required for early seedling development, but is not required for AtRGS1-mediated sugar signaling.

Cellobiose, a disaccharide obtained by the partial hydrolysis of cellulose, consists of two D-glucopyranoses joined by a 1,4′,-β-glycoside bond (Fig. 2). Cellobiose cannot be digested by humans and cannot be fermented by yeast. However, it is metabolizable in plants (Loreti *et al.*, 2000). Cellobiose at 200 mM also has an inhibitory effect on seedling growth and development (Fig. 3) and has an equal effect on wild-type and *Atrgs1* mutants. Melibiose is composed of galactose and glucose (Fig. 2). In barley embryos, melibiose is a nonmetabolizable disaccharide (Loreti *et al.*, 2000). Melibiose at 200 mM inhibits Arabidopsis seedling growth and development to a similar extent in wild-type and *Atrgs1* mutants (Fig. 3).

This structure–function analysis of disaccharides indicates that a fructose moiety is needed for disaccharide sensing. Alteration of the fructosyl moiety, such as in turanose and palatinose, or replacing the fructose moiety with glucose or galactose, such as in cellobiose and melibiose, results in significant inhibition of seedling growth and development and masks the sugar insensitivity or hyposensitivity observed in *Atrgs1* mutants.

Taken together, *Atrgs1* mutants are insensitive to D-glucose and less sensitive to fructose and sucrose. Because sugar metabolism and phosphorylation by HXK are not required for AtRGS1-mediated signal signaling, AtRGS1 most likely functions in an HXK-independent glucose signaling pathway.

Acknowledgments

Work in A.M.J.'s laboratory on the Arabidopsis G protein is supported by the NIH (GM65989-01) and NSF (MCB-0209711).

References

Assmann, S. M. (2002). Heterotrimeric and unconventional GTP binding proteins in plant cell signaling. *Plant Cell.* **14**(Suppl.), S355–S373.

Bechtold, N., and Pelletier, G. (1998). In planta Agrobacterium-mediated transformation of adult *Arabidopsis thaliana* plants by vacuum infiltration. *Methods Mol. Biol.* **82,** 259–266.

Chen, J. G., Willard, F. S., Huang, J., Liang, J., Chasse, S. A., Jones, A. M., and Siderovski, D. P. (2003). A seven-transmembrane RGS protein that modulates plant cell proliferation. *Science* **301,** 1728–1731.

Chevalier, C., LeQuerrec, F., and Raymond, P. (1996). Sugar levels regulate the expression of ribosomal protein genes encoding protein S28 and ubiquitin-fused protein S27a in maize primary root tips. *Plant Sci.* **117,** 95–105.

Ciereszko, I., Johansson, H., and Kleczkowski, L. A. (2001). Sucrose and light regulation of a cold-inducible UDP-glucose pyrophosphorylase gene via a hexokinase-independent and abscisic acid-insensitive pathway in Arabidopsis. *Biochem. J.* **354,** 67–72.

Colombo, S., Ma, P., Cauwenberg, L., Winderickx, J., Crauwels, M., Teunissen, A., Nauwelaers, D., de Winde, J. H., Gorwa, M. F., Colavizza, D., and Thevelein, J. M. (1998). Involvement of distinct G-proteins, Gpa2 and Ras, in glucose- and intracellular acidification-induced cAMP signalling in the yeast *Saccharomyces cerevisiae*. *EMBO J.* **17,** 3326–3341.

Forsberg, H., and Ljungdahl, P. O. (2001). Sensors of extracellular nutrients in *Saccharomyces cerevisiae*. *Curr. Genet.* **40,** 91–109.

Gogarten, J. P., and Bentrup, F. W. (1989). Substrate specificity of the hexose carrier in the plasma membrane of Chenopodium suspension cells probes by transmembrane exchange diffusion. *Planta* **178,** 52–60.

Jang, J. C., Leon, P., Zhou, L., and Sheen, J. (1997). Hexokinase as a sugar sensor in higher plants. *Plant Cell* **9,** 5–19.

Jones, A. M. (2002). G-protein-coupled signaling in Arabidopsis. *Curr. Opin. Plant Biol.* **5,** 402–407.

Karimi, M., Inze, D., and Depicker, A. (2002). GATEWAY vectors for Agrobacterium-mediated plant transformation. *Trends Plant Sci.* **7,** 193–195.

Koch, K. E., Ying, Z., Wu, Y., and Avigne, W. T. (2000). Multiple paths of sugar-sensing and a sugar/oxygen overlap for genes of sucrose and ethanol metabolism. *J. Exp. Bot.* **51**(GMP Special issue), 417–427.

Kraakman, L., Lemaire, K., Ma, P., Teunissen, A. W., Donaton, M. C., Van Dijck, P., Winderickx, J., de Winde, J. H., and Thevelein, J. M. (1999). A *Saccharomyces cerevisiae* G-protein coupled receptor, Gpr1, is specifically required for glucose activation of the cAMP pathway during the transition to growth on glucose. *Mol. Microbiol.* **32**, 1002–1012.

Loreti, E., Alpi, A., and Perata, P. (2000). Glucose and disaccharide-sensing mechanisms modulate the expression of α-amylase in barley embryos. *Plant Physiol.* **123**, 939–948.

Martin, T., Hellman, H., Schmidt, R., Willmitzer, L., and Frommer, W. B. (1997). Identification of mutants in metabolically regulated gene expression. *Plant J.* **11**, 53–62.

Mita, S., Murano, N., Akaike, M., and Nakamura, K. (1997). Mutants of *Arabidopsis thaliana* with pleiotropic effects on the expression of the gene for β-amylase and on the accumulation of anthocyanin that are inducible by sugars. *Plant J.* **11**, 841–851.

Moore, B., Zhou, L., Rolland, F., Hall, Q., Cheng, W. H., Liu, Y. X., Hwang, I., Jones, T., and Sheen, J. (2003). Role of the Arabidopsis glucose sensor HXK1 in nutrient, light, and hormonal signaling. *Science* **300**, 332–336.

Neubig, R. R., and Siderovski, D. P. (2002). Regulators of G-protein signalling as new central nervous system drug targets. *Nature Rev. Drug Discov.* **1**, 187–197.

Oliveira, J., Tavares, R. M., and Geros, H. (2002). Utilization and transport of glucose in *Olea europaea* cell suspensions. *Plant Cell Physiol.* **43**, 1510–1517.

Pego, J. V., Weisbeek, P. J., and Smeekens, S. C. M. (1999). Mannose inhibits Arabidopsis germination via a hexokinase-mediated step. *Plant Physiol.* **119**, 1017–1023.

Price, J., Li, T. C., Kang, S. G., Na, J. K., and Jang, J. C. (2003). Mechanisms of glucose signaling during germination of Arabidopsis. *Plant Physiol.* **132**, 1424–1438.

Rolland, F., De Winde, J. H., Lemaire, K., Boles, E., Thevelein, J. M., and Winderickx, J. (2000). Glucose-induced cAMP signalling in yeast requires both a G-protein coupled receptor system for extracellular glucose detection and a separable hexose kinase-dependent sensing process. *Mol. Microbiol.* **38**, 348–358.

Rolland, F., Moore, B., and Sheen, J. (2002). Sugar sensing and signaling in plants. *Plant Cell* **14**(Suppl.), S185–S205.

Rolland, F., Winderickx, J., and Thevelein, J. M. (2001). Glucose-sensing mechanisms in eukaryotic cells. *Trends Biochem. Sci.* **26**, 310–317.

Sinha, A. K., Hofmann, M. G., Romer, U., Kockenberger, W., Elling, L., and Roitsch, T. (2002). Metabolizable and non-metabolizable sugars activate different signal transduction pathways in tomato. *Plant Physiol.* **128**, 1480–1489.

Smeekens, S. (2000). Sugar-induced signal transduction in plants. *Annu. Rev. Plant Physiol. Plant Mol. Biol.* **51**, 49–81.

Tiessen, A., Prescha, K., Branscheid, A., Palacios, N., McKibbin, R., Halford, N. G., and Geigenberger, P. (2003). Evidence that SNF1-related kinase and hexokinase are involved in separate sugar-signalling pathways modulating post-translational redox activation of ADP-glucose pyrophosphorylase in potato tubers. *Plant J.* **35**, 490–500.

Ullah, H., Chen, J. G., Temple, B., Boyes, D. C., Alonso, J. M., Davis, K. R., Ecker, J. R., and Jones, A. M. (2003). The β-subunit of the Arabidopsis G protein negatively regulates auxin-induced cell division and affects multiple developmental processes. *Plant Cell* **15**, 393–409.

Ullah, H., Chen, J. G., Wang, S., and Jones, A. M. (2002). Role of a heterotrimeric G protein in regulation of Arabidopsis seed germination. *Plant Physiol.* **129**, 897–907.

Ullah, H., Chen, J. G., Young, J. C., Im, K. H., Sussman, M. R., and Jones, A. M. (2001). Modulation of cell proliferation by heterotrimeric G protein in *Arabidopsis*. *Science* **292**, 2066–2069.

Vanderbeld, B., and Kelly, G. M. (2000). New thoughts on the role of beta-gamma subunit in G-protein signal transduction. *Biochem. Cell Biol.* **78**, 537–550.

Versele, M., Lemaire, K., and Thevelein, J. M. (2001). Sex and sugar in yeast: Two distinct GPCR systems. *EMBO Report* **21**, 574–579.

Versele, M., Winde, J., and Thevelein, J. M. (1999). A novel regulator of G protein signalling in yeast, Rgs2, downregulates glucose-activation of the cAMP pathway through direct inhibition of Gpa2. *EMBO J.* **18**, 5577–5591.

Xiao, W., Sheen, J., and Jang, J. C. (2000). The role of hexokinase in plant sugar signal transduction and growth and development. *Plant Mol. Biol.* **44**, 451–461.

[21] Identification and Functional Analysis of the *Drosophila* Gene *loco*

By Sebastian Granderath and Christian Klämbt

Abstract

In contrast to vertebrates, the fruit fly *Drosophila melanogaster* contains only a small number of regulator of G-protein signaling (RGS) domain genes. This article reviews current knowledge on these genes. Although the fruit fly is particularly amenable to genetic analysis and manipulation, not much is known about the functions and mechanisms of action. The best-studied RGS gene in *Drosophila* is *loco*, a member of the D/R12 subfamily. The four different protein isoforms all contain RGS, GoLoco, and RBD domains. This article describes the identification and functional analyses of *loco* in the *Drosophila* system and discusses some mechanistic models that may underlie *loco* function.

Introduction

One of the many pathways cells used to couple cell surface-bound receptors to intracellular signaling systems are heterotrimeric G proteins. These membrane-associated complexes comprise a GTP-hydrolyzing subunit called $G\alpha$ and a $G\beta/G\gamma$ heterodimer. In its inactive form, $G\alpha$ is bound to GDP and forms a complex with the $G\beta/G\gamma$ subunit. Following activation, GTP displaces GDP and $G\alpha$GTP dissociates from the trimer. Both the free $G\alpha$ subunit and the $G\beta\gamma$ complex are capable of activating specific downstream signaling components. Intrinsic GTPase of $G\alpha$ activity hydrolyzes GTP to GDP and inorganic phosphate; the subsequent reassociation of $G\alpha$ and $G\beta\gamma$ subunits terminates signal transduction.

The regulation of trimeric G-protein signaling has for a long time remained elusive. The intrinsic GTPase activity alone could not account for the kinetics of deactivation *in vivo*. GTPase-activating proteins (GAPs) acting on trimeric G proteins were identified in the early 1990s; subsequent analyses showed that they belonged to an evolutionary conserved protein family. The hallmark of these so-called regulator of G-protein signaling (RGS) proteins is a conserved domain of approximately 120 amino acids. This RGS domain is sufficient for both binding of Gα and activation of GTPase activity (De Vries and Gist Farquhar, 1999; De Vries *et al.*, 1995; Koelle and Horvitz, 1996; Neubig and Siderovski, 2002; Siderovski *et al.*, 1996). RGS domain proteins often contain additional sequence motifs, which also aid in the regulation of G-protein function. In particular, the 19 amino acid long GoLoco motif is frequently associated with G-protein regulators (Granderath *et al.*, 1999; Siderovski *et al.*, 1999). The GoLoco motif acts as a guanine nucleotide dissociation inhibitor (GDI), inhibiting GDP dissociation from Gα_i (De Vries *et al.*, 2000; Natochin *et al.*, 2000). The structure of the GoLoco–Gαi·GDP complex suggests that sequences adjacent to the GoLoco motif are involved in controlling the specificity of interactions with distinct Gα proteins (Kimple *et al.*, 2002).

RGS Proteins in *Drosophila*

In contrast to the great diversity of RGS proteins in mice and humans, which express at least 21 functional RGS proteins (Sierra *et al.*, 2002), a genome-wide survey revealed only nine genes encoding RGS proteins in *Drosophila* (Table I). To date, only three of these genes have been analyzed genetically. *axin* encodes the best-studied protein based on its function in antagonizing Wingless signaling (Jones and Bejsovec, 2003). Axin contains binding sites for a large number of proteins [β-catenin, GSK3β, APC, phosphatase 2A (PP2A) and Dishevelled] and thus appears to act as a scaffold for the assembly of the protein degradation complex required for the negative regulation of *wingless* signaling. The function of the RGS domain of Axin itself has only been analyzed in vertebrates; there, it mediates APC binding (Spink *et al.*, 2000) but does not regulate any of the known G proteins (Mao *et al.*, 1998).

The *Drosophila* G-protein-coupled receptor kinase 2 encoded by *gprk2* also contains an RGS domain and is required for normal cAMP levels. Mutations in *gprk2* cause defects in the anterior patterning during embryonic development (Lannutti and Schneider, 2001; Schneider and Spradling, 1997a).

TABLE I
RGS DOMAIN PROTEINS IN *Drosophila*[a]

Name	Map position	Mutants	Phenotype	References
axin (axn)	99D	Yes	Interacts with *wnt* pathway members	Bienz (1999), Willert *et al.* (1999)
CG1514	7C	No	n.d.	—
CG5036	54E	No	n.d.	Egger *et al.* (2002)
G-protein-coupled receptor kinase 1 (Gprk1)	41B	No	n.d.	Cassill *et al.* (1991)
G-protein-coupled receptor kinase 2 (Gprk2)	100C	Yes	Anterior patterning, cAMP levels	Fan and Schneider (2003), Lannutti and Schneider (2001), Schneider and Spradling (1997b)
protein kinase A anchor protein (pkaab)	35F	No	n.d.	Bolshakov *et al.* (2002)
CG7095 (RGS6)	17E	No	n.d.	—
RGS7	15B	No	n.d.	Elmore *et al.* (1998)
locomotion defects (loco)	94B	Yes	P1:glial differentiation P2/P3:oogenesis	Granderath *et al.* (1999)

[a] No reference is given when the sequence is just predicted by the genome project (BDGP).

The *Drosophila* gene *loco* represents the third member of the RGS gene family that has been studied in *Drosophila*. It was first recognized due to its function in the so-called lateral glial cells in the *Drosophila* embryo, but is also required for peripheral nervous system (PNS) development and oogenesis (Granderath and Klämbt, 1999; Pathirana *et al.*, 2001).

Enhancer Trap Technique to Identify Genes

Development of the enhancer-trap technique (Bellen *et al.*, 1990; O'Kane and Gehring, 1987) provided an efficient tool for the selection of interesting genes based solely on their expression patterns, obviating the need for a predefined or even observable mutant phenotype. At the same time, the enhancer-trap element provides a convenient entry point for subsequent molecular analyses. Most importantly, a P-element integration (described later) allows the generation of a mutant allele by remobilizing the inserted transposon. Thus mutations in the locus of interest can be generated relatively easily—paving the way for detailed phenotypic analyses of the *in vivo* function of the gene product(s) of interest. Today, large collections of P-element insertion mutants are established[1] that eventually may allow the direct analysis of most of the *Drosophila* genes.

The enhancer-trap technique makes use of the so-called P-element, a transposon naturally occurring in about 40 copies in the *Drosophila* genome (Engels, 1989). The P-element carries a transposase encoding gene, flanked by inverted repeats of 31 bp, which serve as recognition sites for the transposase. By genetic engineering, the original transposase gene was replaced by a "marker gene" that allows easy selection of animals carrying the enhancer-trap construct (usually a copy of the *white*$^+$ or the *rosy*$^+$ gene, which allow selection by eye color of the adult flies) and a reporter gene (the bacterial *lacZ* coding sequence inserted downstream of a weak general promoter). Expression of the β-galactosidase reporter can be assessed easily in the embryo by *in situ* histology. After germline transformation, stable transformants can be obtained. The construct will integrate at random positions in the genome, with a notable integration tendency in the 5' region of genes (Liao *et al.*, 2000). The reporter gene will, in most cases, be under the control of the same enhancer(s) regulating the neighboring gene. By assaying β-galactosidase activity, one can thus "trap" these genes easily—merely based on their cell- or tissue-specific expression patterns—without any prior knowledge of their sequence or function.

Eventually, the enhancer-trap construct can be remobilized. This is achieved by crossing in a chromosome that harbors a P-element that encodes a constitutively active transposase but is unable to jump itself due to mutations in the inverted repeats (Robertson *et al.*, 1988). In most cases, remobilization leads to precise excision (followed by either reintegration at a different site or loss of the construct). Occasionally, imprecise excisions cause the deletion of sequences flanking the original insertion point. Such an imprecise "hop out" represents an effective mutagen for the tagged gene(s), opening the way for detailed phenotypical and functional analyses.

Enhancer-trap elements also provide a convenient tool for deciphering transcriptional regulatory networks. Their reporter gene activity can be assayed easily in strains homozygous for mutations in possible regulator genes. Altered reporter gene expression (level, tissue specificity, and/or timing) can provide important cues on the regulation of the tagged gene.

---

[1]Databases used to retrieve P-element insertions: Flybase (http://fbserver.gen.cam.ac.uk:7081/), genome project at Baylor (http://flypush.imgen.bcm.tmc.edu/pscreen/), and Genexcel (commercial) (http://genexel.kaist.ac.kr/mapview3/).

Mobile DNA elements can also be used for direct detection of protein expression (Morin *et al.*, 2001). Furthermore, P-elements can be exchanged relatively easily through recombination; it is thus possible to replace the *lacZ* reporter gene with, for example, the yeast transcriptional activator GAL4. The GAL4 construct then allows tissue-specific expression from a second transgenic element, carrying an effector gene under control of the yeast UAS sequence. This binary system can be used to manipulate specific cells for a wide variety of purposes, for example, cell fate changes, rescue experiments, or targeted cell ablation (Brand and Perrimon, 1993; Sepp and Auld, 1999). In addition to these approaches, one can utilize *in vivo* RNAi techniques where a given gene of interest can be inactivated in a temporally and spatially controlled manner (Lee and Carthew, 2003).

Identification of *loco*

The ventral nerve cord of *Drosophila* consists of 13 segmentally repeated neuromeres, each of which comprises about 650 neurons and only about 65 quite well characterized glial cells (Granderath and Klämbt, 1999). The initial specification of the glial cells has been studied extensively and depends on the presence of transcription factors such as glial cells missing (Gcm), reversed polarity (Repo), and different variants of the ETS transcription factor pointed (Pnt) (reviewed by Granderath and Klämbt, 1999; Jones, 2001; Van De Bor and Giangrande, 2002). We set out to identify possible targets of these transcription factors using the enhancer trap approach described earlier.

The *loco* gene was identified by a number of P-element enhancer trap insertions that are located upstream of the P1 promoter (Fig. 1). Cloning of the gene and analysis of the completed *Drosophila* genome sequence revealed the presence of at least three additional promoter elements, which direct the expression of at least four transcripts. The transcripts encode four different protein isoforms named Loco P1–Loco P4. Conceptual translation reveals proteins of 829, 1175, 872, and 1541 amino acids (Fig. 1). While the N termini of the proteins differ, all four isoforms share the same 828 amino acid C-terminal "backbone," encoded by the exons 2, 3, and 4.

The Loco P1 encoding mRNA is the shortest of the four transcripts. The P1-specific exon 1 merely encodes a translation start codon, which is then fused to the shared exons 2,3, and 4, resulting in a protein comprising 829 amino acids (Granderath *et al.*, 1999; Pathirana *et al.* 2001).

The "backbone" of all four isoforms clearly classifies Loco as a member of the D/R12 subfamily of RGS proteins (Hollinger and Hepler, 2002; Ross and Wilkie, 2000; Sierra *et al.*, 2002). First, the Loco RGS domain shows

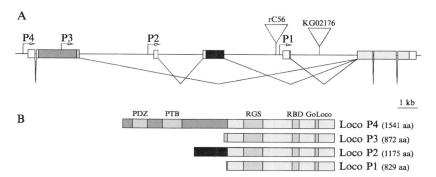

FIG. 1. The *loco* gene. (A) Schematic drawing of the *loco* gene. Transcription is from left to right. Transcriptional start sites and P-element insertion sites are indicated. Open reading frames are indicated by shading. (B) Domain structure of the four deduced Loco proteins.

much higher similarity to the corresponding regions in vertebrate RGS12 and RGS14 than to other members of the RGS family. In addition, the Loco C-terminal RBD and GoLoco domains are arranged in the same relative order as in RGS12 and RGS14. However, the inclusion of Loco in the D/R12 subfamily becomes even more evident when analyzing the longest isoform (Loco P4, 1541 amino acids). The region N-terminal of the Loco backbone contains two additional sequence modules: a PDZ motif and a PTB phosphotyrosine-binding domain. The PDZ domain of approximately 90 amino acids [named after the first three PDZ containing proteins: PSD-95, Discs-large, ZO-1, (Ponting, 1997)] mediates protein–protein interactions and is involved in the assembly of large complexes that are frequently associated with the plasma membrane (Hung and Sheng, 2002). Thus a PDZ domain might help target RGS proteins to specific G-protein-coupled receptor systems (Snow *et al.*, 1998). The PTB domain may confer binding to phosphotyrosine, but studies suggest that PTB domains may also mediate binding to target sequences independent of phosphorylation (Schiff *et al.*, 2000; Yan *et al.*, 2002).

Both these domains are present in the same order in the longest RGS12 isoform. It is interesting to note that at the level of the gene, a similarly complex alternative splicing pattern has been described for the RGS12 gene (Snow *et al.*, 1998) as well as for *loco* (e.g., isoforms with N-terminal PDZ and PTB domains and C-terminal RGS and GoLoco domains, as well as isoforms that lack PDZ and PTB domains and carry only the RGS and GoLoco domains).

loco Expression during Development

The reporter gene activity of the original P-element insertion that led to the identification of *loco* is within the nervous system confined specifically to glial cells. It was subsequently shown to depend on the function of the glial transcription factors Pointed and Gcm (Granderath *et al.*, 2000; Granderath *et al.*, 1999). During oogenesis, at least two different *loco* transcripts are expressed. The Loco P2 encoding RNA is found in anterior-dorsal follicle cells, whereas the Loco P3/4 encoding RNA is expressed by nurse cells (Pathirana *et al.*, 2001). During zygotic development, Loco P2 RNA is expressed in subsets of cells in the lateral ectoderm, which correspond to PNS progenitor cells. However, expression is transient and disappears shortly after the onset of neuronal differentiation (Granderath *et al.*, 1999; Klämbt, unpublished result). Loco P3/4 is expressed in a uniform manner in all ectodermal cells. Loco P1 is expressed in the central nervous system, where most, if not all, of the lateral glial cells initiate *loco* expression. *loco* is expressed continually throughout development and can be also detected in larval and adult stages (Klämbt, unpublished result).

Sequence analysis of the *loco* P1 promoter region (Granderath *et al.*, 2000) revealed the presence of putative binding sites for both Gcm and Pointed [which is itself downstream of Gcm (Giesen *et al.*, 1997)]. Reporter gene constructs carrying different combinations of either wild-type or mutated binding sites were used to confirm the functional significance of these sites. Glial expression of *loco* is apparently a (at least) two-step process: GCM initiates the expression of *loco* in glial cells during earliest glial development. In subsequent stages of gliogenesis, Pointed P1—whose expression is under control of *gcm*—then cooperates with Gcm to maintain and possibly upregulate *loco* expression (Granderath *et al.*, 2000).

However, additional and as yet unknown regulators must operate on the *loco* P1 promoter in glial cells, and virtually nothing is known about the regulation of *loco* expression in nonglial cells. In particular, *loco* was found in screens for genes affecting dorsal-ventral patterning in the *Drosophila* embryo (Pathirana *et al.*, 2001; Stathopoulos *et al.*, 2002) and circadian regulation of gene expression in the *Drosophila* head (Claridge-Chang *et al.*, 2001)—in both cases, it is unlikely that expression is under the control of the Gcm/Pointed machinery.

The *loco* Mutant Phenotype

To further analyze the function of *loco* we generated specific mutants (Granderath *et al.*, 1999, 2000). Utilizing the P-element insertion into the Loco P1 promoter, two chromosomal deficiencies were generated: *loco*$^{\Delta/13}$

and $loco^{\Delta/293}$. Both deficiency chromosomes led to lethality when homozygous or in *trans* to larger deficiencies of the chromosomal region. In order to obtain "additional and better-defined" mutations, we generated EMS induced *loco* alleles, assuming that they should led to lethality. After screening 3088 mutagenized chromosomes for lethality in *trans* to $loco^{\Delta/13}$, we isolated 5 EMS induced lethal *loco* mutants. Alleles *F1* and *L1* appeared to be the strongest alleles, whereas $loco^{M1}$ could be identified as a temperature-sensitive mutation that allowed the survival of paralyzed flies at the permissive temperature.

Within the CNS the *loco* loss of function phenotype resembles the *pointed* phenotype: longitudinal glial cells fail to enwrap the connectives properly and do not establish the intimate glial-glial-cell contacts, leading to a disruption of the blood–brain barrier (Granderath *et al.*, 1999; Klaes *et al.*, 1994). Similar phenotypes were observed in neurexin and gliotactin mutants (Baumgartner *et al.*, 1996; Schulte *et al.*, 2003).

It is likely that *loco* expression does not reflect the terminal step in glial differentiation but participates in earlier regulatory decisions. *loco* mutants do not fully activate expression of the late glial gene *M84*. It is thus conceivable that *loco* acts in a pathway that signals directly to the nucleus to modulate gene transcription. Alternatively, this phenotype might reflect relatively late differentiation defects, where loss of cell–cell contacts results in subsequent aberrations of the gene expression program.

Flies carrying weak *loco* alleles survive to the adult stage, but these flies have a drastically reduced life span and severely impaired locomotor capabilities, which lead to the name—*locomotion defective* (*loco*) (Granderath *et al.*, 1999).

Loco P2 and Loco P3/4 are both expressed during *Drosophila* oogenesis (Pathirana *et al.*, 2001). Initially, both can be detected in the so-called germarium. Later, they are expressed in different domains: while *loco* P2 is detectable in the anterior-dorsal cells of the follicular epithelium, *loco* P3/4 is localized to the nurse cells. *loco* P2 disruption leads to eggs with defects in anterodorsal patterning. If P3/4 function in the nurse cells is disrupted, the so-called "dumpless" phenotype develops, where smaller eggs are being laid (Pathirana *et al.*, 2001). These oogenesis phenotypes resemble those reported for mutation in other components of the G-protein signal transduction pathway. Pathirana and colleagues (2001) also observed occasional defects in the dorsoventral patterning of *loco* mutant embryos following antisense RNA expression. However, the time window of sensitivity to a reduction in *loco* function seems to be rather small and, therefore, the postulated function of *loco* in dorsoventral patterning has to be corroborated by further experiments.

"GAP" or "Trap": Molecular Function(s) of *loco*?

How might the mutant *loco* phenotype be explained in molecular terms? The presence of the RGS domain suggests that *loco* exerts its function by acting as a GAP for $G\alpha_i$ Indeed, it appears that Loco P1 is always coexpressed with $G\alpha_i$—in the lateral glial cells and in the oocyte (Granderath *et al.*, 1999; Pathirana *et al.*, 2001; Wolfgang *et al.*, 1991). Furthermore, the phenotype of $G\alpha_i$ mutations is consistent with a direct interaction of the two proteins. Homozygous $G\alpha_i$ mutant flies are viable and display locomotion defects, similar to the phenotype of hypomorphic *loco* mutations (Yu *et al.*, 2003). Only when both maternal and zygotic $G\alpha_i$ transcripts are removed do $G\alpha_i$ mutations lead to lethality.

Direct protein–protein interaction between Loco and $G\alpha_i$ was demonstrated in a yeast two-hybrid screen for binding partners of $G\alpha_i$ It was therefore postulated that Loco acts as a terminator of $G\alpha_i$ signaling via GTPase activation through the RGS domain (Granderath *et al.*, 1999).

However, the precise function of Loco in the signaling cascade has yet to be determined. The yeast two-hybrid assay also showed that binding of Loco to $G\alpha_i$ does not require the RGS domain. This interaction can also be mediated by C-terminal fragments containing three small adjacent stretches of high homology initially termed B, C, and D (Granderath *et al.*, 1999). Region D was later identified as the GoLoco domain. The GoLoco domain, which binds specifically to $G\alpha_i$ subunits, has been shown to serve as a GDP dissociation inhibitor (GDI) for $G\alpha_i$ proteins (De Vries *et al.*, 2000; Kimple *et al.*, 2001; Kimple *et al.*, 2002).

The *Drosophila* protein Pins (Partner of inscuteable) is a GoLoco protein without an RGS domain. With similarity to $G\alpha_i$, Pins plays an essential role in asymmetric cell divisions in neural precursor cells. Interestingly, Pins can compete with $G\alpha_i$ for binding to the $G\beta\gamma$ subunit. Furthermore, in precursors of the central (but not the peripheral) nervous system, overexpression of a constitutively active form of $G\alpha_i$ has no effect on asymmetric cell division, whereas expression of the wild-type isoform does. Apparently, the GoLoco protein Pins binds "inactive" GDP-$G\alpha_i$ through its GoLoco domain, releasing it from the $G\beta\gamma$ subunit in the absence of an extracellular signal (Schaefer *et al.*, 2001).

Does Loco perhaps act in a related manner? The combination of GDI (GoLoco), GAP (RGS), and putative Ras-binding (RBD) domains may indicate that Loco does not simply act as a terminator of a GTP-$G\alpha_i$ signal. It is tempting to speculate that the protein is instead responsible for the initiation and/or prolongation of a $G\beta\gamma$-dependent signaling pathway. The RGS domain would, in this model, merely serve as a "backup" to ensure that no active GTP-$G\alpha_i$ is present. Interestingly, preliminary experiments show that the expression of constitutively active $G\alpha_i$ isoforms in glial cells has no apparent effects on nervous system development or viability (Granderath, unpublished results). If this finding can be confirmed and extended, it would argue against a "GAP-only" mode of action of *loco* P1 and instead support the second, "GAP-and-trap" model.

Open Questions

While it is quite clear that Loco does interact with $G\alpha_i$, the precise nature of the signaling cascade(s) is entirely obscure. Are extracellular signals and transmembrane receptor(s) required to initiate the $G\alpha_i$/Loco signaling cascade in the glia and in the oocyte and, if yes, what is their nature? The expression of the *Drosophila* G-protein kinase 2 (gprk2) in lateral glial cells may hint toward a receptor-dependent mechanism. The *Drosophila* FGFR2 homolog Heartless (Htl) represents a candidate receptor; it is expressed in glial cells, and *htl* and *loco* mutants show strikingly similar glial phenotypes (Granderath *et al.*, 1999; Shishido *et al.*, 1997). However, experiments to unravel interactions among *heartless, loco, gprk2,* and other G-protein signaling genes required in gliogenesis have yet to be carried out. Neither have any downstream components of the putative "*loco* pathway(s)" been identified.

Finally, functional differences among the different Loco isoforms remain to be analyzed. The role of the different N termini and their interaction partners—although certainly involved in transmission and/or modulation of the signal—is at present completely obscure. It is even possible that the different Loco isoforms may be used to regulate aspects of asymmetric cell divisions in the nervous system. Loco P1 is expressed from the early stages of gliogenesis in the CNS, in the progeny of the so-called lateral glioblast. The glioblast divides symmetrically, in contrast to neuroblasts, which give rise to neuron-only or mixed neuron-glial lineages. Embryonic expression of the longer P2 isoform is confined to precursor cells of the peripheral nervous system, which are known to divide asymmetrically (Bellaiche and Schweisguth, 2001; Knoblich, 2001). One may thus speculate that Loco P1 serves to suppress asymmetry during cellular division in gliogenesis, whereas the longer P2 isoform regulates or augments asymmetry during the earliest steps of PNS development.

In summary, first functional analyses of a *Drosophila* RGS protein have posed many more new questions than answered old ones. What is the nature of the signal and the receptor involved? Does Loco serve to terminate a $G\alpha_i$ signal, to support a $G\beta\gamma$-dependent pathway, or does it act as an effector itself—or is its role a combination of these mechanisms? What is the nature of the other interaction partners of Loco? How do the different Loco isoforms exert their functions on the cellular level—what processes other than cellular extension (as seen in the wrapping phenotype) are involved? Is there indeed a connection to the regulation of asymmetric cell divisions?

In principle, the numerous genetic tools available in *Drosophila* should allow one to tackle these questions efficiently. In the eye of the classical geneticist, the sole drawback is the lack of a "convenient" phenotype—one that can conveniently be identified and then be used to screen genetic modifiers, i.e., enhancers or suppressors of *loco*. Once such a phenotype is available, it will be highly interesting to use *loco* as a starting point to a further understanding of its cellular functions and to shed further light onto this exciting gene family.

References

Baumgartner, S., Littleton, J. T., Broadie, K., Bhat, M. A., Harbecke, R., Lengyel, J. A., Chiquet-Ehrismann, R., Prokop, A., and Bellen, H. J. (1996). A Drosophila neurexin is required for septate junction and blood nerve barrier formation and function. *Cell* **87**, 1059–1068.

Bellaiche, Y., and Schweisguth, F. (2001). Lineage diversity in the Drosophila nervous system. *Curr. Opin. Genet. Dev.* **11**, 418–423.

Bellen, H. J., Wilson, C., and Gehring, W. J. (1990). Dissecting the complexity of the nervous system by enhancer detection. *Bioessays* **12**, 199–204.

Bienz, M. (1999). APC: The plot thickens. *Curr. Opin. Genet. Dev.* **9**, 595–603.

Bolshakov, V. N., Topalis, P., Blass, C., Kokoza, E., della Torre, A., Kafatos, F. C., and Louis, C. (2002). A comparative genomic analysis of two distant diptera, the fruit fly, *Drosophila melanogaster*, and the malaria mosquito, *Anopheles gambiae. Genome Res.* **12**, 57–66.

Brand, A. H., and Perrimon, N. (1993). Targeted gene expression as a means of altering cell fates and generating dominant phenotypes. *Development* **118**, 401–415.

Cassill, J. A., Whitney, M., Joazeiro, C. A., Becker, A., and Zuker, C. S. (1991). Isolation of Drosophila genes encoding G protein-coupled receptor kinases. *Proc. Natl. Acad. Sci. USA* **88**, 11067–11070.

Claridge-Chang, A., Wijnen, H., Naef, F., Boothroyd, C., Rajewsky, N., and Young, M. W. (2001). Circadian regulation of gene expression systems in the Drosophila head. *Neuron* **32**, 657–671.

De Vries, L., and Gist Farquhar, M. (1999). RGS proteins: More than just GAPs for heterotrimeric G proteins. *Trends Cell Biol.* **9**, 138–144.

De Vries, L., Fischer, T., Tronchere, H., Brothers, G. M., Strockbine, B., Siderovski, D. P., and Farquhar, M. G. (2000). Activator of G protein signaling 3 is a guanine dissociation inhibitor for Galpha i subunits. *Proc. Natl. Acad. Sci. USA* **97**, 14364–14369.

De Vries, L., Mousli, M., Wurmser, A., and Farquhar, M. G. (1995). GAIP, a protein that specifically interacts with the trimeric G protein G alpha i3, is a member of a protein family with a highly conserved core domain. *Proc. Natl. Acad. Sci. USA* **92,** 11916–11920.

Egger, B., Leemans, R., Loop, T., Kammermeier, L., Fan, Y., Radimerski, T., Strahm, M. C., Certa, U., and Reichert, H. (2002). Gliogenesis in Drosophila: Genome-wide analysis of downstream genes of glial cells missing in the embryonic nervous system. *Development* **129,** 3295–3309.

Elmore, T., Rodriguez, A., and Smith, D. P. (1998). dRGS7 encodes a Drosophila homolog of EGL-10 and vertebrate RGS7. *DNA Cell Biol.* **17,** 983–989.

Engels, W. R. (1989). P-elements in Drosophila. *In* "Mobile DNA" (M. M., Howe, ed.), pp. 437–484. American Society for Microbiology, Washington, DC.

Fan, S., and Schneider, L. E. (2003). The role of maternal and zygotic Gprk2 expression in Drosophila development. *Biochem. Biophys. Res. Commun.* **301,** 127–135.

Giesen, K., Hummel, T., Stollewerk, A., Harrison, S., Travers, A., and Klämbt, C. (1997). Glial development in the Drosophila CNS requires concomitant activation of glial and repression of neuronal differentiation genes. *Development* **124,** 2307–2316.

Granderath, S., Bunse, I., and Klämbt, C. (2000). gcm and pointed synergistically control glial transcription of the drosophila gene loco. *Mech. Dev.* **91,** 197–208.

Granderath, S., and Klämbt, C. (1999). Glia development in the embryonic CNS of *Drosophila. Curr. Opin. Neurobiol.* **9,** 531–536.

Granderath, S., Stollewerk, A., Greig, S., Goodman, C. S., O'Kane, C., and Klämbt, C. (1999). The *pointed* target gene *loco* encodes an RGS protein required for *Drosophila* glial differentiation. *Development* **126,** 1781–1791.

Hollinger, S., and Hepler, J. R. (2002). Cellular regulation of RGS proteins: Modulators and integrators of G protein signaling. *Pharmacol. Rev.* **54,** 527–559.

Hung, A. Y., and Sheng, M. (2002). PDZ domains: Structural modules for protein complex assembly. *J. Biol. Chem.* **277,** 5699–5702.

Jones, B. W. (2001). Glial cell development in the Drosophila embryo. *Bioessays* **23,** 877–887.

Jones, W. M., and Bejsovec, A. (2003). Wingless signaling: An axin to grind. *Curr. Biol.* **13,** R479–R481.

Kimple, R. J., De Vries, L., Tronchere, H., Behe, C. I., Morris, R. A., Gist Farquhar, M., and Siderovski, D. P. (2001). RGS12 and RGS14 GoLoco motifs are G alpha(i) interaction sites with guanine nucleotide dissociation inhibitor activity. *J. Biol. Chem.* **276,** 29275–29281.

Kimple, R. J., Kimple, M. E., Betts, L., Sondek, J., and Siderovski, D. P. (2002). Structural determinants for GoLoco-induced inhibition of nucleotide release by Galpha subunits. *Nature* **416,** 878–881.

Klaes, A., Menne, T., Stollewerk, A., Scholz, H., and Klämbt, C. (1994). The Ets transcription factors encoded by the Drosophila gene pointed direct glial cell differentiation in the embryonic CNS. *Cell* **78,** 149–160.

Knoblich, J. A. (2001). Asymmetric cell division during animal development. *Nature Rev. Mol. Cell Biol.* **2,** 11–20.

Koelle, M. R., and Horvitz, H. R. (1996). EGL 10 regulates G protein signaling in the C. elegans nervous system and shares a conserved domain with many mammalian proteins. *Cell* **84,** 115–125.

Lannutti, B. J., and Schneider, L. E. (2001). Gprk2 controls cAMP levels in Drosophila development. *Dev. Biol.* **233,** 174–185.

Lee, Y. S., and Carthew, R. W. (2003). Making a better RNAi vector for Drosophila: Use of intron spacers. *Methods* **30,** 322–329.

Liao, G. C., Rehm, E. J., and Rubin, G. M. (2000). Insertion site preferences of the P transposable element in *Drosophila melanogaster. Proc. Natl. Acad. Sci. USA* **97,** 3347–3351.

Mao, J., Yuan, H., Xie, W., Simon, M. I., and Wu, D. (1998). Specific involvement of G proteins in regulation of serum response factor-mediated gene transcription by different receptors. *J. Biol. Chem.* **273,** 27118–27123.

Morin, X., Daneman, R., Zavortink, M., and Chia, W. (2001). A protein trap strategy to detect GFP-tagged proteins expressed from their endogenous loci in Drosophila. *Proc. Natl. Acad. Sci. USA* **98,** 15050–15055.

Natochin, M., Lester, B., Peterson, Y. K., Bernard, M. L., Lanier, S. M., and Artemyev, N. O. (2000). AGS3 inhibits GDP dissociation from galpha subunits of the Gi family and rhodopsin-dependent activation of transducin. *J. Biol. Chem.* **275,** 40981–40985.

Neubig, R. R., and Siderovski, D. P. (2002). Regulators of G-protein signalling as new central nervous system drug targets. *Nature Rev. Drug Discov.* **1,** 187–197.

O'Kane, C., and Gehring, W. (1987). Detection in situ of genomic regulatory elements in Drosophila. *Proc. Natl. Acad. Sci. USA* **84,** 9123–9127.

Pathirana, S., Zhao, D., and Bownes, M. (2001). The Drosophila RGS protein Loco is required for dorsal/ventral axis formation of the egg and embryo, and nurse cell dumping. *Mech. Dev.* **109,** 137–150.

Ponting, C. P. (1997). Evidence for PDZ domains in bacteria, yeast, and plants. *Protein Sci.* **6,** 464–468.

Robertson, H. M., Preston, C. R., Phillis, R. W., Johnson-Schlitz, D. M., Benz, W. K., and Engels, W. R. (1988). A stable source of P-element transposase in *Drosophila melanogaster. Genetics* **118,** 6341–6351.

Ross, E. M., and Wilkie, T. M. (2000). GTPase-activating proteins for heterotrimeric G proteins: Regulators of G protein signaling (RGS) and RGS-like proteins. *Annu. Rev. Biochem.* **69,** 795–827.

Schaefer, M., Petronczki, M., Dorner, D., Forte, M., and Knoblich, J. A. (2001). Heterotrimeric G proteins direct two modes of asymmetric cell division in the Drosophila nervous system. *Cell* **107,** 183–194.

Schiff, M. L., Siderovski, D. P., Jordan, J. D., Brothers, G., Snow, B., De Vries, L., Ortiz, D. F., and Diverse-Pierluissi, M. (2000). Tyrosine-kinase-dependent recruitment of RGS12 to the N-type calcium channel. *Nature* **408,** 723–727.

Schneider, L. E., and Spradling, A. C. (1997a). The Drosophila G protein coupled receptor kinase homologue Gprk2 is required for egg morphogenesis. *Development* **124,** 2591–2602.

Schneider, L. E., and Spradling, A. C. (1997b). The Drosophila G-protein-coupled receptor kinase homologue Gprk2 is required for egg morphogenesis. *Development* **124,** 2591–2602.

Schulte, J., Tepass, U., and Auld, V. J. (2003). Gliotactin, a novel marker of tricellular junctions, is necessary for septate junction development in Drosophila. *J. Cell Biol.* **161,** 991–1000.

Sepp, K. J., and Auld, V. J. (1999). Conversion of lacZ enhancer trap lines to GAL4 lines using targeted transposition in *Drosophila melanogaster. Genetics* **151,** 1093–1101.

Shishido, E., Ono, N., Kojima, T., and Saigo, K. (1997). Requirements of DFR1/Heartless, a mesoderm specific Drosophila FGF receptor, for the formation of heart, visceral and somatic muscles, and ensheathing of longitudinal axon tracts in CNS. *Development* **124,** 2119–2128.

Siderovski, D. P., Diverse-Pierluissi, M., and De Vries, L. (1999). The GoLoco motif: A Galphai/o binding motif and potential guanine-nucleotide exchange factor. *Trends Biochem. Sci.* **24**, 340–341.

Siderovski, D. P., Hessel, A., Chung, S., Mak, T. W., and Tyers, M. (1996). A new family of regulators of G-protein-coupled receptors? *Curr. Biol.* **6**, 211–212.

Sierra, D. A., Gilbert, D. J., Householder, D., Grishin, N. V., Yu, K., Ukidwe, P., Barker, S. A., He, W., Wensel, T. G., Otero, G. *et al.* (2002). Evolution of the regulators of G-protein signaling multigene family in mouse and human. *Genomics* **79**, 177–185.

Snow, B. E., Hall, R. A., Krumins, A. M., Brothers, G. M., Bouchard, D., Brothers, C. A., Chung, S., Mangion, J., Gilman, A. G., Lefkowitz, R. J., and Siderovski, D. P. (1998). GTPase activating specificity of RGS12 and binding specificity of an alternatively spliced PDZ (PSD-95/Dlg/ZO-1) domain. *J. Biol. Chem.* **273**, 17749–17755.

Spink, K. E., Polakis, P., and Weis, W. I. (2000). Structural basis of the Axin-adenomatous polyposis coli interaction. *EMBO J.* **19**, 2270–2279.

Stathopoulos, A., Van Drenth, M., Erives, A., Markstein, M., and Levine, M. (2002). Whole-genome analysis of dorsal-ventral patterning in the Drosophila embryo. *Cell* **111**, 687–701.

Van De Bor, V., and Giangrande, A. (2002). glide/gcm: At the crossroads between neurons and glia. *Curr. Opin. Genet. Dev.* **12**, 465–472.

Willert, K., Logan, C. Y., Arora, A., Fish, M., and Nusse, R. (1999). A Drosophila Axin homolog, Daxin, inhibits Wnt signaling. *Development* **126**, 4165–4173.

Wolfgang, W. J., Quan, F., Thambi, N., and Forte, M. (1991). Restricted spatial and temporal expression of G-protein α subunits during *Drosophila* embryogenesis. *Development* **113**, 527–538.

Yan, K. S., Kuti, M., and Zhou, M. M. (2002). PTB or not PTB—that is the question. *FEBS Lett.* **513**, 67–70.

Yu, F., Cai, Y., Kaushik, R., Yang, X., and Chia, W. (2003). Distinct roles of Galphai and Gbeta13F subunits of the heterotrimeric G protein complex in the mediation of Drosophila neuroblast asymmetric divisions. *J. Cell Biol.* **162**, 623–633.

[22] Analysis of the Roles of Pins and Heterotrimeric G Proteins in Asymmetric Division of *Drosophila* Neuroblasts

By FENGWEI YU

Abstract

Drosophila neuroblasts divide asymmetrically to give rise to two daughter cells of distinct fates and sizes. A number of studies in asymmetric cell division have begun to elucidate the molecular mechanisms underlying asymmetric protein/RNA localization, spindle orientation, and spindle asymmetry. This article describes methods of analyzing the role of Pins and heterotrimeric G proteins in *Drosophila* neuroblast asymmetric division.

Introduction

During the development of the *Drosophila* central nervous system (CNS), neural progenitors or "neuroblasts" (NBs) delaminate from an epithelium of neuroectodermal cells and subsequently undergo asymmetric cell division along the apical–basal axis, generating two daughter cells of unequal sizes and distinct fates. The larger apical daughter cell remains as a NB and divides repeatedly in a stem cell-like mode while the smaller basal one becomes a ganglion mother cell (GMC), which divides terminally to give rise to two postmitotic neurons or glia (Campos-Ortega, 1995). The distinct cell fates of the two daughter cells are specified by the asymmetric localization and segregation of cell fate determinants (e.g., Prospero and Numb) into the basal daughter cells during NB division (Jan and Jan, 2001; Chia and Yang, 2002). Asymmetric segregation of the determinants requires the coordination of mitotic spindle orientation and the basal localization of determinants. The mitotic spindle is oriented to allow cells to divide along the apical–basal axis (for reviews, see Doe and Bowerman, 2001; Knoblich, 2001). The spindle displacement and asymmetry contribute to asymmetric cytokinesis, which leads to the generation of two daughter cells of different sizes (Cai *et al.*, 2003; Kaltschmidt *et al.*, 2000). Another novel aspect of NB division is the asymmetric formation of astral microtubules (MTs): astral microtubules at the apical side grow robustly whereas astral MTs at the basal side undergo shrinkage (Fuse *et al.*, 2003; Yu *et al.*, 2003).

A complex of proteins that are localized to the apical cortex of NBs are required to coordinate spindle orientation with basal localization of cell fate determinants. This apical complex consists of Inscuteable (Insc), Partner of Inscuteable (Pins), $G\alpha_i$, Bazooka (Baz), DmPar6, and DaPKC. Insc may bridge Pins/$G\alpha_i$ and Bazooka/DaPKC/DmPar6 complexes at the apical cortex of NBs, as Insc is able to interact directly with both Pins and Bazooka. Mutations in any of these genes lead to the mislocalization of cell fate determinants and misoriented mitotic spindles (Kuchinke *et al.*, 1998; Petronczki and Knoblich, 2001; Kraut *et al.*, 1996; Schober *et al.*, 1999; Wodarz *et al.*, 1999, 2000; Yu *et al.*, 2000, 2002). These apical complexes can be subdivided into two pathways, the Pins/$G\alpha_i$ pathway and the Insc/Baz/DmPar6/DaPKC pathway, which redundantly regulate the unequal sizes of daughter cells and suppress the formation of basal astral microtubules in NBs. Double mutants disrupting both pathways simultaneously show a high penetrance of equal size division, whereas each single mutant does not (Cai *et al.*, 2003; Fuse *et al.*, 2003; Yu *et al.*, 2003).

Pins, a GoLoco motif-containing protein, has been shown to play a crucial role in the asymmetric division of *Drosophila* NBs (Parmentier *et al.*, 2000; Schaefer *et al.*, 2000; Siderovski *et al.*, 1999; Yu *et al.*, 2000). Pins was originally isolated from a yeast two-hybrid screen using the asymmetric localization domain of Insc and protein complex purification (Schaefer *et al.*, 2000; Yu *et al.*, 2000). Pins colocalizes with Insc at the apical cortex to coordinate the basal localization of cell fate determinants and orientation of the mitotic spindle. Pins interacts with Insc through its N-terminal tetratricopeptide repeats (TPRs) which provide the apical cue for the apical localization of Pins. Pins also interacts with the GDP-bound form of $G\alpha_i$ through its C-terminal GoLoco motifs, which recruit Pins to the plasma membrane (Yu *et al.*, 2002). Pins acts as a guanine nucleotide dissociation inhibitor (GDI) for GDP-$G\alpha_i$ and prevent GDP-$G\alpha_i$ from binding the $G\beta\gamma$ complex (Schaefer *et al.*, 2001). While $G\alpha_i$ colocalizes with Pins at the apical cortex and functions together with Pins in the apical complex, $G\beta\gamma$ is localized throughout the cell cortex. $G\beta\gamma$ acts upstream of the two apical pathways (Pins/$G\alpha_i$ and Baz/DaPKC/DmPar6/Insc) to control daughter cell size difference and suppress the formation of basal astral MTs in NBs (Yu *et al.*, 2003). This article describes methods of analyzing the molecular mechanisms underlying the asymmetric cell division of *Drosophila* NBs, with particular emphasis on the roles of Pins and heterotrimeric G proteins.

Yeast Two-Hybrid Screen with the Functional Domain of Insc

Insc, the first identified apical protein, plays a critical role in the asymmetric cell division of *Drosophila* NBs. Insc is localized at the apical stalk in delaminating NBs and subsequently at the apical cortex during late interphase and mitosis (Kraut *et al.*, 1996; Li *et al.*, 1997). Insc is required for the basal localization of cell fate determinants, orientation of the mitotic spindle along the apical/basal axis, and coordination of these processes in mitotic NBs. Domain dissection analyses of the Insc protein revealed that the central 218 amino acid region of Insc (asymmetric localization domain) is necessary and sufficient for apical cortical localization and for mitotic spindle orientation along the apical–basal axis (Tio *et al.*, 1999). To identify molecules that may direct apical localization of Insc, we searched for proteins that interact with the asymmetric localization domain of Insc in a yeast two-hybrid screen (Yu *et al.*, 2000).

The yeast two-hybrid system is utilized to identify protein–protein interactions as it has been proven to be a powerful technique to isolate interacting proteins starting with a protein of interest (Bartel *et al.*, 1995). In the MATCHMAKER System 2 (Clontech), a bait protein is expressed as a fusion to the GAL4 DNA-binding domain (DNA-BD), while the library is expressed as a fusion to the GAL4 activation domain (AD). When the bait and library fusion proteins interact, the DNA-BD and AD are brought into a complex, thus activating the transcription of reporter genes. This section describes the application of this method to identify novel interacting partners of Insc using Insc as a bait protein.

Solutions and Materials

YPD medium: 20 g/liter Difco peptone, 10 g/liter yeast extract, 20 g/liter agar, adjust pH to 5.8 with HCl, autoclave and cool to 55°, and add 40% glucose stock to 2% final concentration

Synthetic dropout (SD) medium: 6.7 g/liter Difco yeast nitrogen base without amino acids, 25 g/liter agar, add proper dropout powder. Adjust pH to 5.8 with NaOH, autoclave and cool to 55°, and add glucose to 2% final concentration.

1 M 3-amino-1,2,4-triazole (3-AT)

11 mg/ml salmon sperm DNA

Z buffer: 16.1 g/liter $Na_2HPO_4 \cdot 7H_2O$, 5.50 g/liter $NaH_2PO_4 \cdot H_2O$, 0.75 g/liter KCl, 0.246 g/liter $MgSO_4 \cdot 7H_2O$, adjust to pH 7.0 with NaOH and autoclave

Z buffer/X-gal solution: 100 ml Z buffer, 0.27 ml β-mercaptoethanol, 1.67 ml 20 mg/ml X-gal (in DMF) stock solution

Yeast lysis solution: 2% Triton X-100, 1% sodium dodecyl sulfate (SDS), 10 mM NaCl, 10 mM Tris (pH8.0), 1.0 mM EDTA

10× TE: 0.1 M Tris–HCl, 10 mM EDTA, pH 7.5

10× LiAc: 1 M lithium acetate, adjust pH to 7.5 with acetic acid

TE/LiAc/PEG4000: 40% PEG, 1× TE, 1× LiAc

Library Transformation and Screening

The library used for yeast two-hybrid screens is a 0- to 21-h embryonic *Drosophila* Matchmaker cDNA library cloned in the pACT2 vector containing the Gal4 activation domain (Clontech). The various baits used for screening are cloned in pAS2–1 (Clontech) and include the full-length Insc (Insc-1), different lengths of the central region of Insc containing Insc asymmetric localization domain (Insc-1 to Insc-5), and various truncated forms of Insc lacking the central asymmetric localization domain (Insc-6 to Insc-8), as shown in Fig. 1. The bait pAS-Insc-5 is utilized for the yeast two-hybrid screen as follows.

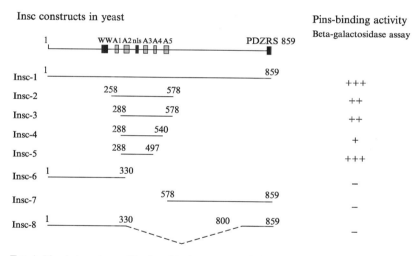

FIG. 1. Pins interacts specifically with the asymmetric localization domain of Insc in yeast two-hybrid assays. The top line is a schematic illustration of the full-length Insc protein. Insc-1 to Insc-8, containing various portions of the Insc coding region, were fused in frame with the Gal4 DNA-binding domain in the pAS2-1 vector. Binding activities between various Insc regions and full-length Pins were semiquantitated based on the time taken for colonies to turn blue in the X-gal filter lift assay and are shown on the right: +++, 30 min; ++, 30–90 min; +, >120 min; −, no significant staining. The five shaded boxes (A1–A5) represent regions that, according to Kraut and Campos-Ortega (1996), share homology with ankyrin repeats. WW denotes location of WW domain target site, and nls indicates a nuclear-location signal. A PDZ recognition site (PDZRS) exists at the end of the Insc protein. Note that all bait constructs, which include the Insc asymmetric localization domain from amino acids 288 to 497, show a positive interaction with Pins. Reproduced with permission from Yu *et al.* (2000), © 2000 Cell Press.

1. Inoculate 1 ml of YPD or SD medium with four to six colonies (2–3 mm in diameter) of the yeast strain Y190 (Clontech). Shake vigorously at 30° overnight.
2. Transfer cells to a flask containing 200 ml medium. Shake at 30° overnight to $OD_{600} = 2$–3.
3. Transfer overnight culture to 1 liter medium at 1:10 dilution to obtain an $OD_{600} = 0.2$–0.3.
4. Incubate at 30° for 5–6 h with shaking of 250 rpm to allow cell growth to $OD_{600} = 0.5$–0.7.
5. Centrifuge at 1000g for 5 min at room temperature. Discard the supernatant and resuspend the cell pellet by vortexing in 1 liter of sterile water.
6. Centrifuge as described previously and wash with 50 ml freshly prepared $1 \times$ TE/LiAc. Pool cells in a 50-ml falcon tube. Centrifuge at 1000g for 5 min at room temperature in a tabletop centrifuge.
7. Decant the supernatant and resuspend pellet in 4 ml $1 \times$ TE/LiAc. Prepare PEG/TE/LiAc solution.
8. Combine 100 μg of library DNA and 10 mg of salmon sperm DNA in a 1.5-ml microfuge tube and mix well.
9. Add DNA mixture to competent yeast cells and vortex well. Add 30 ml PEG/TE/LiAc solution and vortex to mix.
10. Incubate at 30° for 30 min with shaking of 200 rpm.
11. Add 3.5 ml dimethyl sulfoxide (DMSO) and mix well by gentle inversion or swirling. Do not vortex.
12. Heat shock for 15 min in a 42° water bath. Swirl occasionally to mix.
13. Chill cells on ice for 2 min. Centrifuge cells at 1000g for 5 min. Remove the supernatant.
14. Resuspend cells in 4 ml $1 \times$ TE and plate cells onto SD/-Trp/-Leu/-His + 25 mM 3-AT plates.
15. Save 100 μl of a 1:1000 and 1:10,000 dilution of cells to spread onto SD/-Trp/-Leu plates to estimate the transformation efficiency of library DNA.
16. Incubate plates, upside down, at 30° for 1 week.
17. Transfer colonies of 2–3 mm diameter to a SD/-Trp/-Leu/-His+3-AT master plate. Incubate colony patches for 2 days.
18. Replicate the master plate onto another SD/-Trp/-Leu/-His+3-AT plate for backup. One of them is subjected to the colony-lift filter assay as described next.

Colony-Lift Filter Assay

1. Prepare Z buffer/X-gal solution as described in buffer table.
2. Presoak a sterile Whatman filter by placing it in 3 ml Z buffer/X-gal solution in a clean 100-mm petri dish.
3. Place a clean, dry filter over the surface of colony patches to be assayed.
4. Poke holes through the filter into the agar in three asymmetric locations to orient the filter to the agar.
5. Carefully lift the filter off the agar plate with forceps and transfer it to a pool of liquid nitrogen with colonies facing up.
6. After completely frozen for 1 min, remove the filter from liquid nitrogen and allow to thaw at room temperature.
7. Place the filter with colony patches right side up on a presoaked filter. Avoid air bubbles under filter.
8. Incubate filters at 30° and check periodically for the appearance of blues colonies.

Plasmid Isolation from Yeast

1. Inoculate single, separate yeast colonies into 3 ml of SD/-Leu/-Trp medium. Incubate at 30° for at least 20 h with shaking of 250 rpm.
2. Pellet by centrifugation of 1.5 ml of overnight yeast culture in a 1.5-ml microfuge tube. Discard supernatant and resuspend the pellet in the residual liquid by vortexing.
3. Add 0.2 ml of yeast lysis solution.
4. Add 0.2 ml of phenol/chloroform (1:1) and 0.3 g of acid-washed glass beads (Sigma).
5. Vortex for 2 min.
6. Centrifuge at 14,000 rpm for 5 min at room temperature. Transfer the supernatant to a clean microcentrifuge tube and precipitate DNA by adding the same volume of isopropanol. Pellet precipitated DNA by centrifugation at 14,000 rpm for 25 min at room temperature.
7. Wash the pellet with 70% ethanol and air dry.
8. Resuspend DNA pellet in 10 μl sterile water.
9. Add 2 μl of yeast DNA prep into electrocompetent *Escherichia coli* DH5α cells in an electroporation cuvette on ice.
10. Perform the electroporation as per manufacturer's protocol (Bio-Rad).
11. Add 200 μl Luria broth (LB) to allow cells to recover at 37° for 1 h.
12. Spread cells on a selective medium plate (e.g., for Matchmaker 2 library clones, select colonies on Ampicillin).

13. Isolate plasmid DNA from *E. coli* DH5α transformants.
14. Verify plasmid by restriction enzyme digestion with *Eco*RI and *Xho*I.
15. Sequence the insert of the rescued plasmid.
16. Analyze and assemble the clone sequence using the DNAstar program and analyze the encoded protein sequence by SMART.

Verification of Positive Two-Hybrid Interactions

About 200 clones showed up from our primary screen with the bait pAS-Insc-5. To eliminate false-positive clones, rescued plasmids were re-transformed into various baits, including five positive baits containing the middle part of Insc and three negative control baits excluding the Insc functional domain. Various transformants were tested by the colony-lift filter assay. We selected clones that interacted with all positive baits but could not interact with three negative baits. False positives that did not fit these criteria were removed from further analysis. Five out of 200 clones that showed specific interaction with the asymmetric localization domain of Insc in the screen contained sequences from the N-terminal region of Pins (Yu *et al.*, 2000).

In Vitro Protein Interaction Assay

The identification of Pins by its interaction with Insc in a yeast two-hybrid screen suggested a direct interaction of these two proteins. However, yeast two-hybrid analyses may not reveal real interactions in all the cases. Alternatively, potential interactions can be tested further in an *in vitro* interaction assay. The *in vitro* protein interaction assay is a standard method used to test the direct physical interaction of two proteins. The most commonly used method is the pull-down assay in which a "bait" protein is fused to glutathione-*S*-transferase (GST) and the fusion protein is bound to an immobilized glutathione support while the "prey" protein binds to the immobilized "bait" protein. The protein complex is then eluted and analyzed using SDS–polyacrylamide gel electrophoresis (SDS–PAGE). We used this method to provide further evidence that Pins interacts with Insc, in addition to their interaction in a yeast two-hybrid system.

Procedure

Constructs containing the entire Pins coding region, N-terminal Pins (amino acids 1–378) and C-terminal Pins (amino acids 364–658) are generated by cloning polymerase chain reaction (PCR) fragments into pET15b (Novagen). All PCR-generated sequences are checked by DNA sequencing.

Various ^{35}S-labeled Pins translation products are produced using the TNT-*in vitro* coupled transcription and translation kit (Promega) and T7 polymerase. For each binding assay, 5 μl of mixture containing the translated ^{35}S-labeled protein is mixed with beads containing about 4 μg of the various GST-Insc fusion proteins in 0.5 ml phosphate-buffered saline (PBS) for 1 h and washed five times with PBS + 1% Triton X-100. Proteins bound to beads are boiled and eluted in 2× SDS loading buffer and are resolved by electrophoresis on SDS–PAGE gels, which are in turn dried and autoradiographed.

The full-length Pins is precipitated by various GST-Insc proteins containing the middle portion of Insc, suggesting that Pins is able to interact directly with the asymmetric localization domain of Insc (Fig. 2A). Full-length and N-terminal Pins could bind to GST-Insc but not the GST-alone control, whereas C-terminal Pins could not bind to GST-Insc (Fig. 2B). Thus, the N-terminal region of Pins (containing seven TPR domains) is necessary and sufficient for direct interaction with the Insc asymmetric localization domain.

Protein Complex Purification, Coimmunoprecipitation, and Western Blotting

If the physical interaction between Pins and Insc reflects a real *in vivo* interaction, Pins should form a complex with Insc *in vivo*. In order to purify this complex, we used a transgenic fly strain that ubiquitously expresses a FLAG-tagged version of Insc under the control of the *heat shock 70* promoter (hs59NheI in Tio *et al.*, 1999). This version of Insc is fully functional as its expression can rescue the defects associated with the *insc* loss-of-function mutation. Protein extracts are prepared from heat-shocked transgenic embryos; extracts prepared from nonheat-shocked transgenic embryos processed in parallel are used as controls. An anti-FLAG immunoaffinity column is used for both control and experimental extracts. Pins is specifically copurified with the extract prepared from the heat-shocked transgenic embryos containing the FLAG-tagged Insc and not with the control extract (Fig. 2C). These results indicate that Insc and Pins interact either directly or indirectly *in vivo*.

Solutions

Lysis buffer (for crude extraction): 20 mM Tris (pH 7.5), 150 mM NaCl, 1.0% Triton X-100, 1:10 complete protease inhibitors (Roche) (final concentration: 2.5×) (stock: one tablet dissolved in 2 ml H_2O is equivalent to 25× concentration)

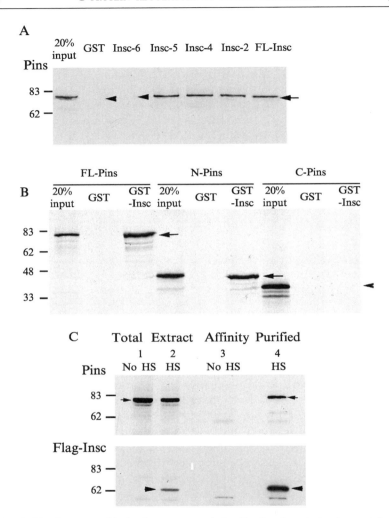

Fig. 2. Pins interacts with the Insc asymmetric localization domain *in vitro* and forms a complex with Insc *in vivo*. (A) *In vitro* interaction between ^{35}S-labeled full-length Pins protein and various GST-Insc fusion proteins (see Fig. 1 for nomenclature). (B) The N-terminal region of Pins containing TPR repeats interacts with Insc *in vitro*. ^{35}S-labeled full-length Pins (FL-Pins), N-terminal Pins (N-Pins, amino acids 1–378), or C-terminal Pins (C-Pins, amino acids 364–658) was tested for interaction with GST (GST alone) or a full-length GST-Insc fusion protein (GST-Insc). (C) Pins and Insc complex *in vivo*. Embryonic extracts were prepared from a fly strain homozygous for a FLAG-tagged *insc* transgene under the control of the *hsp70* promoter. Total extracts from nonheat-shocked (lane 1) and heat-shocked (lane 2) embryos, as well as FLAG affinity-purified samples from nonheat-shocked (lane 3) and heat-shocked (lane 4) embryos, were electrophoresed and the Pins protein (arrow) was visualized by Western blotting with the anti-Pins antibody. The filter was stripped and reblotted with an anti-FLAG antibody to visualize the FLAG-tagged Insc band (arrowhead, lower panel). Reproduced with permission from Yu *et al.* (2000), © 2000 Cell Press.

Wash buffer: 20 mM Tris (pH 7.5), 150 mM NaCl, 0.5% Triton X-100
Elution buffer: 0.1 M glycine (pH 3.5), 0.5% Triton X-100
Anti-FLAG immunoaffinity purification is carried out as follows.

1. Collect embryos at 12 h intervals and treat with heat shock at 37° for 15 min, followed by 1-h recovery in a moisture chamber at room temperature and store at −70°.
2. Grind frozen embryos in liquid nitrogen and mix with 5× volumes of lysis buffer for 30 min in cold room.
3. Centrifuge the embryo lysate at maximum speed in a microfuge tube for 20 min at 4°.
4. Apply the supernatant (embryo extracts) to an anti-FLAG affinity column (Sigma, 1-ml bed column) preequilibrated with the lysis buffer.
5. Wash the column with five aliquots of 3 ml of the wash buffer and elute proteins from the column with elution buffer.
6. Readjust the eluted fractions to neutral pH with Tris buffer and concentrate with a Centricon P10 cartridge (Amicon).
7. Apply approximately 20-μg samples of affinity-purified protein to each lane of an 8% polyacrylamide SDS gel for electrophoresis and Western blot analysis.

Alternatively, one can also carry out the coimmunoprecipitation experiments as follows.

Solution

Lysis buffer: 20 mM Tris–HCl, pH 7.5, 150 mM NaCl, 1 mM EDTA, 1 mM EGTA, 1% Triton X-100, 2.5 mM sodium pyrophosphate (optional), 1 mM sodium orthovanadate, protease inhibitors

Procedure

1. Dechorionate embryos for 5 min with 50% bleach/50% PBT (PBS+0.1%Triton X-100), and then rinse them three times with PBT and then twice with water.
2. After the last wash, remove as much water as possible, leaving embryos at the bottom.
3. Add 5 times the bed volume of lysis buffer (add protease inhibitor just before use).
4. Grind embryos using the blue grinder (KONTES) for about 1 min.
5. Pass the extract 10 times through a 21-gauge needle with a 1-ml syringe and collect supernatant.
6. Add 50 μl protein A/G beads (Pharmacia) to the supernatant and preclear the extract for 1 h.
7. Quick spin down the beads and recover supernatant.

8. Add anti-FLAG antibody (m2, Sigma) to supernatant and incubate for 2 h at room temperature to overnight at 4°.
9. Add 50 μl protein A/G beads and continue incubation for 3–4 h.
10. Pellet and wash beads with lysis buffer five times for 5–10 min each.
11. Add 2× SDS loading sample buffer and boil protein complex with beads for 5 min prior to SDS–PAGE and Western blotting. The rest of the sample can be stored at −20°.

RNA *In Situ* Hybridization Assay

Similar to the Insc protein, *insc* mRNA is expressed in NBs and localized to the apical cortex of NBs (Li *et al.*, 1997). In order to assess the presence of *pins* transcripts in NBs, we carried out an RNA *in situ* hybridization assay to analyze if the *pins* transcript is expressed and localized asymmetrically in NBs, similar to *insc* mRNA. Results indicated that the *pins* transcript is present as early as the syncytial blastoderm stage, suggesting a strong maternal contribution of *pins* mRNA. Zygotic *pins* mRNA is expressed ubiquitously in most of the embryonic tissues, including epithelial cells and NBs. This protocol describes RNA *in situ* hybridization in *Drosophila* embryos.

RNA Probe Labeling (RNA Labeling Kit from Roche)

1. Add the following reagents to a sterile, RNase-free microfuge tube (on ice) in the following order: 1 μg gel purified template of linearized DNA, 2 μl NTP labeling mixture (10×), 2 μl transcription buffer (10×), 1 μl RNase inhibitor and add RNase-free H_2O to a final volume of 18 μl, and 2 μl RNA polymerase (T7).
2. Mix gently and centrifuge briefly. Incubate for 2 h at 37°.
3. Add 2 μl DNase I, RNase free and incubate for 15 min at 37° to remove the DNA template.
4. Add 2 μl EDTA solution to stop the reaction.
5. Add 2.5 μl 4 *M* LiCl and 75 μl prechilled 100% ethanol. Mix well and leave for at least 30 min at −70°. Centrifuge for 15 min at 12,000g and wash the pellet with 50 μl 70% cold ethanol. Air dry briefly and dissolve in 10 μl of RNase-free water and add 1 μl RNase inhibitor for long-term storage.

Preparation of Tissue

1. Rehydrate fixed embryos in 3:1 methanol:4% paraformaldehyde/PBS for 2 min and then in 1:3 methanol:4% paraformaldehyde/PBS for 5 min.

2. Fix for 10 min in 4% paraformaldehyde/PBS.
3. Rinse three times with PBS+0.1% Tween 20.

Hybridization

Solutions

20× SSC: 175.3 g of NaCl, 27.6 g of $NaH_2PO_4 \cdot H_2O$, and 7.4 g EDTA in 800 ml of H_2O and adjust pH to 7.4 with NaOH, 50× Denhardt's reagent: 1% (w/v) Ficoll 40, 1% (w/v) polyvinylpyrrollidone, and 1% (w/v) bovine serum albumin (BSA; Sigma, fraction V)

Hybridization buffer: 50% deionized formamide, 4× SSC, 1× Denhardt's, 250 μg/ml tRNA (Sigma), 250 μg/ml boiled ssDNA (Sigma), 50 μg/ml heparin (Sigma), 0.1% Tween 20, 5% dextran sulfate (Sigma)

1. Add 0.5 ml hybridization buffer to 50- to 100-μl embryos.
2. Prehybridize at 52° for 1 h with rocking.
3. Add 3–5 μl labeled probe.
4. Hybridize at 52° for 12–16 h with rocking (place tube on its side).
5. Rinse with wash buffer (50% formamide, 2× SSC, 0.1% Tween 20).
6. Wash at 52° for 6–18 h with at least five changes of wash buffer with rocking.
7. Rinse with PBT (PBS + 0.1% Tween 20) two to three times.
8. Wash at room temperature for 30 min in PBT.
9. Incubate at room temperature for 1.5 h with a 1:2000 dilution of antidigoxigenin–alkaline phosphatase conjugate in PBT+5% normal goat serum.
10. Rinse with PBT and then wash four times with PBT, 20 min each wash.
11. Rinse twice with AP buffer and then wash for 5 min in AP buffer (100 mM Tris, pH 9.5, 100 mM NaCl, 50 mM MgCl$_2$, 0.1% Tween 20). Make stock of solution without Tween 20 and add Tween 20 just before use.
12. Add 0.3 ml AP buffer containing 2.7 μl NBT (75 mg/ml in DMF) and 2.1 μl BCIP.
13. Incubate with rocking until desired color development is achieved (from 20 min to overnight).
14. Rinse three times with PBT.
15. Remove as much PBT as possible and then add 0.5 ml of 70% glycerol.
16. Embryos are ready to mount or dissect once they have settled to the bottom of the tube.

Immunohistochemistry

To examine the subcellular localization of Pins in *Drosophila* NBs, we carried out immunostaining assays for Pins. Given its *in vivo* interaction with Insc, Pins might colocalize with Insc at the apical cortex of NBs to regulate asymmetric cell division. Immunostaining using the anti-Pins antibody (Yu *et al.*, 2000) was carried out to examine the subcellular distribution of Pins. Data suggested that Pins forms an apical crescent from early prophase to telophase, colocalizing with Insc during mitosis (Fig. 3A and B, at metaphase). However, unlike Insc localizing in the apical stalk of delaminating NB, Pins is absent from the apical stalk, suggesting that Pins expression is slightly later than Insc expression (Yu *et al.*, 2000). As in asymmetrically dividing neuroblasts, spindle asymmetry and displacement are prominent, anti-α-tubulin and anti-β-tubulin antibodies were used to visualize the mitotic spindle and astral microtubules in neuroblasts, respectively, from

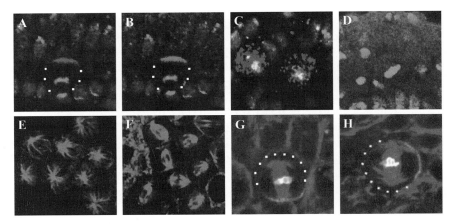

Fig. 3. Pins and Insc localization in wild type (WT), *pins*, and $G\alpha_i$ mutants and spindle morphology. (A–D) Parasagittal optical sections showing NBs from embryos of various genotypes indicated in each panel; apical is up. (A and B) WT embryos triple labeled with anti-Pins (green), anti-Insc (red), and DNA stain (blue); anti-Insc plus DNA staining is shown in A and anti-Pins plus DNA staining is shown in B. Note that Pins colocalizes with Insc at metaphase. (C) *pins* mutant embryos stained with anti-Insc (red) and DNA (blue). Note that Insc is cytoplasmic at metaphase in the absence of Pins. (D) $G\alpha_i$ mutant embryos stained with anti-Pins (green) and DNA (blue). Pins is released from the membrane in both NBs and epithelial cells. (E and F) β-Tubulin staining (in green) in mitotic domain 9 (Kraut *et al.*, 1996) cells of WT and *pins* mutants. In WT, the mitotic spindle is oriented along the apical–basal axis (E) while the spindle is parallel to the surface in the *pins* mutant (F). (G and H) α-Tubulin staining (in red) in WT NBs and epithelial cells. Note that astral MTs grow only at the apical side of NB whereas astral MTs shrink at the basal side. In contrast, astral MTs can grow at both sides of epithelial cells. Reproduced with permission from Yu *et al.* (2000), © 2000 Cell Press and Yu (2003), © 2003 The Rockefeller University Press. (See color insert.)

both wild-type (WT) flies and various mutants. In WT NBs and cells from motitic domain 9 (Kraut *et al.*, 1996), the mitotic spindle is oriented along the apical–basal axis (Fig. 3E). In *pins* mutants, mitotic spindle orientation fails to align perpendicular to the embryo surface (Fig. 3F). Anti-α-tubulin staining also shows that astral microtubules undergo growth at the apical side of NBs, whereas astral MTs undergo catastrophe at the basal side from metaphase to telophase (Fig. 3G). In contrast to NBs, in epithelial cells, astral MTs form at both sides (Fig. 3H).

Solutions

PBT (PBS + 0.1% Triton X-100), 50% bleach, 20% paraformalde-hyde (20 g paraformaldehyde and 250 μl 10 M NaOH in 100 ml PBS, dissolve at $65°$, and filter it before storing at $4°$), heptane, 40% formaldehyde, methanol, ethanol

Fixation of Embryos

1. Rinse embryos in PBT and dechorionate them with 50% bleach/50% PBT for 2 min. Rinse three times in PBT.
2. Transfer to a scintillation vial containing 5 ml 4% paraformaldehyde and 5 ml heptane. Fix by shaking vigorously for 12 min.
3. Remove lower fixative phase and add 5 ml methanol. Shake vigorously for 30 s to devitellinize embryos. The devitellinized embryos sink to the bottom. Embryos are collected and washed in three changes of absolute ethanol. Embryos in ethanol can be stored at $-20°$.
4. Prior to immunostaining, embryos are rehydrated by washing them three times in PBT for 10 min.

Quick Fixation Method Used for Anti-β-Tubulin Antibody Staining

1. Permeabilize dechorionated and washed embryos in 5 ml octane for 30 s.
2. Add the same volume (5 ml) of 40% formaldehyde (unbuffered).
3. Shake vigorously for 3 min and remove the aqueous phase.
4. Add 1 volume (or slightly less) of methanol and devitellinize. Wash three times in absolute ethanol as described earlier.

Quick Fixation Method Used for Anti-α-Tubulin Antibody Staining

1. Dechorionate with 50% bleach/PBT for 2 min. Wash embryos with PBT four times. Remove excess PBT.
2. Transfer embryos into a glass vial containing 5 ml heptane + 5 ml 40% formaldehyde.

3. Swing vial gently (embryos will rotate at the heptane/formaldehyde boundary) for exactly 1 min. Remove formaldehyde within 20 s.
4. Add 5 ml methanol into the vial. Shake vigorously. Transfer the embryos to a new tube.
5. Wash embryos three times with methanol.
6. Wash embryos three times with PBT 10 min for each wash.

Fixation of Other Tissues

To test Pins localization in larval NBs, brains of third instar larvae are dissected in a drop of PBS on a petri dish and fixed in 4% paraformaldehyde in PBS on ice for 10–20 min. To examine Pins localization in pI cells, 15-h APF pupa nota are dissected in PBS and fixed in 4% paraformaldehyde in PBS on ice for 30 min. Testes and ovaries are dissected out of anesthetized adult flies in a drop of PBS and fixed in 4% paraformaldehyde in PBS for 10–20 min. Fixation is performed in 1.5-ml microfuge tubes. These tissues are washed in five changes of PBS.

Immunofluorescent Staining of Fixed Tissues

1. Block fixed tissues in PBT, 3% BSA for at least 30 min.
2. Incubate with primary antibody in PBT, 3% BSA for 2 h at room temperature or overnight at $4°$.
3. Wash three times in PBT, each wash for 10 min.
4. Incubate with secondary antibody in PBT, 3% BSA for 2 h at room temperature.
5. Wash three times in PBT for 10 min.
6. Add To-Pro3 (Molecular Probes Inc.) (1:4000–5000) in last change of PBT and incubate for 15 min.
7. Mount fluorescently labeled samples in Vectashield (Vector).

Microscopy and Image Analysis

Images of fluorescently labeled tissue samples are obtained by confocal microscopy using MRC1024 or Zeiss LSM510 laser-scanning microscopes from Bio-Rad. All digital images are processed using Adobe Photoshop.

Generation of pins and $G\alpha_i$ Mutants

The interaction and colocalization of Pins with Insc suggest that Pins might be required for the asymmetric cell division of *Drosophila* NBs. If Pins is required for the asymmetric division of NBs, *pins* loss-of function mutants should show defects in localizing cell fate determinants and/or orienting the mitotic spindle. To access the function of Pins, *pins* mutants were generated by taking advantage of a derivative P-element insertion line EP3559 (Rorth, 1996).

EP3559 flies carry a P-element derivative that contains the *white* gene inserted about 700 bp upstream of the *pins* transcription unit at the cytological location 98A-B. The P-element in this stock was mobilized using the immobile element P[ry$^+$ Δ2–3] (99B) as a transposase source. One hundred fifty independent *w* revertant lines were established. About 15 lethal or semilethal lines were analyzed on Southern blots using various portions of *pins* cDNA as hybridization probes. Several small deletion lines, which remove part or all of the *pins* coding region, were recovered. Based on the same approach, we also mobilized the P-element insertion KG1907 to isolate several $G\alpha_i$ mutants (Yu *et al.*, 2003).

Immunohistochemical analysis of *pins* and $G\alpha_i$ mutants indicated that Insc is released from the apical cortex of NB and becomes cytoplasmic in *pins* mutants and that Pins is also cytosolic rather than localized apically in NBs (Fig. 3C and D).

RNA Interference

The introduction of dsRNA corresponding to either a portion or the entire coding region of a particular gene into embryonic cells can interfere with the function of the endogenous gene to give a phenotype essentially equivalent to the known genetic mutation. We injected *in vitro*-synthesized *pins* dsRNA to pre-blastoderm embryos to silence *pins* gene function.

Synthesis of dsRNA

The synthesis of dsRNA is based primarily on the T7 RiboMAX express RNAi system from Promega.

Microinjections

1. Dissolve the dsRNA in RNase-free water to a final concentration no greater than 2.5 μM.

2. Pull needles in Sutter P-30 puller and break a tip to a diameter of 0.5–2.5 μm.

3. Back-fill the needle with 1 μl of dsRNA solution.

4. Collect preblastodermal embryos every 1 min at 25° and dechorinate in 50% bleach for 2 min. After rinsing embryos with water, attach them to a coverslip coated with tape glue.

5. Align embryos and cover in Voltalef oil 10 s (Elf Atochem).

6. The injection location is typically on the ventral side in the posterior domain extending from 50–70% egg length. The average injection volume

is 65–110 pL. Approximately 0.2 fmol of dsRNA per embryo should produce the interference effect.

Immunohistochemical Staining

1. Incubate embryos in a moisture chamber at 25°.
2. At the appropriate stage, collect the embryos for staining. With a razor blade, scrape excess oil from the coverslip. Remove as much oil as possible.
3. Take coverslip off of slide and hold coverslip over a 60-mm dish of heptane. With a pipette, wash the embryos on the slide with heptane until the oil is washed into the dish.
4. With a tip-cut 1000-μl pipette tip, transfer the embryos from the dish to a 1.5-ml microfuge tube. Add 0.5 ml heptane and 0.5 ml fixation buffer (to make 5 ml fixation buffer, add 3.5 ml PBS, 0.5 ml 1 M HEPES, and 1 ml 20% paraformaldehyde). Shake for 12 min at room temperature.
5. Remove fixation buffer (bottom). Add 1 ml methanol; vortex for 30 s to 1 min. Remove solution.
6. Wash embryos three times with ethanol and store at −20°.
7. Continue standard immunostaining.

Germline Transformation and Rescue Experiment

Germline transformation is a useful method to generate transgenic flies for either complementation studies or overexpression analysis. Complementation experiments are necessary to confirm that a gene of interest is responsible for the mutations analyzed and one can rescue phenotypes associated with the mutants. Overexpression of Pins leads to the accumulation of Pins in the cytoplasm of NBs, which does not show any defect in NB asymmetric division, whereas overexpression of Gα_i in NB leads to the generation of two equal-sized daughter cells (Schaefer *et al.*, 2001; Yu *et al.*, 2003).

Germline Transformation

pins and *Gα_i* embryos display phenotypes similar to those observed in *insc* mutants (Yu *et al.*, 2000, 2003). Miranda/Prospero and Numb/Partner of Numb (Pon) often show defective localization, either mislocalized as crescents or localized cortically around the cell cortex (Yu *et al.*, 2000, 2003). Mitotic spindle orientation is also defective. To confirm that these phenotypes were caused by a deficiency in *pins* or *Gα_i* locus, *UAS-pins* and *UAS-Gα_i* transgenes were generated, respectively, as described (O'Connor and Chia, 1993).

Full-length *pins* cDNA is isolated from a 4- to 8-h cDNA library and is cloned into pUAST to fuse *pins* to *uas*. Two micrograms pUAST-*pins* is mixed with 1 μg of carrier DNA and resuspended in microinjection buffer. The DNA mixture is filtered through a 0.45-μm syringe and injected into dechorionated embryos, which are covered with Voltalef oil. To obtain germline transformants, individual adult flies developed from microinjected embryos are verified for the marker present in the transformation vector pUAST (Brand and Perrimon, 1993).

Rescue Experiments

To confirm that the phenotypes of *pins* or $G\alpha_i$ mutants are derived from loss-of-function mutations to these genes, we carried out rescue experiments by driving the expression of the *UAS-pins* transgene in NBs with a *scabrous-gal4* driver (the expression of Gal4 under the control of scabrous promoter) in either *pins* mutant or $G\alpha_i$ mutant background (Brand and Perrimon, 1993). The introduction of ectopically expressed Pins and $G\alpha_i$ is able to rescue the defects observed in *pins* and $G\alpha_i$ mutants, respectively, indicating that *pins* and $G\alpha_i$ are responsible for phenotypes of *pins* and $G\alpha_i$ mutants, respectively (Yu *et al.*, 2000, 2003).

Acknowledgments

We thank Dr. Hongyan Wang and Professor William Chia for critical reading and comments. This work is supported by Temasek Life Sciences Laboratory, Singapore.

References

Bartel, P. L., and Field, S. (1995). Analyzing protein–protein interactions using two-hybrid system. *Methods Enzymol.* **254,** 241–263.

Brand, A. H., and Perrimon, N. (1993). Targeted gene expression as a means of altering cell fates and generating dominant phenotypes. *Development* **118,** 401–415.

Cai, Y., Yu, F., Lin, S., Chia, W., and Yang, X. (2003). Apical complex genes control mitotic spindle geometry and relative size of daughter cells in Drosophila neuroblast and pI asymmetric divisions. *Cell* **112,** 51–62.

Campos-Ortega, J. A. (1995). Genetic mechanisms of early neurogenesis in *Drosophila melanogaster. Mol. Neurobiol.* **10,** 75–89.

Chia, W., and Yang, X. (2002). Asymmetric division of Drosophila neural progenitors. *Curr. Opin. Genet. Dev.* **12,** 459–464.

Doe, C. Q., and Bowerman, B. (2001). Asymmetric cell division: Fly neuroblast meets worm zygote. *Curr. Opin. Cell Biol.* **13,** 68–75.

Fuse, N., Hisata, K., Katzen, L. A., and Matsuzaki, F. (2003). Heterotrimeric G proteins regulate daughter cell size asymmetry in Drosophila neuroblast division. *Curr. Biol.* **13,** 947–954.

Jan, Y. N., and Jan, L. Y. (2001). Asymmetric cell division in the Drosophila nervous system. *Nature Rev. Neurosci.* **2,** 772–779.

Kaltschmidt, J. A., Davidson, C. M., Brown, N. H., and Brand, A. H. (2000). Rotation and asymmetry of the mitotic spindle direct asymmetric cell division in the developing central nervous system. *Nature Cell Biol.* **2,** 7–12.

Knoblich, J. A. (2001). Asymmetric cell division during animal development. *Nature Rev. Mol. Cell Biol.* **2,** 11–20.

Kraut, R., and Campos-Ortega, J. A. (1996). Inscuteable, a neural precursor gene of Drosophila, encodes a candidate for a cytoskeleton adaptor protein. *Dev. Biol.* **174,** 65–81.

Kraut, R., Chia, W., Jan, L. Y., Jan, Y. N., and Knoblich, J. A. (1996). Role of inscuteable in orienting asymmetric cell divisions in Drosophila. *Nature* **383,** 50–55.

Kuchinke, U., Grawe, F., and Knust, E. (1998). Control of spindle orientation in Drosophila by the Par-3-related PDZ-domain protein Bazooka. *Curr. Biol.* **8,** 1357–1365.

Li, P., Yang, X., Wasser, M., Cai, Y., and Chia, W. (1997). Inscuteable and Staufen mediate asymmetric localization and segregation of prospero RNA during Drosophila neuroblast cell divisions. *Cell* **90,** 437–447.

O'Connor, M., and Chia, W. (1993). Transgenesis techniques: Principle and Protocols. *In* "Methods in Molecular Biology" (D., Murphy and D. A., Carter, eds.), pp. 75–85. Humana Press Inc., Totowa, NJ.

Parmentier, M. L., Woods, D., Greig, S., Phan, P. G., Radovic, A., Bryant, P., and O'Kane, C. J. (2000). Rapsynoid/partner of inscuteable controls asymmetric division of larval neuroblasts in Drosophila. *J. Neurosci.* **20,** RC84[online].

Petronczki, M., and Knoblich, J. A. (2001). DmPAR-6 directs epithelial polarity and asymmetric cell division of neuroblasts in Drosophila. *Nature Cell Biol.* **3,** 43–49.

Rorth, P. (1996). A modular misexpression screen in Drosophila detecting tissue-specific phenotypes. *Proc. Natl. Acad. Sci. USA* **93,** 12418–12422.

Schaefer, M., Petronczki, M., Dorner, D., Forte, M., and Knoblich, J. A. (2001). Heterotrimeric G proteins direct two modes of asymmetric cell division in the Drosophila nervous system. *Cell* **107,** 183–194.

Schaefer, M., Shevchenko, A., and Knoblich, J. A. (2000). A protein complex containing Inscuteable and the Galpha-binding protein Pins orients asymmetric cell divisions in Drosophila. *Curr. Biol.* **10,** 353–362.

Schober, M., Schaefer, M., and Knoblich, J. A. (1999). Bazooka recruits Inscuteable to orient asymmetric cell divisions in Drosophila neuroblasts. *Nature* **402,** 548–551.

Siderovski, D. P., Diverse-Pierluissi, M., and De Vries, L. (1999). The GoLoco motif: A Galphai/o binding motif and potential guanine-nucleotide exchange factor. *Trends Biochem. Sci.* **24,** 340–341.

Tio, M., Zavortink, M., Yang, X., and Chia, W. (1999). A functional analyses of inscuteable and its roles during Drosophila asymmetric cell divisions. *J. Cell Sci.* **112,** 1541–1551.

Wodarz, A., Ramrath, A., Grimm, A., and Knust, E. (2000). Drosophila atypical protein kinase C associates with Bazooka and controls polarity of epithelia and neuroblasts. *J. Cell Biol.* **150,** 1361–1374.

Wodarz, A., Ramrath, A., Kuchinke, U., and Knust, E. (1999). Bazooka provides an apical cue for Inscuteable localization in Drosophila neuroblasts. *Nature* **402,** 544–547.

Yu, F., Cai, Y., Kaushik, R., Yang, X., and Chia, W. (2003). Distinct roles of Galpha-i and Gbeta-13F subunits of the heterotrimeric G protein complex in the mediation of Drosophila neuroblast asymmetric divisions. *J. Cell Biol.* **162,** 623–633.

Yu, F., Morin, X., Cai, Y., Yang, X., and Chia, W. (2000). Analysis of partner of inscuteable, a novel player of Drosophila asymmetric divisions, reveals two distinct steps in inscuteable apical localization. *Cell* **100,** 399–409.

Yu, F., Ong, C. T., Chia, W., and Yang, X. (2002). Membrane targeting and asymmetric localization of Drosophila partner of inscuteable are discrete steps controlled by distinct regions of the protein. *Mol. Cell. Biol.* **22,** 4230–4240.

[23] Mathematical Modeling of RGS and G-Protein Regulation in Yeast

By Necmettin Yildirim, Nan Hao, Henrik G. Dohlman, and Timothy C. Elston

Abstract

G-protein-activated signaling pathways are capable of adapting to a persistent external stimulus. Desensitization is thought to occur at the receptor level as well as through negative feedback by a family of proteins called regulators of G-protein signaling (RGS). The pheromone response pathway in yeast is a typical example of such a system, and the relative simplicity of this pathway makes it an attractive system in investigating the regulatory role of RGS proteins. Two studies have used computational modeling to gain insight into how this pathway is regulated (Hao *et al.*, 2003; Yi *et al.*, 2003). This article provides an introduction to computational analysis of signaling pathways by developing a mathematical model of the pheromone response pathway that synthesizes the results of these two investigations. Our model qualitatively captures many features of the pathway and suggests an additional mechanism for pathway inactivation. It also illustrates that a complete understanding of signaling pathways requires an investigation of their time-dependent behavior.

Introduction

The pheromone response pathway in yeast is perhaps the best characterized of any known signaling system and has established some important paradigms for G-protein-mediated processes in other organisms. In yeast, haploid **a** and α cell types communicate by secreting type-specific peptide pheromones, which trigger a series of events leading to cell fusion (mating) and formation of an **a**/α diploid. As is typical of such pathways, the pheromone pathway of yeast has three major upstream components: a stimulus (α-factor pheromone is the most commonly studied), a receptor (Ste2 is the α-factor receptor), and a G-protein heterotrimer, consisting of α (Gpa1), β (Ste4), and γ (Ste18) subunits.

Much is known about the proteins that transmit the pheromone signal, as well as about the mechanisms by which events at the cell surface are linked to subsequent biochemical changes in the cytoplasm and nucleus [reviewed in Dohlman and Thorner (2001)]. Less is known about how the signaling components are assembled and regulated over time. One important property of any signal-response pathway is the ability to adapt, or desensitize, to chronic stimulation. Studies in mammalian cells revealed that adaptation can occur at the receptor level through rapid alterations of the receptor, including phosphorylation and uncoupling from the G protein. Studies in yeast revealed that adaptation can also occur at the level of the G protein through rapid induction of the "regulator of G-protein signaling" (RGS) protein Sst2. In general, RGS proteins trigger negative feedback by accelerating the intrinsic GTPase activity of the G-protein α subunit [i.e., acting as GTPase-accelerating proteins (GAPs)] (Berman and Gilman, 1998; Dohlman and Thorner, 1997, 2001).

Having described the essential components and events in G-protein-coupled receptor signaling, an emerging goal is now to develop mathematical models that describe their behavior over time. This should be most feasible for the yeast pheromone signaling pathway, given that there are relatively few components (one receptor, one G protein, one RGS protein, etc.), their activity can be manipulated easily (e.g., gene deletions and gene replacements are constructed easily), and their activity can be measured readily using simple and quantitative functional assays (Hao et al., 2003; Yi et al., 2003). A combination of experimental and computational modeling approaches can provide new insights about (1) how cells communicate and respond to an external stimulus and (2) how those responses change over time.

Two papers on the yeast pheromone response pathway that combined experimental and computational analyses have been published. Yi et al. (2003) modeled the dynamics of receptor and G-protein activation. They used fluorescence resonance energy transfer (FRET) experiments to measure many of the model parameters. Because their study focused on the early response, it was assumed that RGS protein concentration remained constant after pheromone induction. Hao et al. (2003) observed an increase in RGS protein levels after exposure to pheromone and constructed a model to investigate the role of RGS protein induction on the dynamics of the pathway. This article develops a model that combines the results of these two investigations. We describe how the model is constructed from the underlying biochemical steps that make up the pathway. The model consists of rate equations or ordinary differential equations that describe the dynamics (time-dependent behavior) of the various protein concentrations. The behavior of this model is compared with experimental results. Because G protein and MAPK signaling cascades are conserved in all eukaryotes, successful modeling of the yeast pheromone pathway should lead to improved models of signaling events in more complex organisms. Ultimately this approach promises to improve our understanding of how cellular changes in disease states can be predicted and managed.

The Yeast Pheromone Response Pathway

The pheromone response pathway requires activation of a receptor and G protein, a MAP kinase cascade, and a transcriptional regulator. In this study we are interested in regulation at the receptor/G-protein level and, therefore, do not explicitly model the MAP kinase portion of the pathway. A schematic diagram of the system is given in Fig. 1. The pathway functions in the following manner. When inactive, the G protein forms a heterotrimer consisting of Gα (Gpa1), Gβ (Ste4), and Gγ (Ste18). Upon agonist binding, the receptor (Ste2) undergoes a conformational change that causes the formation of a ternary complex of ligand, receptor, and G protein. GDP on Gα is next exchanged with GTP, which in turn causes Gα to dissociate from G$\beta\gamma$. G$\beta\gamma$ then initiates a MAPK cascade, which culminates with phosphorylation of the transcription factor Ste12. The yeast RGS protein Sst2 is one of the proteins upregulated by Ste12 activity. The RGS domain of Sst2 increases the GTPase activity of Gα by approximately 25-fold (Yi et al., 2003). The pathway is turned off when GDP-bound Gα recombines with G$\beta\gamma$. Thus, RGS protein induction provides a negative feedback mechanism. It has been demonstrated that Sst2 is ubiquitinated and degraded in a pheromone-dependent manner (Hao et al., 2003), thereby providing positive feedback to counteract eventually the negative effects of RGS protein induction. It has been shown that the pheromone also induces the synthesis of Gα (Hao et al., 2003). As suggested by Yi et al. (2003), we additionally consider the pheromone-dependent activation of G$\beta\gamma$. In Fig. 1, this is indicated by the pheromone-induced synthesis of Gβ. Increased G$\beta\gamma$ activity could also be accomplished by the pheromone-induced phosphorylation of G$\beta\gamma$ or redistribution of G$\beta\gamma$ within the membrane (Kim et al., 2000).

Model Development

The model of the pathway consists of seven independent state variables representing the time-dependent concentration of free receptor ([R]), agonist-bound receptor ([R $\cdot$ L]), inactive G protein ([G$^{GDP}\alpha\beta\gamma$]), active G protein ([G$\beta\gamma$]), GTP-bound Gα subunit ([G$^{GTP}\alpha$]), GDP-bound Gα subunit ([G$^{GDP}\alpha$]), and RGS protein ([Sst2]). Dose–response data reported in Hao et al. (2003) were measured using β–galactosidase as a reporter. Therefore, the model also includes a state variable for the concentration of β–galactosidase ([βgal]).

The time-dependent behavior of the state variables is determined by a set of eight rate equations. Each rate equation has a similar form:

$$\frac{d[C]}{dt} = \text{input chemical fluxes} - \text{output chemical fluxes}$$

where the time derivative (instantaneous rate of change in time) of the concentration, [C], is equated to the sum of the gain terms (input chemical fluxes) that cause the concentration to increase minus the sum of the loss terms (output chemical fluxes) that act to decrease the concentration. In general, each process depicted in Fig. 1 can be broken down into elementary biochemical steps. In this case, the law of mass action can be used to determine the form of the chemical fluxes required in the rate equations. Under appropriate conditions, it is possible to reduce the number of equations needed to model the system by combining elementary biochemical steps into higher order effective kinetics (e.g., Michaelis–Menten or Hill kinetics) (Fall, 2002). In general, the chemical fluxes are nonlinear functions of the concentrations. Therefore, the rate equations represent a set of coupled nonlinear ordinary differential equations, whose solutions can only be determined by numerical integration. A good introduction to the study of nonlinear dynamics is the text by Strogatz (1994).

Our model of the pheromone response pathway consists of eight rate equations. The equation for the free receptor concentration is

$$\frac{d[R]}{dt} = k_1 + k_3[R \cdot L] - k_2[R][L] - k_5[R] \tag{1}$$

where the first two terms on the right-hand side of this equation are input fluxes. The first term k_1 models the constitutive production of receptors, and the second term, $k_3 [R \cdot L]$, describes the rate at which agonist is liberated from the receptor. The third and fourth terms in Eq. (1) are output fluxes; $k_2 [R][L]$ models agonist binding and $k_5 [R]$ models receptor degradation.

The rate equation for the concentration of agonist-bound receptors is

$$\frac{d[R \cdot L]}{dt} = k_2[R][L] - k_3[R \cdot L] - k_4[R \cdot L] \tag{2}$$

where the terms $k_2 [R][L]$ and $k_3 [R \cdot L]$ enter as input and output fluxes of agonist-bound receptor $[R \cdot L]$, respectively. The last term, $k_4 [R \cdot L]$, represents degradation of the agonist-bound receptor. The model assumes that the rate constants k_5 and k_4 for the degradation of free and agonist-bound receptor, respectively, are different. This assumption is based on

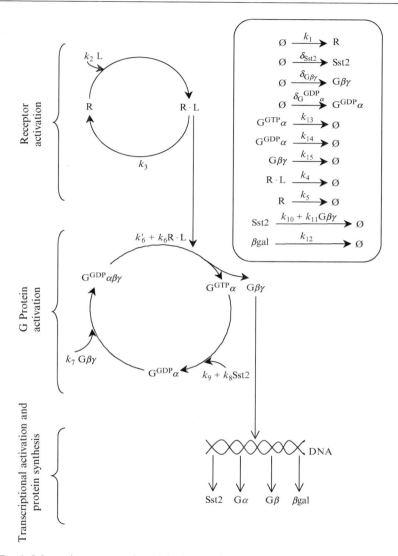

FIG. 1. Schematic representation of the heterotrimeric G-protein activation cycle and rate constants for each biochemical process. Receptor activation occurs when the pheromone (L) binds to the receptor (R). The agonist/receptor complex causes dissociation of the G-protein heterotrimer producing $G\beta\gamma$ and $G^{GTP}\alpha$. $G\beta\gamma$ then induces the synthesis of $G\beta$, $G\alpha$, Sst2, and the reporter β–galactosidase. Sst2 regulates the pathway negatively by increasing the rate at which $G^{GTP}\alpha$ is converted to $G^{GDP}\alpha$. The pathway is inactivated when $G^{GDP}\alpha$ and $G\beta\gamma$ reassociate. Circled reactions represent constitutive protein synthesis and degradation. Note that the model assumes that the heterotrimeric form of the G protein is protected from degradation.

the experimental observation of Yi *et al.* (2003) that k_4 is approximately 10 times greater than k_5.

The next three equations describe the dynamics of the various forms of G protein. The equation for the concentration of inactive G protein is

$$\frac{d[G^{GDP}\alpha\beta\gamma]}{dt} = k_7[G\beta\gamma][G^{GDP}\alpha] - (k_6[R \cdot L] + k_6')[G^{GDP}\alpha\beta\gamma] \qquad (3)$$

where $k_7[G\beta\gamma][G^{GDP}\alpha]$ represents a gain in $[G^{GDP}\alpha\beta\gamma]$ from the reassociation of $G\beta\gamma$ and $G^{GDP}\alpha$ and $(k_6[R \cdot L] + k_6')[G^{GDP}\alpha\beta\gamma]$ is an output flux due to the conversion of $G^{GDP}\alpha\beta\gamma$ to $G^{GTP}\alpha$ and $G\beta\gamma$ by the guanine nucleotide exchange function of the active receptor. In the absence of agonist, the rate constant for this process is k_6' and, in the presence of pheromone, the rate constant is increased by the amount $k_6[R \cdot L]$. We assumed that $G^{GDP}\alpha\beta\gamma$ is stable and, therefore, do not include a degradation term in Eq. (3) (Finley, 1993).

The equation for $[G^{GTP}\alpha]$ is

$$\frac{d[G^{GTP}\alpha]}{dt} = (k_6[R \cdot L] + k_6')[G^{GDP}\alpha\beta\gamma] + \frac{u_1[G\beta\gamma]}{K_1 + [G\beta\gamma]}$$
$$- (k_8[Sst2] + k_9)[G^{GTP}\alpha] - k_{13}[G^{GTP}\alpha] \qquad (4)$$

where $(k_6[R \cdot L] + k_6')[G^{GDP}\alpha\beta\gamma]$ appears as an input flux. There is experimental evidence of a 50% increase in the $G\alpha$ level upon exposure to pheromone (Hao *et al.*, 2003). Rather than explicitly model all the downstream components of the pathway (e.g., MAPK cascade, transcription, and translation), this effect was investigated by including the phenomenological term $u_1[G\beta\gamma]/(K_1 + [G\beta\gamma])$. The important feature of this term is that the synthesis rate saturates at high $G\beta\gamma$ concentration at a value of u_1. The Michaelis constant K_1 is the concentration of $G\beta\gamma$ that produces half the maximum synthesis rate. In general, pheromone induction could follow higher-order (Hill) kinetics. However, because the best agreement between the model and experimental results was achieved when the value of u_1 was very small, higher-order kinetics were not considered. In fact, for the results presented later, $u_1 = 0$, that is, pheromone-induced $G\alpha$ synthesis is not included in the model. This apparent discrepancy with the experimental observation of Hao *et al.* (2003) can be reconciled if one considers that receptor and $G\beta\gamma$, but not $G\alpha$, become concentrated within a subdomain of the plasma membrane following exposure to the pheromone (Dohlman and Thorner, 2001). Because $G\alpha$ is not redistributed, its increased synthesis after exposure to pheromone does not significantly change the concentration near the receptors. The GTPase activity of $G\alpha$ is modeled as $(k_8[\text{Sst2}] + k_9)[G^{\text{GTP}}\alpha]$, where k_9 is the rate constant for the process in the absence of Sst2 (i.e., intrinsic GTPase activity) and $k_8[\text{Sst2}]$ models the increase in GTPase activity in the presence of Sst2. The last term in Eq. (4), $k_{13}[G^{\text{GTP}}\alpha]$, models the degradation of $G^{\text{GTP}}\alpha$.

The equation for $[G^{\text{GDP}}\alpha]$ is

$$\frac{d[G^{\text{GDP}}\alpha]}{dt} = \delta_{G^{\text{GDP}}\alpha} + (k_8[\text{Sst2}] + k_9)[G^{\text{GTP}}\alpha] - k_7[G^{\text{GDP}}\alpha][G\beta\gamma]$$
$$- k_{14}[G^{\text{GDP}}\alpha] \tag{5}$$

where $\delta_{G^{\text{GDP}}\alpha}$, on the right-hand-side of this equation, models the constitutive synthesis of $G\alpha$. Our rationale for including constitutive synthesis in this equation rather than in Eq. (4) for $[G^{\text{GTP}}\alpha]$ is the following. The constitutive $G\alpha$ synthesis rate of $G\alpha$ is an important parameter in determining the steady-state level of this protein in the absence of pheromone. In the cell, GTP is more abundant than GDP. Consequently, newly synthesized $G\alpha$ is more likely to bind GTP initially. However, it seems reasonable to assume that $G^{\text{GDP}}\alpha$ has a lower free energy than $G^{\text{GTP}}\alpha$, so that, at steady state, $G^{\text{GDP}}\alpha$ is the dominant form of $G\alpha$ ensuring that the pathway is inactive. Because the model does not explicitly take into account GTP and GDP binding and dissociation, the only mechanism to convert $G^{\text{GTP}}\alpha$ to $G^{\text{GDP}}\alpha$ is through GTP hydrolysis. In the absence of RGS protein-mediated GTPase acceleration, this process is very slow; consequently, if the constitutive synthesis rate is placed in the equation for $[G^{\text{GTP}}\alpha]$, this concentration becomes artificially high and the pathway shows little response to the pheromone. As discussed earlier, the second term in Eq. (5) represents a source of $G^{\text{GDP}}\alpha$ from GTP hydrolysis. The last two terms in Eq. (5) are loss terms. The term $k_7[G^{\text{GDP}}\alpha][G\beta\gamma]$ models the loss of $G^{\text{GDP}}\alpha$ due to the reformation of $G^{\text{GDP}}\alpha\beta\gamma$ from $G^{\text{GDP}}\alpha$ and $G\beta\gamma$. The last term $k_{14}[G^{\text{GDP}}\alpha]$ models the degradation of $G^{\text{GDP}}\alpha$. There is some evidence that $G^{\text{GDP}}\alpha$ is degraded more slowly than $G^{\text{GTP}}\alpha$ (Schauber et al., 1998). Therefore, we take $k_{14} < k_{13}$.

The $G\beta\gamma$ concentration evolves according to

$$\frac{d[G\beta\gamma]}{dt} = \delta_{G\beta\gamma} + (k_6[\text{R} \cdot \text{L}] + k_6')[G^{\text{GDP}}\alpha\beta\gamma] + \frac{u_2[G\beta\gamma]}{K_2 + [G\beta\gamma]}$$
$$- k_7[G\beta\gamma][G^{\text{GDP}}\alpha] - k_{15}[G\beta\gamma] \tag{6}$$

where $\delta_{G\beta\gamma}$ represents the constitutive synthesis rate. The second term in this equation models the constitutive and agonist-dependent production of

G$\beta\gamma$ from G$^{GDP}\alpha\beta\gamma$. To explain their experimental results, Yi *et al.* (2003) speculated that G$\beta\gamma$ synthesis is induced after exposure to the pheromone. As pointed out previously, other potential mechanisms for increased G$\beta\gamma$ activity are pheromone-induced phosphorylation of G$\beta\gamma$ or redistribution of G$\beta\gamma$ within the membrane (Kim *et al.*, 2000). To model the pheromone-dependent increase in G$\beta\gamma$ activity, we include the term $u_2[G\beta\gamma]/(K_2 + [G\beta\gamma])$ in Eq. (6), where we again assume Michaelis–Menten kinetics with respect to G$\beta\gamma$. For clarity, we refer to this term as pheromone-induced G$\beta\gamma$ synthesis. However, it should be kept in mind that either of the other two mechanisms could be modeled in the same way. The fourth term in this equation, $k_7[G\beta\gamma][G^{GDP}\alpha]$, represents the loss in G$\beta\gamma$ due to the reformation of inactive G-protein heterotrimer G$^{GDP}\alpha\beta\gamma$. Finally, the last term, $k_{15}[G\beta\gamma]$, models the degradation of G$\beta\gamma$.

The equation for the Sst2 concentration is

$$\frac{d[Sst2]}{dt} = \delta_{Sst2} + \frac{u_3[G\beta\gamma]}{K_3 + [G\beta\gamma]} - (k_{10} + k_{11}[G\beta\gamma])[Sst2] \tag{7}$$

Where the first term, δ_{Sst2}, represents the constitutive production of Sst2 and the second term models the induction of Sst2 by pheromone. Again, we have assumed Michaelis–Menten kinetics for this process. It has been determined that Sst2 is ubiquitinated and degraded in a pheromone-dependent manner (Hao *et al.*, 2003). To model this effect, we use an output flux of the form $(k_{10} + k_{11}[G\beta\gamma])[Sst2]$. The parameter k_{10} is the constitutive degradation rate and $k_{11}[G\beta\gamma]$ represents the pheromone-induced degradation rate. We note that the accelerated rate of GTP hydrolysis produced by Sst2 provides negative feedback in this system, whereas the increased degradation of Sst2 produces positive feedback.

Finally, the equation for the β-galactosidase concentration, the reporter used in the experiments of Hao *et al.* (2003), is

$$\frac{d[\beta gal]}{dt} = \frac{u_4[G\beta\gamma]}{K_4 + [G\beta\gamma]} - k_{12}[\beta gal] \tag{8}$$

Where the first term $u_4[G\beta\gamma]/(K_4 + [G\beta\gamma])$ represents the pheromone-induced production of β-galactosidase and the second term $k_{12}[\beta gal]$ models the degradation of this protein.

Results

Equations (1)–(8) were solved numerically using MATLAB (The Math Works), and the results were compared with the experimental studies of Hao *et al.* (2003) and Yi *et al.* (2003). Table I lists the experimentally

TABLE I
MEASURED PROTEIN ABUNDANCES AND VALUES USED IN SIMULATIONS

Protein	Molecules/Cell	Source	Value used (molecules)
α factor receptor (Ste2)	8000	Jenness et al. (1986)	8000
RGS (Sst2)	2000–5980	Ghaemmaghami et al. (2003); Hao et al. (2003)	2000
Gα (Gpa1)	8000–12000	Ghaemmaghami et al. (2003); Hao et al. (2003)	8000
Gβ (Ste4)	2050–10000	Ghaemmaghami et al. (2003); Hao et al. (2003)	8000
Gγ (Ste18)	5550–10000	Ghaemmaghami et al. (2003); Hao et al. (2003); Yi et al. (2003)	8000

measured protein abundances. Table II is a list of all model parameters and the numerical values used in the simulations. As indicated in Table II, many of the model parameters have been measured experimentally. The constitutive degradation rate of the RGS protein Sst2 was measured by Hao et al. (2003). All other constitutive degradation and synthesis rates were chosen so that the steady-state level of each protein was consistent with the measurement listed in Table I. For the cases in which an experimental value of a model parameter did not exist, the parameter value was chosen so that the behavior of the model was consistent with experimental results.

Initially, Eqs. (1)–(8) were run to steady state in the absence of pheromone. At time $t = 0$, a pheromone concentration of 3×10^{-6} M was added. This concentration is slightly higher than the EC$_{50}$. Figure 2 shows a time series for the value of [G$\beta\gamma$]. Immediately after exposure to the pheromone, there is a sharp increase in [G$\beta\gamma$] followed by a short decrease and then a slow rise in concentration. This early time dynamics (<20 min) is in agreement with the experimental results of Yi et al. (2003) obtained using FRET. Our model illustrates that the second increase in [G$\beta\gamma$] can be explained by the pheromone-induced protein synthesis of G$\beta\gamma$. As mentioned earlier, the same effect would result from pheromone-dependent activation or redistribution of G$\beta\gamma$. Figure 3 shows that, after 80 min, [G$\beta\gamma$] continues to decrease slowly.

Hao et al. (2003) studied three strains of yeast. The first strain expressed normal levels of Sst2. The second expressed double ("2X") the normal amount of Sst2, whereas the third was an sst2Δ mutant. The 2XSST2 strain

TABLE II

Model Parameters and Numerical Values Used in Simulations

Parameter	Description	Literature values	Source	Value used
k_1	Receptor synthesis rate	4 molec s^{-1}	Yi et al. (2003)	4 molec s^{-1}
k_2	Agonist receptor association rate	2×10^6 M^{-1} s^{-1}	Yi et al. (2003)	1.067×10^3 M^{-1} s^{-1}
k_3	Agonist receptor dissociation rate	10^{-2} s^{-1}	Yi et al. (2003)	10^{-2} s^{-1}
k_4	Degradation rate of the agonist-bound receptor	4×10^{-3} s^{-1}	Yi et al. (2003)	4×10^{-3} s^{-1}
k_5	Degradation rate of the free receptor	4×10^{-4} s^{-1}	Yi et al. (2003)	5×10^{-4} s^{-1}
k_6'	Constitutive G-protein activation rate constant	—	Estimated according to steady-state values given in Table I	1×10^{-4} molec^{-1} s^{-1}
k_6	Pheromone-induced G-protein activation rate constant	10^{-5}–1.5×10^{-5} (molec)$^{-1}$ s^{-1}	Mukhopadhyay and Ross (1999); Yi et al. (2003)	10^{-5} s^{-1}
k_7	Association rate constant for G$^{GDP}\alpha$ and G$\beta\gamma$	1 (molec)$^{-1}$ s^{-1}	Yi et al. (2003)	5.8×10^{-4} molec^{-1} s^{-1}
k_8	Sst2-dependent GTP hydrolysis rate	0.11 s^{-1}–24 s^{-1}	Mukhopadhyay and Ross (1999); Yi et al. (2003)	1.35×10^{-5} molec^{-1} s^{-1}
k_9	Hydrolysis rate of G$^{GTP}\alpha$ to G$^{GDP}\alpha$ in absence of Sst2	4×10^{-3} s^{-1}	Yi et al. (2003)	10^{-3} s^{-1}
k_{10}	Degradation rate of Sst2	8×10^{-4} s^{-1}	Hao et al. (2003)	8×10^{-4} s^{-1}
δ_{Sst2}	Constitutive Sst2 synthesis rate	1.6 molec s^{-1}	Determined by measured Sst2 abundance and k_{10}	1.6 molec s^{-1}
u_2	Maximum G$\beta\gamma$ synthesis rate	—		3.5 molec s^{-1}
u_3	Maximum Sst2 synthesis rate	—		11.2 molec s^{-1}
u_4	Maximum βgal synthesis rate	—		11.2 molec s^{-1}
K_2	Michaelis constant for the G$\beta\gamma$ synthesis rate	—		1×10^4 molec
K_3	Michaelis constant for the Sst2 synthesis rate	—		1.5×10^3 molec
K_4	Michaelis constant for the βgal synthesis rate	—		1.5×10^3 molec
k_{11}	Pheromone-induced Sst2 degradation rate constant	—		1×10^{-7} molec^{-1} s^{-1}
k_{12}	βgal degradation rate constant	—		1×10^{-4} s^{-1}
k_{13}	G$^{GTP}\alpha$ degradation rate constant	—		1.6×10^{-3} s^{-1}
k_{14}	G$^{GDP}\alpha$ degradation rate constant	—		4×10^{-4} s^{-1}
k_{15}	G$\beta\gamma$ degradation rate constant	—		4×10^{-4} s^{-1}
$\delta_{G\beta\gamma}$	G$\beta\gamma$ constitutive production rate	—		8×10^{-5} molec s^{-1}
$\delta_{G^{GDP}\alpha}$	G$^{GDP}\alpha$ constitutive production rate	—		2×10^{-1} molec s^{-1}

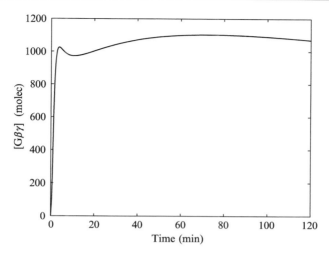

FIG. 2. Time course for $[G\beta\gamma]$ following pheromone induction at time 0. The α-factor concentration is 3×10^{-6} M, which is slightly higher than the EC_{50}. There is an early peak immediately following exposure and then a decline due to receptor degradation. Pheromone-induced synthesis of $G\beta$ causes a second increase in concentration to occur around 10 min.

was modeled by doubling both constitutive and pheromone-dependent production rates, δ_{Sst2} and u_3, and for the $sst2\Delta$ mutant, both of these rates were taken to be zero. Time series of $[G\beta\gamma]$ for both yeast strains are shown in Fig. 3. The $sst2\Delta$ mutant shows a greater maximum response and the second rise in $[G\beta\gamma]$ is missing. Surprisingly, however, the $sst2\Delta$ mutant shows a steady decline in $[G\beta\gamma]$ with its final value being less than the 2XSST2 strain. If desensitization can be equated with a decrease in $[G\beta\gamma]$, then the model suggests an additional mechanism that contributes to desensitization beyond the Sst2-accelerated reformation of $G^{GDP}\alpha\beta\gamma$. In the absence of pheromone, most $G\beta\gamma$ is tied up in the $G^{GDP}\alpha\beta\gamma$ complex and is protected from degradation. Exposure to pheromone causes the heterotrimer to dissociate, allowing the liberated $G\beta\gamma$ to be degraded slowly.

Figure 4 shows a time series for [Sst2]. After exposure to the pheromone, [Sst2] increases slowly for the first 100 min and then begins to decline slowly. This behavior is consistent with the experimental results of Hao $et\ al.$ (2003). The model predicts that the slow decrease in [Sst2] is mainly due to the decline in $[G\beta\gamma]$ that drives $de\ novo$ Sst2 transcription and is largely unattributable to the negative effects of RGS domain GAP activity.

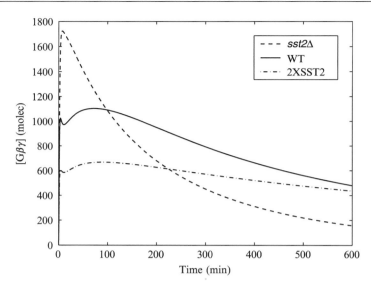

FIG. 3. Time course of $[G\beta\gamma]$ for WT (solid line), 2XSST2 (dot-dashed line), and $sst2\Delta$ (dashed line) strains following pheromone induction (3×10^{-6} M). In all cases the response becomes inactivated after long times (>80 min). The $sst2\Delta$ mutant shows a higher maximal response. RGS protein overexpression decreases the maximum response and increases the time required for the system to return to steady state. In the absence of RGS protein (dashed line), the second increase in $[G\beta\gamma]$ is lost. The decrease of $[G\beta\gamma]$ after long times for the $sst2\Delta$ strain indicates that $G\beta\gamma$ degradation contributes to pathway inactivation.

Figure 5A shows the pheromone dose–response curve measured by Hao *et al.* (2003) for the three different yeast strains (WT, 2XSST2, and $sst2\Delta$). Note that deletion of *SST2* results in a 100-fold decrease in the concentration of α factor necessary to achieve 50% of the maximum agonist response, but does not alter the maximum response. However, a twofold overexpression of *SST2* leads to a reduction in the maximum response by 27%. To test if the model can account for these results, Eqs. (1)–(8) were solved numerically, and the values of β-galactosidase concentration were recorded at 120 min. Figure 5B shows the results. As can be seen, the model nicely reproduces the high sensitivity at low pheromone concentration for the $sst2\Delta$ mutant as well as capturing the reduction in maximum response of the 2XSST2 strain.

The reduction in the maximum response seen in the 2XSST2 strain can be understood by looking at the time series of $[G\beta\gamma]$ at saturating pheromone levels for the three different yeast strains. Figure 6 shows results for a pheromone concentration of 10^{-4} M, which is considerably

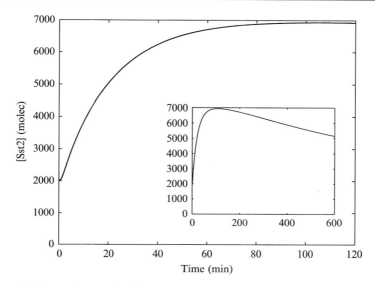

FIG. 4. Time series of [Sst2] changes on pheromone stimulation. The α-factor concentration is again $3 \times 10^{-6}\ M$, which is slightly higher than the EC_{50} value for WT. [Sst2] shows a steady increase at early times due to pheromone-induced synthesis. (Inset) At later times, the decrease in $G\beta\gamma$ concentration also produces a decrease in RGS protein concentration.

higher than the value of $3 \times 10^{-6}\ M$ used in Figs. 2 and 3. Surprisingly, we see that the maximum $G\beta\gamma$ concentration for the 2XSST2 strain is larger than the $sst2\Delta$ mutant. However, the lack of RGS protein GAP activity in the $sst2\Delta$ mutant produces a much slower decrease in the $G\beta\gamma$ concentration immediately after the maximum response. The wild-type $[G\beta\gamma]$ behaves similarly to the 2XSST2 strain in that it decreases rapidly after the maximum response. Also, in both these cases, the $G\beta\gamma$ concentration remains fairly constant between 20 and 120 min. However, the wild-type $[G\beta\gamma]$ is always higher than the 2XSST2 strain. β-Galactosidase degrades very slowly. Therefore, the β-galactosidase concentration measured at 120 min (Fig. 5) represents the cumulative amount of protein synthesized from the time the cells are exposed to pheromone, that is, the total β-galactosidase concentration is roughly proportional to the area under the curves shown in Fig. 6. Therefore, the slow decrease of the concentration in the $sst2\Delta$ strain compensates for its diminished maximum response. The reason for this diminished maximum response is due to the assumption that $G^{GDP}\alpha\beta\gamma$ is protected from degradation. In the absence of pheromone, the

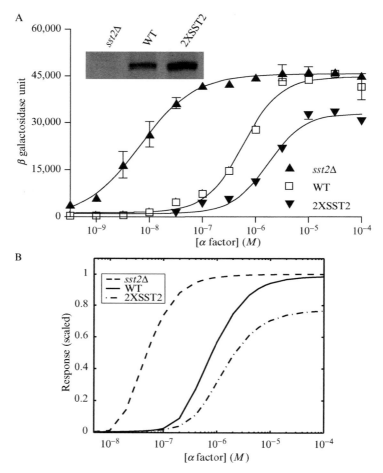

Fig. 5. (A) Pheromone-dependent transcriptional induction (from Hao *et al.*, 2003) for wild-type (WT), 2XSST2, and *sst2Δ* yeast strains. (Inset) The expression level of Sst2 protein in the three strains, as measured by immunoblotting. (B) Predicted pheromone-dependent transcriptional induction for the three strains shown in A. The model captures the decrease in EC_{50} seen in the *sst2Δ* strain (dashed line) relative to the wild type strain (solid line) and the decrease in the maximal response seen in the 2XSST2 strain (dot-dashed line). The response is scaled relative to the maximum response in wild-type cells.

sst2Δ strain cannot convert $G^{GTP}\alpha$ to $G^{GDP}\alpha$ efficiently. Therefore, the steady-state level of $G^{GDP}\alpha\beta\gamma$ for the *sst2Δ* strain is less than for the other two yeast strains.

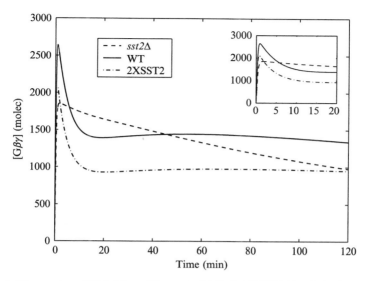

FIG. 6. Time series of [G$\beta\gamma$] for wild-type (solid line), 2XSST2 (dot-dashed curve), and *sst2Δ* (dashed curve) strains at saturating pheromone concentration (10^{-4} M). In this case, the maximal response in the *sst2Δ* strain is less than the wild-type and 2XSST2 strains (inset). However, expression of the RGS protein Sst2 causes wild-type and 2XSST2 strain concentrations to decrease rapidly immediately following the maximum response. Because β-galactosidase is degraded very slowly, its concentration is roughly proportional to the area under the curves. Therefore, the 2XSST2 strain is predicted to show a diminished maximum response (see Fig. 5).

Conclusions

We developed a mathematical model of G-protein signaling in yeast that consists of a set of nonlinear rate equations. The model revealed that the degradation of liberated G$\beta\gamma$ after pheromone exposure contributes to pathway inactivation. It also highlighted the need to consider the time-dependent response of signaling pathways when trying to interpret experimental results. All the mathematical techniques applied in developing and analyzing the model are directly applicable to other organisms.

As with all scientific methods, it is important to understand the limitations of this modeling approach. The validity of the rate equations are based on two important assumptions. The first is that the spatial heterogeneity of protein concentrations does not significantly affect the properties of the pathway. If spatial effects are important, more sophisticated modeling approaches are required. These include compartmental models or using reaction–diffusion equations. The second assumption is that the stochastic nature of the underlying biochemical reactions can be ignored. The rate equations should be interpreted as equations for the average concentration. Their validity depends on concentration fluctuations being small in comparison with the mean concentration. This is not always true, particularly in the case of transcriptional regulation in which mRNA numbers can be quite small. It is possible to construct stochastic models of signaling pathways. However, the analysis and simulation of these models are considerably more difficult. Even with these caveats, the modeling techniques presented in this article provided an important tool for analyzing signaling pathways.

References

Berman, D. M., and Gilman, A. G. (1998). Mammalian RGS proteins: Barbarians at the gate. *J. Biol. Chem.* **273,** 1269–1272.

Dohlman, H. G., and Thorner, J. (1997). RGS proteins and signaling by heterotrimeric G proteins. *J. Biol. Chem.* **272,** 3871–3874.

Dohlman, H. G., and Thorner, J. W. (2001). Regulation of G protein-initiated signal transduction in yeast: Paradigms and principles. *Annu. Rev. Biochem.* **70,** 703–754.

Fall, C. P. (2002). "Computational Cell Biology." Springer, New York.

Finley, D. (1993). The yeast ubiquitin system. *In* "The Molecular and Cellular Biology of the Yeast Saccharomyces" (E. W. Jones, J. R. Pringle, and J. R. Broach, eds.). Cold Spring Harbor Laboratory Press, Cold Spring Harbor, NY.

Ghaemmaghami, S., Huh, W. K., Bower, K., Howson, R. W., Belle, A., Dephoure, N., O'Shea, E. K., and Weissman, J. S. (2003). Global analysis of protein expression in yeast. *Nature* **425,** 737–741.

Hao, N., Yildirim, N., Wang, Y., Elston, T. C., and Dohlman, H. G. (2003). Regulators of G protein signaling and transient activation of signaling: Experimental and computational analysis reveals negative and positive feedback controls on G protein activity. *J. Biol. Chem.* **278,** 46506–46515.

Jenness, D. D., Burkholder, A. C., and Hartwell, L. H. (1986). Binding of alpha-factor pheromone to *Saccharomyces cerevisiae* a cells: Dissociation constant and number of binding sites. *Mol. Cell. Biol.* **6,** 318–320.

Kim, J., Bortz, E., Zhong, H., Leeuw, T., Leberer, E., Vershon, A. K., and Hirsch, J. P. (2000). Localization and signaling of G(beta) subunit Ste4p are controlled by a-factor receptor and the a-specific protein Asg7p. *Mol. Cell. Biol.* **20**(23), 8826–8835.

Mukhopadhyay, S., and Ross, E. M. (1999). Rapid GTP binding and hydrolysis by G(q) promoted by receptor and GTPase-activating proteins. *Proc. Natl. Acad. Sci. USA* **96,** 9539–9544.

Schauber, C., Chen, L., Tongaonkar, P., Vega, I., and Madura, K. (1998). Sequence elements that contribute to the degradation of yeast G alpha. *Genes Cells* **3,** 307–319.

Strogatz, S. H. (1994). "Nonlinear Dynamics and Chaos: With Applications to Physics, Biology, Chemistry, and Engineering." Addison-Wesley, Reading, MA.

Yi, T. M., Kitano, H., and Simon, M. I. (2003). A quantitative characterization of the yeast heterotrimeric G protein cycle. *Proc. Natl. Acad. Sci. USA* **100,** 10764–10769.

[24]   Identification of Yeast Pheromone Pathway Modulators by High-Throughput Agonist Response Profiling of a Yeast Gene Knockout Strain Collection

By Scott A. Chasse and Henrik G. Dohlman

Abstract

Gene deletion analysis is a powerful tool for resolving the contributions of individual open reading frames to the physiology of cells. Analysis of deletion phenotypes in conjunction with a specific pathway reporter can identify constituents of a physiological pathway and reveal potential effectors that regulate the pathway by quantifying the phenotypic responses of the mutant cells. This article describes a high-throughput method of analyzing a yeast gene deletion library for novel G-protein signaling modulators using a yeast pheromone pathway-specific reporter.

Background

Over the years, the pheromone signaling pathway of the budding yeast *Saccharomyces cerevisiae* (hereafter, "yeast") has been studied intensively. Investigations have shown that pheromone acts via a G-protein-coupled receptor (GPCR)-directed protein kinase signaling cascade, which ultimately results in growth arrest, new gene transcription, and mating of haploid cells. Genetic studies and deletion analysis have identified all the components necessary and sufficient for signaling, and most are well characterized. However, components that modulate the signal strength and efficiency have been more difficult to identify (by virtue of the more subtle effects they have on signaling). A notable exception is the prototypical regulator of G-protein signaling (RGS) protein Sst2 (supersensitivity to pheromone) due to its robust deletion phenotype. To identify subtler signal modulators and quantify their effects on signaling, we generated a comprehensive series of pheromone dose–response curves of 4349 of the available 4847 nonessential gene deletions. These data quantify the effect of each gene deletion on the yeast pheromone signaling pathway and, in so doing, identify modulating elements. This article describes the high-throughput approaches used to collect these data. A comprehensive discussion of the results is presented elsewhere (Chasse *et al.*, 2004).

Two indispensable components of this whole genome screening are the yeast nonessential gene deletion mutant collection and the pathway-specific reporter. The deletion collection is available commercially from Research Genetics (Invitrogen). In addition, a *FUS1-lacZ* reporter construct, which is a fusion of the pheromone-responsive *FUS1* promoter and the β-galactosidase coding sequence, has been used extensively by many laboratories (see Hoffman *et al.*, 2000). Historically, it has proven to be a highly specific, effective, and reliable reporter of the yeast pheromone/ GPCR pathway.

Library Maintenance

Because of the large number of knockout strains, it is crucial that they be cataloged and maintained properly. Strains arrive as frozen cell suspensions in 52 × 96-well plates. The first task that should be performed is to prepare a secondary frozen stock to be stored at −80°. This is to ensure the viability of the strains on the primary, or master, plates by minimizing freeze–thaw cycles. Once a secondary frozen stock is prepared, rethawing of the primary plates should be done as a last resort.

To prepare secondary frozen stocks, each source plate should be thawed at room temperature. A 96-well pinning tool (Denville Scientific Inc.) is used to transfer cells from the wells of the source plate to a fresh 96-well assay plate containing 100 μl per well of yeast extract (Difco)/ Peptone (Difco)/dextrose (Sigma) (1/2/2%, w/w) (YPD) broth. The plates should be incubated for 3–5 days at 30° until growth is well established at the bottom of the wells. Rare nonsurviving strains should be noted. Final preparation for −80° storage involves exchanging the spent media with fresh YPD by centrifuging the cells to pellets at 2000g at 4° for 20 min, resuspending each pellet into 100 μl of fresh medium, and mixing each cell suspension with an equal volume of sterile glycerol (30% diluted in YPD) and flash freezing at −80°. Prior to freezing, however, stab plates for shorter-term storage at 4° may be prepared on 150-mm petri plates containing 150 ml of YPD 2% agar supplemented with G418 (Clontech) at 300 μg/ml. Stabs are prepared by using a 96-well pinning tool to transfer small amounts of material from each well to the petri plate. The pinning tool should be stabbed only partially into the agar surface. Pins inadvertently driven too far may cause the strains to grow together at the underside of the agar surface. Upon pinning, the plates are incubated for 3 days at 30°. Once grown the plates are wrapped in the original plastic bags and refrigerated at 4°. Stab plates remain useful for ∼6 months by which time colonies begin to grow together. Finally, it should also be noted that the refrigerator stock plates contain G418 simply as a prophylactic against contaminating microbial growth, rather than for superfluous maintenance of the genomically integrated selection marker.

96-Well Format Pheromone Response Assay

The β-galactosidase pheromone response assay has been a mainstay of this laboratory for many years and has been performed in the same general way throughout this time. Briefly, a yeast strain is transformed chemically with a pheromone-response pathway-specific reporter plasmid containing *FUS1-lacZ*. One or more individual colonies are selected randomly, streaked onto a fresh selection plate, and incubated. Next, lone colonies are inoculated into selective liquid culture media and grown to saturation. This culture is used to inoculate fresh medium. The cells are grown until midlog phase growth (defined later). Samples of the culture are then transferred to wells of a 96-well plate each containing α-factor pheromone, and the plate is incubated. A cocktail containing a fluorescence-developing β-galactosidase substrate is added and the reaction is allowed to progress. After a time, color development is halted with an aliquot of Na_2CO_3, and the fluorescence of each well is measured by spectroscopy. The measurements are normalized for cell density and results are plotted.

Typically, the assay is performed simultaneously on four to eight strains at one time. However, performing the same experiments on a scale sufficient to collect data on nearly 5000 strains in a reasonable time frame requires a 40-fold throughput increase. This scale up entails significant adaptation and optimization of the procedures.

Strain Preparation

During the course of screening, it is usually necessary to return to the source plates for fresh yeast material. Therefore, in order to reduce contamination of the refrigerated stab plates and to conserve the small amount of material occluding the stabs, it is convenient to prepare working patch plates from the stab plates. These patch plates should be prepared on YPD agar in 150-mm petri plates using the pinning tool. They are prepared in a similar manner as the stab plates, but instead use the pinning tool to "paint" small, discrete, circular patches of yeast onto the plate. These plates should be incubated 24–48 h at 30° until the patches are well established. These plates can be stored up to 30 days before overgrowth of the patches occurs. Further, in our screening it was necessary to include on each plate an internal wild-type (WT) control strain (BY4741). Therefore, when the patch plates were prepared, the WT strain was included at position H2 (a vacant position in every BY4741 *MAT **a*** library collection plate). However, because the WT strain has no integrated antibiotic resistance marker, the working patch plate media can contain no G418.

Once the experimental strain arrays have been prepared as patch plates, the strains must be transformed with the reporter plasmid. The following is an adaptation of the transformation (TRAFO) protocol reported by Schiestl *et al.* (1993). All solutions listed in the following procedure (except DNA) should be prepared from filter-sterilized stock solutions. The patch plates are used to inoculate 96-well V-bottom plates (Nunc) containing 250 μl of YPD broth per well. These plates are incubated 24 h at 30° without shaking. The plates are then centrifuged at 2000g at 4°, and the liquid media are aspirated with a multichannel pipettor, using fresh tips for each row to avoid cross-contamination. Each pellet is then resuspended thoroughly into 50 μl of 100 mM lithium acetate. Each well then receives 150 μl of a cocktail containing 42% (w/w) polyethylene glycol 3350 (Sigma), 100 mM lithium acetate (Sigma), 210 μg/ml of single-stranded salmon sperm carrier DNA (Eppendorf), and 6.7 μg/ml of reporter plasmid DNA. The contents of the wells should be mixed thoroughly, and the plates are incubated without shaking for 30 min at 30° followed by 30 min at 42°. The plates are then centrifuged at 2000g at 4° for 20 min, and the supernatant is aspirated. Finally, the pellets are each resuspended into 250 μl of YPD broth and incubated for 4 h at 30°. The plates are then centrifuged at 2000g at room temperature for 20 min, and the pellets are resuspended into 100 μl of synthetic complete dextrose histidine dropout (SCD-his) broth (1.7 g yeast nitrogen base [without amino acids or $(NH_4)_2SO_4$] (Difco), 0.1 g NaOH (J. T. Baker), 30 mg adenine (Sigma), 5 g $(NH_4)_2SO_4$ (Fisher), 20 g dextrose (Sigma), and complete supplement mixture minus histidine (CSM–his) (Bio101)) and plated individually onto 6-well plates containing 5 ml SCD-his agar per well. Histidine dropout medium is used in this particular application to select for cells (*his3* background) containing the reporter plasmid that includes the *HIS3* selection marker. Plates should be incubated for 3–4 days at 30° until the transformed colonies are well established. Finally, a single colony from each well should be transferred via loop, patch plated onto a 150-mm plate of SCD-his agar as an 8 × 12 array matching the original source plate, and grown for 36 h or until the patches of growth are well established.

β-*Galactosidase Activity Assay*

The following procedure was adapted for this high-throughput application from a protocol described in detail by Hoffman *et al.* (2002). Transformed strain arrays are used to inoculate a 96-well assay plate containing 250 μl per well of SCD-his broth via the pinning tool and incubated for 48 h at 30° without shaking. From these cultures, 50-fold dilutions of each strain should be prepared by transferring 35 μl per well to a 2-ml × 96-well growth block (Nunc) containing 1.72 ml per well of SCD-his broth. It was determined empirically (data not shown) that for this strain and reporter background (at a 50-fold dilution) the ideal incubation time to achieve near log phase growth was 8 h. Therefore, after 7 h the optical densities (OD) of the strains should be monitored by an optical plate reader (SpectraMax 340PC, Molecular Devices) until the largest majority of the cultures achieve midlog phase growth. Optical density readings are determined empirically at a given volume for a given plate to correspond to between 0.5 and 0.9 $OD_{600\ nm}$ in a standard cuvette with a 1-cm path length.

Assay plates (96-well) containing 10 μl of α factor (Sigma or any commercial peptide synthesis source) as 10× stock solutions (0, 3.0, 10, 30, 100, 1000 μM; one concentration per plate; three replicates per concentration) should be mixed with 90 μl of the cell suspensions using a liquid-handling robot (Hydra-96, Matrix Technologies) and incubated at 30° for 90 min. A freshly prepared substrate cocktail (20 μl) containing 0.5 μM fluorescein di-β-D-galactopyranoside (FDG) (Molecular Probes), 0.025% Triton X-100, and 130 mM PIPES, pH 7.2, is added to each well using the liquid-handling robot and incubated for 60 min at 37°. The developing reactions are halted by adding 20 μl of 1.0 M Na_2CO_3 using the robot. The fluorescence of each well should then be measured (485 nm excitation/ 530 nm emission) using a fluorescence plate reader (CytoFlor, PerSeptive Biosystems).

Data Analysis

Analysis of raw data is theoretically straightforward. As described in Eq. (1), a raw fluorescence reading must be adjusted to correct for differences in cell density as measured by $OD_{600\ nm}$ and normalized to some arbitrary standard $OD_{600\ nm}$:

$$F_{normalized} = F_{raw} \times \frac{0.800}{OD_{600\ nm}} \qquad (1)$$

Our experience with this assay has demonstrated that the most consistent data are collected from cultures at an $OD_{600\ nm} = 0.800$. We therefore chose to normalize all raw fluorescence readings obtained from cultures within the acceptable range of $OD_{600\ nm}$ to this value. Once these readings have been normalized, dose–response curves can be plotted and various curve parameters calculated. Figure 1 shows a representative dose–response curve from data obtained in the screening experiment. For our purposes, we were interested in changes (relative to WT) in the maximum response to pheromone (efficacy), the concentration of pheromone at which effect is half-maximal EC_{50} (sensitivity or potency), and/or an increase in the baseline activity (in the absence of pheromone, suggesting a constitutive activation of the pheromone pathway). These values are

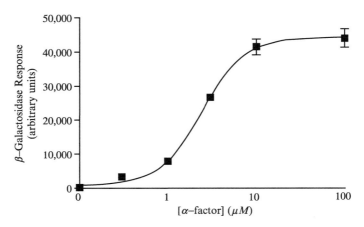

FIG. 1. Pheromone dose–response curve. The β-galactosidase response to the pheromone of BY4741 (WT) obtained during the screening is plotted as a function of log(pheromone concentration). Data were fit to a theoretical dose response (variable slope) by the GraphPad Prism application. Calculated parameters are: baseline 780 arbitrary units, maximum response 44,300 arbitrary units, and EC_{50} 2.4 μM.

calculated easily by commercial software such as GraphPad Prism (Graph-Pad Software Inc.). Finally, each dose–response parameter calculated was presented as a ratio of the corresponding value calculated for the WT strain in each plate. This allowed a direct comparison of data collected on different days and from different plates.

For a small number of deletion strain source plates, simple spreadsheet analysis of the data would be manageable but tedious. However, with a large number (52 library plates), as with the BY4741 *MAT a* collection, the manipulations can become quite unwieldy in terms of the time required to format data, calculate dose–response parameters, sort, analyze, and record the results. Therefore, it is useful to automate the data analysis using scripts and macros. A Microsoft Excel spreadsheet was programmed to accept the 19 sets of data (6 pheromone concentrations in triplicate = 18 readings plus $OD_{600 \text{ nm}}$ readings) and macros written in Visual Basic (Excel editor) to format data, normalize the readings, and prepare it for export to GraphPad Prism for dose–response curve parameter calculation. Further, because both Microsoft Excel and GraphPad Prism are fully scriptable using Macintosh AppleScript (MacOS), the manipulations between programs can be, in large part, automated. Therefore, once data from the experiment are formatted and normalized properly, it is copied and pasted into a GraphPad Prism template file prepared to plot data as individual dose–response curves with associated standard errors. The program is then directed to calculate the desired parameters, plot data, and prepare it for export back to Excel. Data are then pasted back into Excel where they are sorted, and standard deviations are calculated. Scripts and macros are very useful but application specific. For any adaptations of this procedure it will be necessary to develop a customized set of programs for data formatting and manipulation.

For this screening we felt that data obtained justified setting a selection cutoff for each parameter at twofold above or below the control WT value, and at least two standard deviations from the mean for each plate. The exception to this was to set the threshold for baseline activity to fivefold above WT. This was because the prohibitively large number of strains with a twofold or better baseline shift dictated that a higher threshold should be chosen. These arbitrarily set limits served to allow for the selection of the most interesting hits (greatest magnitude) while simultaneously minimizing the clutter of false positives. To ensure that these thresholds would not result in false negatives, we examined known pathway component knockouts and confirmed that these were not excluded erroneously. Table I lists

TABLE I
GENES KNOWN TO AFFECT THE PHEROMONE PATHWAY[a]

Common name	Systematic name	Identification
SST1	YIL015W	Hit
SST2	YLR452C	Hit
STE02	YFL026W	Hit
STE03	YKL178C	**a**-factor binding
STE04	YOR212W	Hit
STE05	YDR103W	Hit
STE07	YDL159W	Hit
STE08	YLR442C	Hit
STE09	YDR227W	Hit
STE11	YLR362W	Hit
STE13	YOR219C	**a**-factor processing
STE14	YDR410C	OOR
STE16	YDL090C	**a**-factor processing
STE20	YHL007C	OOR
STE21	YDR335W	Hit
STE22	YPR122W	**a**-factor processing
STE23	YLR389C	**a**-factor processing
STE24	YJR117W	Hit
STE50	YCL032W	Hit

[a] Those identified as hits by this α-factor screening are labeled. Strains not identified as hits because they failed to grow are identified as out of range (OOR). Components whose function is to process or bind mating pheromones are identified as such.

the known pathway components included in the deletion collection library and the results from the original screening of each. Of the 19 components listed, all but 7 were identified by the screen. Five of these 7 are only required for **a**-factor pheromone processing (Ste16, Ste22, Ste23) or binding (Ste3), α-factor processing (Ste13), and are not involved directly in α-factor signal transduction. The remaining two consistently failed to reach the suitable $OD_{600\ nm}$ range for the pheromone response assay. Based on this, we calculate an estimated false-negative rate of $\sim$1%. Finally, *ste6Δ*, *ste12Δ*, and *ste18Δ* were absent from the haploid strain collection. These strains were not included in the false-negative calculation.

False positives are less a concern for data quality than one of convenience. Therefore, we intentionally set selection thresholds low to reduce the possibility of false negatives. In our screening, using the selection thresholds outlined earlier, 167 hits were initially recorded over all 52 plates. Following a second round of transformation and a β-galactosidase assay of these candidates, 98 were confirmed to significantly affect pheromone signaling in the same manner and to a sufficient degree as first demonstrated. However, the lack of reproducibility was more a function of using a wider range of optical densities than usual to facilitate the screening (discussed later).

High-Throughput Considerations

The first major concern when scaling up the reporter assay was the need to perform a large number of uniform and timely reagent additions. On a small scale, any given reagent addition of the assay procedure is performed easily and accurately with a standard electronic 8-channel pipettor. However, the logistics of liquid handling in a 96-well format, in numbers sufficient to generate full dose–response curves (six pheromone concentrations), complete with error analysis (three replicates per concentration) for even a single library source plate, would require manipulating over 1700 discrete aliquots per reagent addition. Using a standard 8- or 12-channel pipettor would be tiring, time-consuming, and a potential source of human error.

There are many advantages of employing a liquid-handling robot. First, reagent addition volumes are standardized by the mechanical action of the Hamilton-type syringes and are therefore more precise than repeated, manual manipulations. Further, the simultaneous addition of a reagent to an entire plate allows one to better standardize the timing of reagent addition to time-sensitive, color-developing reactions than can be achieved with 8 or 12 serial additions via a multichannel pipettor. Finally, robots are programmable and these programs can be stored, reducing one source of human error.

The screening was performed by a single individual over the course of 4 months. A subsequent screen performed in our laboratory by two individuals has reduced the time required to complete the entire procedure to 3 months. The procedure can be subdivided into three phases: preparing the strains for analysis, generating data and data analysis, and hit confirmation. A flowchart illustrating an overview of the approach is provided in Fig. 2. Using this procedure, each strain is given at least two opportunities to transform with the reporter plasmid and subsequently two opportunities to produce acceptable β-galactosidase assay results. Using the transformation procedures described herein, less than 1% (40 of 4847) of the strains consistently failed to transform with the reporter plasmid. Of the remaining 4807 strains, 4349 strains produced full dose–response curves suitable for analysis. This translates to coverage of 90% of all available strains and 75% of all known yeast ORFs.

In general, one individual can, in a single day, process two plates; either as transformations or β-galactosidase assays. Assays should be performed as soon as possible after transformation. Use of older transformed strains increases the likelihood that any given strain will not respond to the pheromone at all. We found it convenient to use a staggered schedule where 10 plates are transformed (two per day over 5 days) before β-galactosidase assays were performed at the same rate over the following 5 days. Once data were analyzed, it was then possible to consolidate and retest strains that failed to produce acceptable data. Strains consolidated from each set of 10 plates were reassayed twice: once with a slightly shorter growth time and once with a slightly longer growth time to accommodate those strains that grew at a slightly higher and lower growth rate, respectively. In general, the entire cycle for completing 10 plates (960 strains and 17,280 data points) took 3 weeks. It should be noted that the few strains that initially failed to transform were retransformed only after work on the entire strain collection was completed and all nontransforming strains identified.

Generally, there were two reasons data from transformed strains were unacceptable. The first was if there was no apparent response. The second and most common reason was if, at the time of the β-galactosidase assay, the density of the culture was not within the defined range of 0.5–0.9 $OD_{600\ nm}$ units. Typically, when we perform small-scale β-galactosidase assays, great care is taken to grow all cultures to an $OD_{600\ nm} = 0.8$ prior to performing an experiment. However, when performing the reporter assay on 188 strains simultaneously, it is unrealistic to expect all strains to reach this optical density simultaneously. For instance, differences in strain growth rates or initial inoculum cell density can affect the culture cell density at the time of assay. Further, cell size can affect the apparent cell density, increasing or decreasing the optical density for larger or smaller cells, respectively.

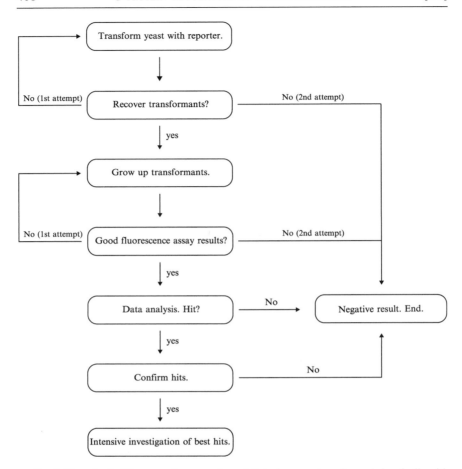

FIG. 2. Flowchart of the screening procedure. All strains were transformed chemically with a pheromone response reporter construct. Strains that failed to transform were later subjected to a second round of transformation. Those that failed a second attempt were eliminated from the screening. Randomly selected transformants from each deletion strain were grown to midlog phase as determined by optical density at 600 nm and were assayed by the liquid β-galactosidase response assay as described in the text. Each transformed strain was given two chances to produce acceptable results. Those that failed to yield acceptable results after two attempts were eliminated from the screening. Once the screening was completed, all initially identified hits were retransformed and reassayed to confirm the mutant phenotypes. Those that failed to confirm were eliminated from the screening. Those that remained after these analyses were considered hits for a more comprehensive analysis.

However, the alternative was to count manually the cell density for each of the 188 strains being assayed, which would be prohibitively time-consuming. It is for these reasons we chose to use a range of $OD_{600\ nm}$. Preliminary experiments with WT strains (data not shown) demonstrated that cultures within this range could be normalized to $OD_{600\ nm} = 0.8$ with little effect on maximum response and no effect on EC_{50}. For this reason, maximum response was the main source of false positives. However, we felt this was a compromise necessary to facilitate the screening.

With such a large number of strains to examine, it was necessary to minimize the number of agonist concentrations tested. For instance, each additional concentration examined would have required three more assay plates. Further, we were limited by the amount of culture contained in each growth block well. Each growth block well could contain only 1.77 ml of culture and 90 μl was required for each assay well. This means that no more than 18 data points can be collected for a given strain. We chose three replicates of six concentrations rather than two replicates of nine concentrations to allow calculation of a standard error at each point. Next, we assumed that most gene knockouts would not affect the sensitivity of the strain to the pheromone. Therefore, we chose to examine a range of concentrations that was centered at the EC_{50} of the WT strain (1–3 μM). Considering these criteria, we chose 0, 0.3, 1.0, 3.0, 10, and 100 μM. A concentration of 10 μM would have been sufficient to activate the pathway maximally. The 100 μM concentration was included to ascertain if any mutant had an altered response at a very high agonist concentration.

Finally, once all data had been collected and all initial hits identified, these strains were consolidated onto two plates, retransformed as before, and assayed one final time to confirm the mutant phenotype. This ended the high-throughput phase of the nonessential gene knockout collection analysis. As with all high-throughput candidate analyses, the identification of hits was only the initial step. The hits identified can now be characterized in greater depth as part of a hypothesis-driven research program to characterize new proteins that regulate the yeast pheromone pathway.

References

Chasse, S. A., Parnell, S. C., Zeller, C. E., and Dohlman, H. G. (2004). Manuscript in preparation.

Hoffman, G., Garrison, T. R., and Dohlman, H. G. (2000). Endoproteolytic processing of Sst2, a multidomain regulator of G protein signaling in yeast. *JBC* **275**(48), 37533–37541.

Hoffman, G., Garrison, T. R., and Dohlman, H. G. (2002). Analysis of RGS proteins in *Saccharomyces cerevisiae*. *Methods Enzymol.* **344**, 617–631.

Schiestl, R. H., Manivasakam, P., Woods, R. A., and Gietz, R. D. (1993). Introducing DNA into yeast by transformation. *Methods* **5**, 79–85.

Author Index

A

Abogadie, F. C., 183
Adami, G. R., 290
Adams, B., 151
Addicks, K., 239
Aebersold, R., 120, 289
Ahn, W., 98, 119, 123
Ahringer, J., 308
Ajit, S., 277
Akaike, M., 340
Akaike, N., 132
Akil, H., 158, 160
Alexander, R. W., 77
Alonso, J. M., 339
Alpi, A., 348
Alt, A., 162
Altman, S., 245
Amarzguioui, M., 247, 262
An, F., 290
Anisimov, S. V., 237
Ansel, K. M., 16
Apaniesk, D., 231
Apanovitch, D. M., 48, 65, 172, 191, 231
Aronovitz, M., 230
Arora, A., 352
Arshavsky, V. Y., 50, 230
Artemyev, N. O., 173, 191, 234, 351
Arya, S., 74
Asano, T., 133
Aspenstrom, P., 247
Assmann, S. M., 339
Asundi, J., 245, 249, 254
Auld, V. J., 354, 357
Avery, L., 307, 311
Avigne, W. T., 345
Avruch, J., 327
Aymen, A. S., 245

B

Babaie, E., 262
Backlund, P. S., 36, 37

Backlund, P. S., Jr., 35
Bahia, D. S., 172, 188, 202
Bai, D., 230
Baird, G. S., 58
Balasubramanian, N., 40
Baleux, F., 23
Barker, S. A., 4, 22, 91, 95–97, 279, 306–308, 351, 355
Barnes, C. A., 77, 78, 80, 90, 230
Barrot, M., 73, 231
Bartel, P. L., 366
Bartz, S. R., 261
Bashar, M., 157
Bateman, A., 323
Baumgartner, S., 357
Baylor, D. A., 120, 133, 151, 190
Bean, B. P., 132, 137, 184
Bechtold, N., 341
Becker, A., 352
Behe, C. I., 57, 58, 62, 67, 246, 266, 327, 329, 330, 358
Behr, J. P., 199
Bejsovec, A., 352
Bellaiche, Y., 360
Belle, A., 391
Bellen, H. J., 357, 363
Bender, C., 277
Benovic, J. L., 77
Bentrup, F. W., 345
Benz, W. K., 354
Bera, A. K., 4, 127, 128, 130
Bergson, C. M., 249
Berker, S. A., 143
Berman, D. M., 41, 57, 58, 62, 78, 80, 90, 129, 132, 157, 191, 205, 231, 232, 246, 266–268, 322, 326, 327, 384
Bernard, J. L., 280
Bernard, M. L., 351
Bernstein, L. S., 37–38, 94, 97, 280
Berridge, M. J., 72
Berstein, G., 119
Berthiaume, L., 43

411

M

Subject Index

A

Actin, F-actin polymerization in chemokine-activated lymphocytes, 25–26

Adenylyl cyclase, assay for regulators of G-protein signaling-insensitive Gα subunit studies, 164

Akt, phosphorylation in lymphocytes after chemokine activation, 23–25

Antisense, clinical limitations, 244–245

Arabidopsis thaliana regulator of G-protein signaling-1
bioinformatic identification and analysis, 321, 323
Gα subunit interactions
coprecipitation, 326–327
surface plasmon resonance of nucleotide-dependent interactions, 327–330
G-protein types, 320–321
GTPase-accelerating protein assays
fluorescent guanine nucleotides as substrates, 333–334, 336
single-turnover assays using *Escherichia coli* phosphate-binding protein, 330–333
purification of recombinant proteins
Gα subunit, 323–326
regulator of G-protein signaling, 326
structure, 320–322, 338
sugar signaling studies
germination inhibition at high sugar concentration, 343, 345
hexokinase in signaling, 340, 346–347
materials, 340–341, 343
mutant resistance to germination inhibition, 345–348
overview, 338–340
sugars and analogs, 344–345
sugar type effects, 347–348

AtRGS1, *see Arabidopsis thaliana* regulator of G-protein signaling-1

B

Behavioral analysis, *see Caenorhabditis elegans* regulators of G-protein signaling; Transgenic mice, regulators of G-protein signaling

Boyden chamber, chemotaxis assay of chemokine-activated lymphocytes, 26–28

C

Caenorhabditis elegans regulators of G-protein signaling
Gα subunit interactions, 310–311
G-protein types, 306
mutant studies of function, 307–308
rationale for study, 305–306
transgenic expression studies
approaches, 308–309
egg-laying behavior assays
freshly laid egg assay, 317–319
unlaid egg assay, 314–317
locomotion assay, 311–314
microinjection of plasmids, 309–310
types, 306–307

Calcium channel, N-type calcium channel studies using regulators of G-protein signaling-insensitive Gα subunit in neurons
electrophysiology studies, 185–187
reconstitution, 183

Calcium flux
assay for regulators of G-protein signaling-insensitive Gα subunit studies, 166–167
lymphocyte measurement after chemokine stimulation, 22–23
regulators of G-protein signaling studies of agonist-evoked calcium signaling
experimental system selection, 121
permeabilized pancreatic acinar cell studies

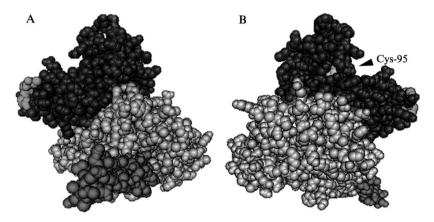

JONES, CHAPTER 3, FIG. 3. Structure of the RGS box binding $G\alpha_i$. A space-filling model of the RGS box of RGS4 (dark green) binding $G\alpha_{i1}$ (gray) based on the crystal structure of this complex (Tesmer *et al.*, 1997). (A) The presumed membrane-facing surface of the complex. (B) Rotation by about $120°$ through the vertical axis of the structure shown in A. Cys-95 (yellow), which corresponds to Cys-98 in RGS16, is only visible on this surface. The amino-terminal end of the RGS box, which starts at residue 51, is shown in aqua, and the amino- and carboxyl-terminal ends of $G\alpha_{i1}$ are shown in pink. Reprinted with permission from Hiol *et al.* (2003).

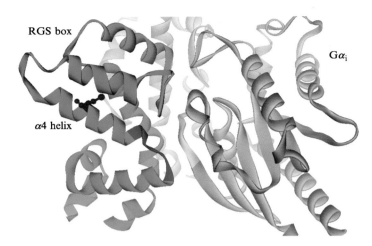

JONES, CHAPTER 3, FIG. 5. Relationship between internal palmitoylation and the RGS–G protein interface. A ribbon representation of the crystal structure of RGS4 (green) bound to $G\alpha_i$ (gray) demonstrating the close proximity of Cys-95 in RGS4 (brown) to the residues between the $\alpha3$ and the $\alpha4$ helices of RGS4 (purple) that interact with switch region I (yellow) of $G\alpha_i$. The membrane-facing side of the complex is predicted to be in the foreground. Reprinted with permission from Osterhout *et al.* (2003).

muscarinic m2 receptor. (B) Alignment and movement of perfusion barrels for rapid solution exchange. (*Top*) The three-barrel array positioned with the patch-clamped cell (see patch electrode) being superfused with 5 mM K^+ solution. Flow through both the middle and the right barrels is continuous and gravity driven, whereas the left barrel is typically not used in these experiments. (*Bottom*) The position of the barrels following computer-controlled movement (700 μm), where the 25 mM K^+ barrel is now aligned with the recorded cell. (C) Electrophysiological recording of a CHO-K1 cell voltage clamped at -100 mV (Vm) during superfusion with the 5 mM K^+ solution and during the triggered application (15 s) and washout of the 25 mM K^+ solution. Voltage ramps (-100 to 50 mV, 500 ms) were evoked before and during 25 mM K^+ solution application to evaluate the voltage dependence of receptor-independent currents. Changes in the whole cell current (I_m) in response to the voltage ramps and solution change are shown; the horizontal line indicates the zero current level. (D) Current–voltage plots derived from the voltage ramps evoked before (5 mM K^+) and during the 25 mM K^+ solution application. The reversal potential shift is consistent with a K-selective conductance having steep inward rectification; both are signature features of GIRK currents. (E) The time course for changes in I_m with 25 mM K^+ solution application and washout was fit with a single exponential function (red lines) to derive time constants (τ) for solution exchange. Note the delay in the onset, which is attributed to the distance of the perfusion barrels from the recorded cells and the associated volume dead space. (F) Plots of the solution exchange time constant obtained from applications and washouts of 25 mM K^+ are nearly equivalent. Data are means $\pm$ SEM ($n = 3$).

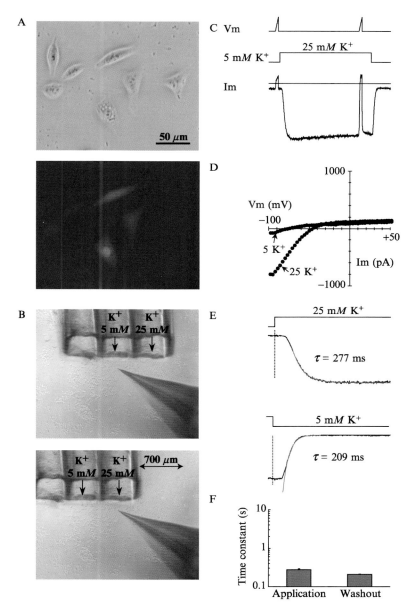

DOUPNIK *ET AL.*, CHAPTER 9, FIG. 1. Determining the limiting rates of solution exchange for CHO-K1 cells under whole cell patch-clamp recording. (A) CHO-K1 cells 24 h after DNA transfection visualized under phase-contrast (*top*) and epifluorescence (*bottom*) microscopy. Cells were transfected with EGFP, GIRK channel subunits (Kir3.1/Kir3.2a), and the

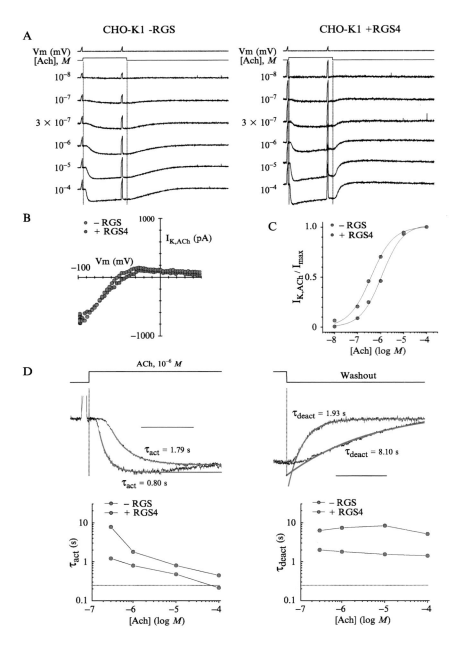

A

CHO-K1 -RGS

Vm (mV)
[Ach], M

10^{-8}

10^{-7}

3×10^{-7}

10^{-6}

10^{-5}

10^{-4}

CHO-K1 +RGS4

Vm (mV)
[Ach], M

10^{-8}

10^{-7}

3×10^{-7}

10^{-6}

10^{-5}

10^{-4}

B

- $-$ RGS
- $+$ RGS4

1000

$I_{K,ACh}$ (pA)

-100

Vm (mV)

-1000

C

1.0

- $-$ RGS
- $+$ RGS4

$I_{K,ACh} / I_{max}$

0.5

0

-8 -7 -6 -5 -4

[Ach] (log M)

D

ACh, 10^{-6} M

τ_{act} = 1.79 s

τ_{act} = 0.80 s

Washout

τ_{deact} = 1.93 s

τ_{deact} = 8.10 s

- $-$ RGS
- $+$ RGS4

10

τ_{act} (s)

1

0.1

-7 -6 -5 -4

[Ach] (log M)

- $-$ RGS
- $+$ RGS4

10

τ_{deact} (s)

1

0.1

-7 -6 -5 -4

[Ach] (log M)

Doupnik *et al.*, Chapter 9, Fig. 2. Quantitative analysis of RGS-modulated GIRK current kinetics in CHO-K1 cells. (A) Acetylcholine (ACh)-activated whole cell GIRK currents recorded from CHO-K1 cells cotransfected with the muscarinic m2 receptor, Kir3.1 and Kir3.2 subunits, and EGFP, with (*right*) or without (*left*) rat RGS4. Families of currents from each cell are from different ACh concentrations applied for 15 s from a holding potential of -100 mV as described in Fig. 1. Vertical dotted reference lines indicate triggering times for the movement of the perfusion barrels for agonist application and washout. (B) Current–voltage plots for $I_{K,ACh}$ derived from RGS4 (green circles) and non-RGS-transfected CHO-K1 cells in A. Plots were obtained by subtracting the ramp-evoked current prior to ACh application from the ramp-evoked current obtained during 1 μM ACh application. Both I–V plots display steep inward rectification and similar reversal potential near the K^+ equilibrium potential (-40 mV), indicative of receptor-evoked GIRK currents. (C) Steady-state dose–response relations for ACh-evoked GIRK currents ($I_{K,ACh}$) in the absence (red circles) and presence of coexpressed RGS4 (green circles). The peak $I_{K,ACh}$ amplitude for each ACh concentration was normalized to the maximal amplitude elicited by 100 μM ACh from each cell. The normalized $I_{K,ACh}$ amplitudes were then fit with a Hill function ($I_{K,ACh}/I_{Max} = 1/\{1+[ACh]/(EC_{50})^{[nH]}\}$) to derive the effective ACh concentration evoking a half-maximal response (EC_{50} value) and the Hill coefficient value (nH). RGS4 expression causes a small rightward shift of the ACh dose–response curve (EC_{50} values: $-$RGS, 0.82 ± 0.16 μM ACh, $n = 10$; $+$RGS4, 1.39 ± 0.30 μM ACh, $n = 6$), contrary to earlier observations (Doupnik *et al.*, 1997). (D) Time constants for ACh-evoked GIRK current activation (τ_{act}) and deactivation (τ_{deact}) were derived from single exponential fits to the currents. The fitted exponential functions are superimposed on current data ($-$RGS, red; RGS4, green) from GIRK currents elicited by 1 μM ACh. The current amplitudes have been normalized for comparison. The horizontal line in each panel represents a 5-s scale bar. (*Bottom*) Plots of τ_{act} and τ_{deact} derived from currents elicited by the different ACh concentrations, illustrating RGS4 effects on both kinetic parameters.

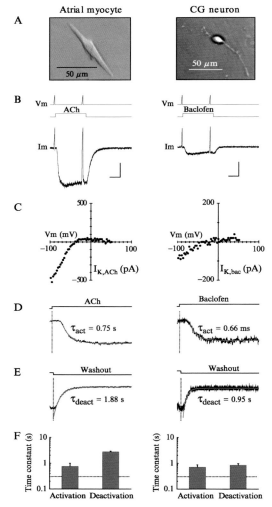

DOUPNIK *ET AL.*, CHAPTER 9, FIG. 3. Quantitative analysis of native GIRK currents recorded from rat atrial myocytes and rat cerebellar granule (CG) neurons. (A) Phase-contrast images of a typical neonatal rat atrial myocyte and a rat CG neuron maintained in primary culture and selected for electrophysiological recordings. (B) ACh (10 μM) and the GABA$_B$ receptor agonist baclofen (100 μM) evoke characteristic GIRK currents from atrial myocytes and CG neurons, respectively. Both display steep inward rectification (C) and rapid activation (D, F) and deactivation kinetics (E, F), with deactivation of baclofen-activated GIRK currents being slightly faster than ACh-activated GIRK currents. Data are means ± SEM. Dashed lines in F refer to time constants for solution exchange from Fig. 1 and represent the limit of resolving kinetic events any faster using the described methods.

```
mGNAoA    1 MGCTLSAEERAALERSKAIEKNLKEDGISAAKDVKLLLLLGAGESGKSTIVKQMKIIHEDGFSGEDVKQYKPVVYSNTIQS  80
mGNAoB    1 MGCTLSAEERAALERSKAIEKNLKEDGISAAKDVKLLLLLGAGESGKSTIVKQMKIIHEDGFSGEDVKQYKPVVYSNTIQS  80
rGNAi1    1 MGCTLSAEDKAAVERSKMIDRNLREDGEKAAREVKLLLLLGAGESGKSTIVKQMKIIHEAGYSEEECKQYKAVVYSNTIQS  80
rGNAi2    1 MGCTVSAEDKAAAERSKMIDKNLREDGEKAAREVKLLLLLGAGESGKSTIVKQMKIIHEDGYSEEECRQYRAVVYSNTIQS  80
rGNAi3    1 MGCTLSAEDKAAVERSKMIDRNLREDGEKAAREVKLLLLLGAGESGKSTIVKQMKIIHEDGYSEDECKQYKVVVYSNTIQS  80
            ****:***::** **** *::**:*** **:*************************** *:* :: :**: *********

mGNAoA   81 LAAIVRAMDTLGVEYGDKERKTDSKMVCDVVSRMEDTEPFSAELLSAMMRLWGDSGIQECFNRSREYQLNDSAKYYLDSL  160
mGNAoB   81 LAAIVRAMDTLGVEYGDKERKTDSKMVCDVVSRMEDTEPFSAELLSAMMRLWGDSGIQECFNRSREYQLNDSAKYYLDSL  160
rGNAi1   81 IIAIIRAMGRLKIDFGDAARADDARQLFVLAGAAEEG-FMTAELAGVIKRLWKDSGVQACFNRSREYQLNDSAAYYLNDL  159
rGNAi2   81 IMAIVKAMGNLQIDFADPQRADDARQLFALSCAAEEQGMLPEDLSGVIRRLWADHGVQACFGRSREYQLNDSAAYYLNDL  160
rGNAi3   81 IIAIIRAMGRLKIDFGEAARADDARQLFVLAGSAEEG-VMTSELAGVIKRLWRDGGVQACFSRSREYQLNDSASYYLNDL  159
            : **::**  *  ::: :  * *:: : :  *:  :*  :* : *** * *:* ** *********** *** *

mGNAoA  161 DRIGAGDYQPTEQDILRTRVKTTGIVETHFTFKNLHFRLFDVGGQRSERKKWIHCFEDVTAIIFCVALSGYDQVLHEDET  240
mGNAoB  161 DRIGAGDYQPTEQDILRTRVKTTGIVETHFTFKNLHFRLFDVGGQRSERKKWIHCFEDVTAIIFCVALSGYDQVLHEDET  240
rGNAi1  160 DRIAQPNYIPTQQDVLRTRVKTTGIVETHFTFKDLHFKMFDVGGQRSERKKWIHCFEGVTAIIFCVALSDYDLVLAEDEE  239
rGNAi2  161 ERIAQSDYIPTQQDVLRTRVKTTGIVETHFTFKDLHFKMFDVGGQRSERKKWIHCFEGVTAIIFCVALSAYDLVLAEDEE  240
rGNAi3  160 DRISQTNYIPTQQDVLRTRVKTTGIVETHFTFKELYFKMFDVGGQRSERKKWIHCFEGVTAIIFCVALSDYDLVLAEDEE  239
            :**  * **:*:*:*********** * *::**********************  * :: *** ** ***** * **

mGNAoA  241 TNRMHESLMLFDSICNNKFFIDTSIILFLNKKDLFGEKIKKSPLTICFPEYPGSNTYEDAAAYIQTQFESKNR-SPNKEI  319
mGNAoB  241 TNRMHESLKLFDSICNNKWFTDTSIILFLNKKDIFEEKIKKSPLTICFPEYTGPSAFTEAVAHIQGQYESKNK-SAHKEV  319
rGNAi1  240 MNRMHESMKLFDSICNNKWFTDTSIILFLNKKDLFEEKIKKSPLTICYPEYAGSNTYEEAAAYIQCQFEDLNKRKDTKEI  319
rGNAi2  241 MNRMHESMKLFDSICNNKWFTDTSIILFLNKKDLFEEKITQSPLTICFPEYTGANKYDEAASYIQSKFEDLNKRKDTKEI  320
rGNAi3  240 MNRMHESMKLFDSICNNKWFTDTSIILFLNKKDLFEEKIKRSPLTICYPEYTGSNTYEEAAAYIQCQFEDLNRRKDTKEV  319
            *****: ********* * ************* ** :*****:*** *    : :* : ** ::*  *:   **:

mGNAoA  320 YCHMTCATDTNNIQVVFDAVTDIIIANNLRGCGLY*  355
mGNAoB  320 YSHVTCATDTNNIQFVFDAVTDVIIAKNLRGCGLY*  355
rGNAi1  320 YTHFTCATDTKNVQFVFDAVTDVIIKNNLKDCGLF*  355
rGNAi2  321 YTHFTCATDTKNVQFVFDAVTDVIIKNNLKDCGLF*  356
rGNAi3  320 YTHFTCATDTKNVQFVFDAVTDVIIKNNLKECGLY*  355
            *:* ****** *:* ********:**  **: ***:*
```

IKEDA AND JEONG, CHAPTER 11, FIG. 1. Mutations conferring *Bordetella pertussis* toxin (PTX) and RGS protein insensitivity to mammalian G-protein α subunits. Clustal W alignment of rodent Gα subunits of the $G_{i/o}$ class. Amino acid identity is indicated by an asterisk and conserved substitution by a colon. Residues conferring RGS insensitivity when mutated are outlined in green and PTX insensitivity in yellow. Genbank accession numbers: mouse $G\alpha_{oA}$ (M36777), mouse $G\alpha_{oB}$ (M36778), rat $G\alpha_{i1}$ (NM_013145), rat $G\alpha_{i2}$ (NM_031035), and rat $G\alpha_{i3}$ (NM_013106).

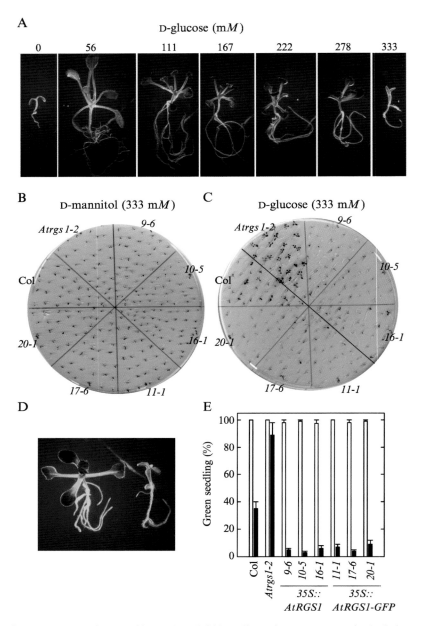

A

D-glucose (mM)

0   56   111   167   222   278   333

B   D-mannitol (333 mM)

Atrgs 1-2   9-6

Col   10-5

20-1   16-1

17-6   11-1

C   D-glucose (333 mM)

Atrgs 1-2   9-6

Col   10-5

20-1   16-1

17-6   11-1

D

E

CHEN AND JONES, CHAPTER 20, FIG. 1. AtRGS1 mediates glucose responses in *Arabidopsis*. (A) Glucose dose-dependent developmental responses of wild-type (Col-0) *Arabidopsis* seedlings. Seedlings were photographed after 14 days grown at 23°, constant light. (B) *Atrgs1* mutants have a wild-type response to D-mannitol. T-DNA insertion mutant *Atrgs1-2* and transgenic lines overexpressing *AtRGS1* (lines *9-6*, *10-5*, and *16-1*) or *AtRGS1-GFP* (lines *11-1*, *17-6*, and *20-1*) were sown on 1/2 MS plates containing 333 mM (6%) D-mannitol. In transgenic lines, expression of *AtRGS1* or *AtRGS1-GFP* was driven by the 35S promoter of cauliflower mosaic virus. (C) *Atrgs1* mutants are insensitive to high concentration of D-glucose. Seeds in B were sown on 1/2 MS plates containing 333 mM (6%) D-glucose. Seedlings in B and C were photographed after 10 days grown at 23°, constant light. (D) Green seedling (*left*) and development arrested seedling (*right*) on 333 mM (6%) D-glucose. Seedlings were photographed after 10 days grown at 23°, constant light. (E) Quantitative analyses of sugar sensitivities in *Atrgs1-2* mutants and transgenic lines overexpressing *AtRGS1* or *AtRGS1-GFP*. Open bar: 333 mM D-mannitol. Closed bar: 333 mM D-glucose. The percentages of green seedlings were scored after 10 days grown at 23°, constant light. Shown are means ± SE of three replicates.

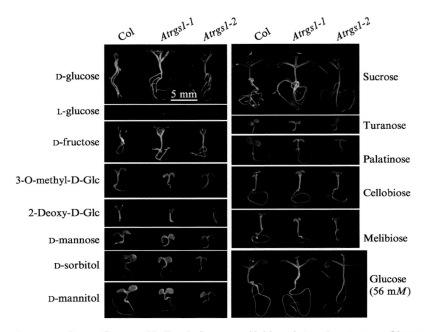

CHEN AND JONES, CHAPTER 20, FIG. 3. Sugar specificities of *Atrgs1* mutants. D-Glucose, L-glucose, D-fructose, D-sorbitol, and D-mannitol were used at 333 m*M*. 3-*O*-Methyl-D-glucose (3-O-methyl-D-Glc) was used at 50 m*M*. 2-Deoxy-D-glucose (2-Deoxy-D-Glc) and D-mannose were used at 1 m*M*. All disaccharides (sucrose, turanose, palatinose, cellobiose, and melibiose) were used at 200 m*M*. Wild-type (Col) and *Atrgs1* mutants grown at 56 m*M* D-glucose were used as controls. In all these experiments, the particular sugar/disaccharide or analogs were the sole sugar source. Seedlings were photographed after 10 days grown at 23°, constant light. Seeds on 333 m*M* L-glucose did not germinate.

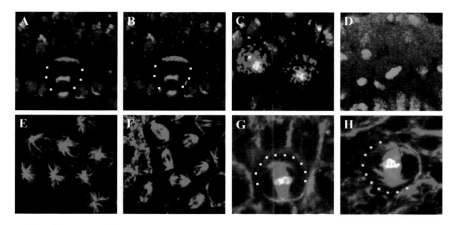

Yu, Chapter 22, Fig. 3. Pins and Insc localization in wild type (WT), *pins*, and $G\alpha_i$ mutants and spindle morphology. (A–D) Parasagittal optical sections showing NBs from embryos of various genotypes indicated in each panel; apical is up. (A and B) WT embryos triple labeled with anti-Pins (green), anti-Insc (red), and DNA stain (blue); anti-Insc plus DNA staining is shown in A and anti-Pins plus DNA staining is shown in B Note that Pins colocalizes with Insc at metaphase. (C) *pins* mutant embryos stained with anti-Insc (red) and DNA (blue). Note that Insc is cytoplasmic at metaphase in the absence of Pins. (D) $G\alpha_i$ mutant embryos stained with anti-Pins (green) and DNA (blue). Pins is released from the membrane in both NBs and epithelial cells. (E and F) β-Tubulin staining (in green) in mitotic domain 9 (Kraut *et al.*, 1996). Cells of WT and *pins* mutants. In WT, the mitotic spindle is oriented along the apical–basal axis (E) while the spindle is parallel to the surface in the *pins* mutant (F). (G and H) α-Tubulin staining (in red) in WT NBs and epithelial cells. Note that astral MTs grow only at the apical side of NB whereas astral MTs shrink at the basal side. In contrast, astral MTs can grow at both sides of epithelial cells. Reproduced with permission from Yu *et al.* (2000), © 2000 Cell Press and Yu (2003), © 2003 The Rockefeller University Press.